ELECTRONICS FOR TECHNICIAN ENGINEERS

The purpose of this book is to provide the trainee technician engineer with a broad insight into a diverse range of electronic components and circuits.

Both thermionic valves and semiconductors are discussed and their application in electronic circuits. Both large signal (graphical) and small signal (equivalent circuit) techniques are covered in detail.

Mathematics are kept to a minimum and for those readers with a limited mathematical ability, graphs and tables are included which will enable them to cover the majority of the work successfully.

The book is not intended to cover a particular course of study but should provide some very useful material for readers who are taking electronics at ordinary and advanced certificate or diploma level and for trainee technician engineers undergoing their training in engineering training centres or firms where the training includes circuit design work.

Some very elementary material is included for home-study readers with an interest in electronics.

To Paul and Judith

ELECTRONICS
FOR
TECHNICIAN ENGINEERS

W. W. SMITH

Area Manager, London and South East Region,

Engineering Industry Training Board.

HUTCHINSON EDUCATIONAL

HUTCHINSON EDUCATIONAL LTD
178–202 Great Portland Street, London W.1

London Melbourne Sydney
Auckland Bombay Toronto
Johannesburg New York

★

First published August 1970

© W.W. Smith 1970

Illustrated by David Hoxley.
This book has been set in cold type by E.W.C. Wilkins and Associates Ltd.,
printed in Great Britain by Anchor Press, and bound by Wm. Brendon,
both of Tiptree, Essex

CONTENTS

Author's note.
Introduction.

Chapter 1.		Electrical networks and graphs.	1
	1.1.	Ohm's law.	1
	1.2.	Voltage/current graphs.	4
	1.3.	Current/voltage graphs.	4
	1.4.	Composite current/voltage characteristics.	5
	1.5.	Series load resistors.	6
	1.6.	Shunt load resistors.	7
	1.7.	Introduction to load lines.	8
	1.8.	Voltage distribution in a series circuit.	9
	1.9.	Non-linear characteristics.	10
	1.10.	Plotting the points for positioning a load line.	12
	1.11.	Drawing load lines on restricted graphs.	12
Chapter 2.		**Further networks and simple theorems.**	15
	2.1.	Internal resistance.	15
	2.2.	Effective input resistance.	16
	2.3.	Four terminal devices.	18
	2.4.	Voltage and current generators.	20
	2.5.	Input resistance, current operated devices.	21
	2.6.	Simple theorems.	23
	2.7.	Kirchoff's laws.	24
	2.8.	Derivation of a formula.	27
	2.9.	Superposition theorem.	28
	2.10.	Reciprocity theorem.	29
	2.11.	Thevinin's theorem.	29
	2.12.	Norton's theorem.	29
	2.13.	Comparison of theorems.	29
	2.14.	'Pi' to 'tee' transformation.	33
Chapter 3.		**Linear components.**	41
	3.1.	The Resistor.	41
	3.2.	The perfect inductor.	42
	3.3.	Rise of current through an inductor.	42

v

3.4.	The capacitor.	45
3.5.	Capacitors in series.	50
3.6.	Parallel plate capacitors	51

Chapter 4.   Revision of basic a.c. principles   53

4.1.	Alternating current.	53
4.2.	R.M.S. value.	55
4.3.	Mean value.	56
4.4.	a.c. circuits.	56
4.5.	Resonant circuits.	58

Chapter 5.   Diodes, rectification and power supplies.   63

5.1.	The thermionic Diode.	63
5.2.	The half wave rectifier.	65
5.3.	Power supply units.	68
5.4.	The full wave circuit.	72
5.5.	Filter circuits.	75
5.6.	Multi-section filter.	80
5.7.	Parallel tuned filter.	80
5.8.	Choke input filters.	81
5.9.	Diode voltage drop.	86
5.10.	Metal rectifiers.	87
5.11.	Bridge rectifiers.	90
5.12.	Voltage doubling circuit.	91

Chapter 6.   Meters.   93

6.1.	A simple voltmeter.	93
6.2.	Switched range ammeter.	95
6.3.	Universal shunts.	96
6.4.	High impedance voltmeter.	100
6.5.	A.c. ranges, rectification, RMS and average values.	101
6.6.	A simple ohmmeter.	104
6.7.	Simple protection circuits.	107
6.8.	Internal resistance of the meter movement.	108

Chapter 7.   Triode valves, voltage reference tubes and the thyratron.   109

7.1.	The triode valve.	109
7.2.	Triode parameters.	112
7.3.	Ia/Va triode characteristics, common cathode.	116
7.4.	Gas-filled devices.	118
7.5.	Simple stabiliser circuits.	119
7.6.	Stabiliser showing effects of H.T. fluctuations.	120

7.7.	Stabiliser showing effect of load variations.	121
7.8.	The gas-filled Triode.	122
7.9.	Control ratio.	125
7.10.	Grid current.	125
7.11.	Firing points.	126

Chapter 8. Amplifiers. 129

8.1.	The triode valve-simple equivalent circuit.	129
8.2.	Voltage amplification. Load Lines. The operating point.	130
8.3.	Signal amplification.	131
8.4.	Construction of a bias load line.	134
8.5.	Maximum anode dissipation.	136
8.6.	Deriving resistor values for an amplifier. (d.c. considerations).	138
8.7.	Voltage gain (a.c. conditions).	139
8.8.	Maximum power transference.	142
8.9.	Maximum power theorem (d.c.)	143
8.10.	Maximum power theorem (a.c.)	144
8.11.	An inductive loaded amplifier.	146

Chapter 9. Simple transformer coupled output stage. 153

9.1.	Simple concept of transformer action on a resistive load.	153
9.2.	Power equality, input and output.	153
9.3.	Equality of ampere-turns.	154
9.4.	Reflected load.	154
9.5.	Simple transformer output stage.	155
9.6.	Plotting the d.c. load line.	156
9.7.	Plotting the bias load line.	158
9.8.	The operating point.	159
9.9.	A.c. load line.	159
9.10.	Applying a signal.	160

Chapter 10. Miller effect. 165

10.1.	Miller effect in resistance loaded amplifiers.	165
10.2.	Amplifier with capacitive load.	166
10.3.	Amplifier with inductive load.	167
10.4.	Miller timebase.	167
10.5.	Cathode follower input impedance.	167

Chapter 11. The Pentode valve. 171

11.1.	The tetrode and pentode.	171

Chapter 12. Equivalent circuits and large signal considerations. 177
- 12.1. A simple equivalent circuit of a triode valve. 177
- 12.2. Common cathode amplifier. 178
- 12.3. Un-bypassed cathode; common cathode amplifier. 179
- 12.4. The phase splitter or 'concertina' stage. 180
- 12.5. Common-grid amplifier. 182
- 12.6. Input resistance; common-grid amplifier. 183
- 12.7. Common anode amplifier. 184
- 12.8. Output resistance – common anode amplifier. 185
- 12.9. Input resistance – common anode amplifier. 185
- 12.10. Output resistance of anode – common cathode amplifier. 186
- 12.11. Cathode coupled amplifier – Long tailed pairs. 187
- 12.12. Long tailed pair approximations. 192
- 12.13. Graphical analysis – long tailed pair. 194

Chapter 13. Linear analysis. 199
- 13.1. Elementary concept of flow diagrams. 199
- 13.2. Simple amplifier with resistive anode load. 203
- 13.3. Linear analysis of a clipper stage. 210

Chapter 14. Pulse techniques. 221
- 14.1. Waveform identification. 221
- 14.2. Step function inputs applied to C.R. networks. 223
- 14.3. Pulse response of linear circuit components. 227
- 14.4. A simple relaxation oscillator. 230
- 14.5. Simple free running multivibrators. 232
- 14.6. A basic pulse lengthening circuit. 240
- 14.7. The 'charging curve' and its applications. 241

Chapter 15. Further large signal considerations, a Binary counter. 245
- 15.1. A basic long tailed pair. 245
- 15.2. A basic Schmitt trigger circuit. 245
- 15.3. A simple bi-stable circuit. 247
- 15.4. Binary circuits – the Eccles Jordan. 248
- 15.5. A simple binary counter. 253
- 15.6. Feedback in a simple counter. 255
- 15.7. Meter readout for a scale of 10. 258
- 15.8. Design considerations of a simple bi-stable circuit. 261

Chapter 16. Further considerations of pulse and switching circuits. 267
- 16.1. A cathode coupled binary stage. 267
- 16.2. A biassed multivibrator. 271

16.3.	A direct coupled monostable multivibrator.	273
16.4.	Cathode follower; maximum pulse input.	277
16.5.	A phase splitter analysis.	281
16.6.	Linear analysis of a cathode coupled multivibrator.	285
16.7.	The diode pump.	287

Chapter 17. A delay line pulse generator. 291

17.1.	A simple pulse generator.	291
17.2.	Delay line equations.	295
17.3.	A Delay line.	297
17.4.	The thyratron.	302
17.5.	A delay line pulse generator.	302

Chapter 18. Negative feedback and its applications. 307

18.1.	Feedback and its effect upon the input resistance of a single stage amplifier.	308
18.2.	Feedback in multistage amplifiers.	309
18.3.	Composite feedback in a single stage amplifier.	312
18.4.	Effects of feedback on parameters μ and ra due to composite feedback.	314
18.5.	The effects of feedback on output resistance.	315
18.6.	Voltage and current feedback in a phase splitter.	317
18.7.	Voltage series negative feedback – large signal analysis.	321
18.8.	Stabilised power supplies.	329
18.9.	A series regulator.	331
18.10.	A shunt type stabiliser circuit.	334
18.11.	Negative output-resistance.	335
18.12.	A stabilised power supply unit.	336
18.13.	Attenuator compensation.	344
18.14.	Derivation of component values in an impedance convertor.	346

Chapter 19. Locus diagrams and frequency selective networks. 353

19.1.	Introduction to a circle diagram for a series CR circuit.	353
19.2.	Plotting the diagram.	356
19.3.	Resistance.	356
19.4.	Voltage.	358
19.5.	Current.	358
19.6.	Measurements.	360
19.7.	Power factor.	363
19.8.	Power.	364
19.9.	Use of the operator j.	367
19.10.	A series L.R. circuit.	372

	19.11.	Frequency response of a C.R. series circuit.	374
	19.12.	A frequency selective amplifier.	378
	19.13.	The twin tee network.	380

Chapter 20.   Simple mains transformers.   389

20.1.	A simple design. (1).	389
20.2.	Transformer losses.	396
20.3.	A design of a simple transformer (2).	397
20.4.	A simple practical test of a transformer.	401

Chapter 21.   Semiconductors.   411

21.1.	Junction transistors.	411
21.2.	N. type material.	413
21.3.	P. type material.	413
21.4.	Energy level.	413
21.5.	Donor atoms.	414
21.6.	Acceptor atoms.	414
21.7.	P – n junction.	415
21.8.	Reverse bias.	416
21.9.	Forward bias.	417
21.10.	The junction transistor.	418
21.11.	Input and output resistance – the equivalent tee.	420
21.12.	Bias stabilisation.	424
21.13.	The stability factor, K.	426
21.14.	Common emitter protection circuits.	427
21.15.	Input resistance – common emitter.	429
21.16.	Input resistance – common collector.	432
21.17.	Variations in load resistance, common base.	433
21.18.	Output resistance – common collector.	433
21.19.	Expressions incorporating external resistors.	435
21.20.	Voltage gain.	436
21.21.	Power gain.	437
21.22.	Current gain.	438
21.23.	D.C. amplifier.	439
21.24.	Gain controls.	440
21.25.	Simple amplifier considerations.	441
21.26.	Ic/Vc measurements.	443
21.27.	Clamping.	448
21.28.	Small transformer-coupled amplifier.	450

CONTENTS   xi

Chapter 22.   **'h' Parameters.**   457
- 22.1. Equivalent circuits.   457
- 22.2. 'h' parameters and equivalent T circuits.   463
- 22.3. 'h' parameters. Conversion from T network parameters.   467
- 22.4. Measuring 'h' parameters.   467
- 22.5. Input resistance with R_L connected.   470
- 22.6. Current gain.   471
- 22.7. Voltage gain.   471
- 22.8. Output admittance.   471
- 22.9. Power gain.   472

Chapter 23.   **'H' parameters.**   475
- 23.1. Cascade circuit (common base).   475
- 23.2. H_{11} (Common base).   476
- 23.3. H_{12} (Common base).   477
- 23.4. H_{12} (Common collector).   478
- 23.5. H_{21} (Common base).   479
- 23.6. H_{22} (Common base).   480
- 23.7. H_{21} (any configuration).   481

Chapter 24.   **M.O.S.T. Devices.**   483
- 24.1. Introduction to M.O.S.T. devices.   483
- 24.2. A simple amplifier.   489
- 24.3. Analysis of amplifier with positive bias.   490
- 24.4. An amplifier with negative bias.   492

Chapter 25.   **Ladder networks and oscillators.**   499
- 25.1. Simple ladder networks.   499
- 25.2. The wien network.   502
- 25.3. Phase shift oscillators.   507
- 25.4. Analysis of a transistorised 3 stage phase shift network.   511

Chapter 26.   **Zener Diodes.**   515
- 26.1. Operating points.   515
- 26.2. A voltage reference supply.   517
- 26.3. Transistorised stabilised power supply.   520

Chapter 27.   **Composite devices.**   523
- 27.1. Silicon controlled – rectifiers.   523
- 27.2. A super alpha pair.   529
- 27.3. Application of super alpha pair (High input-resistance amplifier).   530

| | 27.4. | Application of super alpha pair (regulated p.s.u.) | 531 |

Chapter 28. Simple logic circuits — 537
| | 28.1. | Transistorised multivibrator circuit. | 537 |
| | 28.2. | Introduction to a simple digital system. | 540 |

Chapter 29. Combined AND/OR gate. — 543
	29.1.	Simple logic circuit.	543
	29.2.	Simple 'AND' gate.	544
	29.3.	Simple 'OR' gate.	545
	29.4.	Coincidence gate.	546
	29.5.	Combined 'AND/OR' gate circuit.	548

Chapter 30. Analogue considerations. — 555
	30.1.	Laplace terminology.	555
	30.2.	Operational amplifiers.	556
	30.3.	Difference amplifiers.	559
	30.4.	Servomechanisms.	560
	30.5.	Summing integrator.	561
	30.6.	Simple analogue computor.	564
	30.7.	Application to a simple servomechanism system.	565
	30.8.	Solving simultaneous differential equations.	568

Chapter 31. Sawtooth generation. — 571
	31.1.	Modified Miller sawtooth generator.	571
	31.2.	Modified miller with suppressor gating.	573
	31.3.	The Miller balance point.	580
	31.4.	Puckle timebase. (1).	582
	31.5.	Puckle timebase. (2).	584

Answers to Problems — 589

Index — 615

AUTHOR'S NOTE

The purpose of this book is to give all technicians, particularly the Technician Engineer, a broad basic appreciation of some of those aspects of electronic components and circuitry that he is likely to meet in his place of work. It is impossible, in a book of this size, to cover every detail of any circuit, in fact a whole volume could be written for almost every topic in this book. An attempt has been made however to cover the necessary detail likely to be generally required by the junior Technician Engineer, whilst the deeper aspects of technology, which often requires a more advanced mathematical ability, have been limited.

The borderline activities between the qualified Technician Engineer and and the Technologist are often very grey. One is likely to find both in a design department. A graduate may often be found doing production design for a year or two, in order to 'cut his teeth' before moving on to a more senior or a completely different post more in keeping with a university education.

Within this grey area however, it is often possible to identify the Technologist and the Technician Engineer, as the latter will usually demonstrate a more practical approach towards a problem in relation to the more mathematical or academic approach by the Technologist.

One of the most important features of the Technician Engineer's abilities is perhaps his ability to 'fault-find', whether in testing production equipment or a first off in a design department. A successful Technician Engineer will fault-find quickly and efficiently because he will be able to estimate likely quantities whilst taking his measurements. He can only demonstrate this ability when he is throughly familiar with a wide range of circuitry. This book attempts to cover a wide range of basic circuits and to show by examples, many alternative methods of approach towards solving technical problems.

It is a very difficult task to decide just where to draw the line when discussing circuits; one could write many more pages for all of the circuits contained in this book. Whether a successful compromise has been reached will be known in time. It is hoped that readers with views on the matter will inform the Author who would be pleased to modify future editions.

For example, a Technician Engineer might be asked to design and build a power supply unit. He would ensure that the ripple content is within the limits prescribed. It is unlikely that he would generally need to consider harmonics other than the fundamental. On the other hand, he would almost certainly be expected to appreciate output resistance and ensure that the circuit was within specification.

He would be expected to know the orders of values of circuit components, to recognise an 'impossible' value of say, a resistor used as a bias resistor in the cathode of a small power amplifier. He should be able to estimate a very close-to-correct value of resistor after looking at the printed valve or transistor characteristics.

He would be expected to design, analyse or test circuits using a wide range of components, motors, generators, resistors, capacitors, inductors, valves, transistors, etc., often working to requirements laid down by someone else, but he would not be expected to design any one of the components themselves, unless he specialised at a later stage. It is with this in mind that this book has been prepared, to show how to use components rather than to spend much time on the design of them.

In the near future electronic circuit design will become more of a question of 'system' design; choosing from a range of 'modules' including microelectronic devices, many of which will be freely available 'off the shelf'. The Technician Engineer, if he is to become involved in the use of modules, should first have a good appreciation of discrete components and how they function in a circuit. He must be familiar with input and output resistance and how feedback can be used to accomplish certain tasks. This book attempts to show him the basic techniques.

Much of the material in this book resulted from a number of successful industrial training schemes for trainee technician engineers, such as the 1st year course for electronic technician engineers at the Crawley Industrial Training Centre. Courses similar to this have been devised and run by the Author since 1961 and this book reflects what he believes to be the general basic requirements of this trainee technician engineer.

The Technician Engineer generally needs to see a practical application for his theory. This book attempts to continually show how to apply this basic theory. Trainees should wherever possible, practice building and testing circuits that they have designed (or analysed) as an academic exercise and convince themselves that their theory really works in practice.

It is hoped that whatever course of further education or training the electronics trainee undertakes, he will find a lot of very useful information in this book.

Many of the examples contain values for components that have been chosen to highlight a particular point. Hence some values may be larger, or smaller, than is met in practice. For example, the meter movement used in the section on Meters, has been given a high internal resistance. This allowed various factors to be emphasised that might have been insignificant with a low value.

There are many levels of technician. Some will study for the H.N.C., others for the Full Technological Certificates of the City and Guilds, these are the technician engineers, others might take the final only of the C & G

AUTHOR'S NOTE

course 57, some the Radio and T.V. Mechanics course, and many others. This book has been prepared with the needs of all technicians in mind, hence the alternative methods shown. Some are more academic, whilst others are non-mathematical and employ graphs, charts and tables. The reader will of course, select the method that applies to him most.

Finally, the needs of the 'home study' student has not been overlooked, some very basic material is included from time to time to enable him to progress through most of the book without too much difficulty.

The Author gratefully acknowledges the advice, assistance and encouragement given to him by Dr. T. Siklos, Principal of Crawley College of Further Education, to Mr. J.R. Bee for his advice and assistance in the checking of examples and in particular, the section on transformer design, and to Mr. A. Cain, Manager of the Electronics Section of the Crawley Industrial Training Centre, for his help and advice and for checking the final draft and for his many suggestions for improving the presentation.

Finally, the Author would like to express his appreciation to Mr. Patrick Moore for his assistance and advice which led to the preparation of the early draft stage of the book.

The Author wishes to acknowledge Mullards Ltd. for their kind permission to reproduce several of their valve and transistor characteristics.

EAST GRINSTEAD.

INTRODUCTION

The technician engineer in the electronic industry is complementary to the chartered engineer and due mainly to the efforts of the I.E.E.T.E. supported by the I.E.E., now has a standing and status in industry as an engineer in his own right. In the near future he might use the designation 'Tech. Eng.' as a complementary term to the graduate's 'C.Eng.'

He is responisble for design, development, planning, estimating, design draughting and servicing of electronic equipment of all types. His is a key post in industry and after suitable formal training and academic attainment of say a Higher Technician Certificate or Diploma in Electronics, Radio or Electrical engineering, carries out many tasks which a few years ago were carried out by graduate engineers.

One of the prime qualities of the technician engineer is the ability to diagnose, to analyse, to approach the solution of technical problems in a true logical and engineering manner. Properly planned training during the early part of his career will assist him to develop these qualities.

This book is written for the potential technician engineer in an attempt to provide him with the means of developing a diagnostic approach towards his technical problems. It should provide him with a substantial broad foundation upon which he can build a later expertise in any of the many branches of electronic engineering.

Every technician engineer should be able to read and understand electronic circuit diagrams and to be thoroughly familiar with the printed characteristics of the numerous devices used in electronic engineering. He should be equally familiar with load line techniques and to be able to produce acceptable answers by means of both printed characterists and small signal analyses using equivalent circuit techniques.

He should be able to provide rapid approximate answers using any one of a number of techniques and to be so well versed in circuitry that he can freely choose whether to accept approximate answers or to obtain precise answers according to the requirements at any given time.

He will find that from time to time, a particular approach towards the solution of a problem, although quite precise, may not necessarily be the quickest manner in which to provide the answer he seeks. He should be able to both recognise the need for, and the ability to use, a different and equally precise approach that will enable him to reach his goal with much less effort. He will be capable of doing this only when he has acquired a very broad knowledge of the basic fundamentals of electronic principles through his studies and a proper training.

The reader should attempt to consolidate his position at each stage as he progresses throughout this book; he should try to practice the theory he learns and more important, he should practice the theory on circuits which differ from those shown as examples throughout these series.

Little mention is made of electron theory and a.c. principles. There are numerous books available which deal with matters such as 'electrons in magnetic and electrostatic fields', and he should refer to one or more of these if he so desires. This book attempts to cover a very wide range of circuit diagrams, covering both the basic design and analysis thus providing a very real and useful background.

With a pass at 'O' level or a good C.S.E. in both mathematics and physics, no reader who is continuing with his studies, should have any real difficulty in progressing throughout this book.

Almost every page contains worked examples, some of which are biassed towards design while others are biassed towards analysis of circuits previously designed by someone else. Some are precise whilst alternative methods by approximation are shown.

The electronics field is rapidly changing and techniques vary almost from one month to the next. The basic principles of electronics shown in this book however, apply now and when considering known components, will be valid for the foreseeable future.

The earlier sections contain a great deal of useful basic theory and practical worked examples. Valves are used, as with these devices, accurate answers are usually obtained in practice. The latter section deals almost solely with semiconductors, and although the same principles apply, worked examples show clearly the allowance one must make for some semiconductors and their effect upon calculated values.

The importance of establishing correct d.c. conditions, as a general rule, before considering a.c. conditions, is stressed throughout. Although there are exceptions to this, the reader is advised to adopt this principle until he has gained sufficient experience to enable him to decide whether this general approach can be varied on the particular occasion.

Some errors are inevitable in a book of this size and although every attempt has been made to reduce these to a minimum, some may occur. The Author would be glad to receive notification of any errors in order to ensure that future editions are amended accordingly.

CHAPTER 1

Electrical networks and graphs

Many devices and theorems are used in electronic engineering in order to facilitate the analysis and discussion of networks, but not all of these are needed by the technician engineer. In this chapter, the basic methods used for dealing with circuits are illustrated by very easy examples involving resistors only, although the same ideas apply, of course, when reactances are introduced at a later stage. Especially important are graphical methods, particularly load line techniques. Later the concepts of the equivalent voltage and current generators are discussed. Much of this early material will not be new to the student, although the techniques discussed are of *paramount importance* and will be extended for more advanced analyses later on in this book.

1.1. Ohm's Law

If V is the voltage across a conductor (potential difference between the ends of the conductor) in volts, I is the current in amperes flowing through the conductor, and R ohms is the resistance of the conductor, then these three are related by Ohm's law. The resistance depends upon the material and dimensions of the conductor and upon the temperature, but in given circumstances will be a constant for a particular resistor.

Examples.

1. If an e.m.f. of 3 volts is applied across a resistor having a value of 2Ω, then a current of 1.5 amps will flow. A circuit diagram is shown in figure 1.1.1.

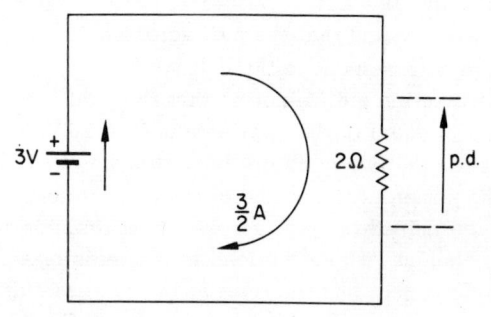

Fig. 1.1.1.

(Note that in the figure, an arrow is used to denote the polarity of the p.d. The arrow head indicates the positive end of the potential.)

The current would flow through the resistor in a clockwise direction if the battery were connected with its positive terminal as marked. If the connections to the battery are reversed, then the current will flow anti-clockwise. The clockwise flow of current is shown on the diagram by an arrow.

The current flowing into the resistor causes a 'voltage drop' or potential difference to be developed across it, this potential being positive at the end of the resistor into which the current is flowing.

In figure 1.1.2. the circuit diagram is given for two resistors connected in series with a battery V.

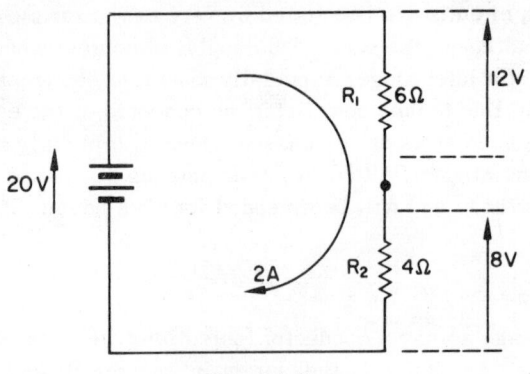

Fig. 1.1.2

Suppose the battery has an e.m.f. of 20 volts, and that R_1 is 6Ω, R_2 is 4Ω. The total resistance to current flow will be the sum of R_1 and R_2 $6 + 4 = 10\Omega$. The current, by Ohm's law, will be $I = V/R = 20/10 = 2A$.

This current, leaving the positive terminal of the battery, flows in a clockwise direction round the circuit and enters R_1 at the top of the resistor. Hence the p.d. developed across R_1 will have such a polarity as to make the top of the resistor positive with respect to the bottom. The magnitude of this p.d. given by Ohm's law, is $I \times R_1 = 2 \times 6 = 12V$.

In similar way, we may find that the p.d. across R_2 is $8V$, the top end being again positive with respect to the bottom.

By adding the two series p.d.'s, noting that they aid one another, we see that the total p.d.'s is equal to the applied e.m.f. of 20V. The applied voltage is shared between the two resistors in such a way that the current in each resistor is the same, as it must be of course, for series connections.

We could have determined the p.d. across either resistor by applying the 'load over total' technique, where the load is the resistor across which the p.d. needs to be determined, and the total is the total circuit resistance.

Example

$$VR_2 = \frac{\text{load}}{\text{total}} \times \text{applied volts.} = \frac{4 \times 20}{10} = 8V.$$

ELECTRICAL NETWORKS AND GRAPHS 3

For the case where two resistors are connected in parallel across a battery as shown in figure 1.1.3., it is clear that the p.d. across each resistor is equal to the battery e.m.f., because each is directly connected to the battery. The separate currents in the resistors I_1 and I_2 add to give the total battery current, I.

$$I = I_1 + I_2 = \frac{V}{R_1} + \frac{V}{R_2} = \frac{V(R_1 + R_2)}{R_1 R_2}$$

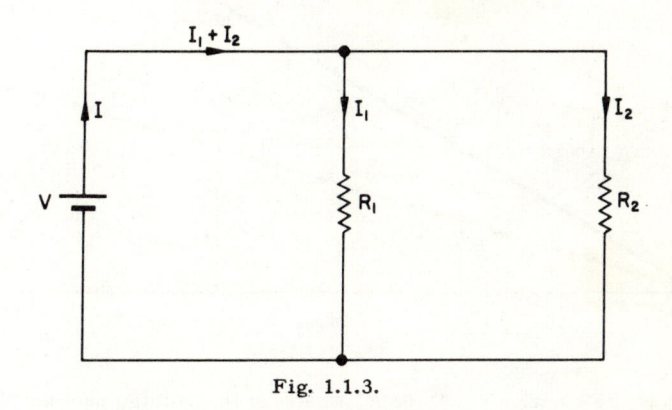

Fig. 1.1.3.

If we write R for the effective resistance of the shunt combination, so that $I = V/R$, then it is clear that

$$R = \frac{R_1 R_2}{R_1 + R_2}.$$

The effective resistance is less than either of the two component resistors, and is given by the usual rule for shunt resistors, product/sum. If we know the current I, and need to evaluate say, I_2, we can by duality use 'load over total' where for currents, the load resistor will be that resistor which has I_1 flowing through it. Therefore

$$I_2 = \frac{I \times \text{load}}{\text{total}} = \frac{I \times R_1}{R_1 + R_2}.$$

In a simple circuit such as 1.1.3.,

$$I_1 = \frac{I \times R_2}{R_1 + R_2}$$

It can be seen that the resistors R_1 and R_2 change position in the formula 'load over total' compared with the same expression for voltage as shown in the example 1.1.2.

1.2. Voltage/current graphs.

Suppose we take a particular resistor, apply different direct voltages to it and note the current which flows for each voltage applied. We may then plot a graph of V against I and the result will be a straight line, say the line $0-A$ in figure 1.2.1. The graph, or curve, must be a straight line because the

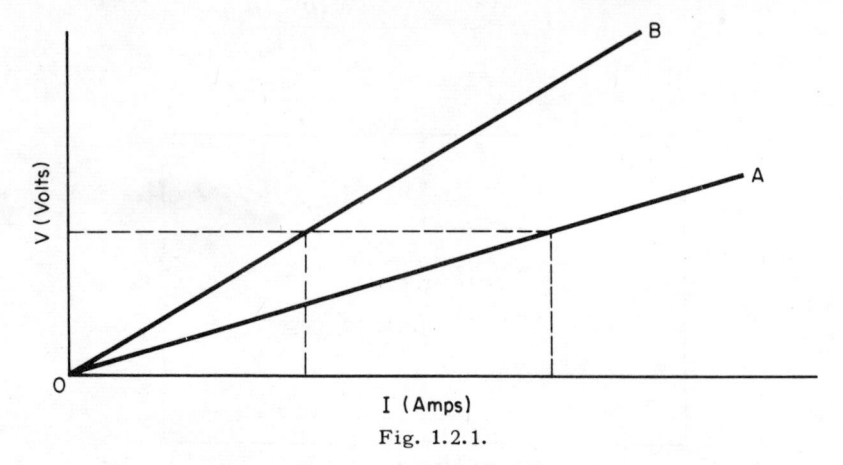

Fig. 1.2.1.

ratio V/I is the same, a constant R, whatever the voltage may be. The slope of the line is a measure of the value of the resistance R, so that if we used a larger resistance we would obtain a line of greater slope such as $0-B$.

We can find the value of resistance by simply selecting a voltage and noting the current flow for that value of voltage. Applying Ohm's law and using the values obtained from the graph will give the answer.

The example shown in figure 1.2.1. will show what is meant.

1.3. Current/voltage graphs.

We are often given information about electronic componets in the form of a graph. Sometimes these graphs are drawn with the current on the Y axis and voltage on the X. Figure 1.3.1. illustrates this.

Now that the axes have been changed over, the slope of the line represents not R but the reciprocal of R, i.e., $1/R$. The reciprocal of R is known as the conductance, G. We will have to get used to recognising graphs with the axes 'inverted' as almost every graph we are likely to see from now on will be of this type. The fact that the slope is a reciprocal need not give rise to concern for we can ignore this fact; our answers will be just as easy to obtain as with the previous types.

Figure 1.3.2. shows a graph of a resistor, but this time, the line is drawn from right to left and actually represents the negative reciprocal of R.

Once again, we will get used to this, particularly as we will be drawing

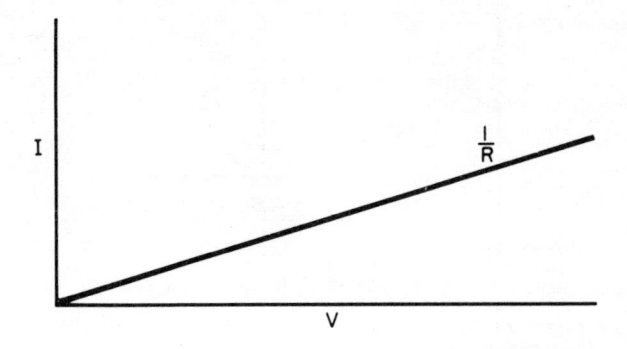

Fig. 1.3.1.

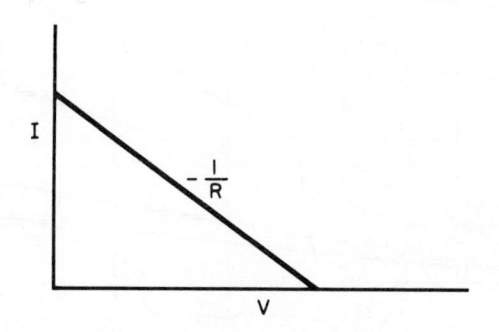

Fig. 1.3.2

many ourselves later on. Again we need not concern ourselves with the negative reciprocal, all we want to know is that we must obey Ohm's law always. We will therefore consider these lines as though they represented positive resistances.

1.4. Composite current/voltage characteristics.

We are going to discuss a circuit similar to that shown in figure 1.1.1, but in this instance, we will be studying a circuit containing a 'device' that has all of those characteristics of a 3Ω resistor. This device is shown ringed in the circuit in figure 1.4.1.

We intend at a later stage, to connect a load resistor in series with the device, and to discuss one method of obtaining composite I/V characteristics for both the device plus its load. Figure 1.4.2. shows the static I/V 'curve' for the device.

The line $0-A$ represents the static characteristics of the device, the slope of which is a measure of its resistance. As before, if we need to know the resistance of the device, we select a voltage, note the corresponding current

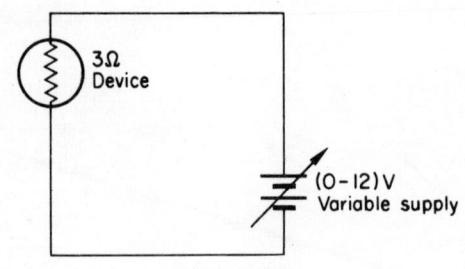

Fig. 1.4.1

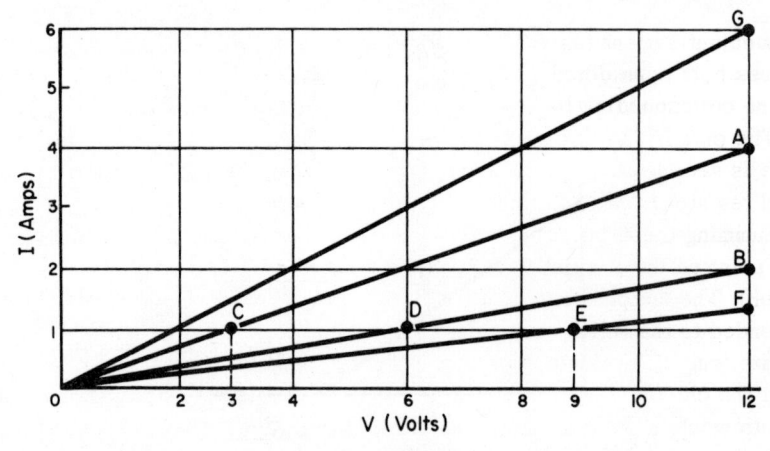

Fig. 1.4.2.

that would flow, and from Ohm's law, determine the resistance at that voltage. In this example, the resistance will be seen to be 3Ω, and as the 'curve' is a straight line, this value would apply for any chosen voltage.

1.5. Series load resistors.

Figure 1.5.1. shows the device with a series load resistor connected.
Figure 1.5.2. shows the device with a shunt load resistor connected.

Before we attempt to deal with the problem of determining the total effective resistance at any given voltage, we need to obtain the I/V graph for the device we intend to use. This is given in figure 1.4.2. by the line 0–A.

If we now refer to the circuit diagram in figure 1.5.1., we will discuss the steps that need to be taken in order to derive a composite I/V curve for the complete circuit, i.e., the device plus its series connected load resistor.

Figure 1.4.2. also contains the composite I/V curve. The line 0–A represents the device alone. The line 0–B represents the 6Ω resistor alone. The technician engineer will need to draw the latter line himself on the given characteristics. As we are considering a series circuit, we may choose any

ELECTRICAL NETWORKS AND GRAPHS

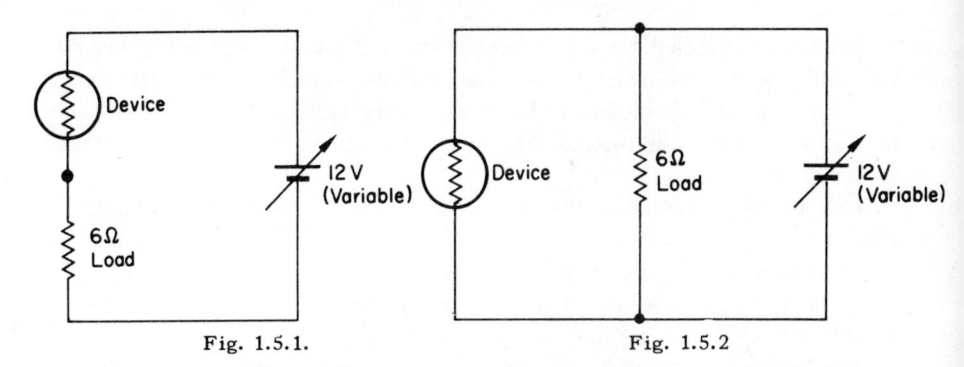

Fig. 1.5.1.

Fig. 1.5.2

convenient current (as this will be common to both components), note the p.d., across both considered separately, sum them and mark a point on the graph on a line corresponding to both the chosen current and the summed p.d.'s.

The current chosen for this example was 1 amp. The p.d. across the device is seen to be $1 \times 3 = 3$ V. (point C on the graph). The drop across the load resistor is seen to be $1 \times 6 = 6$ V. (point D on the graph).

Summing these potentials gives us 9 V. This point is plotted above the 9 V point on the V axis, and on the line representing 1 amp. (point E on the graph). The composite characteristic is drawn from 0 through point E and extended to the full length of the graph. Therefore the line $0-F$ is the composite 'curve' for the device plus its series 6Ω resistor.

If the characteristic is non-linear, as with a Diode, a number of similar points would need to be drawn in order to obtain a composite 'curve'. This technique is very useful when dealing with a Diode having a load resistor series connected and an alternating sinusoidal supply. The load voltage is easily determined with this dynamic characteristic.

1.6. Shunt load resistors.

We will now discuss the method previously described, but in relation to a shunt load. Figure 1.5.2. shows the circuit arrangement we are considering. The load resistor is seen to be connected in parallel with the device. The line $0-A$ represents the device as with the previous example. We have already established that it behaves as a 3Ω resistor. The line $0-B$ represents the load resistor considered alone, as before.

The voltage is common to both components as may be seen from the circuit diagram. The currents through the components will depend upon their relative values and may be determined by the use of Ohm's law.

On this occasion we select a suitable voltage and calculate from the graph, the current flowing through the device and load respectively. If we choose 12 V, we can see that the current through the device is $12V/3\Omega = 4$ A. Similarly, we can see that the current that would flow through the load would be 2 A. These are marked as points A and B respectively.

You will recall that for a common current in the last example, we summed the voltages. In this example, we have chosen a common voltage and must sum the currents. These sum to 6 A for the selected voltage of 12 V. A point is plotted on the graph that corresponds to 6 A on the line above 12 V, this is marked as point G.

The line $0-G$ represents the composite 'curve' for the complete shunt circuit.

The slope of the composite curve is a measure of its resistance. If we select say, 6 $\dot{V}$, we can see that a circuit current of 3 A would flow. The circuit resistance is therefore $6V/3A = 2\Omega$.

The reader is advised to refer to page 3 and to calculate the effective resistance of a 3Ω and 6Ω resistor in parallel and convince himself that he has mastered this technique before proceeding with the section dealing with load lines.

1.7. Introduction to load lines.

In both of the previous examples, the supply voltage was assumed to be variable. This would be the case for instance, if the supply was alternating in a sinusoidal manner. (An alternating supply varies in amplitude in a particular manner and will be discussed at a later stage).

When we have a steady supply voltage (d.c.), it is not normally necessary to derive a composite curve for a circuit containing a device and series load. During this example we will see how to plot a single load line for a load resistor and to 'read off' from the graph, the voltage distributions and circuit current. The circuit we wish to discuss is given in figure 1.7.1.

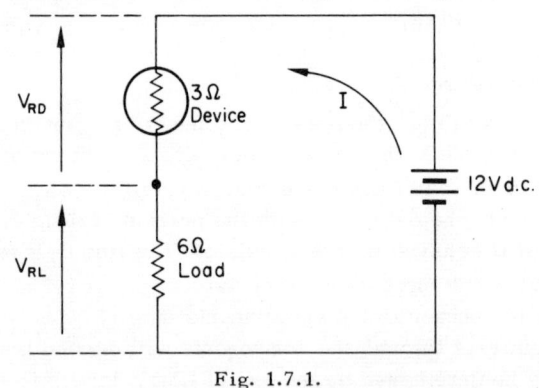

Fig. 1.7.1.

The given characteristic for the device in figure 1.7.1. is shown in figure 1.7.2. by the line $0-A$.

Students should try to appreciate that information on transistors, valves and many other electronic components are often presented in graphical form and that the techniques discussed here are of paramount importance,

ELECTRICAL NETWORKS AND GRAPHS

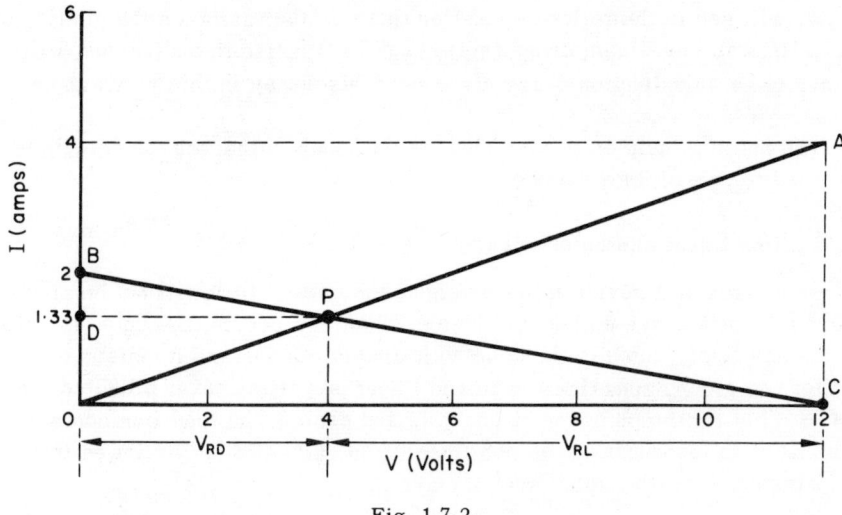

Fig. 1.7.2.

particularly when dealing with the more complex components we will meet later on in this book.

1.8. Voltage distribution in a series circuit.

The device characteristic is shown on the graph in figure 1.7.2. by the line $0-A$. This is often the main information one is likely to be given on printed characteristics. We will see later however, that graphs of more advanced electronics components follow this general principal, but are more complex.

Positioning a load line is a very simple matter and one method is as follows;

1. Ignore the 'curve' of the device. (Line $0-A$).
2. Refer to the circuit diagram — assume that the device resistance is zero.
3. Calculate the current that would flow through R_L.
4. Mark this current on the Y axis of the graph. (Point B).
5. Mark a point on the X axis corresponding to the supply voltage (Point C).
6. Connect points $B - C$ with a straight line.

$B-C$ is the load line for the 6Ω load resistor. The point P identifies the intersection of the device curve and the load line. The dotted line perpendicular to point P is dropped down to the X axis, and terminates at a point seen to be 4 V. The distance 0–4 V gives us the device voltage, V_{RD}. The remainder of the supply voltage, seen in this case to be $12 - 4 = 8$ V, is developed across the load resistor, V_{RL}. The circuit current flowing is seen to be 1.333 A, as indicated by the line drawn from point P to the Y axis.

We will see in the following section that our discussion on the plotting of load lines for non-linear components, such as transistors and valves, will prove to be an extension of the arguements discussed in this very simple case.

The basic principles remain however, although further reasoning will be applied to consolidate the position.

1.9. Non-linear characteristics.

When a curve of a device is not straight, the current flow will not be proportional to the voltage across the device. When this occurs, the curve is said to be non-linear. Such is the case with diodes, valves and transistors. The latter two groups sometimes approach linear proportions over a limited range of use, but for the purposes of analysis and design, the load line approach has much to recommend it as non-linearity in curves is fully allowed for.

Figure 1.9.1. illustrates such a curve.

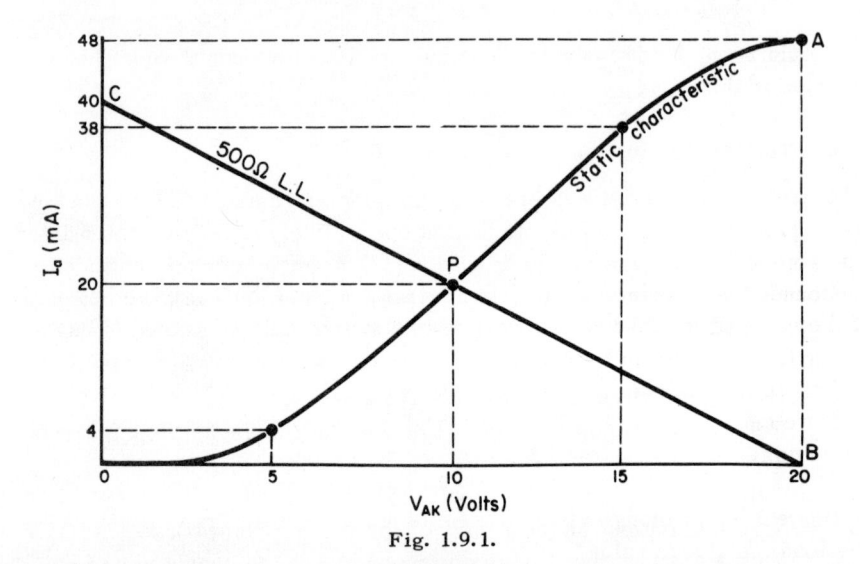

Fig. 1.9.1.

If we were to plot a table of I/V, we would see that the resistance of the device changes with different voltages, This typifies the principles of non-linearity. The d.c. resistance of the device at the voltages chosen is given in the following table.

V	I (mA)	R (Ohms)
20	48	417
15	38	395
10	20	500
5	4	1250

When we require to know the d.c. resistance, we simply select a voltage, note the current, and by Ohm's law, calculate the resistance, $R = V/I$ Ohms.

We will see later how a.c. resistance differs from d.c. resistance, and when it is desirable to know either or both. We will not concern ourselves with this subject at this stage.

Figure 1.9.2. shows the circuit diagram relating to the characteristics shown in figure 1.9.1. Note that for the first time, we now have an actual electronic device and that its resistance is not constant but varies disproportionately to its potential. In a practical circuit, the e.m.f. may be the High Tension (H.T.) supply to the circuit.

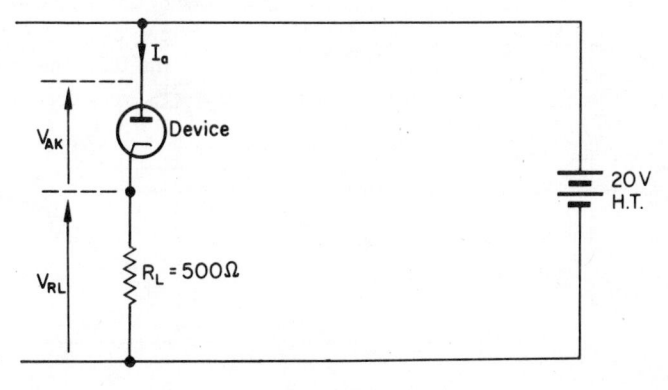

Fig. 1.9.2.

Connecting a series load resistor to this device is quite a common practice. If the value of the load resistor is known, then for a given supply voltage, a simple load line can be drawn thus enabling us to quickly determine the circuit current and the voltage distributions.

If we knew the resistance of the device at a given voltage across it, we could quite easily determine the voltages across the components by using the 'load over total' technique. The problem is, of course, that for the circuit in figure 1.9.2. although we know the d.c. supply voltage (H.T.), and know the load resistance value, we do not know the device resistance. Before we can determine the latter, we need to know either the device voltage or current and then, from the graph, we could easily evaluate the unknowns.

This problem is very much akin to the 'chicken and the egg'.

The load line technique overcomes this problem because as we have seen, we completly ignore the device and its curve, when plotting the load line. The intersection of the curve and the load line will, as before, provide us with our answers without any undue effort.

The load line shown in figure 1.9.1. is positioned in the same manner as before. It is identified as a 500Ω load line as shown.

12   ELECTRONICS FOR TECHNICIAN ENGINEERS

1.10. Plotting the points for positioning a load line.

Before positioning further load lines, we will discuss how to indicate where a point lies on a graph by means of co-ordinates.

Students will remember that a point on a graph may be shown as X, Y.

Example.

The expression (16,84) indicates that the point on a graph lies 16 units to the right of the zero and 84 units above the X axis.

A diode in series with a 5KΩ load is connected to a 200V supply. Figure 1.10.1., shows the device (a) open circuit and (b) short circuit.

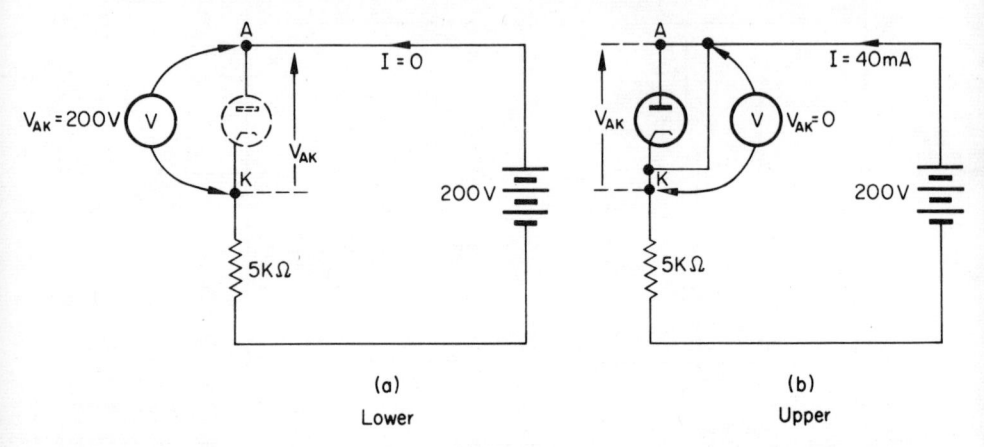

```
         (a)              (b)
        Lower            Upper
```

Fig. 1.10.1

If we assume that the device is open circuit, the current flow will be zero. Hence $Y = 0$.

As the device is open circuit, the potential across the device terminals will be equal to the H.T. (This of course assumes a loss free meter). Hence $X = 200$.

The co-ordinates for the lower end of the load line may be expressed as 200,0. This is shown as point A on the graph in figure 1.11.1.

Conversely, with the device short circuited, as in figure 1.10.1.,b, the current will be of the value H.T./R_L. Hence $Y = 40$. It follows therefore, that $X = 0$.

The co-ordinates for the upper end of the load line will be 0, 40.

This is marked on the graph as point D, in figure 1.11.1.

1.11. Drawing load lines on restricted graphs.

A load line may need to be drawn on a graph for a given resistor and for a supply that has a greater value than that shown on the graph.

ELECTRICAL NETWORKS AND GRAPHS 13

Fig. 1.11.1.

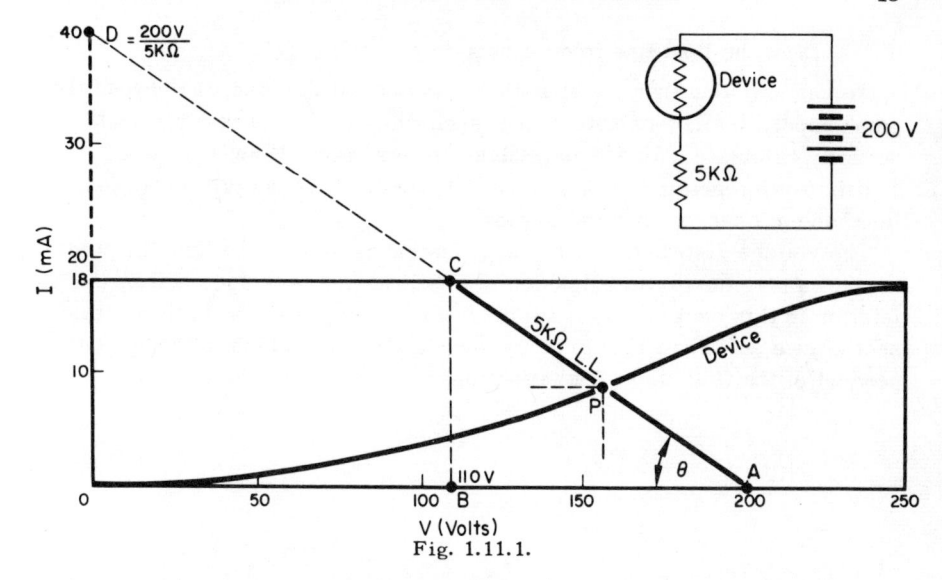

We sometimes meet the problem of plotting a load line which requires an upper point on the graph that has a greater value than that shown on the graph. Figure 1.11.1. illustrates one example where the graph cannot accomodate the upper end of the load line. The circuit diagram is also given.

The H.T. is seen to be 200 V. The load resistor has a value of 5000 Ω. The lower end of the load line will be positioned at the H.T. point 200, 0. The upper end of the load line should terminate at a point H.T./R_L = 40 mA. i.e., 0, 40.

The graph shown has a maximum value of 18mA.

A very important aspect of positioning load lines is that when correctly positioned, the load line will be drawn at a particular angle, θ. Where θ is the angle that the load line makes with the base or voltage line.

This angle is solely dependant upon the resistor value and is unaffected by the H.T. to be employed. (This was seen to be the case in 1.2.1.).

The following steps show just how we can position our load line in these and similarly difficult circumstances.

1. Position point A. This will be at the H.T. potential, i.e. 200, 0.
2. Note the maximum value of current printed on the graph. (18 mA in this case).
3. Multiply this current by the load resistance to produce a voltage Vx. In this example, $Vx = 5000 \times 18/1000 = 90$ V.
4. Position a temporary point on the voltage axis at Vx volts below the H.T. (In this case the point is 200 V − 90 V = 110 V and is marked as point B).
5. Position a temporary point corresponding to (H.T. − Vx) and the max' current. In this example, this point is 110, 18 and is marked as point C.

6. Draw the load line from point A to point C.

We can see from figure 1.11.1. that if we extend the load line beyond the graph paper, it will terminate at the 'short-circuit' current that we would normally calculate with a graph embracing this value. Point D.

It is worth repeating at this stage that the angle of the full length load line has not changed from the original θ.

The voltage distribution and circuit current is determined from the point p as shown previously. An extension to this load line technique will be covered later on in this book when we are discussing Triode valves. In this later section, we will show how to plot a second load line which, although independent of the first, is no less important for rapid design or analysis.

CHAPTER 2

Further networks and simple theorems

2.1. Internal resistance.

In previous examples, the source of e.m.f. (e.g., the battery), has been assumed to maintain its e.m.f. at a steady value irrespective of the load which may be connected to it. In fact, a practical source of e.m.f., always has some internal resistance, and the current which it supplies will always cause some voltage drop within the battery itself.

This has two effects. Firstly, the available voltage for the circuit connected to the battery is reduced by the amount of the voltage drop within the battery and, secondly, the current flowing through the internal resistance dissipates heat, so that part of the chemical energy of the cell is 'lost' as far as the external circuit is concerned.

We may take into account the internal resistance of a source by considering it to be made up of two parts in series, a constant e.m.f., plus a resistance equal to the internal resistance of the source, as shown in figure 2.1.1.

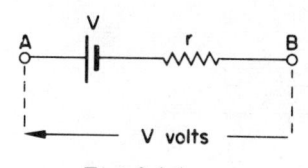

Fig. 2.1.1.

We take as an example a battery of 10 V e.m.f. and a 1 ohm internal resistance, connected to a 9 ohm resistor as shown in figure 2.1.2.

A and B are the actual battery terminals. The current is clearly 1A, and the p.d. across the load resistor is therefore 9 V, this being 1 V less than the battery e.m.f. The other 'lost' volt is accounted for my the drop across the internal resistance.

If more current is taken from the battery, by reducing the value of the load resistance, then the terminal voltage of the battery will fall still more. For example, with a 4 Ω load, the current will be 2 amps and the load voltage will be only 8 volts. It should be clear that if a supply is to have a good 'regulation', that is to say its terminal voltage is not to vary too much with varying loads, then its internal resistance must be very small compared with the load resistances which it is proposed to connect to the supply source. If the source is on 'no load' meaning that the terminals are open, then of course the terminal voltage will be equal to the e.m.f., since there will be no loss of p.d. across the internal resistance when there is no current flowing.

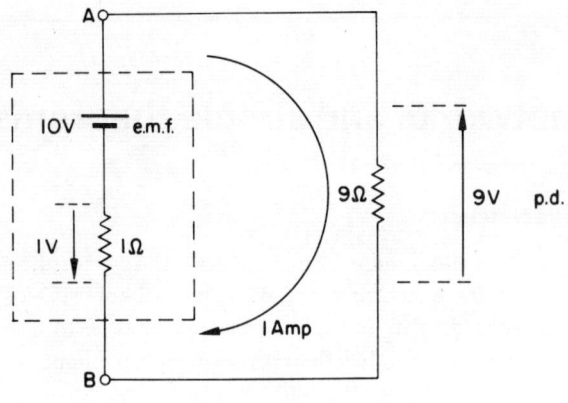

Fig. 2.1.2.

2.2. Effective input resistance.

Suppose we are given a 'black box' which has two input terminals but are told nothing of what is inside. We might try to find out the contents by connecting a known voltage to the terminals and measuring the current flowing into the box. Suppose that on connecting the box to a 10 V battery (with negligible resistance), through an ammeter we found that the current was 10 A. Then by Ohm's law, we could deduce that the box contained a 1 Ω resistor. Figure 2.2.1.

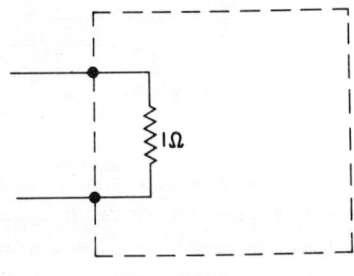

Fig. 2.2.1.

Of course if we opened the box we might well discover that any one of a number of possible circuit arrangements actually existed inside of the box. Two such possible arrangements are shown in figure 2.2.2.

Despite the marked differences in circuit arrangements shown in figure 2.2., there can be no doubt that the *effective* input resistance in each case is indeed 1 Ω. We could have put this another way by stating that either of the circuits could have caused the 'investigator' to believe that each box contained a 1 Ω resistor and further, had either box been connected to any other external circuit, that external circuit would have 'seen' a 1 Ω resistor. There is an important lesson to be learnt at this stage, the actual resistance

FURTHER NETWORKS AND SIMPLE THEOREMS

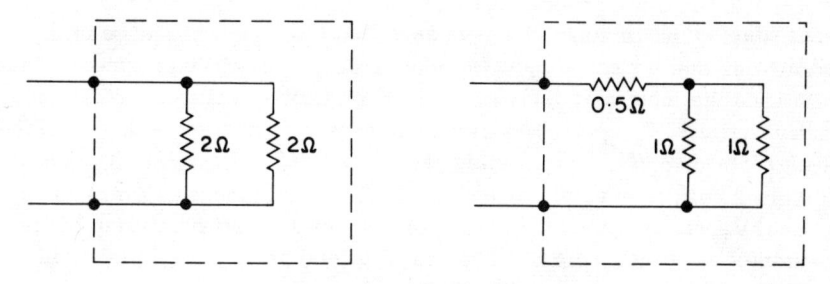

Fig. 2.2.2

inside a 'box' may differ considerably from the 'effective' resistance looking into the input terminals. This difference may cause considerable confusion unless it is carefully studied.

A box is shown in figure 2.2.3. which contains a $0.5\,\Omega$ resistor in series with a 5 V battery. The *actual* resistance is obviously $0.5\,\Omega$. What happens if we try to find out the effective input resistance of this box by means of our little test? The reader will see that since the two batteries are in series opposition, the effective e.m.f. is 5 V and the current measured would be $5/0.5 = 10\text{A}$. This is the same value of current we measured in our previous test and we are lead to believe that the box contains a $1\,\Omega$ resistor. Upon opening the box, however, we would find that it actually contained a $0.5\,\Omega$ resistor.

Suppose that we repeated this test, only this time, we reversed our 10 V battery. In this case, the reader will readily find that the input current would be not 10A but 30A. This time he could hardly be blamed for deducing that the box contained a $1/3\,\Omega$ on this occasion.

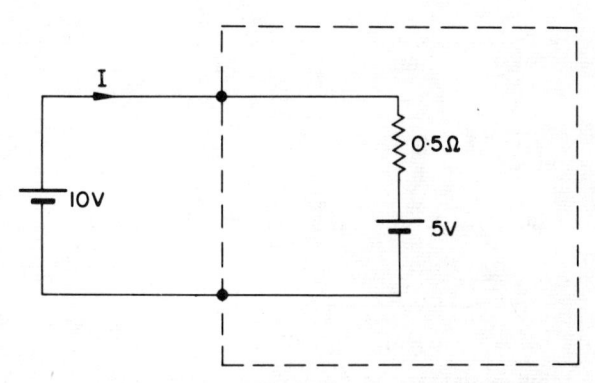

Fig. 2.2.3.

The reader will readily appreciate that if we had used a 5 V battery for our test, no input current would have flowed and he would have been forced to conclude that the resistance inside of the box was infinite.

Summing up our findings, we have seen that the input resistance of the box had a value that depended upon the input voltage we applied during our tests. Sometimes the input resistance was 1 ohm, on another occasion $1/3\,\Omega$ and finally, infinity. The point here is that when we are dealing with a black box which contains *active* components (like the battery in figure 2.2.3,) the input resistance will, in general, depend upon the circumstances in which the box is used. Valves and transistors are examples of devices which may be treated in circuits as if they were black boxes, provided proper care is taken in specifying the conditions under which they are operating.

It may be of interest to mention in passing that for the box in figure 2.2.3, to be a better analogue for say a transistor, it should be assumed that the 5 V battery in the box is only present when an input voltage is applied. However this is a complication which need not concern us at the moment as it in no way affects the principle or the definition of input resistance.

2.3. Four-terminal devices.

Most of the devices used in electronics like valves and transistors are 'four-terminal' devices, with one pair of input terminals and one pair of output terminals. When we say that the effective input resistance of the device depends upon the conditions under which it is used, we must include in these conditions the load which is connected to the output terminals. This applies to both passive and active devices, and the example given below should make this clear.

A four-terminal device has the circuit shown in figure 2.3.1.

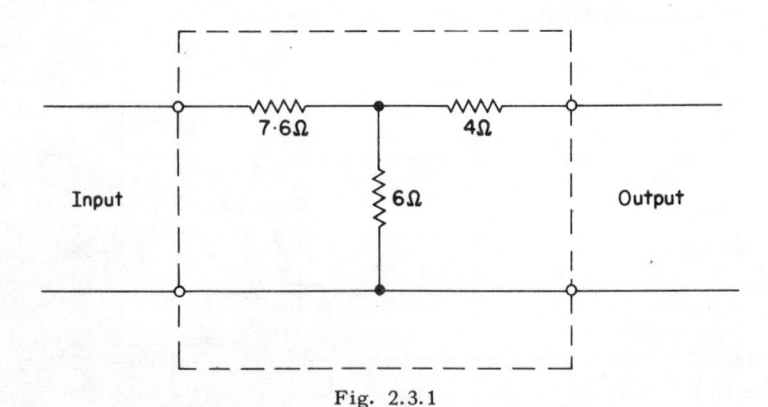

Fig. 2.3.1

What is the input resistance of the box shown in figure 2.3.1. if its output terminals are (a) open-circuited and (b) short-circuited?

On open circuit it is obvious that the input resistance is $7.6 + 6 = 13.6\,\Omega$.

If the output terminals are short circuited together, however, the $4\,\Omega$ and the 6 ohm resistors will be connected in parallel giving a combined resistance of 2.4 Ω. When this is added to the 7.6 Ω, the input resistance of 10 Ω is obtained.

Let us now take this one step further and examine a box shown in figure 2.3.2.

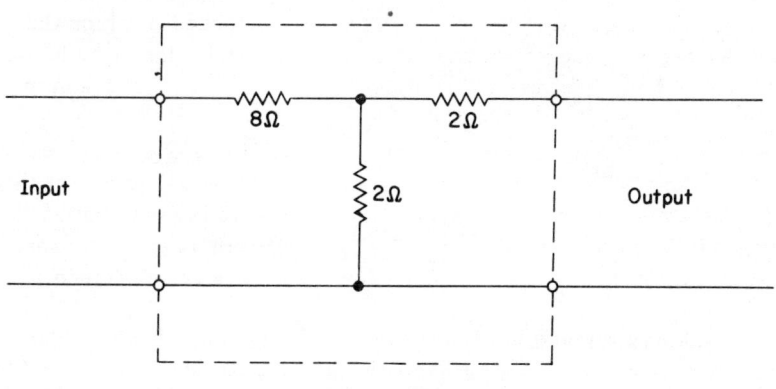

Fig. 2.3.2.

What is the input resistance with

    (a) the output terminals open circuit?

    (b) the output terminals short circuit?

What is the output resistance with

    (c) the input terminals open circuit?

    (d) the input terminals short circuit?

(a) Simply adding the 8 Ω and 2 Ω gives us 10 Ω.

(b) The two, 2 Ω resistors are in parallel giving 1 Ω. Adding this 1 Ω to the 8 Ω gives us 9 Ω.

(c) Looking into the output terminals with the input terminals open circuit we simply add the pair of 2 Ω resistors to give 4 Ω.

(d) In this case, the 8 Ω and the 2 Ω are in parallel = 1.6 Ω.
Adding the 1.6 Ω and the 2 Ω results in an output resistance of 3.6 Ω.

The reader should not assume that these examples are of academic interest only, for in practice unless the technician engineer has a good understanding of these and other basic principles, even very simple design and analytical work may be frustrating in that practical tests may not agree with those results that were theoretically predicted.

We will discuss in the following section, generators other than the simple voltage source discussed so far.

2.4. Voltage and current generators.

Fig. 2.4.1.

(a) (b)

We are able to replace either generator with the other should we so desire. Before we do however, we must know something about the relationship between them. The voltage generator develops an e.m.f. shown in the figure as e. This e.m.f., is an alternating voltage (this is often referred to as an a.c. voltage) and will appear across its terminals A and B in the absence of a load. The internal resistance, r, is shown in precisely the same manner as that of the battery in figure 2.1. As the internal resistance is shown separately from the generator, we can see that the generator itself must have zero resistance.

If we were to apply a simple test to this generator, we would find that when measuring across the terminals A and B, the no-load e.m.f. would be e volts. If we then measured the resistance across the same terminals, we would find a resistance of value r.

The current generator shown in the figure has a generator that develops a current i, and will have a value equal to the current that will flow through the voltage generator should we short-circuit the voltage generator terminals. We can calculate this current quite easily once we know the value of e and r. The current value is given by Ohm's law in the usual manner, $i = e/r$.

We have shown a shunt resistor r which is connected across the current generator terminals, and if we were to apply our simple test once again, only this time, to the current generator, we would find that the resistance across its terminals would be r ohms. It follows then, that the current generator itself must have infinite resistance. Further, the potential across the current generator terminals would be e volts and equal to $i \times r$.

Two important facts have emerged, one is that the voltage generator has zero resistance and the second, the current generator has infinite resistance.

2.5. Input resistance, current operated device.

We discussed in the earlier pages of this chapter how to apply a simple test to a box in order to determine the input resistance. Figure 2.5.1. shows another method of determining input resistance. The reader will see that whereas in previous examples we applied an input voltage, we intend in this little test, to apply a current instead. What will happen when we do so?

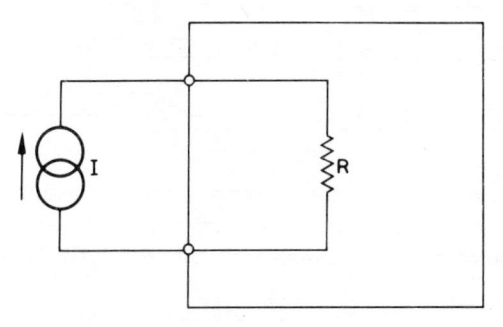

Fig. 2.5.1.

If we apply a current input of known value, the current flowing through the resistance R will produce a p.d. across the input terminals. The input resistance by Ohm's law, will be given by dividing the p.d. by the applied *input current*.

If R has a value of 10Ω, and i is 1 amp, then the p.d. will be $10v$. The input resistance will of course be

$$\frac{10v}{1\,\text{amp}} = \frac{10}{1} = 10\,\Omega$$

This is a very simple example and the reader could not be critiscised for believing that this simple example is untypical of the real life practical problems with which he is likely to come into contact; he would be right.

The example in figure 2.2.4 showed a box that contained a battery that set up an e.m.f. in response to an applied external voltage. Such is the case with the box for which a current input is appropriate. These also may have a generator 'inside' the box, one that responds to an external current applied to the input terminals.

Let us take this one step further and examine a box that does have an internal current generator, one that will generate a current in response to an applied input current. Figure 2.5.2. shows the circuit we intend to discuss.

We have learnt that a current generator has an infinite resistance. It follows therefore, that as this is so, we cannot force a current into it from another source. If i is the applied input current, and i' the internally generated current, then when we apply our input, a p.d. will be developed across

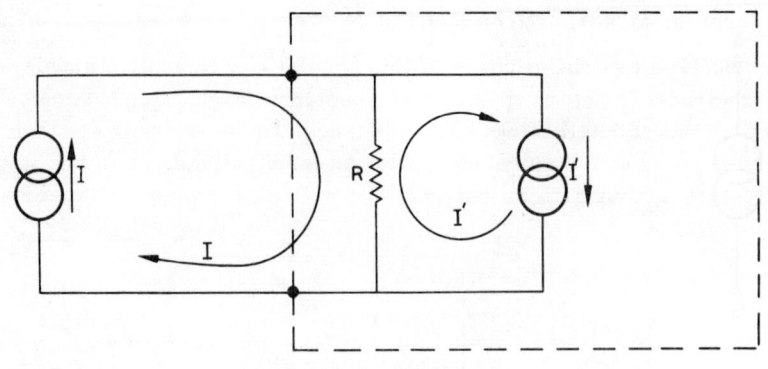

Fig. 2.5.2.

R due to the algebraic sum of the currents i and i' flowing through R. In this example, i' is flowing in such a direction so as to oppose, or subtract from, the input current flowing through R. The p.d. will therefore be

$$(i - i')R = v$$

The input resistance

$$\frac{v \text{ in}}{i \text{ in}} = \frac{v}{i} = \frac{(i-i')R}{i} = \frac{iR - i'R}{i} = R - \frac{i'R}{i} = R\left(1 - \frac{i'}{i}\right)$$

If i is 1 amp, and i' 0.5 amp, and $R = 10\,\Omega$, then the effective input resistance will be

$$10\left(1 - \frac{0.5}{1}\right) = 5\,\Omega.$$

If the internal generator had been reversed, then both currents would have flowed through R in the same direction, and augmented each other.

The total current through R would have been $i + i'$ and the p.d. would have become, $(i + i')R$. Hence the input resistance would be

$$\frac{v}{i} = \frac{(i + i')R}{i} = R\left(1 + \frac{i'}{i}\right)$$

Using the previous values, the input resistance,

$$R \text{ in} = 10\left(1 + \frac{0.5}{1}\right) = 15\,\Omega.$$

The reader was given earlier on, a hint that transistors could often be treated as though they were a black box. We ought to take this a little further now and examine a box, similar to the previous example, that takes us a step nearer to an actual transistor. Figure 2.5.3 is an extension of figure 2.5.2.

The 6 Ω resistor has been added to give a degree of realism to this example. The reason for its presence need not concern us at this stage, but is included so that we can embrace it within our examination of the circuit.

FURTHER NETWORKS AND SIMPLE THEOREMS 23

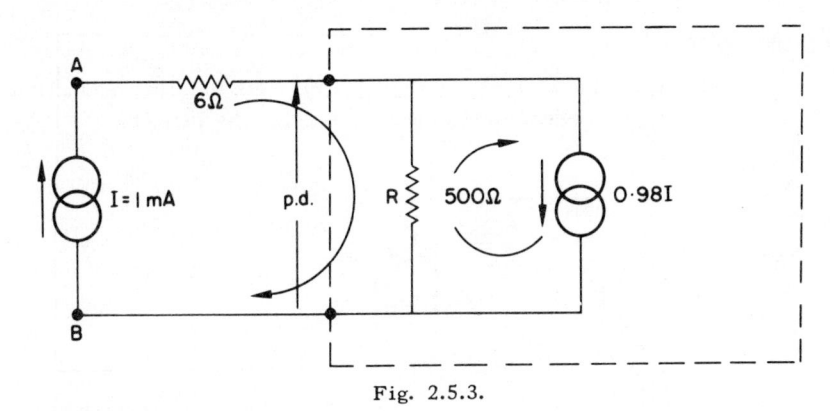

Fig. 2.5.3.

The reader will reason for himself that the only current to flow through the 6Ω resistor outside the box will be the input current i.

We have seen that the effective input resistance is determined by the p.d., set up across the input terminals resulting from both i and $0.98i$. As the potential across the 6Ω will not affect the 'input p.d.', we will discard the 6Ω resistor for now, and replace it once we have established the effective input resistance to the box proper.

When i is applied, the internal generator develops a current having a magnitude of $0.98i$. The effective current flowing through R is therefore $i - 0.98i = 0.02i$.

The p.d. across R becomes $0.02\,\text{mA} \times 500\,\Omega = 0.01\,\text{V}$.

The effective input resistance to the *box* becomes $0.01\,v/1\,\text{mA} = 10\,\Omega$.

It now remains for us to replace the 6Ω resistor to complete the story. The total input resistance to the complete circuit at the terminals A to B is therefore the sum of $10 + 6 = 16\,\Omega$.

2.6. Simple theorems.

The student should understand how to apply the largest number of theorems for his later, more advance, analytical work in electronics. The more theorems he can master, the more alternatives he will have when he considers which approach to use in order to solve circuit problems. He will see that for instance, if he is able to change the form of a circuit to another that is electrically identical, he might complete an analysis in a much reduced time.

For instance, we might attempt to solve a problem using 'voltages'. We should obtain a correct answer of course, but it might take a long time. If we had say, used 'currents' instead, we might have obtained the same answer in less than half the time. The reverse could be equally true, of course.

In order to facilitate the analyses of circuits and networks, we will discuss some of the more common theorems and show how to apply these and compare them with one another.

2.7. Kirchhoff's laws.

Kirchhoff's first law. The algebraic sum of the currents entering a junction is equal to the currents leaving the junction.

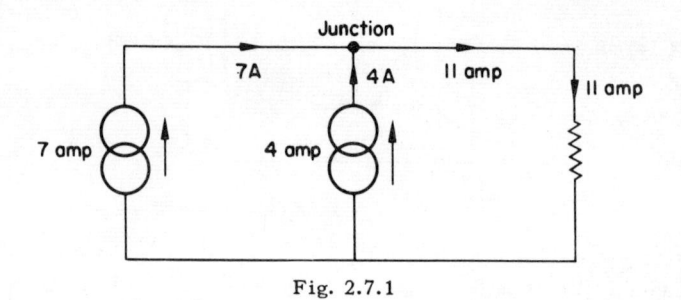

Fig. 2.7.1

Kirchhoff's second law. The algebraic sum of the *I.R.* drops in a closed loop is equal to the effective e.m.f. in the loop.

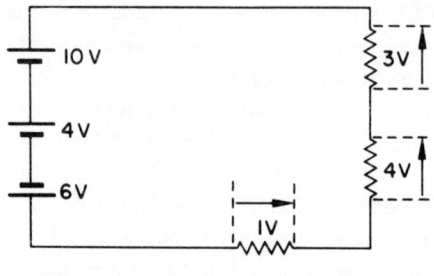

Fig. 2.7.2.

Example.

Calculate the p.d. across the 0.1Ω resistor and all battery currents, in the circuit shown in figure 2.7.3.

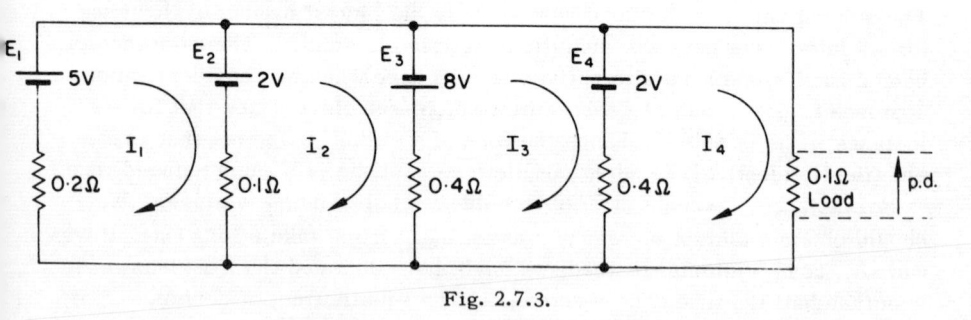

Fig. 2.7.3.

We will deal with this problem by means of Kirchhoff's second law. We have labelled the currents in each loop and have assumed a clockwise

FURTHER NETWORKS AND SIMPLE THEOREMS

rotation for each one. If any of these assumptions are wrong, the correct answer will contain a negative value for the current concerned. It is then a simple matter to note this and, if necessary, correct the sketch.

We have four unknown currents and therefore we must have four equations. We will find an expression for I_1, substitute this in the second equation and find an expression for I_1. Having found an expression for I_2 we substitute this in the third equation and find an expression for I_2. This will give us an expression for I_3 which is finally substituted for I_3 in equation 4 leading to a derivation for I_4.

I_4 then, will have a known numerical value. Once this has been obtained, we substitute it in equation 3 and find a numerical value for I_3.

This numerical value for I_3 is substituted in equation 2 and the value of I_2 is found.

The numerical value of I_2 is similarly substituted in the first equation in order to find a numerical value for I_1.

The actual battery currents are then found by taking the difference between the loop currents flowing through the appropriate battery. The p.d. is found by simply calculating the product of I_4 and the 0.1 ohm load resistor.

$$V_5 = I_4 \times R_5 .$$

$$I_1 \downarrow \quad 3 = 0.3\,(I_1) - 0.1\,(I_2) \quad (1)$$

$$I_2 \downarrow \quad 10 = -0.1\,(I_1) + 0.5\,(I_2) - 0.4\,(I_3) \quad (2)$$

$$I_3 \downarrow \quad -10 = -0.4\,(I_2) + 0.8\,(I_3) - 0.4\,(I_4) \quad (3)$$

$$I_4 \downarrow \quad 2 = -0.4\,(I_3) + 0.5\,(I_4) \quad (4)$$

From equation (1),
$$I_1 = \frac{3 + 0.1(I_2)}{0.3}$$

Substituting for I_1 in equation (2),

$$10 = -0.1\,\frac{[3 + 0.1\,(I_2)]}{0.3} + 0.5\,I_2 - 0.4\,(I_3)$$

Multiplying both sides of the equation by 0.3

$$3 = -0.3 - 0.01\,(I_2) + 0.15\,(I_2) - 0.12\,(I_3)$$

$$\therefore I_2 = \frac{3.3 + 0.12\,I_3}{0.14}$$

Substituting for I_2 in equation (3)

$$-10 = -0.4\,\frac{[3.3 + 0.12\,(I_3)]}{0.14} + 0.8\,I_3 - 0.4\,I_4$$

$$\therefore\ I_3 = \frac{-0.08 + 0.056 I_4}{0.064}\text{, multiplying top and bottom by 100,}$$

then substituting for I_3 in equation (4)

$$2 = -0.4\frac{[-8 + 5.6\,(I_4)]}{6.4} + 0.5 I_4$$

$$\therefore\ I_4 = \frac{9.6}{0.96} = \underline{10\text{ Amps.}}$$

Hence the p.d. across the load = 10×0.1 = 1 V.

We need now to find for I_3

From equation (4) $\quad 2 = -0.4 I_3 + 0.5 I_4$

$$\therefore\ 2 = -0.4 I_3 + 0.5 \times 10$$

$$\therefore\ I = \frac{-3}{-0.4} = \underline{7.5\text{ Amps}}$$

I_2 may now be found. From equation (3),

$$-10 = -0.4 I_2 + 0.8\,(7.5) - 0.4\,(10)$$

$$\therefore\ I_2 = 30\text{ A.}$$

It is left to the reader to verify the remaining loop currents.

The current in the cell $E_4 = I_3 - I_4 = 10 - 7.5 = \underline{2.5\text{ A}}$
The current in the cell $E_3 = I_3 - I_2 = 30 + 7.5 = \underline{-22.5\text{ A}}$
Similarly, cell currents E_1 and E_2 are 20 A and 10 A respectively.

    We are now able to redraw the circuit and label the cells or batteries with the currents that are actually flowing through them, rather than round the loops. This revised circuit is given in figure 2.7.4.

    The reader should note that the battery, or cell, currents in figure 2.7.4. have been labelled numerically so as to coincide with the battery and resistor number. Hence I_1 in *this* circuit does not refer to the *Loop* current in the previous figure, but relates to the *Battery* current given in the answer to the problem.

    The reader is probably already aware that the sum of the battery currents flowing towards the junction (J) is equal to the current leaving the junction and finally flowing down through the load resistor, R_5. This suggests perhaps that we might be able to tackle the original problem by using Kirchhoff's first law. Let us discuss this suggestion, see if we are able to derive a method or formula perhaps, and compare the amount of work it entails in solving the same problem.

FURTHER NETWORKS AND SIMPLE THEOREMS 27

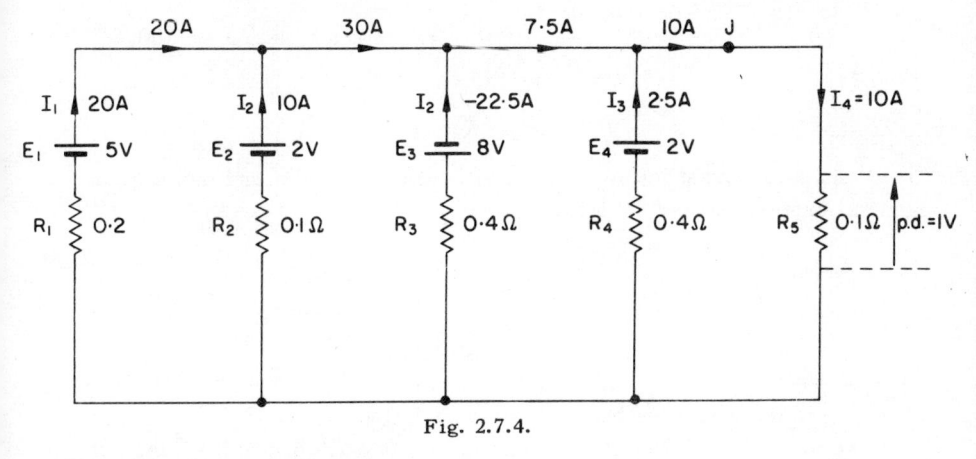

Fig. 2.7.4.

2.8. Derivation of a formula.

It is not necessary to deal with a difficult circuit for this task, and principles we might establish should be valid for the circuit in figure 2.7.4.

Let us examine the simple circuit in figure 2.8.1.

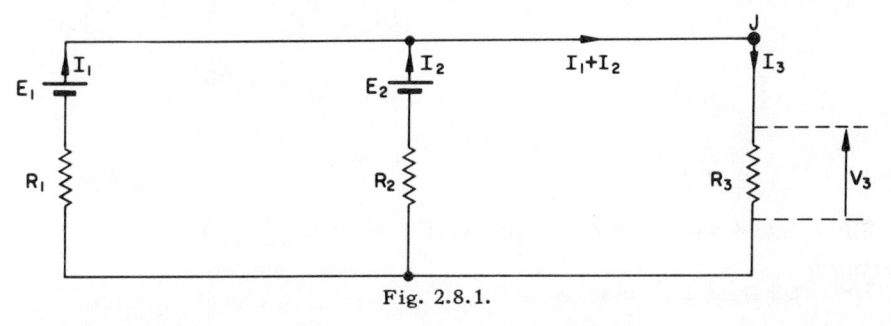

Fig. 2.8.1.

One of the most important, yet basic, rules for design, is to draw a circuit, decide upon the required circuit conditions, assume that they exist, apply Ohm's law and calculate the resistor values which will satisfy the circuit requirements. We have our circuit in figure 2.8.1., we want a p.d. across R_3, and we have decided to label the battery currents, I_1 and I_2.

We can see that

$$I_1 + I_2 = I_3. \quad I_1 = \frac{E_1 - V_3}{R_1}, \quad I_2 = \frac{E_2 - V_3}{R_2}, \quad I_3 = \frac{V_3}{R_3}$$

Hence

$$\frac{E_1 - V_3}{R_1} + \frac{E_2 - V_3}{R_2} = \frac{V_3}{R_3}$$

and collecting V_3 terms,

$$\frac{E_1}{R_1} + \frac{E_2}{R_2} = V_3 \left[\frac{1}{R_1} + \frac{1}{R_2} + \frac{1}{R_3} \right]$$

Hence
$$V_3 = \frac{\frac{E_1}{R_1} + \frac{E_2}{R_2}}{\frac{1}{R_1} + \frac{1}{R_2} + \frac{1}{R_3}}$$

This can be extended for any number of batteries or cells and therefore a general expression may be written as

$$V_{n+1} = \frac{\frac{E_1}{R_1} + \frac{E_2}{R_2} + \ldots \frac{E_n}{R_n}}{\frac{1}{R_1} + \frac{1}{R_2} + \ldots \frac{1}{R_{n+1}}}$$

Where n is the number of batteries or cells in the circuit.

If we now return to the problem shown in figure 2.7.4. and by using our formula, the load p.d.,

$$V_5 = \frac{\frac{5}{0.2} + \frac{2}{0.1} + \frac{(-8)}{0.4} + \frac{2}{0.4}}{\frac{1}{0.2} + \frac{1}{0.1} + \frac{1}{0.4} + \frac{1}{0.4} + \frac{1}{0.1}} = \frac{30}{30} = 1 \text{ V.}$$

$$I_1 = \frac{5-1}{0.2} = 20 \text{ A.} \qquad I_2 = \frac{2-1}{0.1} = 10 \text{ A}$$

$$I_3 = \frac{-8-1}{0.4} = -22.5 \text{ A} \qquad I_4 = \frac{2-1}{0.4} = 2.5 \text{ A.}$$

Which agrees with the answers previously obtained.

2.9. Superposition Theorem.

In any linear network, the current in any branch (or its equivalent p.d. between any two networks) due to a number of voltage (or current) generators connected in any part of the network, is the sum of the currents produced by all of the generators considered one at a time with all others replaced by their internal resistances.

Example.

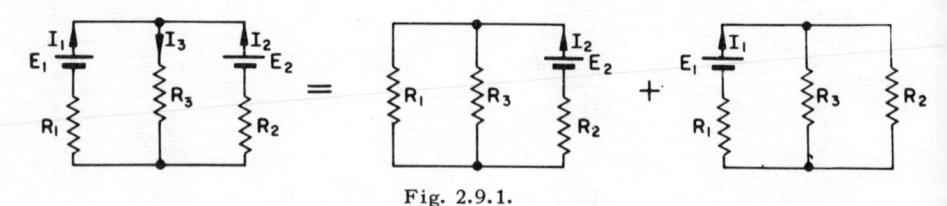

Fig. 2.9.1.

2.10. Reciprocity Theorem.

In any linear network, should a generator produce a current at any point in the network, the same current will flow if the generator and the measuring devices are interchanged.

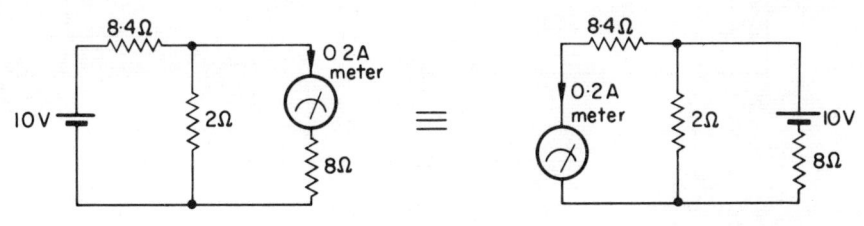

Fig. 2.10.1.

2.11. Thevinin's Theorem.

Any linear network containing voltage (or current) generators may be simplified to one single voltage generator with zero internal resistance in series with an external resistance. The generated voltage will be the same as the open circuit voltage of the complex network.

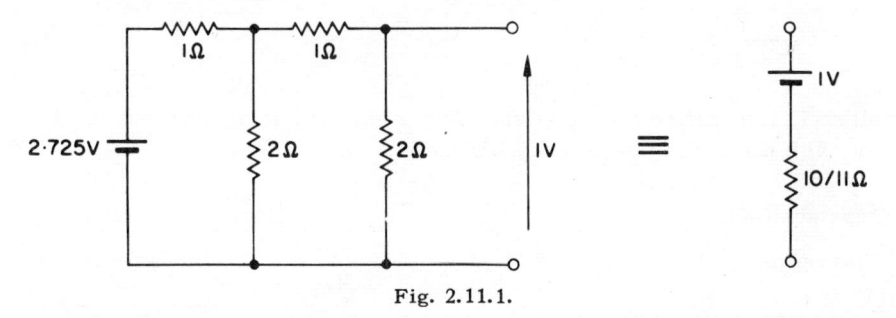

Fig. 2.11.1.

2.12. Norton's Theorem.

Any linear network containing voltage (or current) generators may be simplified to a single current generator of infinite resistance shunted by a resistance. The generator current will equal the short circuit terminal current of the original network, whilst the shunt resistance will equal the resistance seen looking into the original network terminals. (Fig. 2.12.1.)

2.13. Comparison of theorems.

A single example will now be given. We will solve the set problem by using some of the theorems and methods previously discussed. Figure 2.13.1. shows a circuit consisting of two batteries, E_1 and E_2 each with an internal resistance R_1 and R_2 connected in parallel across a load resistor R_3. We will

Fig. 2.12.1.

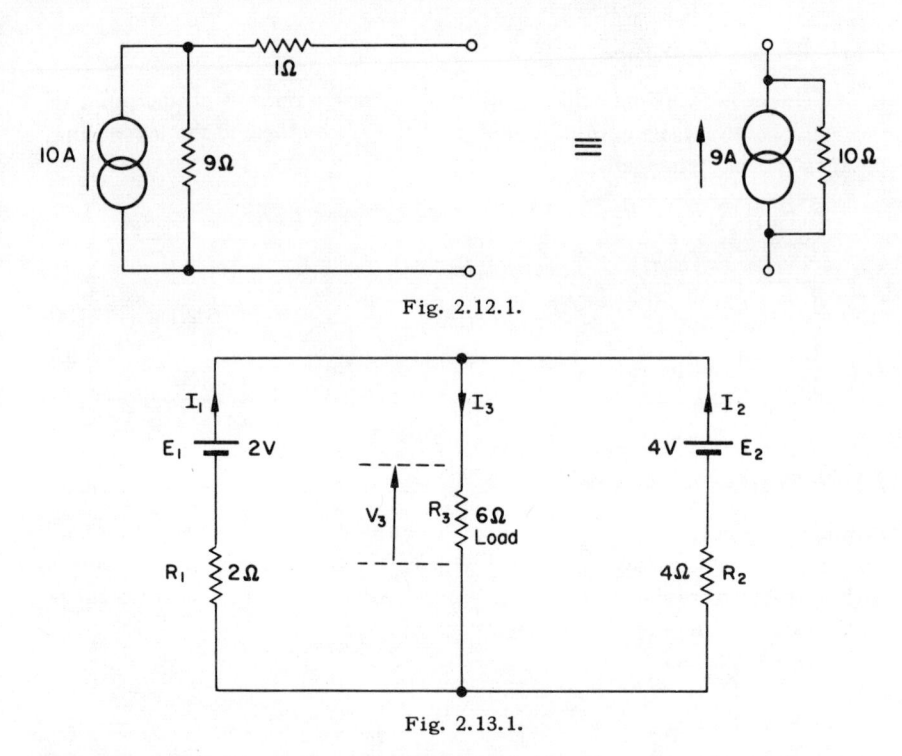

Fig. 2.13.1.

attempt to determine the numerical value of the p.d. across the load resistor, with each of the methods outlined.

Superposition.

The current I_1 due to E_1, with E_2 removed $= \dfrac{E_1}{R_1 + R_2 /\!/ R_3}$

The fraction of I_1 entering $R_3 = \dfrac{E_1}{R_1 + R_2 /\!/ R_3} \times \dfrac{R_2}{R_2 + R_3}$ \hfill (1)

The current I_2 due to E_2, with E_1 removed $= \dfrac{E_2}{R_2 + R_1 /\!/ R_3}$

The fraction of I_2 entering $R_3 = \dfrac{E_2}{R_2 + R_1 /\!/ R_3} \times \dfrac{R_1}{R_1 + R_3}$ \hfill (2)

$I_3 = (1) + (2) = \dfrac{E_1 R_2}{(R_1 + R_2 /\!/ R_3)(R_2 + R_3)} + \dfrac{E_2 R_1}{(R_2 + R_1 /\!/ R_3)(R_1 + R_3)}$

$I_3 = \dfrac{E_1 R_2}{\left[R_1 + \dfrac{R_2 R_3}{R_2 + R_3}\right][R_2 + R_3]} + \dfrac{E_2 R_1}{\left[R_2 + \dfrac{R_1 R_3}{R_1 + R_3}\right][R_1 + R_3]}$

FURTHER NETWORKS AND SIMPLE THEOREMS

$$= \frac{E_1 R_2}{\left[\frac{R_1(R_2+R_3)+R_2 R_3}{R_2+R_3}\right][R_2+R_3]} + \frac{E_2 R_1}{\left[\frac{R_2(R_1+R_3)+R_1 R_3}{R_1+R_3}\right][R_1+R_3]}$$

$$= \frac{E_1 R_2}{R_1 R_2 + R_1 R_3 + R_2 R_3} + \frac{E_2 R_1}{R_2 R_1 + R_2 R_3 + R_1 R_3}$$

$$= \frac{E_1 R_2 + E_2 R_1}{R_1 R_2 + R_1 R_3 + R_2 R_3} = \frac{(2 \times 4) + (4 \times 2)}{8 + 12 + 24}$$

$$\therefore I_3 = \frac{16}{44} = \frac{4}{11} \text{ Amp.}$$

Hence the p.d. $= I_3 R_3 = \frac{4 \times 6}{11} = \frac{24}{11}$ Volts

Thevinin's

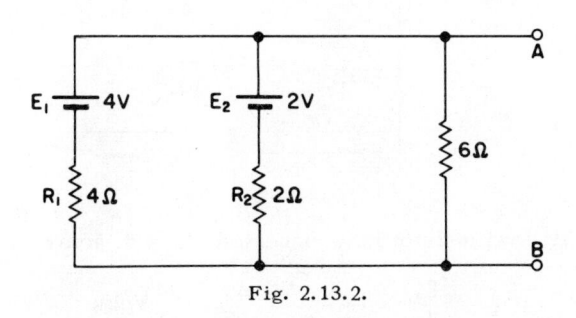

Fig. 2.13.2.

Removing the 6Ω load for now, the circulating current I becomes,

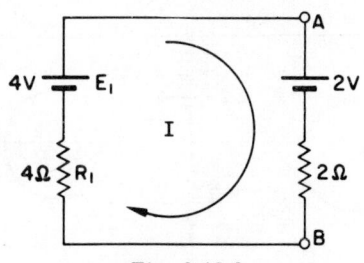

Fig. 2.13.3

$$I = \frac{(4-2)V}{(4+2)\Omega} = \frac{2}{6} = \frac{1}{3} A$$

The potential $A - B = E_1 - IR_1 = 4 - \frac{1}{3} \times 4 = \frac{8}{3}$ V

D

The equivalent resistance with the e.m.f.'s removed becomes,

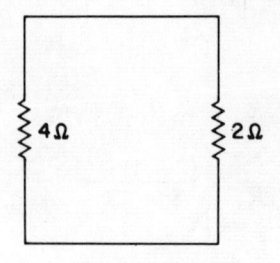

Fig. 2.13.4.

The circuit (less the 6Ω load) now becomes,

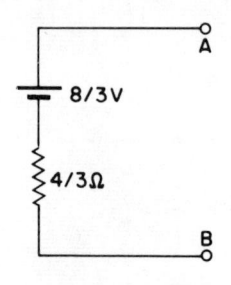

Fig. 2.13.5.

and when the 6Ω load resistor is reconnected, the p.d. across it becomes,

$$\text{p.d.} = \frac{8}{3}V \times \frac{6}{6 + \frac{4}{3}} = \frac{48}{22} = \frac{24}{11} \text{ Volts}$$

Kirchhoff's second law

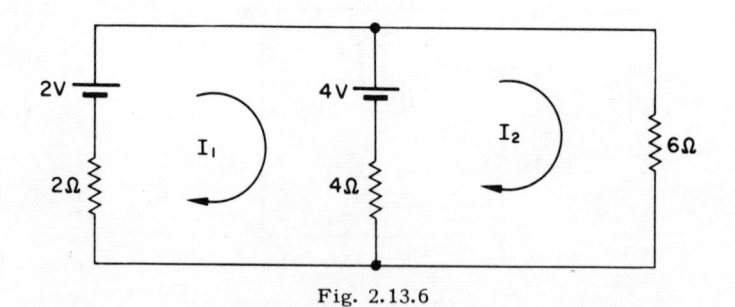

Fig. 2.13.6

I_1) $\quad -2 = 6I_1 - 4I_2$ \hfill (1)

I_2) $\quad 4 = -4I_1 + 10I_2$ \hfill (2)

$$2 \times (1) = -4 = 12I_1 - 8I_2$$
$$3 \times (2) = 12 = -12I_1 + 30I_2$$
$$\text{adding} \quad 8 = 22I_2$$

Hence
$$I_2 = \frac{8}{22} \text{A}.$$

$$V_3 = 6 \times I_2 = \frac{48}{22} = \frac{24}{11} \text{Volts}.$$

Using the formula derived

$$V_3 = \frac{\frac{2}{2} + \frac{4}{4}}{\frac{1}{2} + \frac{1}{4} + \frac{1}{6}} = \frac{2}{\frac{11}{12}} = \frac{24}{11} \text{Volts}$$

We have by no means exhausted the known theorems, neither have we necessarily chosen those of greatest importance. We have however, examined a circuit problem, using a variety of techniques, in an attempt to demonstrate the value of the care that should be taken when deciding which approach should be used.

2.14. 'Pi' to 'Tee' transformation.

It is often advantageous to re draw a circuit in another form thus facilitating an easier solution. One such simplification is discussed here.

Any linear network can be simplified to either an equivalent π network or an equivalent Tee network as shown in their basic forms in figure 2.14.1.

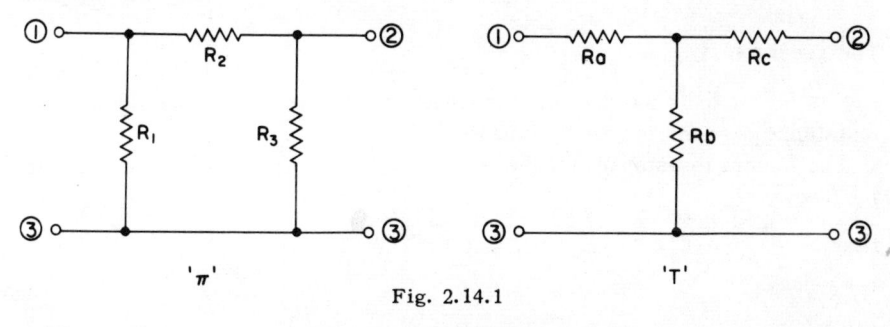

Fig. 2.14.1

If a given network can be reduced to either one of the circuits shown in figure 2.14.1., then there must be a given relationship between them.

The Pi network has components identified as R_1, R_2 and R_3. Let us assume that these are known values whereas the Tee network components labelled, R_a, R_b and R_c, are not of known values. If we had a π network with known values and wished to transform it into a Tee network, and preserve the original functions of the network, we would have to write, for the Tee network, values of R_a, R_b and R_c in terms of the known R_1, R_2 and R_3, such that if tests were carried out on the Tee network, we should obtain precisely the same answers for identical tests to the π network.

We can only do this after we have established the relationship between the two networks.

If we apply some simple tests to the π network and equate the results for identical tests to the Tee network, we will establish the relationship between the two.

There are three unknown components in the Tee network and we will require 3 equations before we can successfully determine the relationship we seek.

The first test.

Suppose we determine the input resistance to both networks, 'looking in' between terminals 1 and 3 for both. Then Rin for the π network
= $R_1 // R_2 + R_3$ and Rin for the Tee = $R_a + R_b$.
(Note that Rc has one end unconnected and plays no part in the expression).

It remains only to equate the two expressions to give us our first equation;

$$R_1 // (R_2 + R_3) = R_a + R_b$$

$$\therefore \frac{R_1(R_2 + R_3)}{R_1 + R_2 + R_3} = R_a + R_b$$

$$\therefore \frac{R_1 R_2 + R_1 R_3}{R_1 + R_2 + R_3} = R_a + R_b \qquad (1)$$

The second test.

Let us repeat the exercise only this time, we will determine the output resistance between terminals 2 and 3.

The 'output resistance' for the π is equated to the 'output resistance' for the Tee.

Hence
$$R_3 // (R_1 + R_2) = R_c + R_b$$

$$\therefore \frac{R_3 R_1 + R_3 R_2}{R_3 + R_1 + R_2} = R_c + R_b \qquad (2)$$

The third test.

Let us now 'look into' the networks between the terminals 1 and 2.

FURTHER NETWORKS AND SIMPLE THEOREMS

When looking into the π, we see an input resistance of $R_2 // (R_1 + R_3)$ and for the Tee, $R_a + R_c$.

and equating these expressions;

$$\frac{R_2(R_1 + R_3)}{R_1 + R_2 + R_3} = R_a + R_c$$

$$\therefore \frac{R_1 R_2 + R_2 R_3}{R_1 + R_2 + R_3} = R_a + R_c \qquad (3)$$

We have now the three necessary equations and by a little manipulation, we can begin to draw up expressions for the 'unknowns' in the Tee in terms of the 'knowns' in the π.

Let us take (2) from (1)

$$\frac{R_1 R_2 + R_1 R_3}{R_1 + R_2 + R_3} = R_a + R_b \qquad (1)$$

$$\frac{R_2 R_3 + R_1 R_3}{R_1 + R_2 + R_3} = R_c + R_b \qquad (2)$$

$$\frac{R_1 R_2 - R_2 R_3}{R_1 + R_2 + R_3} = R_a - R_c \qquad (4)$$

Now let us add equations 3 and 4.

$$(3) \quad \frac{R_1 R_2 + R_2 R_3}{R_1 + R_2 + R_3} = R_a + R_c$$

$$(4) \quad \frac{R_1 R_2 - R_2 R_3}{R_1 + R_2 + R_3} = R_a - R_c$$

$$(3) + (4) \quad \frac{2(R_1 R_2)}{R_1 + R_2 + R_3} = 2R_a$$

Hence $\quad R_a = \dfrac{R_1 R_2}{R_1 + R_2 + R_3}$

R_b and R_c are found in a similar manner.

It is not necessary, or even desirable perhaps, to have to derive these expressions each time they are required. A simple 'aide memoir' is offered, one that will enable the reader to write down expressions for R_a, R_b and R_c in terms of R_1, R_2 and R_3, without resorting to simultaneous equations.

Consider the two networks shown in figure 2.14.2.

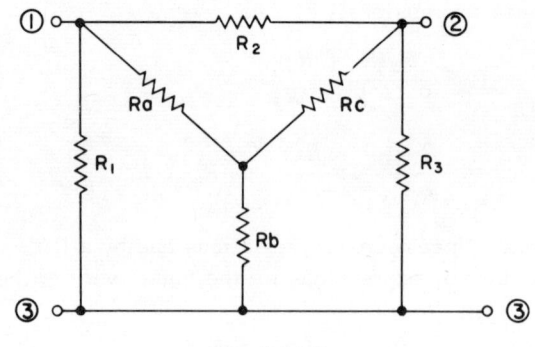

Fig. 2.14.2.

The two networks are superimposed.

The expression derived for $R_a = \dfrac{R_1 R_2}{R_1 + R_2 + R_3}$

may be thought of as $R_a = \dfrac{\text{Straddle}}{\text{Sum}}$, that is, we write down those two known resistors that 'straddle' our unknown, divided by the sum of the knowns.

The reader will readily see that R_a is straddled by R_1, R_2.

Similarly, by using $\dfrac{\text{'straddle'}}{\text{sum}}$,

$$R_b = \dfrac{R_1 R_3}{R_1 + R_2 + R_3} \quad \text{and} \quad R_c = \dfrac{R_2 R_3}{R_1 + R_2 + R_3}$$

Tee to π transformation.

Should we know the values of the components in the Tee network and wish to express the unknowns in the π, in terms of the known values in the Tee, we follow the rules above exactly, except that we write the reciprocal of every resistor in the expression.

Example.

(Where we need to express R_1 in terms of R_a, R_b and R_c.)

$$\dfrac{1}{R_1} = \dfrac{\dfrac{1}{R_a} \times \dfrac{1}{R_b}}{\dfrac{1}{R_a} + \dfrac{1}{R_b} + \dfrac{1}{R_c}}$$

where $\dfrac{1}{R_a} \times \dfrac{1}{R_b}$ straddle the unknown $\dfrac{1}{R_1}$.

FURTHER NETWORKS AND SIMPLE THEOREMS 37

Tee to Pi transformation is covered in detail in later chapters. During this chapter, we will confine our discussion to Pi to Tee transformation only.

Example.

Transform the π network into the equivalent Tee network.

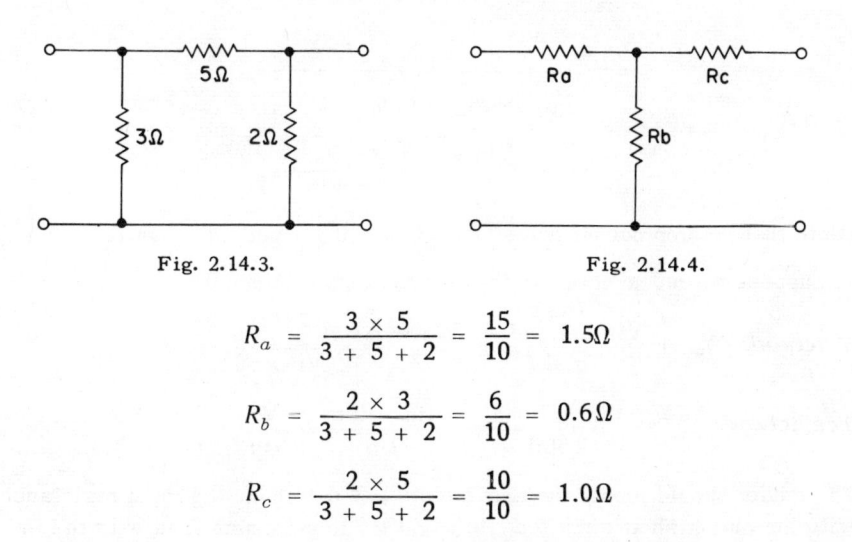

Fig. 2.14.3.   Fig. 2.14.4.

$$R_a = \frac{3 \times 5}{3 + 5 + 2} = \frac{15}{10} = 1.5\Omega$$

$$R_b = \frac{2 \times 3}{3 + 5 + 2} = \frac{6}{10} = 0.6\Omega$$

$$R_c = \frac{2 \times 5}{3 + 5 + 2} = \frac{10}{10} = 1.0\Omega$$

In other words, if we had a Tee network with the values shown, and applied tests to both networks, the results will be identical.

Suppose we applied an input voltage to both networks and calculate the output voltages. Both networks must give the same answer if they are to function in precisely the same manner.

Consider the π network

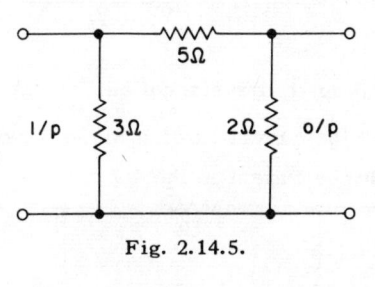

Fig. 2.14.5.

$$V_0 = \frac{V_{in} \times Load}{Total} = \frac{V_{in} \times 2}{7} \quad \therefore \quad \frac{V_0}{V_{in}} = \frac{2}{7}$$

and for the Tee network,

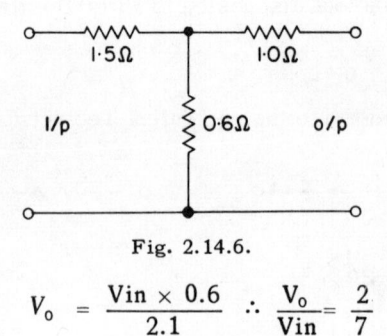

Fig. 2.14.6.

$$V_0 = \frac{V_{in} \times 0.6}{2.1} \quad \therefore \frac{V_0}{V_{in}} = \frac{2}{7}$$

(Note that no drop occurs across Rc when using a perfect voltmeter).

Suppose we repeat this looking into the output terminals.

π network. $V_0 = \dfrac{V_{in} \times \text{Load}}{\text{Total}} = \dfrac{V_{in} \times 3}{8} \quad \therefore \dfrac{V_0}{V_{in}} = \dfrac{3}{8}$

Tee network. $V_0 = \dfrac{V_{in} \times \text{Load}}{\text{Total}} = \dfrac{V_{in} \times 0.6}{1.6} \quad \therefore \dfrac{V_0}{V_{in}} = \dfrac{0.6}{1.6} = \dfrac{3}{8}$

The reader should repeat the above tests and determine the input resistance with the output short circuited. He might try to determine Rout with the input short circuit.

We will discuss at a later stage, more advanced transformations, The resistors (R) will be replaced by impedances (Z). The expression for Za would become

$$Z_a = \frac{Z_1 Z_2}{Z_1 + Z_2 + Z_3}$$

and for Tee to π transformation $Y_1 = \dfrac{Y_a Y_b}{Y_a + Y_b + Y_c}$ where $Y = \dfrac{1}{Z}$.

Example in the use of π to 'T' transformation.

Problem: (a) Calculate the current flowing in the 5 V cell.

(b) Calculate the current in the 4 V cell.

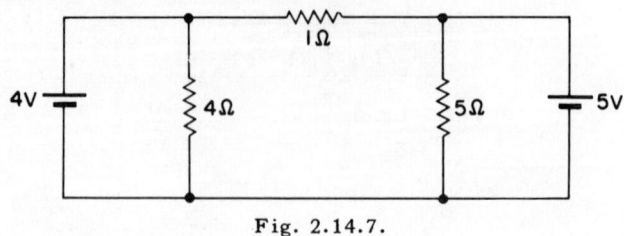

Fig. 2.14.7.

and transforming the π to a Tee,

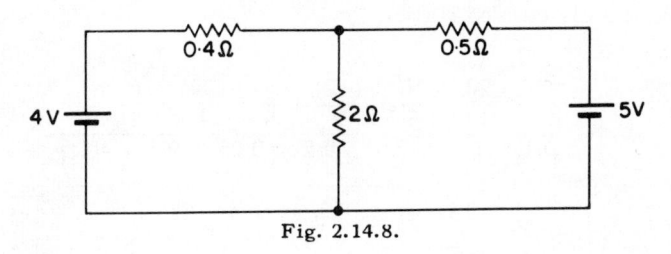

Fig. 2.14.8.

and re-arranging a little,

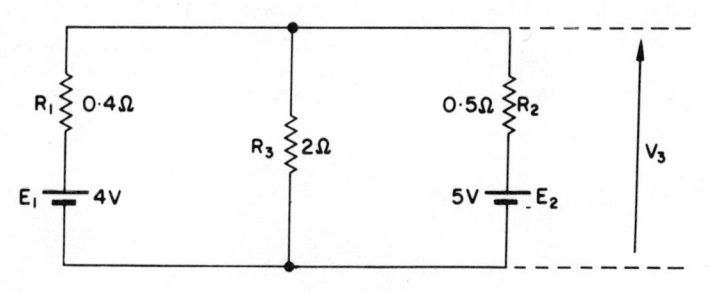

Fig. 2.14.9.

and using the formula derived for this type of circuit earlier on,

$$V_3 = \frac{\dfrac{E_1}{R_1} = \dfrac{E_2}{R_2}}{\dfrac{1}{R_1} + \dfrac{1}{R_2} + \dfrac{1}{R_3}} = \frac{\dfrac{4}{0.4} = \dfrac{5}{0.5}}{\dfrac{1}{0.4} + \dfrac{1}{2} + \dfrac{1}{0.5}} = \frac{10 + 10}{2.5 + 0.5 + 2} = 4 \text{ V}.$$

The current in the 5 V cell $= \dfrac{5 - 4}{0.5} = \dfrac{1}{0.5} = 2$ Amp.

Hence a current of 2 Amp is flowing in the 5 V cell. There is no current flowing in the 4 V cell.

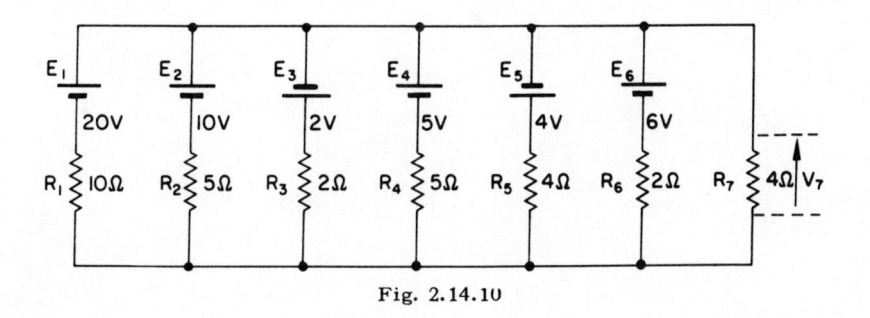

Fig. 2.14.10

A final example is shown in figure 2.14.10. Let us determine the terminal p.d., V_7.

From the formula derived earlier;

$$V_7 = \frac{\frac{20}{10} + \frac{10}{5} + \frac{(-2)}{2} + \frac{5}{5} + \frac{(-4)}{4} + \frac{6}{2}}{0.1 + 0.2 + 0.5 + 0.2 + 0.25 + 0.5 + 0.25}$$

$$V_7 = \frac{6}{2}V = 3V.$$

Individual currents are easily found. The reader might solve this problem by equating e.m.f.'s and $I.R.$ drops in each loop.

CHAPTER 3

Linear components

Components used in electronics fall into two distinct classes, those whose properties (characteristics) do not depend upon the voltage applied or the current flowing through the component and those whose properties change considerably with the voltage or current.

The first class includes resistors, capacitors and inductors which do not have ferromagnetic cores and these are termed *linear* components. It is our intention to summarise the essential facts concerning these components in the present chapter.

On the other hand, certain devices have the basic property that the relationship between voltage applied and current flowing varies according to the magnitude of the voltage and current. Such devices are called *non-linear*, and they include not only valves and transistors but also some inductors which are specially made with ferromagnetic cores for use in magnetic amplifiers.

To make the distinction clear, let us compare a simple capacitor with, say, a transistor. Although the reactance of a capacitor varies with the frequency of the supply to it, the value of its capacitance, and hence of its reactance, does not depend in any way upon the value of the supply voltage. The reactance is the same whether the voltage applied to it is 1 V or 10 mV. On the other hand, the effective input resistance of a transistor will be very different if the input voltage is changed so drastically, even if the frequency of the input is not altered.

Non-linear devices are of vital importance in electronics, but they are discussed in later chapters.

3.1. The resistor.

A component which has pure resistance only (a resistor) is characterised by an opposition to current flow which does not depend at all upon the frequency of that current. It obeys Ohm's law at all times, provided the current is not allowed to rise to such a value that the temperature of the resistor alters appreciably, and there is no question of any storage of electrical energy by a resistor. It is dissipative only, and the power dissipated in the component, given by I^2R or V^2/R, is completely converted into heat energy. It is important in electronic circuits that this heat be allowed to get away by positioning of the component, provision of adequate air circulation or other means. Under no circumstances should the power (or voltage) rating of a resistor be exceeded in any circuit in which it is used.

3.2. The perfect inductor.

Although an actual coil must have some resistance, it is convenient to look at the properties of a perfect inductor (one without resistance) and then regard an actual inductor as a component which is composed of a perfect inductor in series with a resistance equal to the actual resistance of the wire with which it is wound. It is true that this resistance will increase in value at high frequencies but this is usually only of importance at high radio frequencies and this can easily be allowed for in such cases. We shall not find that the so-called 'high frequency resistance' is important in this book.

When d.c. flows through an inductor a steady magnetic field is produced but this has no effect upon the current because e.m.f. is induced only by changing fields. If the magnitude of the field is altered by changing the current magnitude, an e.m.f. is induced which is proportional to the rate of change of current and this e.m.f. opposes the current change. If the current is increasing the induced (or back) e.m.f. will act in such a direction as to prevent it rising, but if the current is falling the e.m.f. will be such as to maintain the current flow. Inductance, then, is the circuit property which tends to oppose any change in the magnitude of the current flowing in the component. If the inductance of a coil is L henries, then the induced e.m.f. is given by e.m.f. $= -L$ (rate of change of current), and is expressed as $-L\, di/dt$, the minus sign indicating the opposition to change. The effect of this when sinusoidal current flows is illustrated in figure 3.2.1. from which it can be seen that the current lags the applied voltage by 90°.

3.3. Rise of current through an inductor.

Figure 3.3.1. shows a 2 henry inductor in series with a 3Ω resistor. The series combination is connected, via a switch S, to either a 12V d.c. supply or a short circuit.

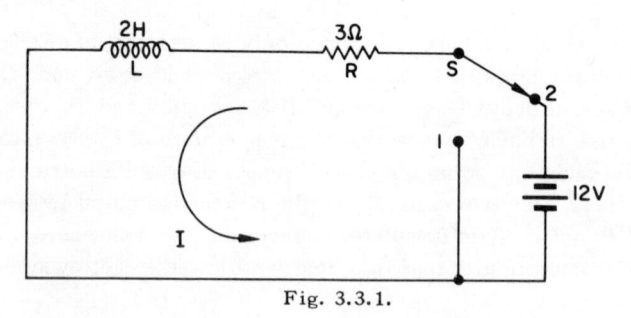

Fig. 3.3.1.

LINEAR COMPONENTS

Assume current is changing as shown.

As it passes from +ve to −ve, rate of change is greatest. At each peak, its change is momentarily zero.

The rate of change is shown..........

This produces a back e.m.f. of opposite phase.

but e.m.f. = $-L.$ (rate of change of current).

The applied voltage v is of opposite phase.

Now compare v with i.

It is seen that the current lags the voltage by 90°.

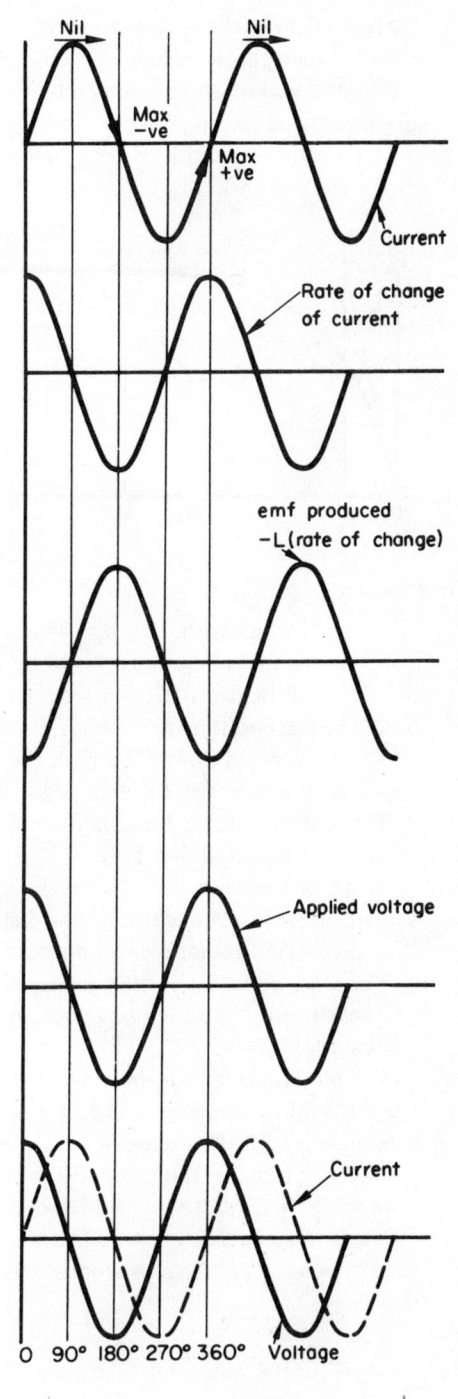

Figure 3.3.2. shows the graph of $I/_t$ where I is the current in ampheres and t is time in seconds.

With the switch S in position 1, a short circuit is connected across the series network and no current flows.

At $t = 0$, the switch is set to position 2 and 12 V d.c. is connected across the network.

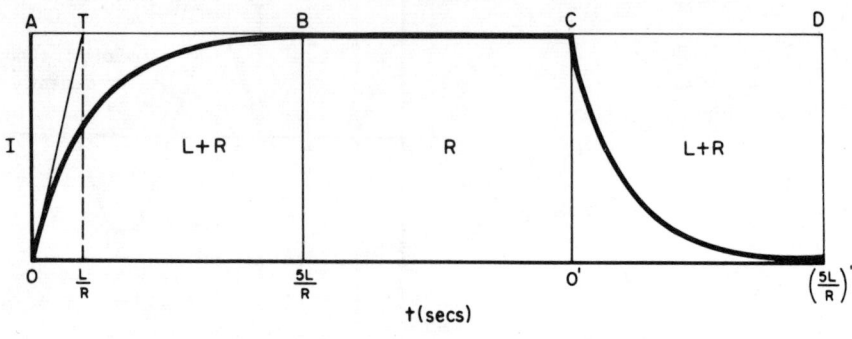

Fig. 3.3.2

At $t = 0$, and with 12 V d.c. connected across the network, no current flows as its formation is opposed by the induced back e.m.f. The circuit current does begin to flow however, and by Lenz's law, this e.m.f. continues to oppose the increase of current.

The 12 V supply provides both the voltage drop across R and the voltage required to neutralise the induced e.m.f.

The initial current flow, if it were to increase at a constant rate, would reach a maximum, given by $I = V/R$, at a time T seconds. The period T seconds is known as the time-constant of the circuit and is given for this circuit as L/R seconds. L is the inductance of the inductor in henries and R is the series resistance in ohms.

The time constant in this example is $\frac{2}{3}$ seconds.

The current continues to increase but at a lessening rate and will reach maximum at $t = \infty$.

For all practical purposes however, the change in amplitude beyond $t = 5(L/R)$ is very small and can be considered as having reached its maximum at $5(L/R)$ seconds.

In this example, the current would reach a maximum of approximately 4 amperes in $(5 \times 2)/3 = 3.34$ seconds.

There may be occasions when we need to know the exact current amplitude beyond $t = 5.L/R$ seconds and this can be evaluated by using the expression

$$i = I\left(1 - e^{\frac{-Rt}{L}}\right),$$

where i is the unknown amplitude, I is the maximum value at $t = \infty$.

It may be seen in figure 3.3.2. that if the switch is set to position 2 at $t = 0$, the current will reach its maximum value at a period $5.L/R$.

The growth of current during the period $A - B$ is a function of L and R. The curve will remain the same whatever the values of L and R but changing values of L and R will change the actual values of both I maximum and the time in seconds.

With the switch left in position 2, the current will remain constant. The figure shows this constant current during the period $B - C$. The value of current during this period is determined by R alone as, with no changing value of current, the inductor behaves as a short circuit.

At $t = 0'$, if the switch is set to position 1, the current will fall during the period $C - D$ as shown in the figure.

If $L = 2H$ and $R = 3\Omega$, the current will fall to approximately zero after a period $(5.L/R)'$ seconds, i.e.,

$$\frac{5 \times 2}{3} = \frac{10}{3} \text{ seconds.}$$

An inductor will resist any attempt to change the current flowing through it from a constant value. It will resist both an increase and decrease in current value.

We will discuss a capacitor in 3.4. and we shall see that by duality, it resists changes in capacitor potential whereas the inductor resists changes in inductor current.

3.4. The capacitor.

A simple capacitor may consist of 2 metal plates, separated a short distance from each other. It has the ability to store electric energy equal to $\frac{1}{2}CV^2$ Joules. Once charged it has a potential gradient across the gap between the plates. The capacitance will be reduced if the gap is increased or if the area of the parallel plates is made smaller. It will be increased if certain materials are inserted into the airgap between the plates.

A capacitor has the ability to store a charge. This charge is measured in Coulombs and has a symbol, Q. If a capacitor is connected across an a.c. supply, as shown in figure 3.4.1., a current, I, will flow for a time, T.

The positive charges that leave the positive pole of the battery, will accumulate on the upper plate of the capacitor. A perfect dielectric is assumed to exist between the plates. An excess positive charge will exist on the upper plate; whenever current flows, any charges lost from the battery must be replaced. As the positive terminal will lose some positive charges, the negative terminal will receive an equivalent charge. This charge can only come from the lower plate of the capacitor. The lower plate of the capacitor having lost some positive charges will be left with a negative charge. (Alternatively one may say that as positive and negative charges flow in opposite directions — and are equal in number — a -ve charge will exist on the lower plate equivalent to the -ve charges on the upper plate)

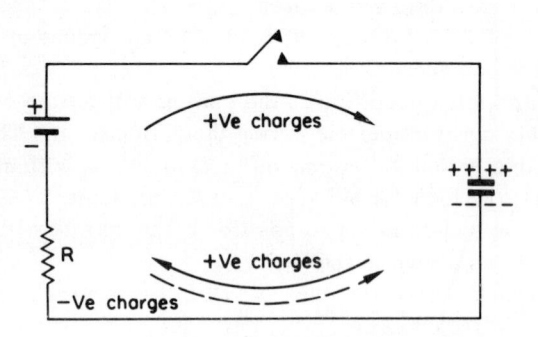

Fig. 3.4.1.

The current will continue to flow until sufficient charges have accumulated on both plates and this will occur at the instant that the capacitor potential is equal to that of the e.m.f. The resultant positive charge on the upper plate and the resultant negative charge on the lower plate, both repel any further charges due to the electrostatic forces which will exist at each plate due to the accumulated charges. When two batteries of equal e.m.f. are connected in parallel, positive terminal to positive terminal, there cannot be any current flow.

The charged capacitor behaves as a second battery, and after a time T, appears to have the same e.m.f. as the supply. The potential difference acquired by the capacitor after a time T, is equal to the e.m.f. of the battery. At the instant that the potential of the capacitor is that of the battery, all current flow ceases. The capacitor is then said to have acquired a charge, $Q = CV$ in Coulombs. If the battery has an internal resistance, then the time taken for equilibrium to take place, is $T = 5.CR$ seconds. Theoretically, if R is zero, the time taken for the current to fall to zero $= 5CR = 5Cx0 = 0$ seconds.

If the cell is reversed, the process is reversed.

If a low frequency sinusoidal voltage is applied across a capacitor, as with the d.c. case, when the sinewave reaches its most positive peak, the capacitor is charged to that value. During the second half cycle, the capacitor is charged in the reverse direction by the negative peak. The current is proportional to the rate of change of charge. For a greater applied e.m.f. at a constant frequency, rate of change of charge will be proportionally greater. Hence the current will be greatest when the rate of change is greatest; the current will be zero when the rate of charge, and e.m.f. is zero.

LINEAR COMPONENTS 47

Applied e.m.f.

Rate of change of current

Current and voltage phase difference

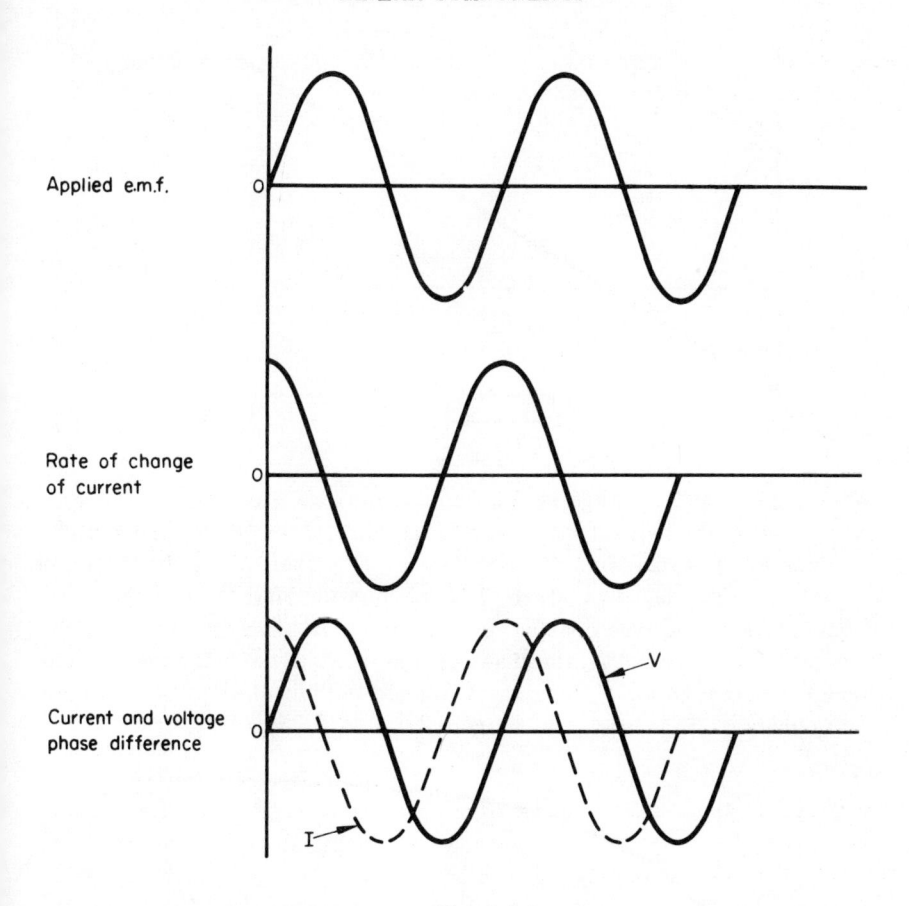

Fig. 3.4.2.

If a constant current was caused to flow into a capacitor, a p.d. would build up across the capacitor. This p.d. would increase at a uniform rate provided the input current remained at a constant value.

Figure 3.4.3. shows the capacitor voltage V_c, increasing in a linear manner, resulting from the constant current I.

During this process, the capacitor will have acquired a charge $Q = I.T.$ where I is the current and T is the duration of time during which the current had flowed into the capacitor.

The capacitor can also acquire a charge $Q = CV$ by connecting a steady voltage across C. Both are expressed in Coulombs and if both charges were of identical value, we can write $Q = CV = I.T.$

This expression shows, for a constant charge, $CV = I.T.$

The latter expression $CV = IT$ is most useful and provides the basis for numerous basic designs and analyses in electronic circuits, many of which are discussed later in this book.

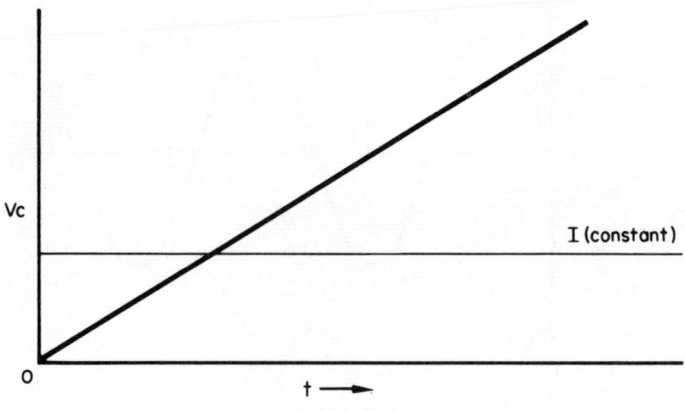

Fig. 3.4.3.

When a d.c. supply voltage is connected across a series $C.R$ circuit as in figure 3.4.4, the capacitor p.d. at any instant t, after closing the switch S, is given by the expression $V_c = V(1 - e^{-t/CR})$. Where V_c is the capacitor potential, V is the applied voltage, t is the duration after the switch is closed and CR is the time constant of the circuit. As with the example for the inductor in figure 3.3.2., the time constant $C.R.$ is defined as the period of time it would take for V_c to equal V should the initial slope of the curve remain constant. This is shown in figure 3.4.4.

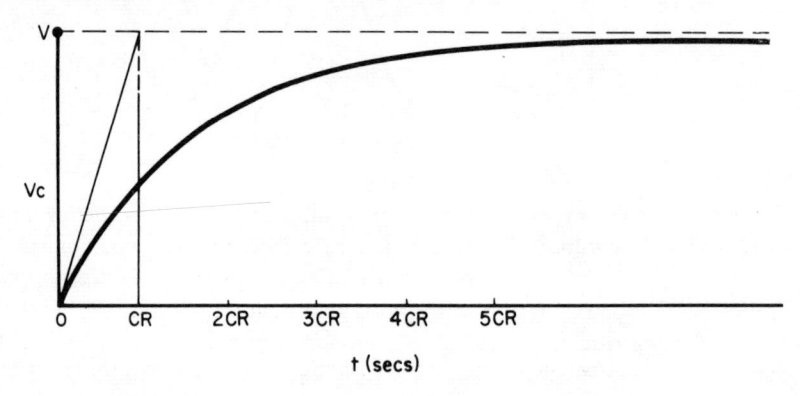

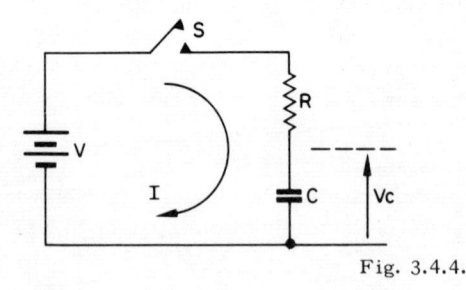

Fig. 3.4.4.

LINEAR COMPONENTS

We will be discussing series C.R. networks in detail at a later stage. For the moment however, we will examine capacitors having no series resistance and see how they may be charged, connected in parallel, and how they may then be regarded as a single charged capacitor.

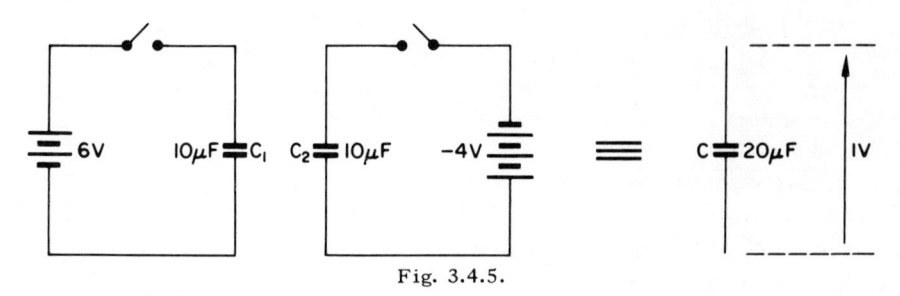

Fig. 3.4.5.

If we were to connect two (or more) capacitors in parallel, we would express the resultant as a single capacitor. $C = C_1 + C_2$. We might also be interested in the resultant charge Q. We obtain Q by simply summing the respective charges $Q1$ and $Q2$.

Hence if a capacitor C_1 of $10\mu F$ having a charge of 20 Coulombs were connected in shunt with a second capacitor C_2 of $20\mu F$ and having a charge of 5 Coulombs, the resultant would be a capacitor $C = 30\mu F$ having a charge of 25 Coulombs.

We can charge a capacitor by connecting a d.c. **voltage supply across it and then removing the supply.**

If we were to charge a $10\mu F$ capacitor with a 6 V supply as shown in figure 3.4.5. and a second $10\mu F$ capacitor with a -4 V supply, and then connected the capacitors in shunt, we can express the resultant as a single capacitor having a resultant charge.

Summing values of respective capacities and charges, the resultant is expressed as a single capacitor of $20\mu F$ having a p.d. of 1V.

The method of deriving these values is given;
The capacitor C_1 has a charge

$$Q_1 = C_1 V_1 = \frac{10}{10^6} \cdot 6 = 60\mu C.$$

The capacitor C_2 has a charge

$$Q_2 = C_2 V_2 \;\; \frac{10(-4)}{10^6} = -40\mu C.$$

(Note the negative sign due to the -4 V supply).

Summing $C_1 + C_2$ gives $C = 20\mu F$ and summing their respective charges, $Q = Q_1 + Q_2 = 60\mu C + (-40\mu C) = 20\mu C$.

Hence C is a $20\mu F$ capacitor having a charge of $20\mu C$.

The resultant p.d. across C_1 from $Q = CV$, is given as

$$V = \frac{Q}{C} = \frac{20\mu C}{20\mu F} = 1V.$$

Capacitors, once charged even via resistance, can be rapidly discharged, with virtually no resistance, in a very short duration.

A very high current can flow during this short period of time. This often forms the basis of a camera flash device. A $10\mu F$ capacitor 'charged' to a potential of 50 V, if discharged in 0.1 mS will provide 5 A into a very low resistance load during the period concerned. Charged capacitors, particularly large capacity paper capacitors, can provide a very nasty, if not lethal, shock, hence care should be taken when handling these components.

3.5. Capacitors in series.

Consider the circuit as shown in figure 3.5.1.

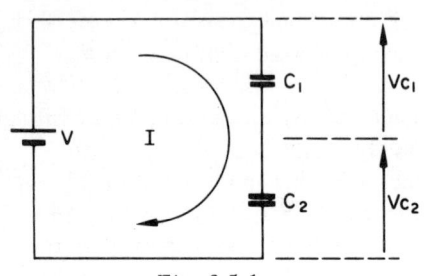

Fig. 3.5.1.

Should a battery having zero internal resistance, be connected as shown, a current I, would flow for a very short period of time.

The current flow would cease when $V_{c_1} + V_{c_2} = V$.

The current I, is seen to 'flow' through both capacitors and from the expression $Q = I.T.$ both capacitors would acquire a charge Q. These charges would be identical irrespective of the capacitor values as may be verified from the expression $Q = I.T.$ (Note that 'C' does not occur in the formula).

The effective capacitance 'presented' to the battery is

$$C' = \frac{C_1 C_2}{C_1 + C_2}$$

The effective capacitance C' must have a charge Q identical to the charge on either capacitor.

LINEAR COMPONENTS 51

then
$$Q = C' \times V$$
$$Q' = \frac{C_1 C_2}{C_1 + C_2} \times V \quad \text{but as} \quad V_{c_1} = \frac{Q}{C_1}$$
$$V_{c_1} = \frac{C_1 C_2}{C_1 + C_2} \times \frac{V}{C_2} = \frac{V C_2}{C_1 + C_2}$$

which may be seen to be similar to the 'load over total' for resistors except that the load is the 'other' capacitor.

Hence,
$$V_{c_2} = \frac{V C_1}{C_1 + C_2},$$

a most useful expression and is analogous to the expression for finding a current through one of two resistors in shunt described earlier.

3.6. Parallel plate capacitors.

An expression relating the capacity of a parallel plate capacitor to its area a, number of plates n, the distance between adjacent plates d, the area of the plate, in various media Er, is given as

$$C = E_0 Er \frac{(n-1)a}{d} \text{ Farads.}$$

where $E_0 = 8.85 \cdot 10^{-12}$ and Er is the relative permittivity (Er of air is 1). n is the number of plates, a is the area in square metres and d is the distance apart in metres.

Example 1.

What is the capacitance of a capacitor that consists of two parallel plates of 1m² in area spaced 10 mm apart in air?

$$C = \frac{E_0 Er (n-1) a}{d} = \frac{8.85 \cdot 10^{-12} (1) 1}{10 \cdot 10^{-3}} = \underline{885 \text{ pF}}.$$

If the plates were to be pulled apart to a distance of 20 mm, what then would be the capacitance. The capacitance is inversely proportional to the distance, therefore as the distance is doubled, the capacity will be halved, the capacity will be 442.5 pF.

If the latter capacitor were to be dipped in oil having an Er of 5, what then will the capacity become? The capacity is proportional to the term Er, therefore as the term Er has been increased by 5, then the capacity will have increased also by 5.

The capacity will be 2212.5 pF.

Example 2.

If a $10\mu F$ capacitor is connected to a 10 V d.c. supply and a $20\mu F$ capacitor is connected to a -10 V supply, then both capacitor terminals connected together, what voltage will exist across the common terminals?

With these problems, the capacitance must be added and the charge should also be added. The charge of the $10\mu F$ capacitor is $100\mu C$. The charge on the $20\mu F$ capacitor is $-200\mu C$. When the capacitors are connected together, the total capacitance is $30\mu F$, and the total charge becomes $(100 - 200)\mu C = -100\mu C$.

The resultant terminal voltage becomes, $\dfrac{-100\mu C}{30\mu F} = -3.33\,V.$

Example 3.

If a capacitor having a capacity of $1\mu F$ is connected to a 1000 V supply, then dipped completely in oil having an Er of 5, what then will the terminal voltage become? If the capacitor is taken out of the oil and discharged in 1 sec, how much current will flow.

The capacitor charges initially to 1000V. It acquires a charge of 1mC. When immersed in oil, its capacitance increases to $5\mu F$. The terminal voltage will fall to 200 V whilst its charge will remain constant. If the capacitor is removed from the oil, the voltage across the terminal will revert to its original value of 1000 V. The charge Q will be 1mC as before. If the capacitor is discharged in 1 second, the current that will flow will be

$$I = \frac{Q}{T} = \frac{1\,mC}{1\,sec} = 1\,mA.$$

Note:

If we connect capacitors in parallel, we add their respective values in the manner in which resistors in series are added. If we connect capacitors in series, the resultant capacitance is evaluated in the same manner in which resistors in parallel are evaluated. For two capacitors in series, the resultant capacitance is obtained by the product/sum rule whilst if there are more than two in series the resultant capacitance becomes

$$\frac{1}{C} = \frac{1}{C_1} + \frac{1}{C_2} + \ldots \frac{1}{C_n}$$

CHAPTER 4

Revision of basic a. c. principles

4.1. Alternating current

The currents and voltages we shall be mainly concerned with have instantaneous values which vary regularly with time in a periodic manner, and the simplest type of variation possible is illustrated in the graph given in figure 4.1.1. It is called a 'sinusoidal' current, and the graph is commonly called a 'sine wave'.

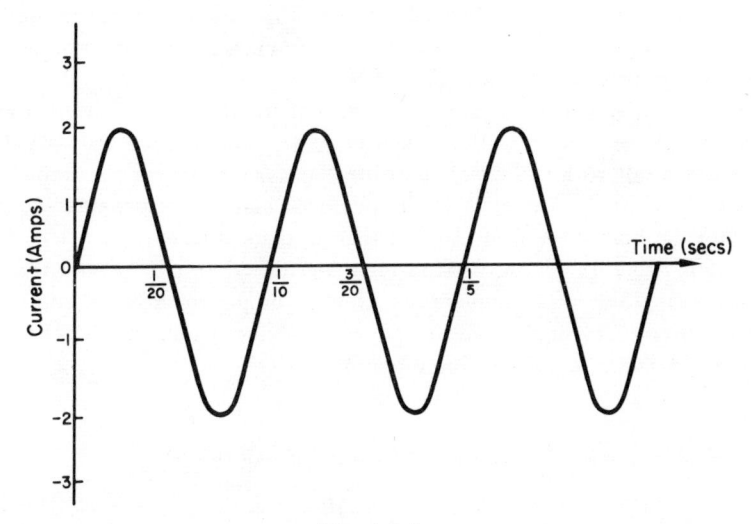

Fig. 4.1.1.

It is the most natural type of variation, and a graph of the voltage induced in a coil rotating at uniform speed in a magnetic field, or that of the displacement of the bob of a simple pendulum, would look exactly like the diagram.

After one complete variation, from zero up to the positive peak, down through zero to the negative peak and back to zero again, the sequence is repeated again and again. One such complete variation is called a 'cycle', and the time taken to complete one cycle is called the 'period' of the wave. The number of periods in one second is called the 'frequency' of the current and is expressed in cycles per second, Hz or c/s. Often in electronics the frequency is so high that multiple units are used:

1 Kilocycle per second (kc/s) = 1000 c/s, or 1 kHz.
1 Megacycle per second (Mc/s) = 1000 kc/s = 10^6 c/s, or 1 MHz.

In figure 4.1.1. the period is one tenth of a second, the frequency ten cycles per second. The 'peak value' is two amperes, so that the current rises to a maximum value of two amperes in the positive direction and falls to a lowest value of two amperes in the negative direction. Sometimes the 'peak-to-peak value' is of interest, and in this example the peak-to-peak value is four amperes.

If this waveform had been produced by an alternator whose coil was rotating in a magnetic field, then one cycle would have been produced for each complete rotation of the coil.

We could therefore plot current against coil position instead of time, each period corresponding to 360° or 2π radians. When discussing general properties of a.c. it is very convenient to use angle instead of time, even if rotating coils are not involved at all, because it is then possible to state general results which are quite independent of the frequency of the particular current or voltage under discussion. It is quite easy to convert from angle to time by noting that one period equals 360°.

The range of frequencies which may be met in electronics varies from a few cycles per second up to thousands of megacycles per second, but the basic ideas dealt with in this chapter are the same whatever the frequency. It is generally true, however, that the physical size of components required is smaller the higher the frequency of the voltages in use.

In calculations involving alternating quantities, we use the fact that the instantaneous value of a sinusoidal current is proportional to the sine of the angle. For example, the current shown in figure 4.1.1, whose peak value is 2A has an instantaneous value given by

$$i = 2\sin\theta,$$

and we can express this in terms of time by noting that

$$\theta = 2\pi \times 10 \times t,$$

t being the time and 10 Hz the frequency. Thus

$$i = 2\sin(20\pi t)$$

Generally, if $\hat{I}$ is the peak value of a current and f the frequency in Hz, then the instantaneous current i is given by

$$i = \hat{I}\sin 2\pi ft.$$

Often the quantity $2\pi f$ is called the 'angular frequency', the unit being radians per second, and a common symbol for angular frequency is ω.

$$\omega = 2\pi f.$$

$$i = \hat{I}\sin\omega t.$$

Throughout this book, lower case letters will be used for instantaneous values.

4.2. R.m.s. value

It is often important to be able to calculate the power dissipated when an alternating current flows through a resistance. As the actual value of the current varies from instant to instant during the cycle, the power will also vary from instant to instant, but these fluctuations of power are not important. What is important is the average power, averaged over a number of complete cycles (or over just one cycle, which gives the same answer).

Suppose a current I is flowing in a resistance R ohms, then the power at any instant is given by I^2R watts. The average power W is then the average or mean of I^2R over a complete cycle. If I is that *direct* current which, flowing in the same resistance, R, would give the same power W, then I is called the 'effective' value of the alternating current. Another name for effective value is 'root-mean-square' or 'r.m.s.' value, because it may be calculated by squaring the instantaneous value, taking the mean of the result over a cycle and then taking the square root.

In other words, the r.m.s. value of an alternating quantity is the value of a direct quantity (voltage or current) which will have the same heating effect in a circuit as the given alternating quantity. In the case of a sine-wave, the r.m.s. value is equal to the peak value divided by $\sqrt{2}$, that is

$$\text{r.m.s. value} = 0.707 \times \text{peak value}.$$

In all cases it should be assumed that stated voltages and currents are given in r.m.s. values unless otherwise indicated. For example if the mains voltage is given as 240 V, 50 Hz, this means 240 V r.m.s., and power calculations can be carried out just as if the supply was 240 V d.c. However, it must be observed that in some applications, it is the *peak* value which is important (for example in deciding whether a capacitor will break down). A 240 V mains supply rises at the peak of its cycle to $240/0.707 = 240 \times 1.414 = 339.4$ V.

Example.

An electric fire has a non inductive resistance of 20Ω and is connected to a 240 V D.C. supply. How much current will flow?

$$I = \frac{V}{R} = \frac{240\,V}{20} = 12\,A.$$

The same electric fire is connected to a 240 V a.c. supply. How much alternating current will flow?

$$I = \frac{V}{R} = \frac{240\,V}{20} = 12\,A.$$

Note that both V and I are expressed in r.m.s. values.

4.3. Mean value.

The average or mean value of a sinusoidal current or voltage, I_{av}, over a complete period is, of course, zero, because the flow is in one direction for half the time and for the other half it is flowing equally in the opposite direction. However, the average over half a cycle is of importance in calculations on rectifiers and meters. The mean value calculated on this basis is the peak value divided by $\pi/2$.

$$\text{Mean value} = \frac{2}{\pi} \times \text{peak} = 0.64 \times \text{peak}.$$

4.4. A.C. circuits.

When an alternating voltage is applied across a non-inductive resistor, the current which flows will be in phase with the voltage, that is to say both the current and voltage will reach their respective positive peaks at the same instant. On the other hand if an alternating voltage is applied to a pure inductor or capacitor, the voltage and current will not be in phase, but will differ in phase by $90°$ ($\pi/2$ radians).

In the case of the inductor, the current *lags* by $90°$ while the current in a capacitor *leads* the voltage by $90°$.

These phase relationships are illustrated graphically in figure 4.4.1.

In order to calculate the magnitude of the current flow through a resistance we merely divide the magnitude of the applied voltage by the resistance (Ohm's Law), but in the cases of inductance and capacitance we divide by the *reactance* of the component. Reactance depends not only upon the value of the component but also upon the frequency of the applied voltage, because higher frequency variations imply a more rapid rate of change of voltage. For an inductor the reactance increases with frequency and is given by

$$XL = 2\pi f L \, \Omega$$
$$= \omega L \, \Omega,$$

where X is the (inductive) reactance, f is the frequency in Hz, and L is the inductance in henries.

On the other hand, the reactance of a capacitor *decreases* as the frequency increases and is given by

$$X_C = \frac{1}{2\pi f C} = 1/\omega C,$$

where C is the capacitance in farads.

Example.

An inductor has an inductance of 31.8 H. Assuming the winding resistance to be zero, how much current would flow if it was connected across a 100 V,

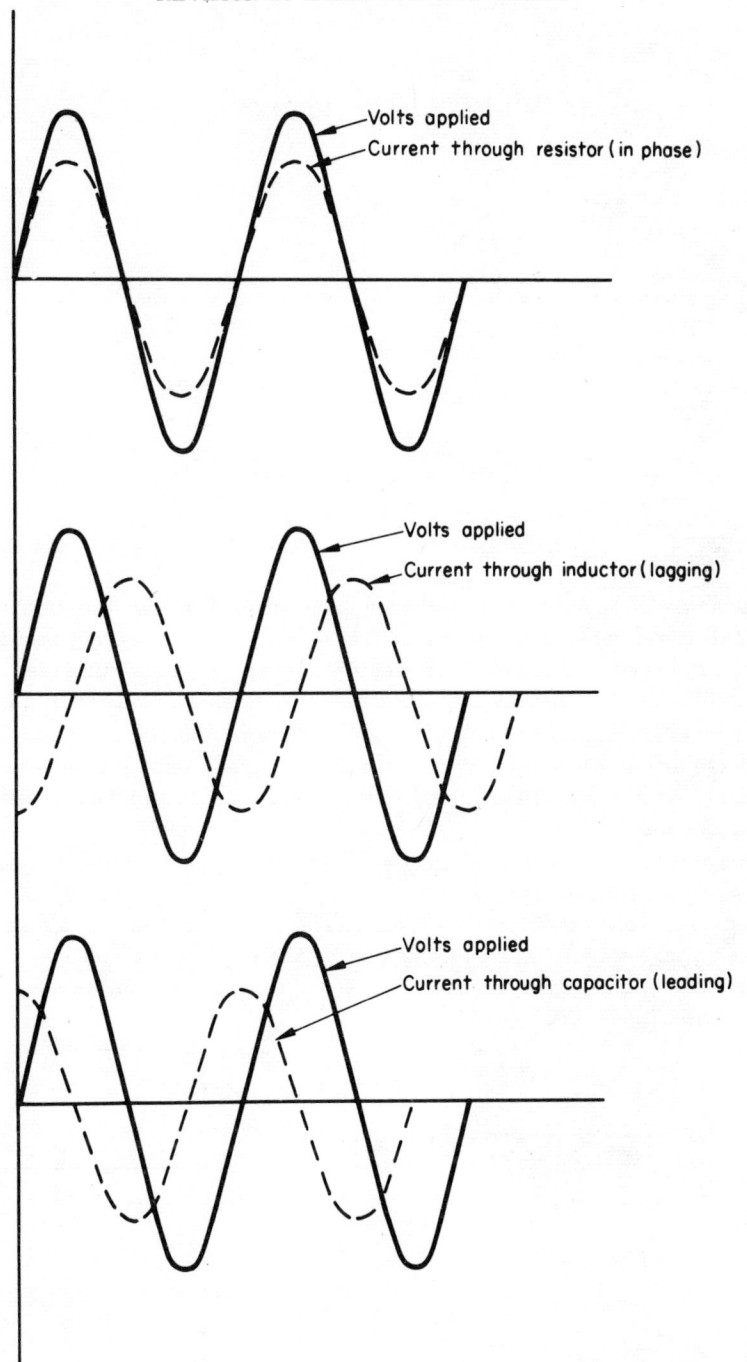

Fig. 4.4.1.

50 Hz supply?

$$XL = 2\pi fL = \frac{31.8 \times 50}{0.159} = 10 \text{ K}\Omega \text{ (Note, } \frac{1}{\pi} = 0.159).$$

$$I = \frac{V}{X_L} = \frac{100 \text{ V}}{10}\text{mA} = 10 \text{ mA}.$$

Example.

A capacitor having a value of $15.9 \mu f$ is connected across a 200 V, 50 Hz supply. What current will flow?

$$X_c = \frac{1}{2\pi fC} = \frac{0.159 \times 10^6}{50 \times 15.9} = 200 \Omega$$

$$I = \frac{V}{X_c} = -\frac{200 \text{ V}}{200 \Omega} = 1 \text{ A}.$$

4.5. Resonant circuits.

The reader will have dealt with elementary examples in which resistance, inductance and capacitance occur in series and parallel combinations. A useful aid to understanding is the 'vector diagram' or 'phasor diagram' in which voltages and currents are represented by lines whose lengths are proportional to magnitude (usually r.m.s.) the angles between lines giving the phase difference between the corresponding quantities. It should be stressed that only quantities having the same frequency may be represented on any one diagram.

We shall be content here to summarise briefly the two important cases of series and parallel resonance.

Consider the following series circuit. The object is to derive a formula for the voltage magnification that occurs at one particular frequency, f_o. The current is common to all components and is therefore chosen as the reference in figure 4.5.1.(b).

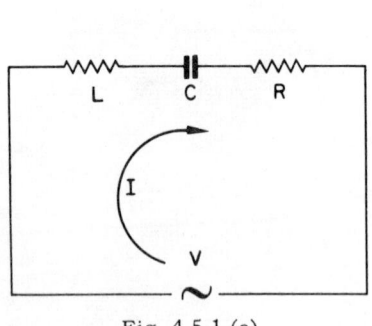

Fig. 4.5.1.(a)   Fig. 4.5.1.(b)

At resonance the capacitive and inductive reactances cancel out leaving R as the effective impedance to current flow.

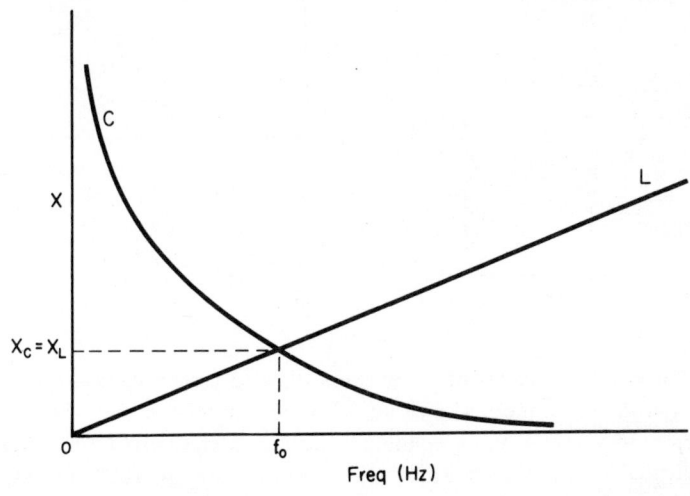

Fig. 4.5.2.

At the particular frequency at which resonance occurs, called the resonant frequency, fo, $XC = XL$ therefore

$$\frac{1}{\omega C} = \omega L \therefore \omega^2 = \frac{1}{LC}, \quad \omega = \frac{1}{\sqrt{LC}} \text{ and } fo = \frac{1}{2\pi\sqrt{LC}}$$

The circuit current is v/R, the instantaneous voltage across

$$L = i \times X_L = \frac{v}{R} \times X_L$$

This voltage is usually much greater than the applied voltage because X_L is usually $\gg R$, and the ratio vL/v is known as the magnification factor of the circuit.

The symbol for the magnification factor is Q.

$$Q = \frac{vL}{v} = \frac{\cancel{v}}{R} \cdot \frac{X_L}{\cancel{v}} = \frac{X_L}{R} = \frac{\omega_0 L}{R}$$

The value of Q may be several hundred, particularly when ferrite cores are used in the inductor. It should be noted that since, at resonance, $X_L = X_C$, an alternative expression for Q is

$$\frac{Xc}{R} = \frac{1}{\omega_0 RC}.$$

Let us consider the following parallel resonant circuit.

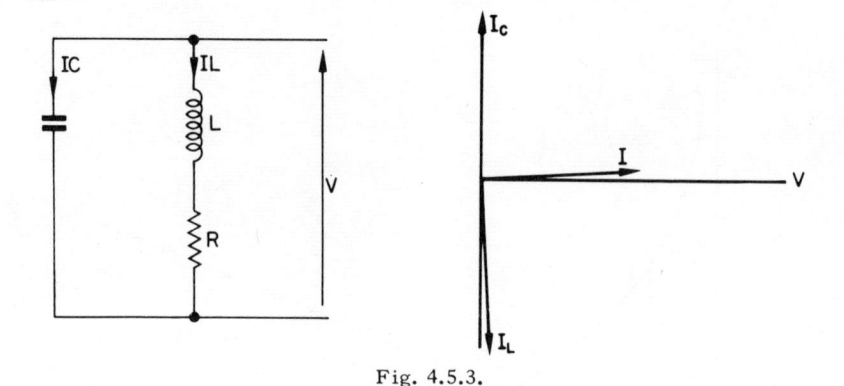

Fig. 4.5.3.

An inductor may consist of many turns of wire wound upon a former. The wire has a d.c. resistance, shown as a resistor in series with the inductance.

The applied voltage, v, is connected across both C and L. In a resonant circuit without coil resistance, $IC = IL$ and as they are 180° out of phase, no supply current would flow.

In practice, when considering the inductor, it has two components, XL and R. This causes a phase shift in the inductive current, IL, which therefore does not quite cancel the capacitive current, IC. I is smaller than Ic or IL whilst if the value of resistance were zero I would become zero. The impedance of the inductor, ZL, is given by $ZL = \sqrt{R^2 + XL^2}$ but if $R \ll XL$ it may be ignored, thus

$$I_L = \frac{V}{X_L} = \frac{V}{\omega L}$$

Similarly,

$$I_C = \frac{V}{X_C} = V\omega C$$

In a perfect loss free parallel circuit, $I_C = I_L$, and therefore $X_C = X_L$.

It is seen then, that if R is very small in a parallel tuned circuit, the formula for resonance is the same as that for a series tuned circuit.

Example.

If a series, or parallel tuned circuit where R is assumed to be zero, consisted of $C = 15.9\,\text{pF}$ and $L = 15.9\,\mu\text{H}$, at what frequency would they resonate?

$$fo = \frac{1}{2\pi\sqrt{LC}} \quad \therefore \quad fo = \frac{0.159}{\sqrt{LC}}$$

hence $fo^2 = \dfrac{0.159 \cdot 0.159}{15.9 \cdot 10^{-6} \cdot 15.9 \cdot 10^{-12}} = 10^{14}$ Hz

$\therefore fo = \sqrt{10^{14}} = 10^7$ Hz $= 10$ MHz.

Example showing the effect of varying 'C'.

A tuned circuit consisting of an inductor L and a variable capacitor 'C' having a maximum value of 500 pF, resonates at 1 MHz when $C = 100$ pF.

We require to tune the circuit to 0.5 MHz — what will be the new value of C?

$$\text{Let } fo_1 = 1.0 \text{ MHz} \quad \text{and} \quad fo_2 = 0.5 \text{ MHz}$$

$$fo_1 = \dfrac{1}{2\pi\sqrt{LC_1}} \quad \text{and} \quad fo_2 = \dfrac{1}{2\pi\sqrt{LC_2}}$$

As 2π and L are constant, let $2\pi\sqrt{L} = K$.

$$\therefore fo_1 = \dfrac{1}{K\sqrt{C_1}} \quad \text{and} \quad fo_2 = \dfrac{1}{K\sqrt{C_2}}$$

$$\text{then } \dfrac{fo_2}{fo_1} = \dfrac{\frac{1}{K\sqrt{C_2}}}{\frac{1}{K\sqrt{C_1}}} \quad \therefore \dfrac{fo_2}{fo_1} = \dfrac{\sqrt{C_1}}{\sqrt{C_2}}$$

$$\text{hence } \dfrac{fo_2}{fo_1} = \sqrt{\dfrac{C_1}{C_2}}$$

Putting in the values for fo_1 and fo_2 and for $C_1 = 100$ pF

$$\dfrac{0.5 \text{ MHz}}{1.0 \text{ MHz}} = \sqrt{\dfrac{100 \text{ pF}}{C_2}} \quad \therefore \dfrac{1}{2} = \sqrt{\dfrac{100 \text{ pF}}{C_2}}$$

$$\therefore \dfrac{1}{4} = \dfrac{100 \text{ pF}}{C_2} \quad C_2 = 400 \text{ pF}.$$

The reader wishing to go into greater detail of the principles of a.c., is advised to read 'Fundamentals of Engineering Science' by G.R.A. Titcomb, Hutchinson Press.

CHAPTER 5

Diodes, rectification and power supplies

In this chapter, we will look at our first non-linear devices, those of which have the property that they are fairly good conductors to current flow in one direction but are very poor conductors, (in some cases, virtually non-conductors), when the direction of the applied voltage is reversed. We may broadly classify these devices as 'diodes', and we shall not only discuss the thermionic diode but also metal, germanium and silicon rectifiers.

All these devices, by virtue of their unidirectional property, are capable of use as rectifiers. A rectifier produces a pulsed d.c. output from an a.c. input, and because the mains supply for electronic equipment is almost invariabley a.c. whilst the valves and transistors in the equipment require a d.c. supply, every piece of electronic equipment requires some form of rectifier circuit for its correct operation.

The basic design principles for 'power supply units' are therefore dealt with at some length in this chapter, and it will be found that one of the main considerations will be the importance of obtaining as 'smooth' a d.c. output as possible from the unit; that is an output which is not only unidirectional, but also as free as possible from a.c. ripple.

When we consider this 'smooth' d.c. output from power supply units, we have in mind, a battery. The output from a battery is of course, completely free from any 'ripple' or alternating component. We are unlikely to be able to design any power supply unit that is as good as the battery, some measure of 'ripple' will invariably be present; we have to make up our mind therefore, just what amount of ripple we can tolerate in each case, and design accordingly. We will consider the fundamental only in the following examples and not concern ourselves with the other harmonics normally present in power supply unit waveforms. We will also assume that all rectifiers have no losses, although in practice, thermionic valve rectifiers have losses but can be allowed for by referring to the published characteristics, thus making the necessary allowances. Further practical examples that allow for losses, using published rectifier data, are included in later chapters.

5.1. The thermionic diode.

A typical diode has a barium oxide coated cathode situated within a nickel anode. When indirectly heated, it will have a heater inserted within the cathode. Figure 5.1.1. shows how these elements may be situated.

F

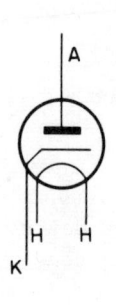

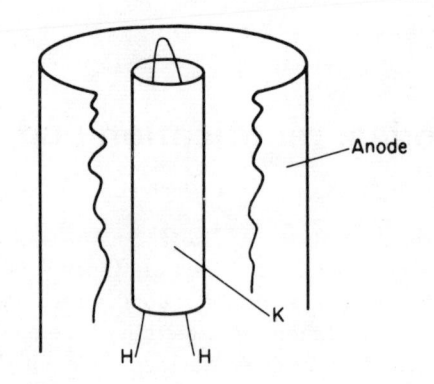

Fig. 5.1.1.

We intend to discuss just one type of thermionic diode in this chapter, as there are so many diodes in use now, that it would be impossible in just one chapter, to do them justice. Figure 5.1.1. also shows the circuit symbol for the diode we intend to discuss.

With a directly heated diode, the heater itself is the cathode. We will concern ourselves for the time being, with the indirectly heated type but, as the heater supply is quite separate from the H.T., supply we will omit the heater on subsequent circuits in order to maintain simplicity. If the heater were connected to an appropriate supply voltage, and with the anode disconnected from any supply, the cathode, due to its close proximity to the heaters, would be heated and electrons would be 'boiled off' from the cathode surface. Many of these 'free' electrons would be 'thrown' from the cathode. After travelling some distance depending upon their initial velocity, these free electrons would fall back to the cathode area, forming a cloud in the vicinity of the cathode. This 'cloud' is called the 'space charge'.

We will soon see that if an attracting potential appears above these electrons in the 'space charge' area, those electrons will be more readily available to be drawn towards the attractive potential than the electrons beneath the cloud nearer the cathode surface. It is therefore, from this space charge that electrons are drawn when the anode is connected to a supply which must be at a greater positive potential than the cathode. When the anode voltage is zero, there is no attraction, but as the anode voltage is increased in a positive going direction, more and more electrons are attracted towards the anode. The electrons lost by the space charge are replaced from the cathode surface. The electrons lost by the cathode are replaced from the negative pole of the supply voltage. The electrons lost by the supply are replaced by those electrons that were attracted to the anode, as the anode is directly connected to the positive pole of the supply. As we increase the supply potential, we can continue to attract more and more electrons until a state is reached where although the supply is further increased, the electron flow will not.

This point is known as 'saturation', and this occurs because the cathode is unable to 'boil off' enough electrons to keep the space charge replenished. We can overcome this to some extent by increasing the heater voltage thus raising the cathode temperature further, but care should be taken when considering this step if damage to the cathode is to be avoided. Figure 5.1.2. shows a typical diode Ia/Va curve.

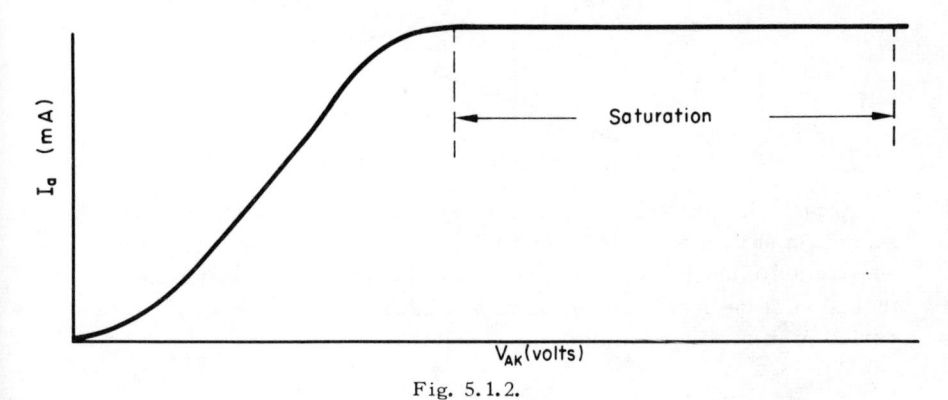

Fig. 5.1.2.

If we were to connect the anode to a negative supply, there would be no attraction for the electrons in the space charge area. The anode, if negative, would repel the electrons and almost all of them would return to the cathode area. Figure 5.1.3. illustrates this effect.

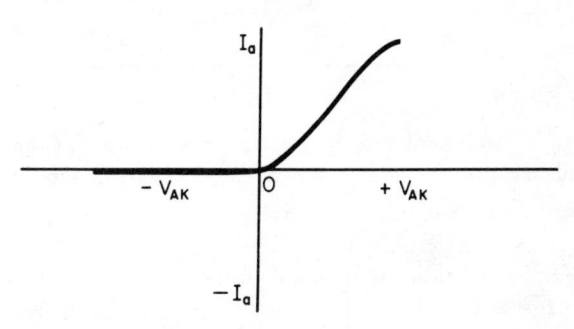

Fig. 5.1.3.

5.2. The half wave rectifier.

If V_{ak} were varying say, in a sinusoidal manner, between V and $-V$, then we can see that the diode would conduct during the positive half cycle but would be cut off during the negative excursion of the supply.

Figure 5.2.1. shows this type of input and the resultant output.

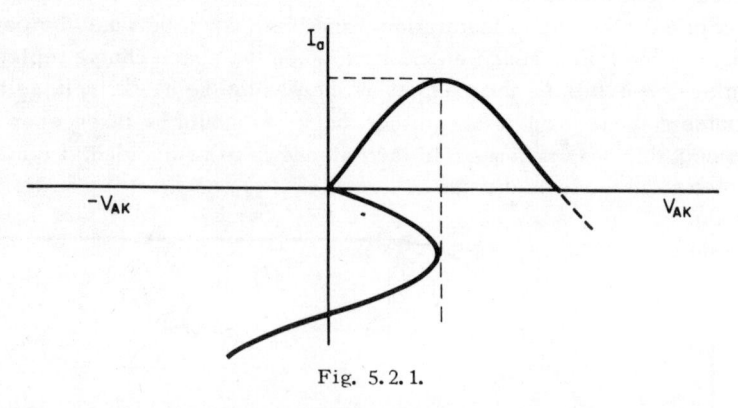

Fig. 5.2.1.

In many circuits, the diode will have a series load resistor connected. We saw in an earlier chapter how static curves may be converted into composite (or dynamic) curves; these resultant curves allow for the inclusion of the load resistor. Figure 5.2.2. gives an example where the device is used with an alternating supply.

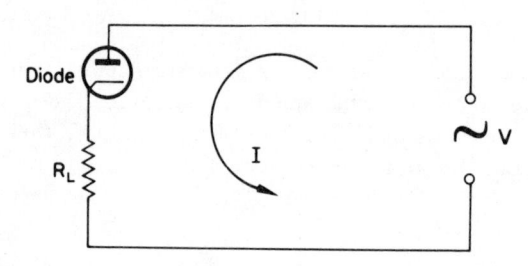

Fig. 5.2.2.

The technique of producing a dynamic characteristic for the diode with its series load resistor connected across an alternating supply is as follows.

1. Select a nominal voltage on the V_{ak} axis, say about 1/10 of the maximum on the graph.
2. Determine the current that would flow through R_L if it were connected across the selected voltage (point A on the graph). The current that would flow is given as point B on the graph.
   Draw a load line for R_L between points $A - B$.

Repeat the foregoing for say, double the voltage previously selected. If voltage $0 - C$ were selected, the current $0 - D$ would flow.
Draw a load line for R_L between points $C - D$.
Repeat a few more times as shown in the figure 5.2.3.

Draw a dotted line from each intersection of each load line and the diode static curve and terminate this exactly above the assumed voltage in each case. For example, the first horizontal dotted line terminates above voltage $0 - A$.

Mark a cross above each assumed voltage at the point each dotted line terminates.

Connect all crosses as shown by the line $0 - H$. This is the dynamic characteristic for the diode and its series load resistor, R_L.

As the voltage varies, perhaps sinusoidally, the current at any instant can be determined from the graph of the dynamic curve in the normal manner.

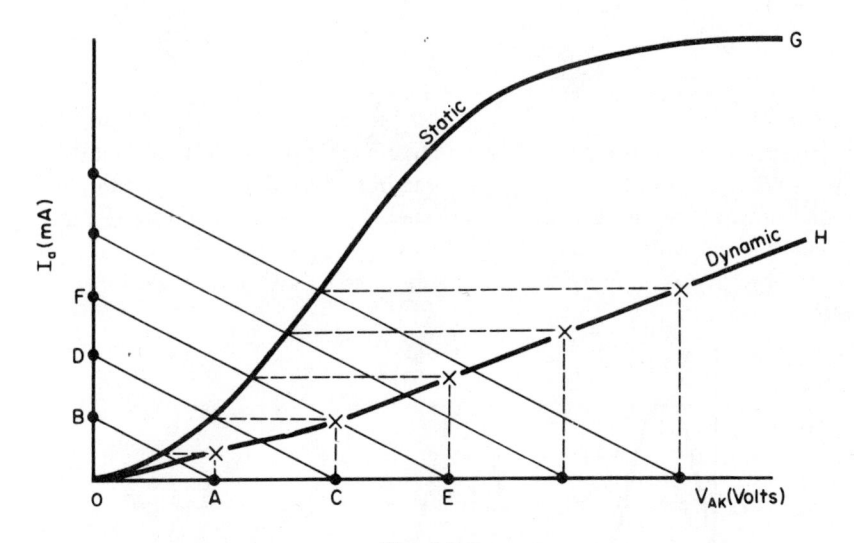

Fig. 5.2.3.

As the current must be undirectional, the circuit forms the basis of a circuit that will give us a d.c. output for an a.c. input. The final circuit will have to be much more ambitious than this, but the refinements are to be discussed in the following pages.

We can see from Figure 5.2.3. that the higher the value of R_L the more horizontal the dynamic curve. This curve also becomes more straight as R_L is increased.

This occurs because R_L begins to 'swamp' the diode resistance and is one example of the subservient role often played by diodes etc, when circuit resistors play a most important role.

We can see that from figure 5.2.3. the I_a/V_a dynamic curve is almost linear. The straight line means that the output current will follow the same law as the applied voltage. Therefore although the output current is unidirectional, it is far from being a steady d.c. output at this stage. The output voltage will be developed across the load resistor and will be of the same shape as the current flowing through it as shown in figure 5.2.4.

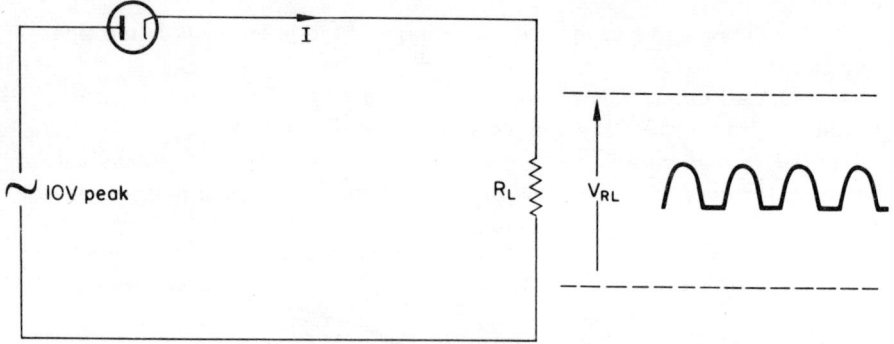

Fig. 5.2.4.

The output d.c. voltage will be partly sinusoidal and will have an average d.c. amplitude of peak input/π therefore, with a peak input of 10 V as shown, the average d.c. output voltage is seen to be $\simeq 3.1$ V.

The average voltage is shown in figure 5.2.5. We can see that due to the unidirectional properties of the diode there is no appreciable negative voltage excursion.

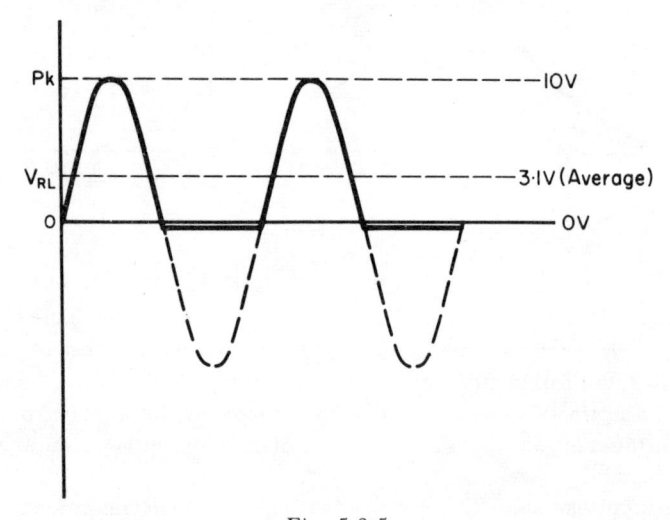

Fig. 5.2.5.

5.3. Power supply units.

A power supply unit should have an output d.c. voltage that should be quite steady, and with as little voltage variation as possible. It should also have

DIODES, RECTIFICATION AND POWER SUPPLIES

a fairly low internal resistance.

We know that our simple rectifier circuit will give us pulsed d.c. and it is clear that the 'rise and fall' across the load would be intolerable for electronic equipment. Should we have such an alternating H.T. the H.T. variations would interfere with the normal circuit operation. The next step then is to consider some means by which we can 'smooth' out these voltage fluctuations yet leaving us with the steady d.c. we require.

Let us take one step towards obtaining this steady d.c. We will add to our simple circuit, a capacitor and connect it as a 'reservoir' as shown in figure 5.3.1.

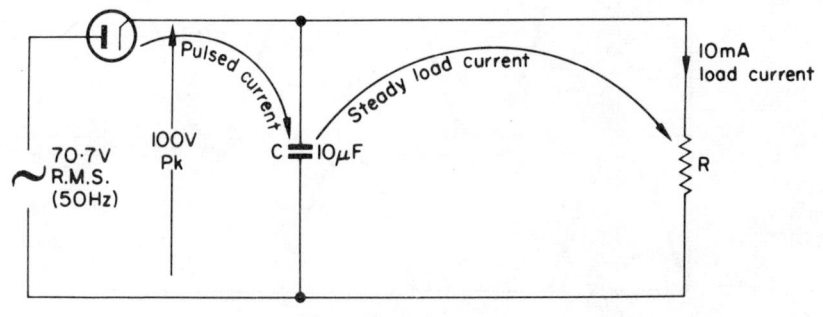

Fig. 5.3.1.

There is a relationship between the charge Q, the voltage V, the current I, and the capacitor value C, and the time T.

$Q = I.T. = C.V.$

C is the capacity of the capacitor in Farads.
V is the voltage across the capacitor.
I is the current flowing out of the capacitor into the load.
T is the periodic time in seconds, during which the diode current is pulsed into the Capacitor.

We do not need to concern ourselves with the charge Q, and as $C.V. = I.T.$, we are able to derive answers for the variables we are likely to meet during the following analytical exercises.

Suppose an alternating voltage of 70.7V r.m.s. were applied to the circuit as shown in figure 5.3.1. Neglecting any diode losses the peak voltage at the diode anode will be 100 V pk. During conduction, the diode will be 'on', and when the input reaches its peak value, the cathode will be at the anode potential of 100 V. The capacitor will charge to the peak value almost instantaneously. We will ignore the time actually taken for this to occur. Once charged, the capacitor will provide a reservoir from which the load current will be taken. As the load is assumed constant, the current it requires will be constant also. As the load current is taken at a steady rate from the reservoir capacitor, the potential across the capacitor will fall linearly. Before the capacitor voltage can fall to zero, the capacitor voltage is 'topped up' from a further current pulse from the rectifier.

The cycle will repeat itself. Figure 5.3.2. illustrates this effect.

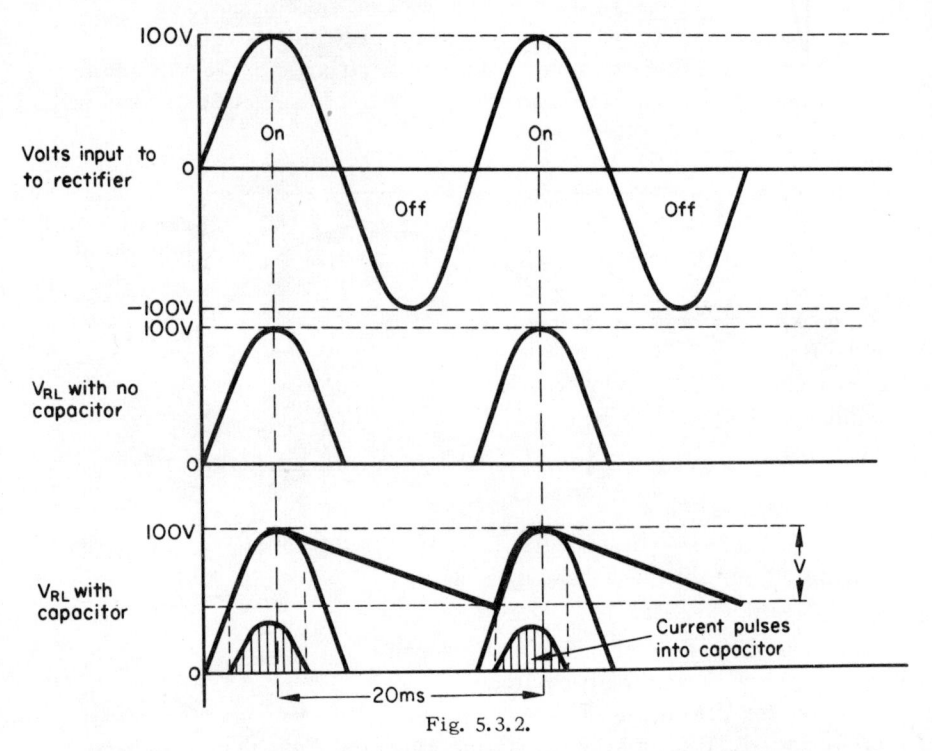

Fig. 5.3.2.

The current pulse from the rectifier returns to zero during the period of time that the rectifier is cut off.

If the frequency of the supply is 50 Hz then the approximate period of time between 'topping up' the capacitor is 20 mS. If we let the fall in capacitor potential be say, V, then $V = I.T./C$. For this example,

$$V = \frac{\frac{10}{1000} \times \frac{20}{1000}}{\frac{10}{1\,000\,000}} = 20\,V$$

This fall in capacitor voltage, V, is seen in an enlarged illustration in figure 5.3.3.

The fall in voltage across the capacitor is then, 20 V.

It is convenient to assume that the average value of the 20 V (shown in

DIODES, RECTIFICATION AND POWER SUPPLIES 71

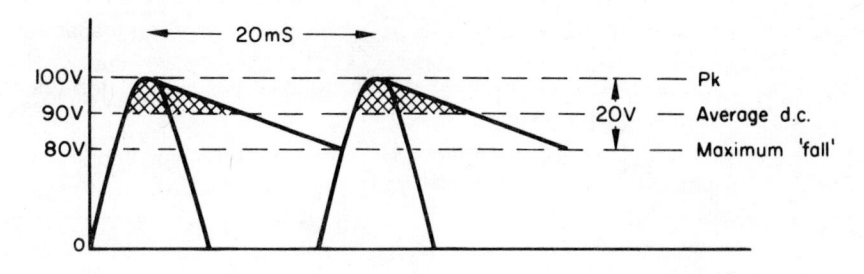

Fig. 5.3.3.

figure 5.3.3. as that voltage between 100 V and 80 V) is midway between the two levels shown. The average d.c. therefore, is seen to be 100 − 20/2 = 90V. This means that the average d.c. output voltage is 90 V. We have now got an H.T. line of 90 V but of course we have also got a ripple voltage superimposed on top. The ripple voltage has a peak value of 100 − 90 = 10 V. pk. The peak ripple is that shaded voltage above the average d.c. line shown in the figure. It can be seen that the ripple voltage approximates to that of a triangular waveform. If we wished to express the ripple as r.m.s., we would have to divide the peak value by $\sqrt{3}$. **The r.m.s. ripple voltage superimposed on our H.T. is therefore**

$$\frac{10}{\sqrt{3}} = 5.78 \text{ V. (r.m.s.)}.$$

If the charge Q is assumed to be constant, the following aide memoir may be useful in these initial stages.

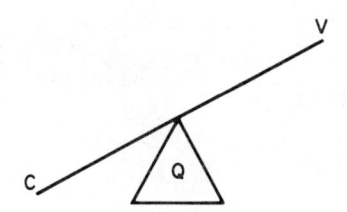

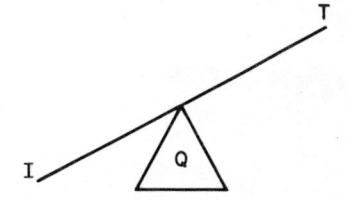

Fig. 5.3.4.

It is seen that as C is increased, V will decrease.

Also if T is decreased, I will increase.

The maximum value of C however is limited to the value given by the rectifier valve manufacture, who will state the recommended maximum value of C.

A very approximate expression for the average d.c. voltage, for a very very small average load current, is

$$V_{av} = \sqrt{2}\, V_{in} \left[1 - \frac{T}{2CR_l}\right] \qquad (1)$$

Where v_{in} is in r.m.s., T is the period between 'topping up' pulses into the capacitor, C, and R_L is the load resistor.

For example, a half-wave circuit has 141v r.m.s. applied, a capacitor of $10\,\mu F$, a load current of 2 mA and a load resistor of 100 KΩ.

$$V_{av} \doteq \sqrt{2}\, V_{in} \left[1 - \frac{T}{2CR_L}\right] = 200 \left[1 - \frac{20.10^{-3}}{2.10^{-6}.10^{5}}\right]$$

$$\doteq 200 \left[1 - \frac{0.02}{2}\right] \doteq 200 - \frac{4}{2} \doteq \underline{198\text{ V average.}}$$

By comparing with the graphical method shown, if the load current is 2 mA average, then

$$V = \frac{I.T}{C} = \frac{2.10^{-3}.20.10^{-3}}{10.10^{-6}} = 4V.$$

The average d.c. across $R_L = 200 - \frac{4}{2} = \underline{198V \text{ average.}}$

In this example, R_L would be, $R_L = \dfrac{198V}{2mA} = 99K\Omega.$

The expression (1) is very approximate and should be used to give a quick indication when the load current is very very small only.

5.4. The full-wave circuit

If two diodes are connected in a rectifier circuit in such a way that one diode conducts during one half-cycle and the other diode conducts during the opposite half-cycle of the a.c. input waveform, then important advantages are obtained over the simpler half-wave circuits just discussed. The capacitor will charge up *twice* in each cycle so that the time of discharge will be halved, and therefore the load voltage will not fall so much before being 'topped up' by the next pulse of charging current. The circuit for the full-wave system is shown in figure 5.4.1.

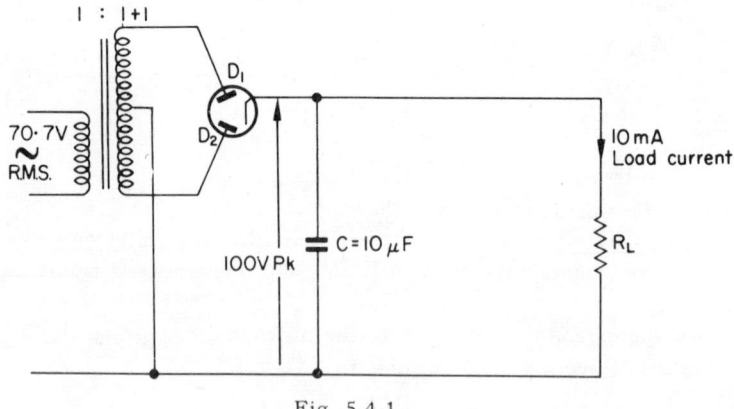

Fig. 5.4.1.

D_1 and D_2 are two identical diodes. C is as before, a reservoir capacitor, from which the load will draw its current. When the anode of D_1 is at its most positive value, then due to the phase difference across the transformer secondary windings, the anode of D_2 will be at its most negative.

During the next half-cycle, the role of the diodes reverse. It should be noted that the current flowing through the load will be unidirectional and that the unidirectional current from both diodes are in the same direction (Fig. 5.4.2.). The load voltage therefore will be of the same polarity as it would be for one diode only.

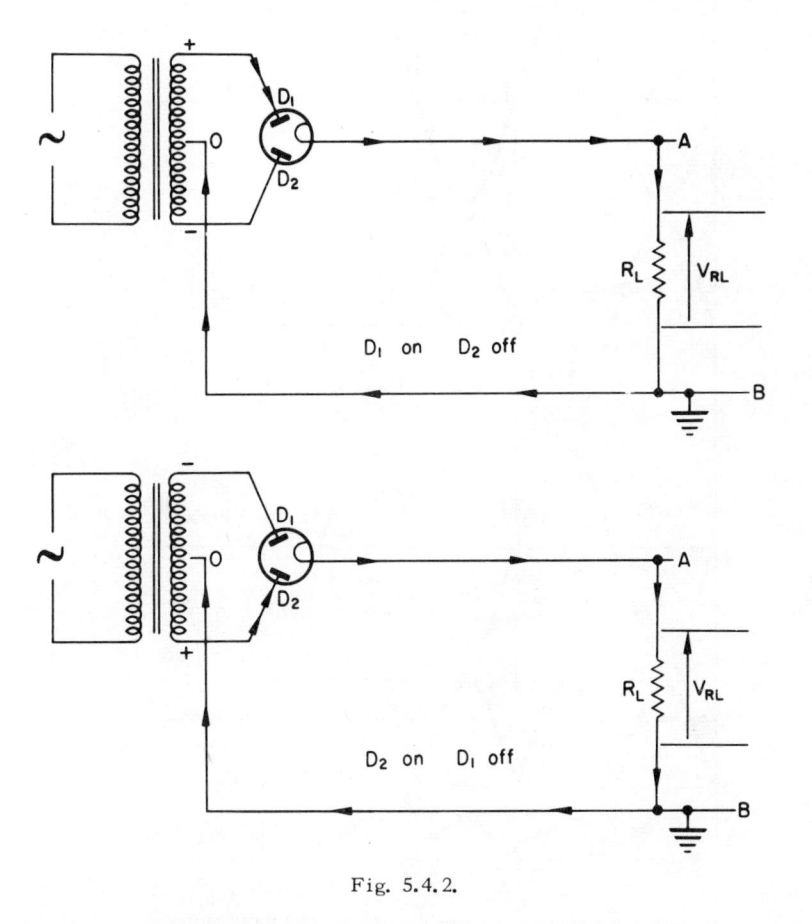

Fig. 5.4.2.

Note that when either diode conducts, the circuit current flowing through the load, flows in one direction only.

Some supplies need to be negative and when this is so, the point (A) may be earthed leaving point (B) as the H.T. supply.

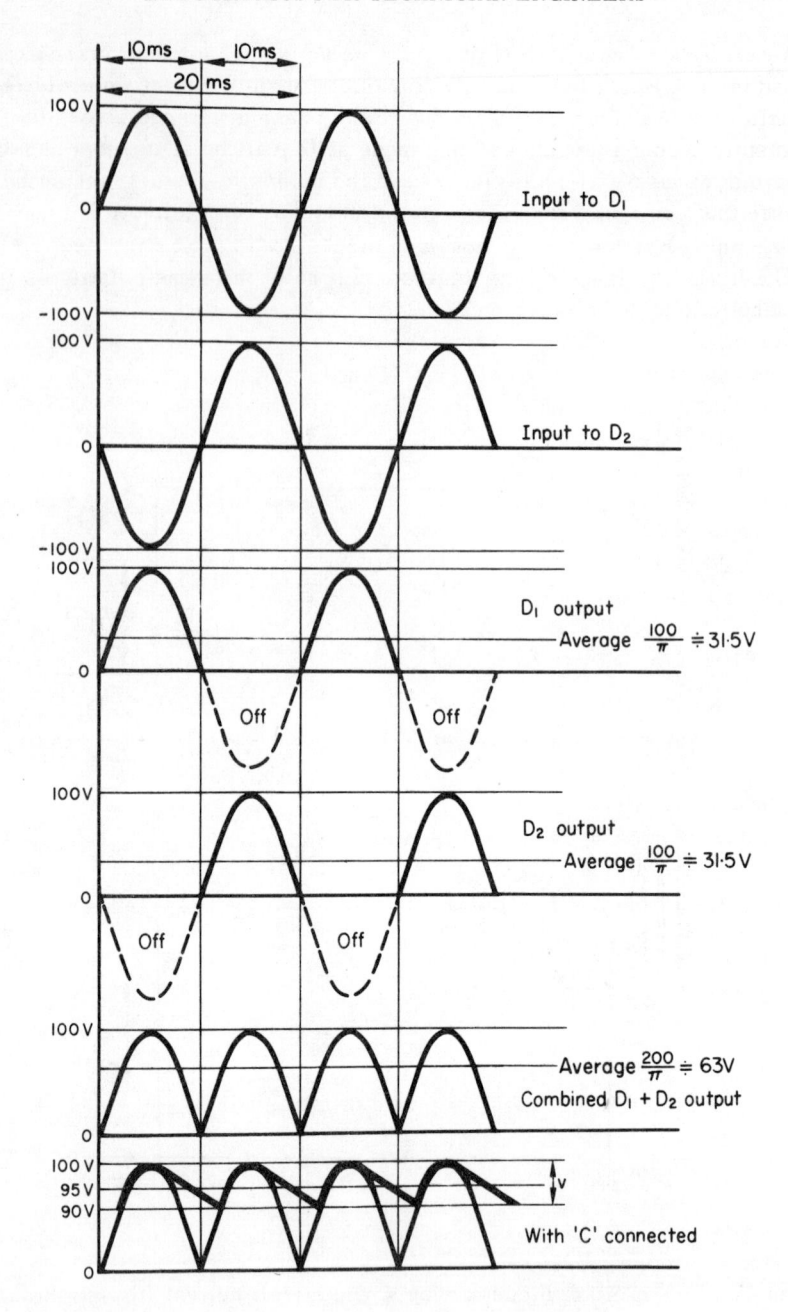

Fig. 5.4.3.

DIODES, RECTIFICATION AND POWER SUPPLIES

We can see from figure 5.4.3. that the average value of the rise and fall across the reservoir capacitor is half of that in a half-wave circuit, the capacitor will 'fall' for 10mS only before it is topped up, therefore the fall in potential is only 10V. If we assume that the average of this rise and fall voltage is a line drawn at half amplitude, as with the half-wave circuit, we can see that the average d.c. voltage is $100 - 5 = 95V$. d.c.

The waveforms for this circuit are given in figure 5.4.3.

The rise and fall voltage will be less for a full-wave circuit, the ripple frequency will be double the a.c. supply frequency to the rectifiers. Half wave rectifier circuits are often confined to loads requiring relatively high voltage and light load currents. The full-wave circuit however, may be used for higher load currents. The ripple voltage is proportional to the load current, and inversely proportional to both the value of the reservoir capacitor value and the number of half wave inputs in the circuit. That is to say, if the ripple was say 30V in a half wave system, then for a full wave system it would be 15V.

If the ripple was say, 25V at a load current of 60mA, then for a load current of 120mA, the ripple would become 50V.

5.5. Filter circuits

We have seen how by varying circuit components, the number of rectifiers and the load current, a different value of rise and fall voltage will exist across the reservoir capacitor. This voltage will normally be far too high to be acceptable for electronic equipment and a means of reducing it to an acceptable level is required. The circuit associated with this 'ripple reduction' is known as a filter. A filter should reduce the ripple or a.c. component of the d.c. output and do so without seriously reducing the d.c. voltage itself.

Let us consider the circuit diagram in figure 5.5.1. This shows a simple resistance – capacitance filter.

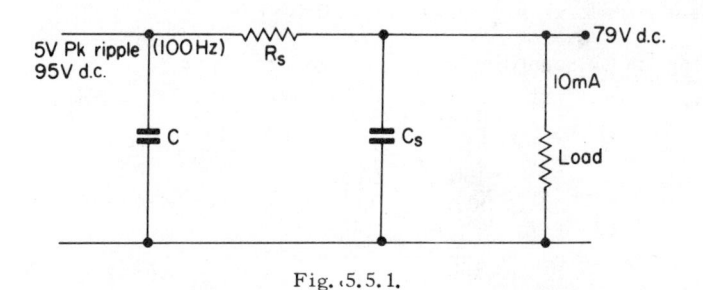

Fig. 5.5.1.

We have assumed 5V peak ripple voltage across the reservoir capacitor, C. Let us assume that this is the 5V ripple superimposed upon our average d.c. level in the previous example. The peak voltage across C is 100V. Suppose we had a load current of 10mA and that at that load current, we need a d.c. output voltage of 79V.

The value of the series filter resistor R_s, needs to be determined first of all. As we require 79V at our output, and as we have 95V mean d.c. across C, we obviously need to 'lose' 16V across R_s. For a load current of 10mA, and a potential of 16V across it, R_s must be 16V/10mA = 1600Ω.

Suppose we decided that for a given equipment, our acceptable output ripple was to be say, 50mV peak. We have already established the value of R_s as 1600Ω. This combined with C_s forms a filter and will attenuate the applied ripple at the input of the filter as follows:-

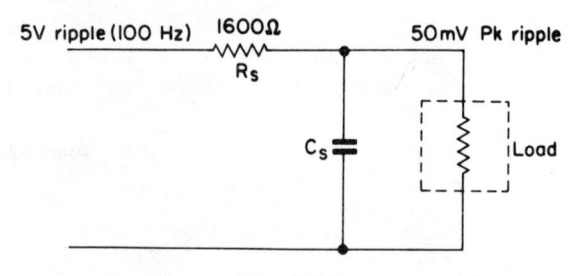

Fig. 5.5.2.

The reactance of C_s is assumed to be very much smaller than the load resistance. The load resistance, as it is larger and is shunted by the reactance of C_s, may be ignored. Figure 5.5.2. shows the circuit we are discussing. The impedance of the series circuit consisting of R_s in series with C_s is given as $Z = \sqrt{R_s^2 + X_{cs}^2}$ where X_{cs} is the reactance of C_s at the input ripple frequency of 100Hz.

The reactance of $C_s = XC_s = 1/2\pi fC$.

As we want only 50mV output ripple, and have 5V input, we need to attenuate that ripple 100 times.

By load over total, we can say that $50\text{mV} = 5\text{V} \cdot \dfrac{\text{Load}}{\text{Total}}$.

Putting known quantities in the expressions above,

$$50\text{mV} = 5000\text{mV} \left(\frac{\text{Load}}{\text{Total}}\right)$$

$$\frac{50}{5000} = \frac{\text{Load}}{\text{Total}}$$

$$\frac{1}{100} = \frac{X_{cs}}{Z} \text{ hence}$$

$$100 = \frac{Z}{X_{cs}} \text{ and if } R_s \gg X_{cs}$$

DIODES, RECTIFICATION AND POWER SUPPLIES

$$100 \div \frac{R_s}{X_{cs}} \quad \text{and as} \quad R_s = 1600\Omega \text{ and}$$

$$X_{cs} \div 16\Omega, \quad \text{hence} \quad C_s = 100\mu\text{F}.$$

The slight approximations are of little consequence and in practice would be most acceptable. Figure 5.5.3. shows the final circuit.

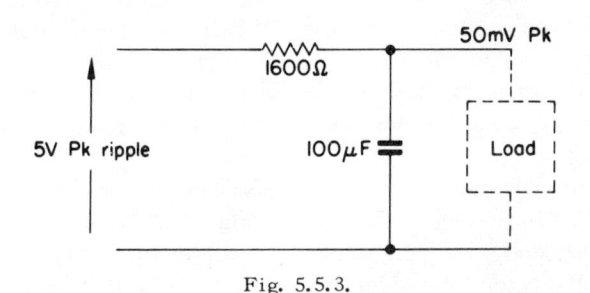

Fig. 5.5.3.

We have reached the stage where by using the foregoing approximations we can examine a complete power supply unit in detail calculating all of the components for both a.c. and d.c. potentials

A complete power supply unit is shown in figure 5.5.4.

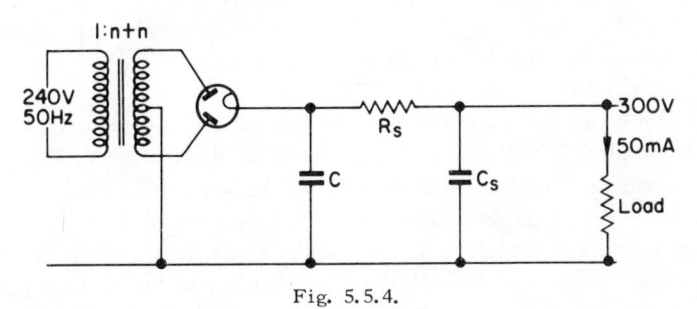

Fig. 5.5.4.

The next step is to design a circuit, having decided upon the electrical specification.

Suppose we need a power supply unit to give us 300V at 50mA load current, with a ripple not greater than 50mV peak.

The first step is to calculate $R_L = 300\text{V}/50\text{mA} = 6\text{K}\Omega$ required to load the circuit during subsequent testing.

Valve manufacturers state the maximum value of reservoir capacitor that we can use with a given rectifier. A common value is $50\mu\text{F}$. This will be the value that we will choose for our circuit. All losses in the rectifier and transformer windings are assumed to be negligible.

A full-wave circuit has a ripple frequency of 100Hz, and the time between 'topping up' the reservoir will be 10mS. The rise and fall voltage across C will be

$$\frac{\frac{50}{1000}A \times \frac{10}{1000}S}{\frac{50}{1000\,000}F} = 10V.$$

We will once more assume that the average value of the ripple or a.c. component superimposed upon the d.c. is half of the complete rise and fall voltage. The average d.c. is therefore, 5V below the peak output from the rectifiers, i.e., the d.c. will have 5V peak ripple superimposed.

The peak ripple across C is 5V. We want only 50mV. As the required attenuation has to be about 100, the reactance of C_s has to be one hundreth of the value of R_s.

Let us arbitrarily choose 50μF for the smoothing capacitor. This has a reactance of 32Ω at 100Hz. The ohmic value of R_s must therefore be 99 times 32Ω to give us the ripple attenuation we require. (This assumes a simple resistive potential divider and is adequate for this basic study). A suitable standard value is 33KΩ.

With 50mA d.c. flowing through R_s, we will develop $50 \times 33 = 165V$ across R_s. If the output is to be 300V, then we must allow for the 165V 'lost' across R_s and at this stage, we can state that the average d.c. across the reservoir capacitor must be $300 + 165 = 465V$.

But we already know that the peak voltage necessary to give us our average voltage must be 5V greater than the average, therefore the peak voltage will need to be $465 + 5V - 470V$ peak.

If we now quote this peak voltage in terms of r.m.s. we can state the secondary winding potential. The r.m.s. is $0.7 \times 470V = 329V$, r.m.s.

Each secondary winding must produce 329V, but at this stage we ought to allow for any voltage drop across the secondary winding resistance. We will allow for this by deciding on say, 350V, r.m.s. from each secondary winding. When connected to a 240V mains supply, the transformer ratio of primary to secondary turns, needs to be $350/240 = 1.45$. The secondary windings will therefore be $1.45 + 1.45$ times that of the primary winding. (Hence $n = 1.45$).

The actual ripple will be $\frac{5V \times 32}{3300} \doteq 50mV$, peak.

In practice, it is quite an easy matter to vary R_s a little, on load, so as to adjust its value in order to obtain the 300V, d.c. at the output.

This simple power supply unit has one main disadvantage where larger load currents are concerned. The larger the load current, the greater the voltage lost across R_s. This will result in a greater voltage needed from the transformer. If we have a larger output from the rectifier, its peak voltage may cause a lot of difficulty when considering the type of reservoir capacitor. More expensive rectifiers may also be needed, as we will have to carefully select a valve that can withstand 2 × Pk voltage (known as the peak inverse voltage, P.I.V.). This occurs when the valve is off. Its anode will have a potential of the peak negative input and its cathode, the peak positive input. We can overcome the loss across R_s by replacing R_s with an inductor. This component, when used in this type of filter circuit, is often referred to as an L.F. Choke. A fairly common value of inductance for this choke would be about 10H.

Inductors were the subjects of an earlier discussion, and we saw that they have two components, resistance and inductance. In our power supply unit, we will use a choke that has low resistance yet a high enough inductance to be suitable in our a.c. filter. The effective improvement over our simple filter containing R_s, is that the d.c. voltage 'lost' across the low resistance in the choke will be small. The reactance of the choke however, will compare with the high resistance value of the previous R_s thereby giving the a.c. attenuation required.

Figure 5.5.5. shows the power supply unit complete with choke.

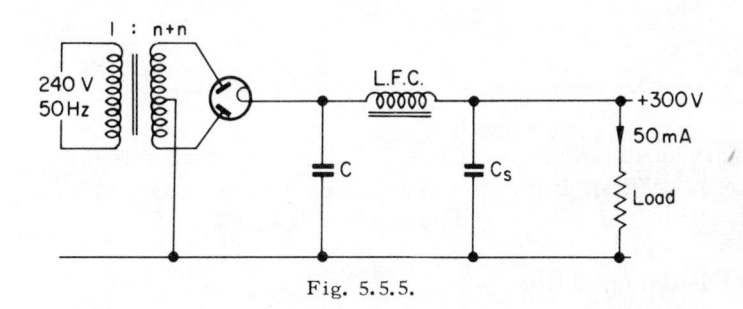

Fig. 5.5.5.

We require 300V, d.c. at a load current of 50mA and a ripple not exceeding 50mA Pk.

Let us assume that the choke has a resistance of 330Ω and an inductance of 10H. The voltage drop across the choke resistance will be 50/1000 × 330 = 16.5V. This is small but can however be allowed for. We require then, on average d.c. across the reservoir, 300 + 16 = 316V d.c. The peak voltage across the reservoir will be 5V greater than the average, as before, therefore the peak voltage across C will be 321V, peak. This corresponds to 321 × 0.7 = 224.7 say 230V, r.m.s.

The transformer ratio needs to be 230/240 = 1:0.96. Hence, $n = 0.96$.
The ripple will be attenuated as follows :-

$$\text{Ripple voltage} = 5V \times \frac{X_c}{Z}, \text{ where } Z = (X_c \sim X_L)$$

X_L at 100Hz = $2\pi f L$ = 6.28 × 100 × 10 = 6280Ω.

The difference between the reactances (which is the circuit impedance) is $6280 - 32 = 6248\Omega$.

Therefore the ripple output is $\dfrac{5000 \times 32}{6248} = 26\text{mV Pk}$.

We can see then, that this is a considerable improvement on our resistance – capacitance filter network.

There are many other types of filter circuits, some of which are very briefly described.

5.6. Multi-section filter.

Figure 5.6.1. shows a typical circuit.

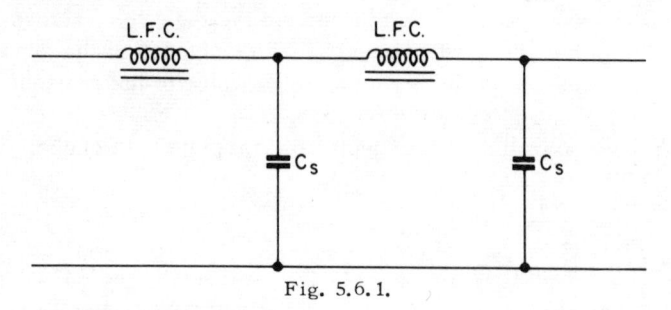

Fig. 5.6.1.

This is similar to the previous $L - C$ filter described except that it has two stages of filter. If the attenuation of each stage is say, 100, then the overall attenuation will be approximately 10 000 (100^2).

5.7. Parallel tuned filter.

Figure 5.7.1. shows such a filter.

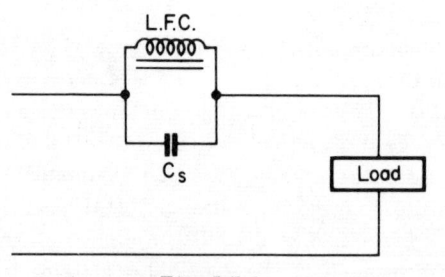

Fig. 5.7.1.

A shunt tuned circuit offers a very high impedance to a.c. at the resonant frequency. If this filter were designed to resonate at 100Hz (for a full-wave power supply unit operating on a 50Hz mains), the attenuation will be quite high.

DIODES, RECTIFICATION AND POWER SUPPLIES 81

All of the power supply units considered so far are mainly for light current loads. The maximum ripple current is stated on the capacitor, and the maximum value of reservoir capacitor for a valve rectifier is given by the valve makers. When we use, and top up, a reservoir capacitor, we subject the rectifier to large current pulses.

5.8. Choke input filters.

The regulation of power supplies is most important and when used in a full wave system, this filter has a good regulation. Figure 5.8.1. shows a circuit of a half wave rectifier feeding an inductive load.

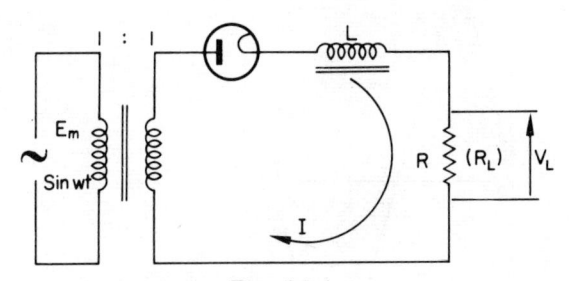

Fig. 5.8.1.

The choke will attempt to keep the current I at a constant value, even when the load current varies, this results in a much steadier voltage across the load. The value of the choke is critical, (depending upon several factors that we are going to discuss), and bears some relationship to the average load current. Some approximations were made with the previous power supply unit calculations and in this example, although we will use further approximations, we will deal with the subject a little more fully.

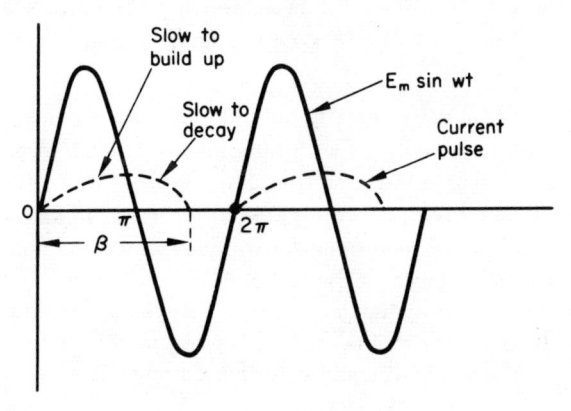

Fig. 5.8.2.

The half wave circuit shown in figure 5.8.1. is not practicable as due to the back e.m.f. developed across the choke, the input voltage for large load currents, will tend to be cancelled. The current will also be slow to build up due to the time constant L/R. The current will continue to flow even when the input voltage has fallen to a negative value. This is again due to the back e.m.f. of the choke attempting to keep the current flowing. This is illustrated in figure 5.8.2.

Figure 5.8.3. shows the period of time β, that the current is flowing through the choke. With a resistive load, $\beta = \pi$ radians.

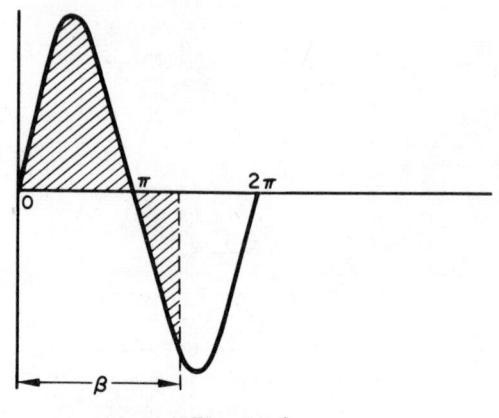

Fig. 5.8.3.

$$\text{average volts} = \frac{1}{2\pi} \int Em \sin \theta \, d\theta = \frac{Em}{2\pi}(1 - \cos \beta)$$

where $\omega t = \theta$ and is an angle when the voltage across the load $(R + L)$ is $Em \sin \omega t$, during conduction.

The regulation of a power supply unit may be shown graphically. We will see later how this is associated with internal resistance. A good regulation will result in a straight line and will be horizontal. This indicates that for an increase in load current, no voltage drop occurs. The half-wave circuit with an inductive load, is notorious for its poor regulation and this, as well as the ideal graph, is shown in figure 5.8.4.

For small average currents, the e.m.f. developed by the inductor is small, but for large currents, the e.m.f. becomes larger and tends to cancel the input voltage. This is discussed further on page 85.

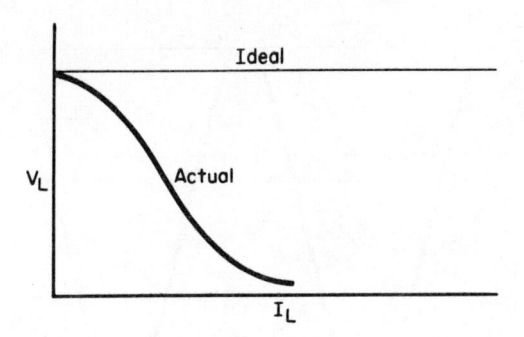

Fig. 5.8.4.

Figure 5.8.5. shows a full-wave circuit. The regulation characteristics are very much better than those of the half-wave circuit.

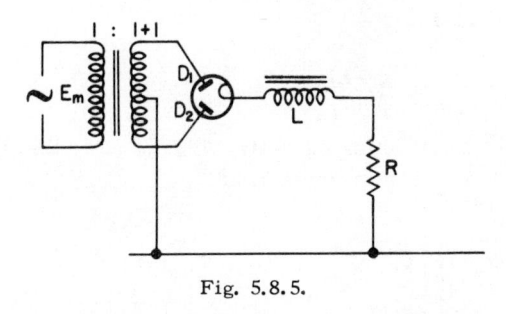

Fig. 5.8.5.

The effective supply to the load is shown in figure 5.8.6.

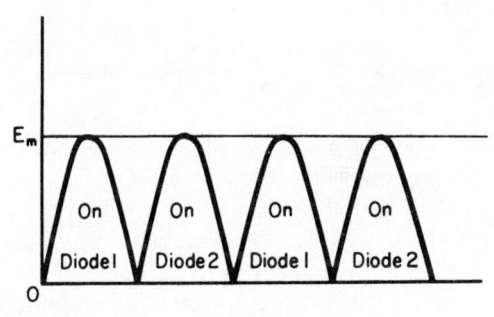

Fig. 5.8.6.

This is seen to consist of a train of rectified pulses from alternate diodes.

The figure 5.8.6. may be slightly rounded off in order to produce a near sinusoidal wave form as shown in figure 5.8.7.

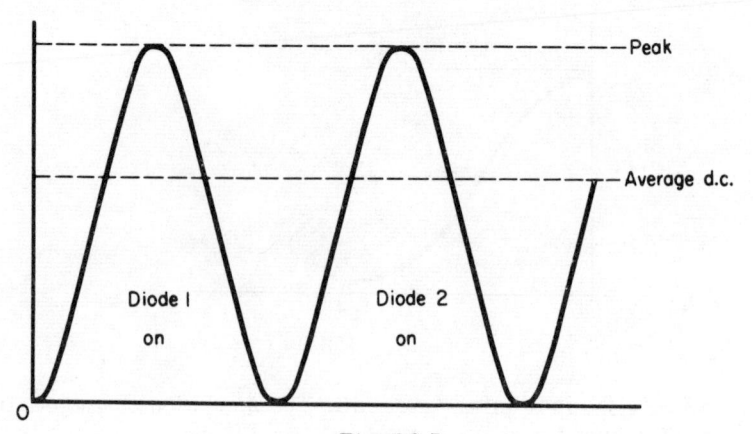

Fig. 5.8.7.

The peak value of a.c. $\doteqdot \dfrac{4 Em}{3\pi}$

and the average d.c. $\doteqdot \dfrac{2 Em}{\pi}$

The addition of a smoothing capacitor will complete this basic power supply unit. This is shown in figure 5.8.8.

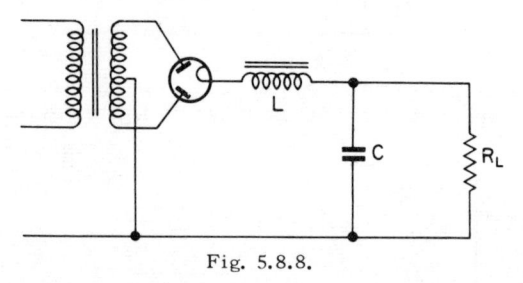

Fig. 5.8.8.

The value of the capacitor should be chosen such that its reactance at ripple frequency will be very much lower than the ohmic value of the load resistor, R_L, thus ensuring that the unwanted ripple current will pass through C instead of the load which is represented by R_L.

If this is the case, then the capacitor reactance will also be very much lower than the reactance of the choke, which in turn should be high. The load resistor can be ignored as it is shunted by the low ohmic value of the capacitor.

The circuit may be re-drawn as in figure 5.8.9.

It is seen that as the X_C is almost short circuit with respect to the X_L, the effective a.c. load across the rectifiers will be the reactance of the choke.

DIODES, RECTIFICATION AND POWER SUPPLIES

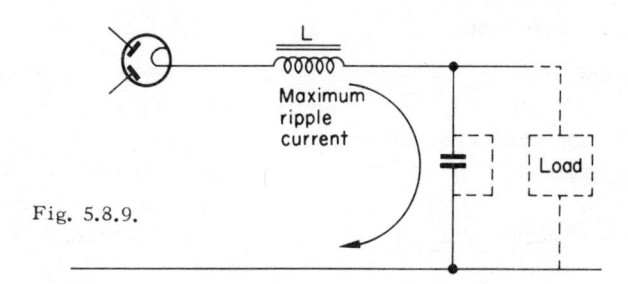

Fig. 5.8.9.

It is necessary to consider the a.c. current that will flow through the choke. The lower the reactance, the greater the current. If the current is too high, it will 'clip' on the extreme negative tips, as shown in figure 5.8.10.

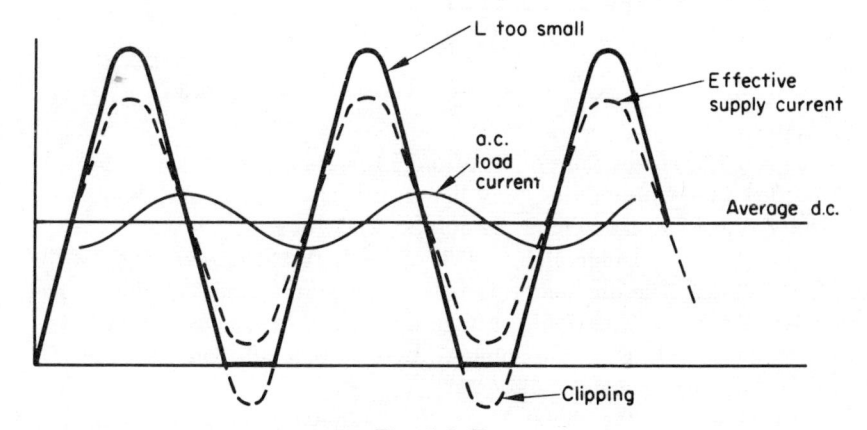

Fig. 5.8.10.

The peak fundamental of the a.c. current component is $4Em/3\pi \omega L$. Where Em is the peak input to the rectifier. The average d.c. component of the current is $2Em/\pi R$. In order to avoid cut off, the average d.c. current must never be less than the 'short circuit' current that will flow through the choke as shown in figure 5.8.9. It follows, therefore that

$$\frac{2Em}{\pi R} > \frac{4Em}{3\pi \omega L} \quad \text{therefore,} \quad \frac{2}{R} > \frac{4}{3\omega L} \quad \text{hence} \quad L > \frac{2R}{3\omega}. \quad \text{Thus} \quad \omega L > \frac{2R}{3}$$

consequently the reactance of the choke must be greater than $2R/3$. R may be a bleeder resistor connected permanently across the output. This will not only ensure that the minimum steady d.c. current flows at all times, but that it will also serve to discharge the capacitor when the supply is switched off, particularly if the load is disconnected prior to switching off the supply.

5.9. Diode voltage drop.

Although with modern diodes there will not be a large voltage drop, the following example allows for such a drop so as to give a more complete picture. The drop across the diode will be determined from the I_a/V_a characteristics given in figure 5.9.2. We will calculate the average load voltage, ripple across the load and to establish the minimum load current if efficient commutation and a good regulation is to be ensured.

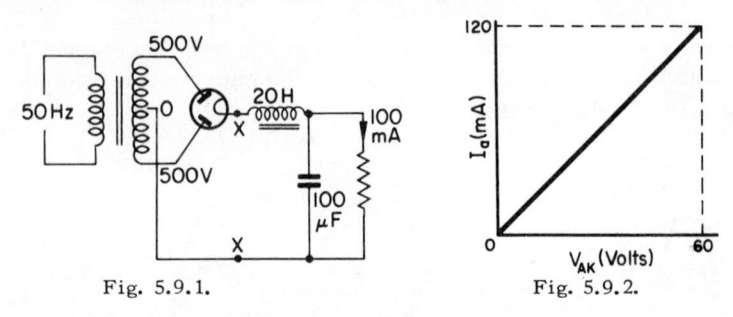

Fig. 5.9.1.

Fig. 5.9.2.

Figure 5.9.1. shows the circuit of the power supply unit, and figure 5.9.2. shows the characteristics of the rectifier.

The average voltage across the points, X, X, is given as
$V_{av} = 500 \sqrt{2}. \, 2/\pi -$ Diode drop. $= 450 - 50 = 400V$. (The resistance of the diode is 500Ω,). The diode drop is the product of the diode resistance and the average current. The ripple voltage across the load is determined after the ripple voltage at the point X, X. is calculated. The ripple at point X. X is approximately 300V Pk.

The reactance of the choke $X_L = 2\pi f L = 12.6 \text{K}\Omega$, (where f is 100Hz). The load resistor is $400V/100\text{mA} = 4\text{K}\Omega$. The reactance of the capacitor at 100Hz is given by $X_C = 1/2\pi f C = 16\Omega$. (The minimum value for L, at $R = 4\text{K}\Omega$, is $\doteqdot$ 4H for this case.).

The reactance of 16Ω is very much lower than the value of the 4K resistor and therefore the resistor may be ignored. The effective reactance across the point X. X is that of the choke, $12.6\text{K}\Omega$. The 'short circuit' current that will flow in the choke will be the peak voltage (ripple) at the point $X.X.$ divided by the reactance of the choke.

This becomes
$$\frac{300}{12.6} \, \text{K}\Omega = 23.8\text{mA Peak.}$$

The peak ripple current will flow through the capacitor. The ripple voltage across the load resistor will be the product of the peak ripple current times the reactance of the capacitor. The load ripple is $23.8\text{mA} \times 16\Omega = 380\text{mV}$ Peak. The r.m.s. value of load ripple will be $380 \times 0.707 = 269\text{mV}$.

DIODES, RECTIFICATION AND POWER SUPPLIES

The minimum average load current in order to ensure efficient commutation must be not less than 23.8mA. Figure 5.9.3. shows the final characteristics of the circuit. The regulation curves are given and particular reference is made to the critical average current which will ensure good commutation of the rectifiers (Point X.). (An 18KΩ bleeder resistor could be connected across the load so as to provide the necessary minimum average current should the load current fall below 23.8mA).

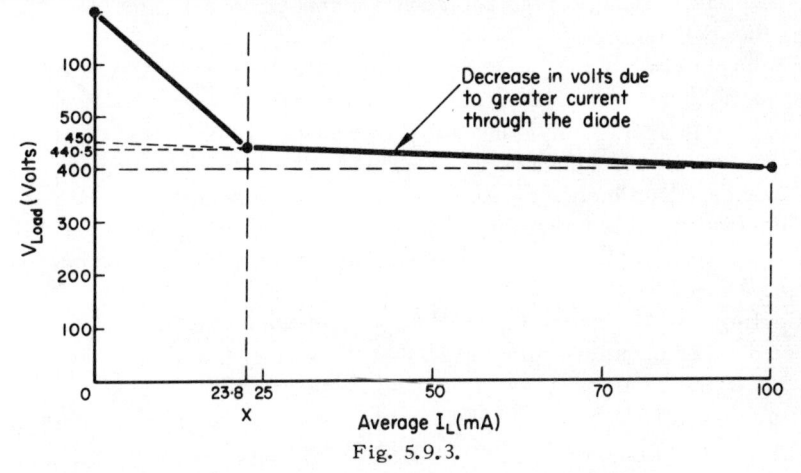

Fig. 5.9.3.

At 100mA, V load = 400V. At I_L = 23.8A, V_L = 440.5V.

Below I_L = 23.8A, the regulation is very poor as may be seen from the figure. Unlike the capacitor filter, the greater the load current, the lower the load resistance, the better the smoothing and the better the regulation.

5.10. Metal rectifiers.

The metal rectifier consists of a metal sheet separated from a semiconductor sheet by a thin layer of insulating material known as the barrier layer.

The uppermost plate will not normally act as a contact to which external connections may be made.

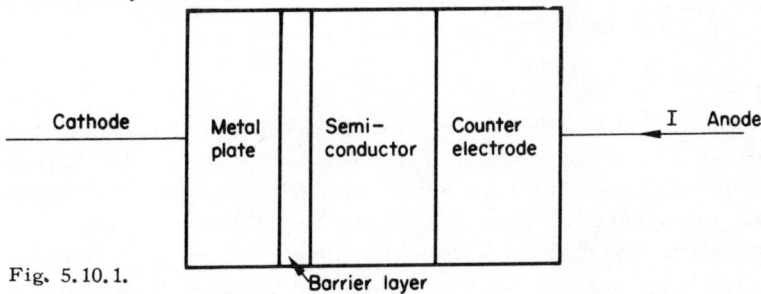

Fig. 5.10.1.

The copper-oxide type. (Type A).

In the type A, the metal is copper. The copper sheet is heated up to a temperature of approximately 1000°C, after which, it is cooled at a controlled rate. An outer skin is formed which, as it will have too high a resistance, is etched away by chemical process. Near the surface, there will exist a layer which has neither too much or too little oxygen. This is known as the blocking layer. The resultant copper/copper oxide junction forms the rectifier.

The upper plate, known as the counter-electrode, may consist of a layer of graphite and finally, a plate of rather soft metal such as lead is pressed firmly against the graphite. The pressure of this plate is critical and should not be loosened as when replaced, the rectifier's properties may change. Each rectifier element may be in the form of a disc thus making assembly of several elements quite easy. An insulating tube will be passed through the centre of the discs and through this will be passed a threaded metal rod.

Cooling fins will be interspersed throughout the assembly as required. These devices should be situated so that the surrounding air may assist the fins in the cooling process by normal convection currents.

Current will flow from semiconductor to the metal, or oxide to copper for type A, and selenium to alloy for type B.

The Selenium type. (Type B).

With this type, the semi-conductor may be formed by melting a layer of selenium on to the iron or steel plate. An insulating surface is formed on the semiconductor by chemical action followed by the connection of the counter electrode, or upper plate. This connection may be made by evaporating a soft metal once more, on to the surface.

As with the copper-oxide type, several discs may be stacked in series according to the voltage it is intended to use.

These discs are very sensitive to light and just one disc may develop quite a potential when connected across a resistor if subjected to, say, sunlight.

Figure 5.10.2. shows the I_a/V_a characteristics of these devices.

Germanium and Silicon types.

Germanium junction diodes are often made by allowing Indium to diffuse into a crystal of n type germanium. The crystal may be mounted in a metal can. This is filled with dry air and then hermetically sealed. This will prevent any moisture from damaging the surface of the germanium.

Silicon junction diodes are not unlike the germanium types. The process is as follows. The crystal is grown by slowly pulling a seed crystal from

DIODES, RECTIFICATION AND POWER SUPPLIES

molten n type silicon. When the seed is of the right thickness, a small amount of p type impurity is added. The result is a p type crystal. When complete, the result is a $p-n$ type junction diode.

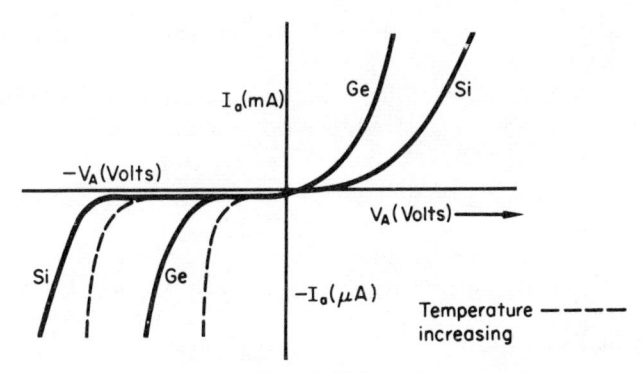

Fig. 5.10.2.

Further types known as point contact types are made from both germanium and silicon. A small wire spring with a fine point, is caused to be in contact with a 'n' type crystal. The whole assembly is mounted in a glass tube.

Figure 5.10.3. shows a typical point contact type.

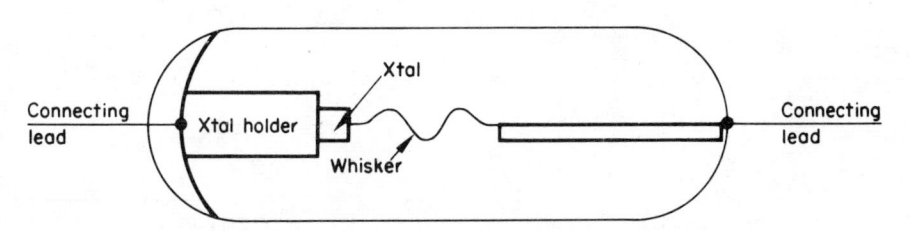

Fig. 5.10.3.

The forward current will be determined by the applied voltage and the external resistance. There will be no appreciable increase in current with an increase in temperature. The reverse voltage must not exceed the manufacturer's rating.

During the time that forward current is flowing, the forward drop must be kept as small as possible in order to keep the power dissipation as low as possible.

During the period of time that the rectifier is cut off, its cathode will be at the d.c. level of the storage load input. The anode will have, as it is cut off, a large negative voltage applied to it. Adding these voltages algebraically, the potential across the device will be approximately twice that of the peak value of the input.

This voltage is known as the 'peak inverse'. The peak inverse voltage for germanium types are in the order of 300V. Selenium types may be capable of withstanding up to 40V. Silicon types may be able to withstand much higher potentials.

If a greater peak inverse is likely to be encountered, then more elements must be placed in series with the others.

The peak current rating of these devices may often be exceeded for a short period of time but this should not be continued for any longer than the period of time stated by the manufacturers. The manufacturers may often quote the peak permissible current averaged over a 20 or 50 millisecond period.

Germanium rectifiers have a low forward drop. Selenium types are slightly higher whilst silicon types have a forward drop of 0.5 to 1.0V.

5.11. Bridge rectifiers.

This circuit could consist of metal rectifiers.

If four rectifiers are connected all facing the same way as shown in figure 5.11.1. and 'stretched' at the junctions of d_1, and d_2, d_3 and d_4, they will become the more familiar circuit of a convential bridge rectifier.

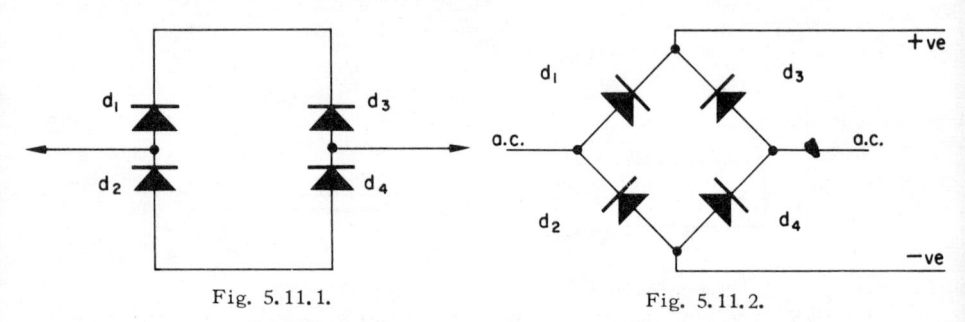

Fig. 5.11.1.

Fig. 5.11.2.

The a.c. input is connected to the points marked; the d.c. output is taken from d_1, d_3 (positive) and d_2, d_4 (negative) as in figure 5.11.2.

This is a full-wave circuit, and, as in the previous full wave circuit, there is a 'double topping up' process. The action is as follows:

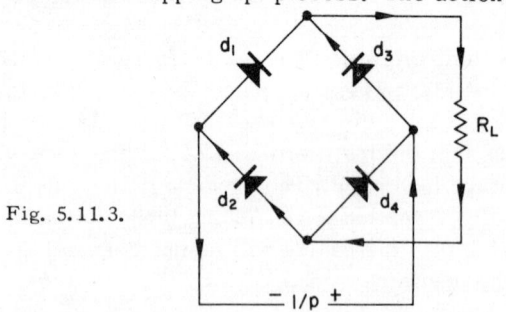

Fig. 5.11.3.

DIODES, RECTIFICATION AND POWER SUPPLIES 91

Consider the instant that the a.c. input is as shown.
D_3 anode is positive (D_3 conducting).
D_4 cathode is positive (D_4 not conducting).
D_1 anode is negative (D_1 not conducting).
D_2 cathode is negative (D_2 conducting).
The current flows as indicated.

During the next half cycle of the alternating input, the roles of all diodes are reversed, and appear as in figure 5.11.4.

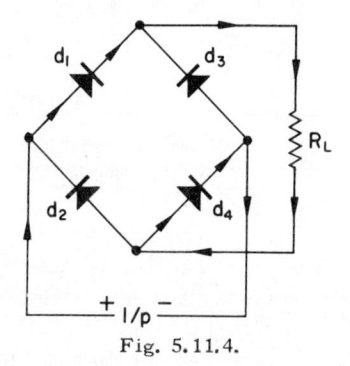

Fig. 5.11.4.

D_1 anode positive (conducting). D_2 cathode positive (non conducting) D_3 anode negative (non conducting). D_4 cathode negative (conducting).

It is clear, therefore, that for each half-cycle of input, a current is passed *in the same direction*, through R_L and behaves in a similar manner to the power supply unit previously described. The addition of a reservoir capacitor, filter resistance and smoothing capacitor (C_s) will complete the picture.

The advantage of this circuit is that the secondary winding of the mains transformer need not be centre-tapped, which may be useful or more economical. Further, current will flow through the winding on each half cycle, thus giving better transformer utilisation. The VA of the transformer winding is reduced by $1/\sqrt{2}$ in this type of circuit.

Where, in the double diode full wave circuit, the average d.c. output is $2/\pi \times$ Pk input from each secondary winding, the output obtained in this circuit is $2/\pi \times$ Pk input from the single secondary winding (represented by the $1/p$ in figure 5.11.4.).

5.12. Voltage doubling circuit.

With R_L connected as shown in figure 5.12.1. one capacitor is 'topped up' every alternate half-cycle, whilst the other is discharging through R_L. The resultant output d.c. across $R_L = 2 \times$ Pk output from the secondary winding. Many variations of this basic circuit are possible, and voltages many times the peak transformer output may be obtained.

A detailed drawing of the voltage doubler is shown in figure 5.12.1.

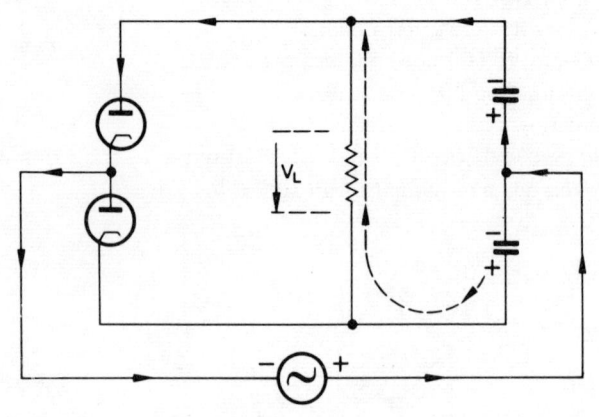

Fig. 5.12.1.

This has been a rather brief introduction to power supply units. We will discuss in a later volume, several other very important factors relating to power supplies.

We will need to examine stabilised supplies and to gain a better understanding of source resistance and to see how it can be controlled.

Before we can discuss either factor in depth, it is important to have a further look at more advanced equivalent circuits. **This subject is discussed in detail in chapter 18.**

CHAPTER 6

Meters

The moving-coil meter is the basic unit employed in almost all the instruments used in practice for the measurement of voltage, current and resistance. It is sensitive, accurate, and when used properly will have negligible effect on the circuit in which the measurement is being made. However, the meter movement itself must be associated with resistances which are connected either in series (for voltmeters) or in shunt (for ammeters) in order to ensure that the full-scale movement of the needle corresponds to the desired range of variation in the circuit being measured. Also if a battery is added the meter may be used for the measurement of resistance. It is common to design so called 'universal' meters or 'multirange instruments' which can be arranged by means of switches to operate as voltmeters, ammeters or ohmmeters over a variety of different ranges.

In this chapter we shall not describe the moving-coil movement itself as this is covered in any elementary book on electricity, but we will describe in detail the design of the multirange meter, discuss how a.c. is measured and also mention meter protection circuits.

6.1. Simple voltmeter.

Given a moving coil milliammeter with a known resistance R_m, a certain current (d.c. flowing in the right direction through the coil) will be just sufficient to cause the needle to move to full scale reading on the associated scale. This current, a characteristic of the movement, is called the full scale deflection current, abbreviated to f.s.d. The movement will be given the following circuit representation, with the marked current signifying f.s.d.

Figure 6.1.1. Movement with 100μA f.s.d.

Fig. 6.1.1.

In order to design a voltmeter, we must place in series with the movement a resistance R_s (often called a multiplier) of such a value that the current flowing in the circuit is equal to f.s.d. when the voltage across the whole circuit is equal to the maximum voltage it is desired to measure.

Suppose the meter has an internal resistance of 50 Ω and f.s.d. is 100μA, and we wish to construct a voltmeter of range 0–1 V. This means that 1 V must be applied to the voltmeter terminals (not the meter terminals) in order to cause a current of 100μA and hence give f.s.d. The required value for R_s is then easily calculated by Ohm's law from the circuit in figure 6.1.2.

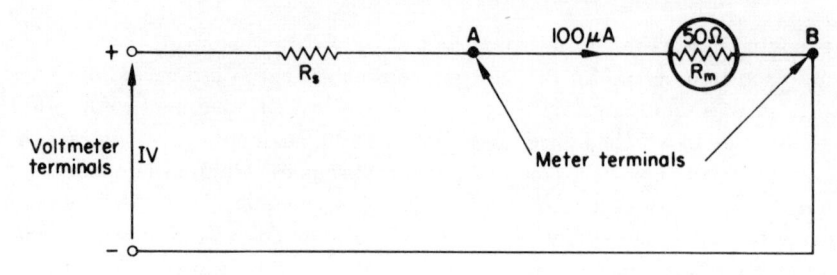

Fig. 6.1.2.

Voltmeter, range 0–1 V.

The first step is to calculate the p.d. across R_m when 100μA is flowing. This p.d. is 100μA × 50 Ω = 5mV. The remainder of the 1 V input is developed across R_s, that is the p.d. across R_s is to be 1 V – 0.005 V = 0.995 V.

Hence R_s = 0.995 V/100μA = 9.95 KΩ.

By a precisely similar procedure, we may calculate the required value of multiplier resistance for other voltage ranges. For example, a 0–10 V voltmeter.

R_s = 9.995 V/100μA = 99.95 KΩ.

while for a 0–100 V voltmeter

R_s = 99.95 V/100μA = 999.5 KΩ.

In practice, we may use a 100 KΩ resistor for the 10 V range and a 1 MΩ resistor for the 100 V range with negligible loss of accuracy.

A multirange voltmeter can be constructed with a switch to select the multiplier appropriate to the selected range, and a circuit diagram for such an instrument with three voltage ranges is shown in figure 6.1.3.

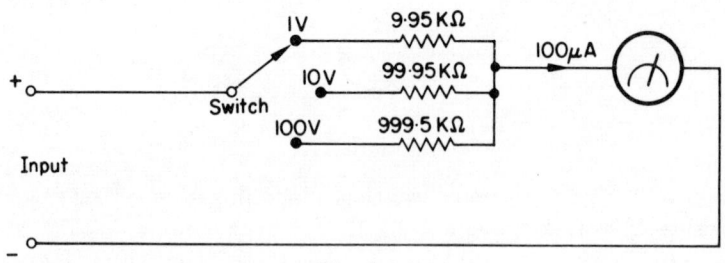

Fig. 6.1.3.

METERS

6.2. Switched-range ammeter.

In the case of an ammeter, a shunt resistance must be connected to the meter so that only a proportion of the main circuit current to be measured flows through the meter and the remainder flows through the shunt. The calculation of the value for the shunt merely involves arranging that when the maximum desired current is flowing into the ammeter, the correct f.s.d. current, in this case 100μA, is flowing in the meter movement.

Let us design an instrument to measure $0-100\mu$A, $0-1$ mA and $0-10$ mA in three ranges, using the same meter with f.s.d. 100μA and resistance 50Ω.

For the 100μA range it is obvious that the meter may be used as it stands, with no modification. We may calculate the required value of shunt resistance, R_{sh}, for the 1 mA range with the aid of figure 6.2.1.

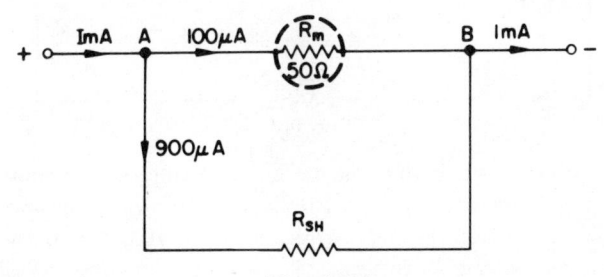

Fig. 6.2.1

If 1 mA enters the positive terminal and we require 100μA to cause f.s.d., then the remaining 900μA must flow through R_{sh}, as marked on the diagram.

The p.d. across R_m is 100μA $\times$ 50Ω = 5 mV. If 5 mV exists across R_m then 5 mV must also exist across R_{sh}, because the two are in parallel. But R_{sh} has 900μA flowing through it.

Hence R_{sh} = 5 mV/0.9 mA = 5.55Ω.

In a similar way, for the 10 mA range,

R_{sh} = 5 mV/9.9 mA = 0.505Ω.

The circuit for the proposed ammeter is shown in figure 6.2.2.

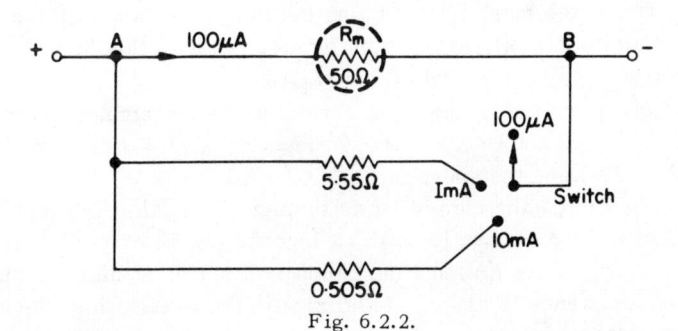

Fig. 6.2.2

H

It is important that the switch is a 'make before break' type in this circuit, so that when changing ranges it is not possible to connect the movement alone in the circuit to be measured even for a moment when the current is much larger than f.s.d.

The shunt resistors, which for large currents have to be very small and of awkward values, may be very difficult to manufacture in practice, and it is common to insert a swamp resistor, as shown in figure 6.2.3., in order to raise the 'effective p.d.' across points $A'.B$.

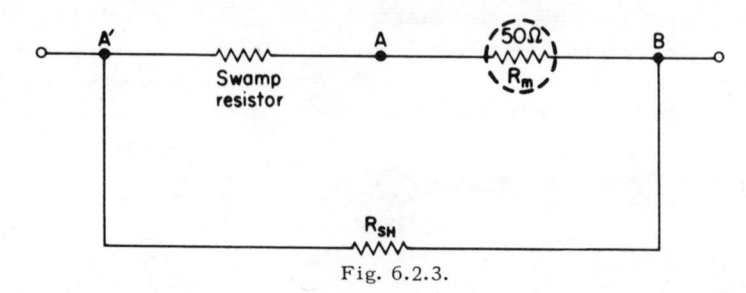

Fig. 6.2.3.

This has the effect of increasing the shunt resistors to a reasonable value so that they may be made without difficulty. If the swamp resistor has a value of, say, 950Ω, then the p.d. across $A'.B$ for 100µA is 0.1 V and the new shunt values become as follows:-

For the 1 mA range, R_{sh} = 0.1V/0.9 mA = 111.1Ω.
For the 10 mA range, R_{sh} = 0.1V/9.9 mA = 10.11Ω.

A disadvantage of using a swamp resistor is that the effective resistance of the meter is increased, giving an increased voltage drop across the meter. This may influence the external circuit too much, so that the current being measured is altered by the presence of the meter itself.

6.3. Universal Shunts.

Many commercial meters have a tapped shunt (R_{sh}) connected permanently across the meter, the input current being fed into the appropriate tap.

Let us 'build' a circuit slowly, step by step, as with this device the design becomes a little more complicated.

With input applied to (1), the input current will flow through Rm and the shunt path ($R1 + R2 + R3 + R4$.) With an input to (2), the current will flow through ($R1 + Rm$) and the shunt path of ($R2 + R3 + R4$).

With an input to (3) the current flows through ($R2 + R1 + Rm$), and the shunt path of ($R3 + R4$). Lastly, with an input to (4) the current flows through ($R3 + R2 + R1 + Rm$) and the shunt path of $R4$. At higher input currents, extra resistance is placed in series with the meter, while the shunt path has a lower resistance.

METERS

Typical circuit diagram.

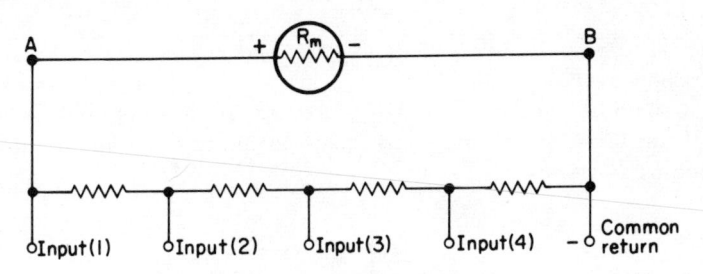

Fig. 6.3.1. Ammeter with universal shunt.

Let us simplify the circuit a little in order to examine the current paths. Let I be the current required for f.s.d. Let NI be the current to be measured where N is a positive integer, i.e. a number such as 1, 2, 5, 10, etc.

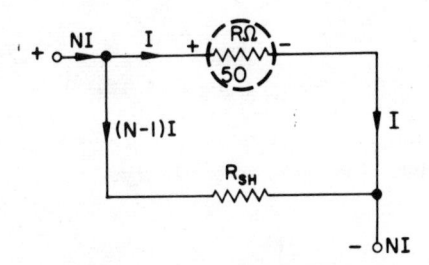

Fig. 6.3.2.

If NI enters the input terminal, and I flows through the meter, then the differenct $(NI - I)$ must flow through R_{sh}. The sum of their currents leaves the common, or negative terminal and totals $= I + (N - 1)I = NI$ once more.

If we now show one tapping point in R_{sh} (for a second range), we need to calculate the exact point at which to connect the tap. Consider the previous circuit with the tap added.

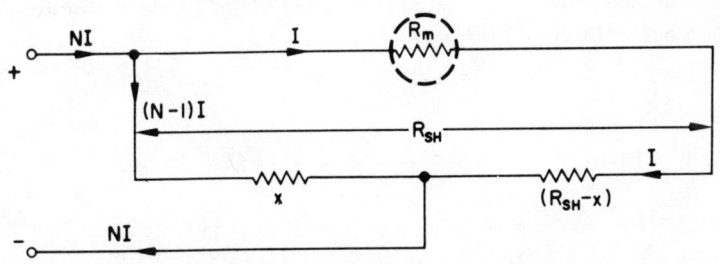

Fig. 6.3.3.

We will use the previous meter, $100\mu A$, f.s.d. with 50Ω resistance. Assume that we choose $200\mu A$ as our lowest and basic range.

R_{sh} must pass 100μA leaving 100μA to flow through R_m, therefore

$$RSH = \frac{5\,mV}{0.1\,mA} = 50\Omega.$$

Having established the value of R_{sh}, as 50Ω, we have to consider the tapping point for a further range. To make matters easier, the circuit may be simplified as follows:-

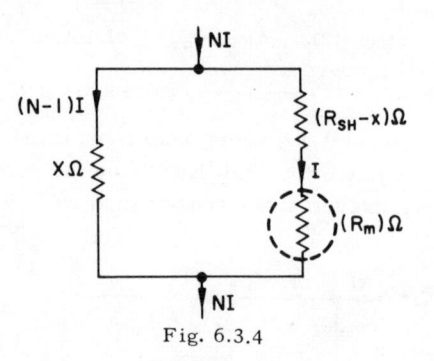

Fig. 6.3.4

It is evident that the meter resistance Rm is in series with $(R_{sh}' - x)$, and is shown as one resistor for the sake of clarity.

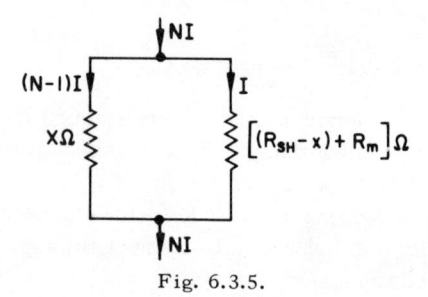

Fig. 6.3.5.

Using the 'load over total' technique for currents in shunt circuits, the current through the resistor x is given as,

$$\frac{(R_{sh} - x + Rm)(NI)}{(R_{sh} - x + Rm) + x} = (N-1)I$$

Tidying up a little

$$\frac{(R_{sh} - x + Rm)NI}{R_{sh} + Rm} = (N-1)I$$

Therefore $(R_{sh} - x + Rm)N = N(R_{sh} + Rm) - (RSH + Rm)$.

and $N(R_{sh} + Rm) - Nx = N(R_{sh} + Rm) - (R_{sh} + Rm)$.

Therefore $-Nx = -(R_{sh} + Rm)$

and $x = \dfrac{R_{sh} + Rm}{N}$

METERS

R_{sh}, Rm and N are all known, and it is now an easy matter to substitute in the formula in order to find x.

It has been assumed that our basic range will be 200μA. R_{sh} has a value of 50Ω. Suppose we decide that the second range should be 1 mA.

1 mA is 10 times that of I (the f.s.d. current); therefore $N = 10$.

Substituting in the formula $\quad x = \dfrac{R_{sh} + Rm}{N}$

$$x = \frac{50 + 50}{10} = 10\Omega.$$

The circuit then becomes as shown. Assume 1 mA input.

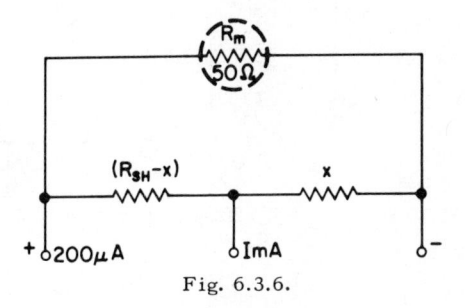

Fig. 6.3.6.

As a check the circuit may be redrawn, the current flowing in both paths may be calculated. Note that Rm is in series with $(R_{sh} - x)$.

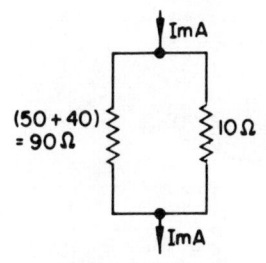

Fig. 6.3.7.

Current through meter $= \dfrac{1\,\text{mA} \times 10}{90 + 10} = \dfrac{1 \times 10\,\text{mA}}{100} = 100\mu\text{A}$

Current through 10Ω $= \dfrac{1\,\text{mA} \times 90}{90 + 10} = \dfrac{1 \times 90}{100} = 900\mu\text{A}$

Further ranges may be calculated in a similar manner. This circuit has, however, the disadvantage that the sensitivity in 'ohms per volt' is reduced, compared with the simpler switched-shunt type.

The basic meter of 100μA (10 KΩ/V) has been modified to that of 200μA (5 KΩ/V) and **consequently** the loading on external circuits is twice that of the original meter, when used as a voltmeter.

It is suggested that for the first attempt at multirange meter design, a plug and socket arrangement should be employed. More ambitious meters can be attempted later, incorporating the rather complicated switching associated with a circuit of this type.

A typical meter may be shown thus:

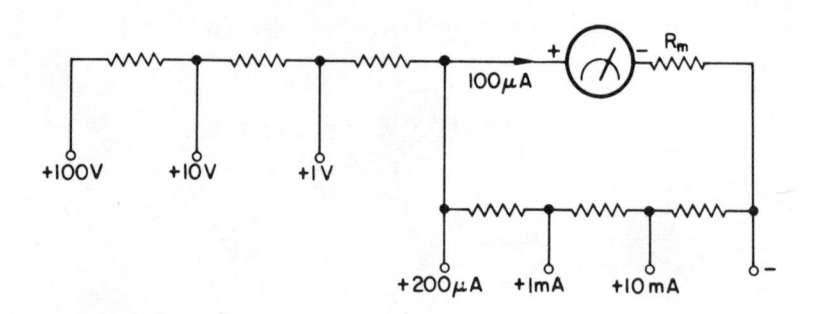

Fig. 6.3.8

6.4. High impedance voltmeter.

Suppose that it is desired to measure the p.d. at point x in the following circuit.

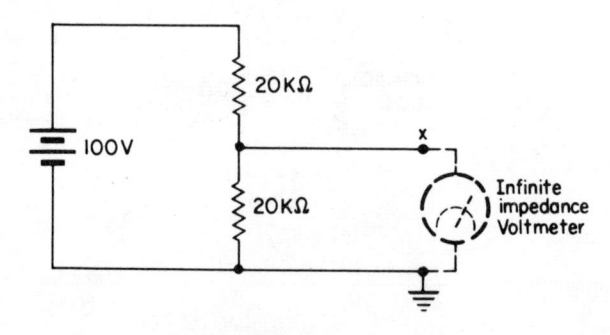

Fig. 6.4.1.

The theoretical value at point $x = \dfrac{20}{40} \times 100 \text{ V} = \underline{50 \text{ V}}.$

Using a 1 KΩ/V meter (on the 100 V range), the circuit will appear as in figure 6.4.2.

METERS

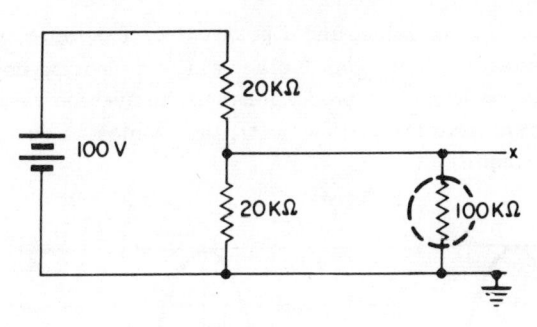

Fig. 6.4.2.

The p.d. $= \dfrac{16.7}{20 + 16.7} \times 100\,\text{V} = 45\,\text{V}$ (where $100\,\text{K}\Omega/20\,\text{K}\Omega = 16.7\,\text{K}\Omega$)

But if a 500 Ω/V meter had been used, the p.d. measured would have been

$$\dfrac{14.3}{34.3} \times 100\,\text{V} = 41.7\,\text{V}.$$

It is important, therefore, to ensure that when measuring potentials in high impedance networks, a very high impedance voltmeter should be used whenever possible.

6.5. A.c. ranges, Rectification, R.M.S. and Average values.

The form factor of an alternating voltage is an indication of its waveshape. The form factor is given as;

$$\text{The form factor} = \dfrac{\text{r.m.s. value}}{\text{Average value}}$$

and for sinusoidal waveform the factor is

$$\dfrac{\sqrt{1/2} \times \text{Pk value}}{2/\pi \times \text{Pk value}} = \dfrac{\pi}{2\sqrt{2}} = \dfrac{\sqrt{2}\pi}{4} = 1.11$$

If the wave form has a form factor less than this value, it indicates that the waveform tends towards a squarewave. If the form factor has a value that is greater than this value, it indicates that the waveform tends towards a triangular type of waveform. A sinusoidal waveform will always have a formfactor of 1.11.

The form factor of a sinusoidal waveform is most important when dealing with the a.c. voltage ranges of a moving coil rectifier type a.c. voltmeter.

Figure 6.5.1. shows a typical sinusoidal waveform, the r.m.s., peak and average values are indicated.

Rectification of the alternating voltages to be measured is often accomplished by the use of a copper-oxide bridge rectifier. The voltage to be measured will have a peak to peak value, and a peak value. The meter will, in general, be calibrated in terms of the r.m.s. value, while the moving coil movement will actually give a deflection proportional to the average value. Once these various parameters have been allowed for, the problem of designing the a.c. ranges will be simplified.

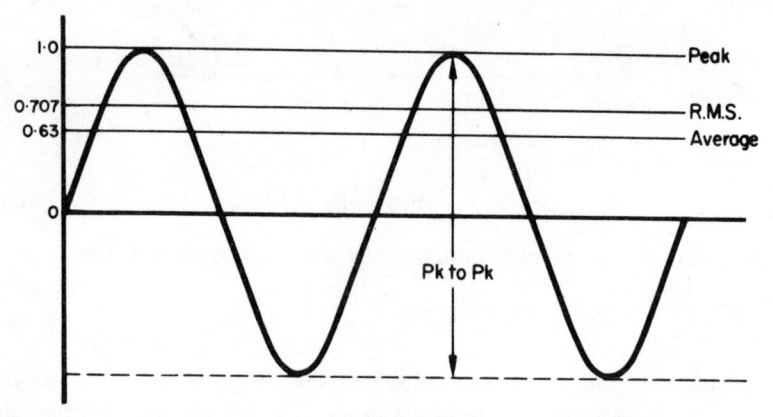

Fig. 6.5.1.

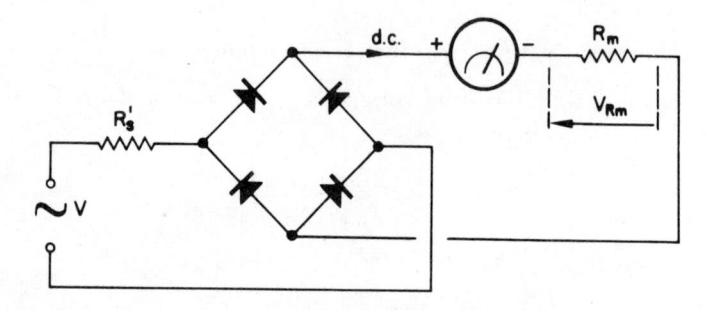

Fig. 6.5.2.

Figure 6.5.2. shows a typical rectifier arrangement.
Figure 6.5.3. shows a typical rectifier output.

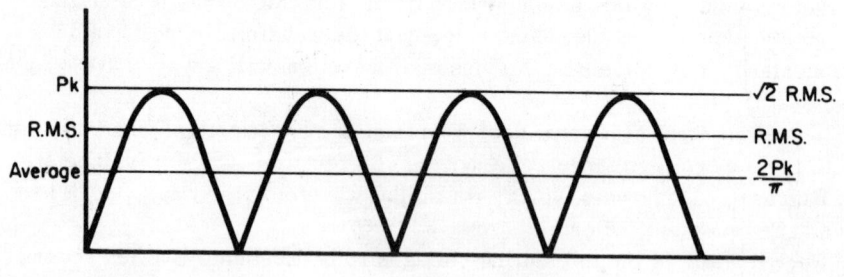

Fig. 6.5.3.

METERS

When a.c. is applied to a full wave bridge, the peak output is applied to the movement.

The movement will register the average of this peak, but the scale must be calibrated in r.m.s.

Hence
$$Im = \frac{[V_{r.m.s.}(\sqrt{2})] \times [2/\pi]}{R_s^1}$$

Where R_s^1 is the series a.c. voltage range resistor.

The meter scale is to be common to both a.c. and d.c. volts, hence
$$V_{d.c.} = V_{r.m.s.}$$

Hence
$$I_m(d.c.) = I_m(a.c.)$$

$$\therefore \quad \frac{V_{d.c.}}{R_s} = \frac{(V_{r.m.s.}\sqrt{2})(2)}{R_s^1 \pi}$$

but as $V_{d.c.} = V_{r.m.s.}$ on the scale, at f.s.d.

$$\frac{1}{R_s} = \frac{2\sqrt{2}}{\pi R_s^1}$$

hence $\quad R_s^1 = \dfrac{2\sqrt{2}}{\pi} \cdot R_s$

$\therefore \quad R_s^1 = 0.9 R_s$

Example.

If the d.c. and a.c. f.s.d. = 100 V, and R_s = 100 KΩ then for a.c., R_s^1 = 0.9 × 100 KΩ = 90 KΩ.

Calibration of a.c. Voltmeter.

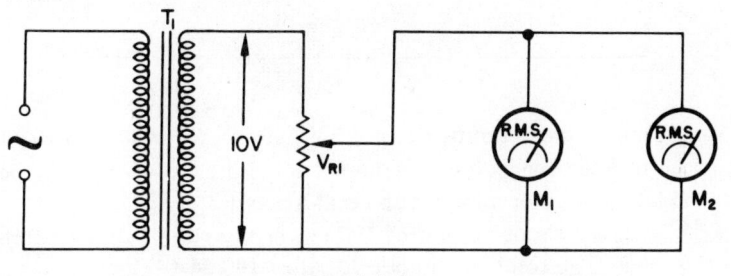

Fig. 6.5.4.

M_1 is the standard or known meter.
M_2 is the meter under construction.
VR_1 is a potentiometer varying the a.c. voltage from 0–10 V a.c.
T_1. is a step down transformer of a suitable ratio.

The constant, 0.90, will be valid for all ranges above 10 V f.s.d. Below this value, the rectifier may not be linear and the forward voltage drop will not be proportional to the applied voltage and current.

The lower ranges, below 10 V, may be calibrated by connecting a known standard meter in shunt with the meter to be built and the scale may be calibrated accordingly (figure 6.5.4.).

The value of Rm must be taken into account on the lower ranges, also.

When R_s is less than 100 times that of Rm, the latter must always be deducted from the total resistance in the entire circuit.

For non-sinusoidal voltages, there will be a different value of form factor. This will result in meter readings that cannot be relied upon for absolute measurements, although changes in level may be reasonably accurate.

6.6. Simple Ohmmeter.

An Ohmmeter may be constructed as an instrument complete in itself; alternatively it may be incorporated in the multi-range instrument previously described.

An Ohmmeter is a device which is used to measure resistance, for example that of resistors, d.c. resistance of transformer windings, short circuit tests, etc.

The basic circuit is shown in figure 6.6.1.

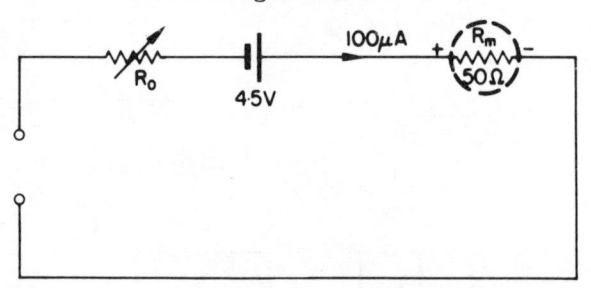

Fig. 6.6.1.

We have chosen, quite arbitrarily, a 4.5 V battery. R_0 is a variable or preset potentiometer connected as a preset resistor; the value of R_0 needs to be found before further design can be attempted.

We will assume a f.s.d. current of $100\mu A$, and assume that we short-circuit the input. The total resistance to allow $100\mu A$ to flow =

$$= \frac{4.5 V}{100 \mu A} = 45 K\Omega$$

R_0 may therefore be a $50 K\Omega$ potentiometer, and set, in practice, to $45 K\Omega - 50\Omega$ for $100\mu A$ through the meter. Rm may now be ignored as, once set, R_0 and Rm may be considered as a single resistor. We will assume that $R_0 = 45\,\text{K}\Omega$. (A finer control will be achieved by using say, a $39\,\text{K}\Omega$ fixed resistor in series with a $10\,\text{K}\Omega$ variable, for R_0.)

R_0 is set to this value with the input terminals short-circuited so that the external resistance is zero. We are now in a position to derive a table which will make it possible to construct an Ohms scale on the meter.

The external resistance to be measured will always be effectively in series with R_0. If we choose a number of different values of external resistance, we can easily calculate the current that will flow through the meter. We can then draw a scale on the meter, for ohms, showing the points that have been chosen.

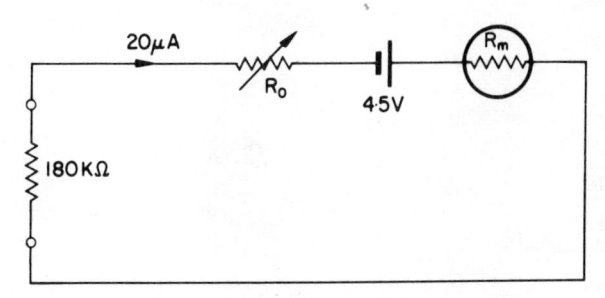

Fig. 6.6.2.

Example.

From figure 6.6.2, $\quad I = \dfrac{V}{R} \quad \therefore \quad I = \dfrac{4.5\,\text{V}}{(180 + 45)\,\text{K}\Omega}$

$$= 20\mu\text{A}.$$

Therefore, at the point corresponding to $20\mu\text{A}$ on the existing scale, we can mark $180\,\text{K}\Omega$. The pointer will indicate this point when a $180\,\text{K}\Omega$ resistor is connected across the input terminals.

The point will be marked on the scale as shown in figure 6.6.3.

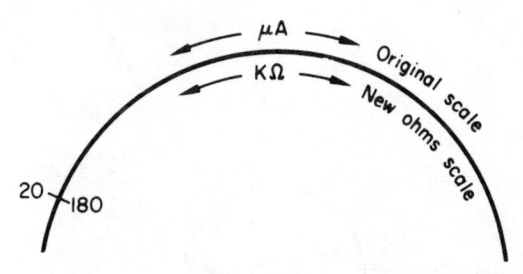

Fig. 6.6.3.

Further points are derived in the following table. Standard values have been chosen, but it is left to the student to decide for himself how many points he will mark when designing his own Ohmmeter. With a few exceptions the values shown are correct to 2 significant figures only.

External Resistor (Ω)	Meter Deflection μA	External Resistor (Ω)	Meter Deflection μA
470K	8.75	22K	67
390K	10	18K	71.5
330K	12	15K	75
270K	14	12K	79
220K	17	10K	82
180K	20	8.2K	84.5
150K	23	6.8K	87
120K	27	5.6K	89
91K	33	4.7K	90.5
82K	35.5	3.9K	92
68K	40	3.3K	93
56K	44.5	2.7K	94
47K	49	2.2K	95.5
39K	53.5	1.8K	96
33K	57.6	1.0K	98
27K	62.5		

A typical meter scale is shown here. Note the original scale for 100μA and the new points for the Ohms scale.

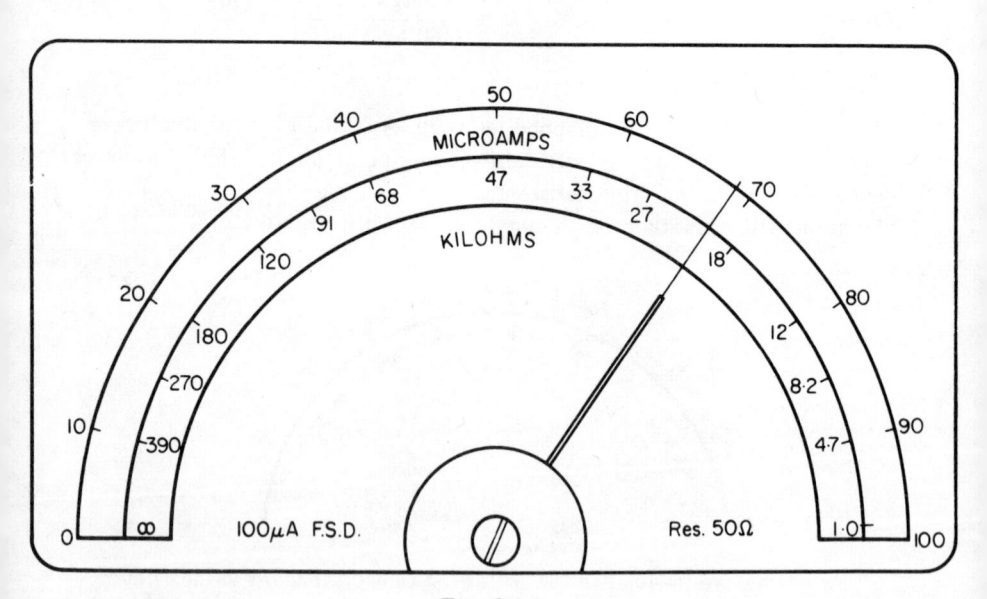

Fig. 6.6.4.

6.7. Simple protection circuits.

Electro-mechanical protection devices are difficult to make. It is fairly simple, however, to design an electronic circuit that will provide reasonable protection for the meter in the event of an accidental overload. The circuit shown here illustrates how a diode connected across the meter enables the voltmeter to withstand a considerable overload without destroying the expensive movement.

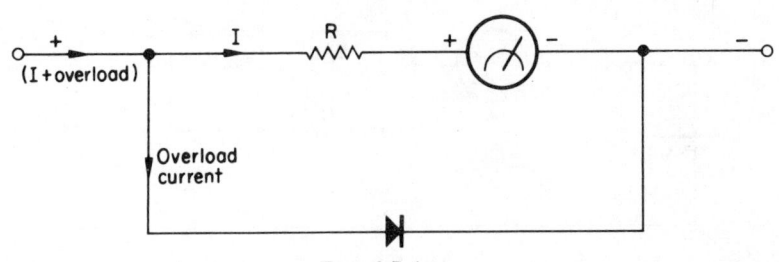

Fig. 6.7.1.

R may be the meter resistance, a swamp resistance or one deliberately positioned there for the purpose of assisting the particular diode to function more effectively. If I is the normal f.s.d., a p.d. $(I \times R)$ will be developed across the diode. The diode is only very slightly forward biassed, and consequently has a fairly high resistance compared with R, so that it does little to modify the circuit. In the event of an accidental overload of, say, five times, the current will become $5I$. The p.d. across the diode will become five times greater and hence the forward resistance will fall rapidly. The new value of diode resistance is now very much smaller than R; consequently the diode diverts most of the overload current in the same manner as a shunt. The diode and R may be determined either by theory or experimentally, but both methods should be attempted in order to demonstrate that the theory is borne out in practice. If a large negative overload occurs, the diode is almost open circuit and does nothing to protect the meter.

A second diode connected the other way round provides a similar protection for reverse overloads. The circuit is given in figure 6.7.2. Silicon diodes are perhaps the best type to use in this circuit.

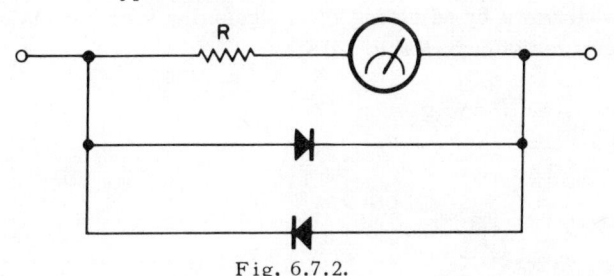

Fig. 6.7.2.

6.8. Measuring the 'internal resistance' of the meter movement.

The internal resistance of a meter cannot be safely measured with ohmmeters. The ohmmeter will have an internal battery that, when applied to the meter terminals, may cause a current, greater than the full scale current, to flow through the meter. A simple method of measuring the internal resistance is to connect a circuit as shown in figure 6.8.1.

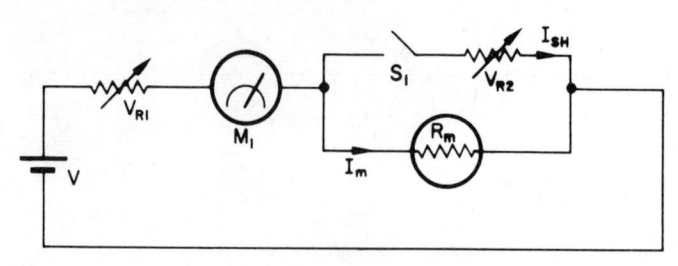

Fig. 6.8.1.

Assume the resistance, Rm to be zero, choose a value for VR_1 such that when V is applied, the current flowing, (Im). in the meter will be a little less than the permissible full scale deflection. Adjust VR_1 so as to cause the current in the meter to give exactly full scale deflection of the pointer. This current is monitored by M_1 and kept constant by adjusting VR_1. Close switch S_1 and adjust the variable resistor VR_2, until the meter reads exactly half full scale deflection. Open the switch S_1 and measure the ohmic value of VR_2. As exactly half of the current had flowed through VR_2, the value of VR_2 must be equal to that of Rm, the meter resistance.

Example.

Assume that Rm is actually 1000Ω. The f.s.d. is 1 mA. Suppose a battery having an e.m.f. of 4.5 V is available. The maximum permissible current is 1 mA, therefore the minimum value of VR_1 must be $4.5\,K\Omega$.

Adjust VR_1 to a value of $3.5\,K\Omega$ and a 1 mA current will flow. The $1\,K\Omega$ value of Rm will be added to the total resistance. Close the switch, S_1, and adjust the variable resistor VR_2 until the meter reads 0.5 mA. Ensure that 1 mA still flows by adjusting VR_1. Measuring VR_2 will indicate that the meter resistance, Rm is $1.0\,K\Omega$.

CHAPTER 7

Triode valves, voltage reference tubes and the thyratron

The triode.

In this chapter we shall be concerned with the properties and characteristics of the thermionic triode valve, and for the first time examine a device which enables the current flowing in one circuit to be controlled by means of the voltage applied in another circuit. This principle introduces the vitally important process of amplification, and the circuit arrangements for triodes will be studied in the following chapter where amplifiers are discussed in more detail.

7.1. The triode valve.

If a third electrode, called the *control grid* or simply the *grid*, is included between cathode and anode in the diode valve, then the externally applied potential between this grid and cathode will exert a great influence upon the anode current flow, particularly if this grid is mounted near to the cathode in the space-charge region. The grid is of mesh-like construction so as not to impede the flow of electrons from cathode to anode and it is normally arranged that the grid does not at any time become positive with respect to the cathode so that the grid does not itself attract electrons and current does not flow in the external circuit between grid and cathode.

The circuit symbol for a triode is shown in figure 7.1.1. together with the symbols used for the voltages and currents which flow when the valve is used in a circuit. These latter are anode current I_a, grid-cathode volts V_{gk}, and anode-cathode volts V_{ak}. The output load resistance R_L is also shown together with the high tension or H.T. supply which is a d.c. used to maintain the anode at a positive potential with respect to the cathode and therefore keep anode current flowing. As mentioned above, the anode current value depends not only upon the anode supply but also on the p.d. V_{gk}.

The arrangement of the electrodes in a triode is shown in figure 7.2.

If the grid potential is zero ($V_{gk} = 0$), then the triode behaves just like the diode, the grid having negligible effect on I_a. Thus I_a is determined only by V_{ak}, and as V_{ak} is gradually increased from zero the anode current rises as shown in the I_a/V_{ak} characteristic of figure 7.1.3.

Now if the grid voltage V_{gk} is made slightly negative, then the attraction of electrons from the cathode produced by the positive anode voltage is offset by a repulsion due to the negative grid, so that the anode current is

Fig. 7.1.1.

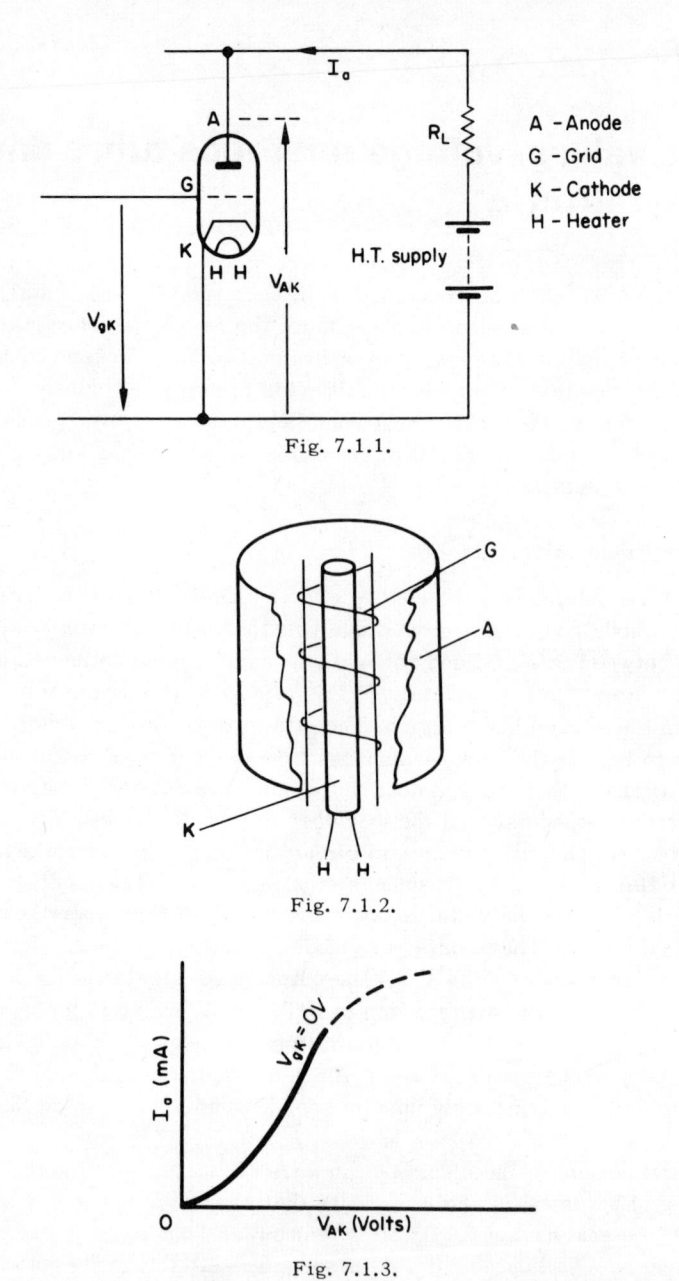

Fig. 7.1.2.

Fig. 7.1.3.

reduced. If V_{gk} is gradually made more and more negative the anode current gradually falls until it ceases altogether. The potential required on the grid to just prevent the flow of anode current (for a given anode voltage V_{ak}) is

TRIODE VALVES, VOLTAGE REFERENCE TUBES AND THE THYRATRON

called the 'cut-off' potential. As an example, suppose V_{ak} is maintained throughout at 100 V, then the anode current might change as the grid is made more negative as shown in the following table:

V_{gk}	I_a
0 V	10 mA
−1	8
−2	6
−3	4
−5 V	0 mA

Fig. 7.1.4.

In this case the cut-off voltage is −5 V.

As there are three variables, I_a, V_{ak} and V_{gk} a graphical representation of the triode characteristics can only be done by plotting two of these variables, one against the other, for a fixed value of the third. For example, we might plot I_a against V_{ak} for a fixed value of grid voltage $V_{gk} - 2V$. This graph is shown in figure 7.1.5. labelled −2. Further graphs can be drawn on the same diagram for other fixed values of V_{gk}, each graph labelled with the corresponding value of grid volts to give a series of curves, and in this way the behaviour of the valve can be exhibited in a convenient form. The characteristics shown in figure 7.1.5. with I_a plotted against V_{ak}, are called the 'anode characteristics' whilst an alternative commonly used involves the so-called 'mutual characteristics' in which I_a is plotted against V_{gk} for different fixed values of V_{ak}.

The two types of diagram are, of course, just two different ways of presenting exactly the same information, and it is a matter purely of

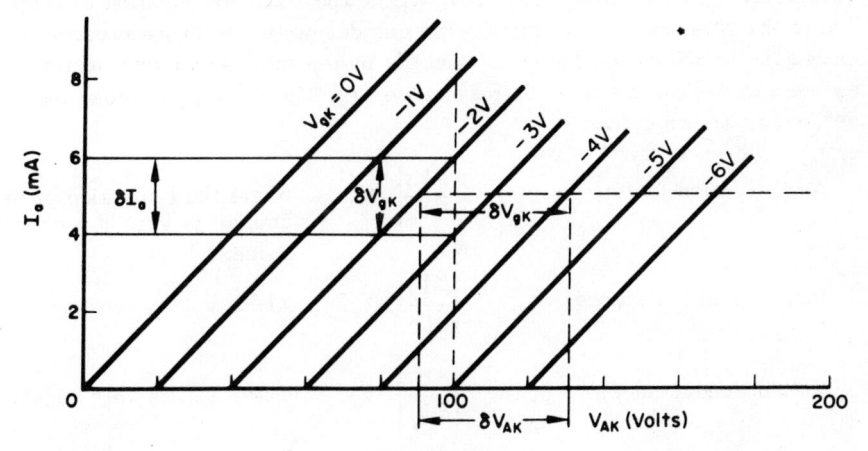

Fig. 7.1.5.

convenience which is employed in any particular application. For example if we wish to see from figure 7.1.5. how I_a varies with V_{gk} for a fixed value of V_{ak} equal to 100 V (the example just given in the table on p.   .), we merely look along the vertical line for 100 V, shown dotted in the diagram, and the reader will find the tabulated values at the points where this dotted line cuts the curves.

The graphs given are idealised to the extent that all curves are straight lines and they are equidistant and parallel. In practice, triodes have characteristics like this except in the region where anode current is very small, and provided we do not enter this region, satisfactory results may be obtained by using the ideal characteristics. True (measured) typical characteristics are more like those shown in figure 7.2.3.

7.2. Triode parameters

The ideal characteristics are linear, so that if one variable is kept fixed then the other two will be proportional to one another over the range between zero grid volts and cut-off, and we may take the constant of proportion as a characteristic parameter for the particular triode.

For example, the previous table (and the graph) show that if V_{ak} is fixed in value, then a 1 V change in grid voltage produces a 2 mA change in anode current, a 2 V change in V_{gk} produces a 4 mA change in I_a, and so on. We therefore say that the triode has a 'mutual conductance' equal to 2 mA per volt.

The symbol used for mutual conductance is g_m, and for this triode,

$$g_m = 2 \text{ mA/V}.$$

Note that for the ideal curves it does not matter what (fixed) value we choose for V_{ak}, the value of g_m will still be the same. By electing to keep one of the other variables constant, we can define two other parameters, called the amplification factor, μ, and the incremental resistance (usually called a.c. resistance or anode resistance), r_a. The three parameters are defined as shown below:

Amplification factor, $\quad \mu = \left| \dfrac{\delta V_a}{\delta V_g} \right| I_a \quad$ Where the I_a outside of the bracket is kept at a constant value.

Incremental resistance, $\quad r_a = \left| \dfrac{\delta V_a}{\delta I_a} \right| V_{gk} \quad$ where V_{gk} is kept constant.

Mutual conductance, $\quad g_m = \left| \dfrac{\delta I_a}{\delta V_g} \right| V_{ak} \quad$ where V_{ak} is kept constant.

TRIODE VALVES, VOLTAGE REFERENCE TUBES AND THE THYRATRON 113

From the definitions it follows immediately that there is a relationship between the three parameters:

$$r_a \times g_m = \mu$$

Suppose we investigate the parameter, μ, using the I_o/V_{ak} characteristics.

$$\mu = \left| \frac{\delta V_a}{\delta V_g} \right| I_a .$$

If we choose a value for I_a, say 5 mA, draw a line as indicated at 5 mA. Mark a point on this line corresponding to a value of V_{gk}, say -2 V. Mark another point on this line corresponding to the value of V_g say -2 V. Then mark a point on the same I_a line at -4 V say. A perpendicular dropped from both points will show that there are two values of V_a. The change in V_{ak} (δV_{ak}) divided by the change in V_{gk} (δV_{gk}). In our example,

$$\mu = \left| \frac{130 - 90}{4 - 2} \right| = \left| \frac{40}{2} \right| = \underline{20}$$

from figure 7.2.1. For a given value of I_a, a change of $V_{gk} (\delta V_{gk})$ would alter I_a. The change of V_{ak} required to restore I_a to its original value is (δV_{ak}).

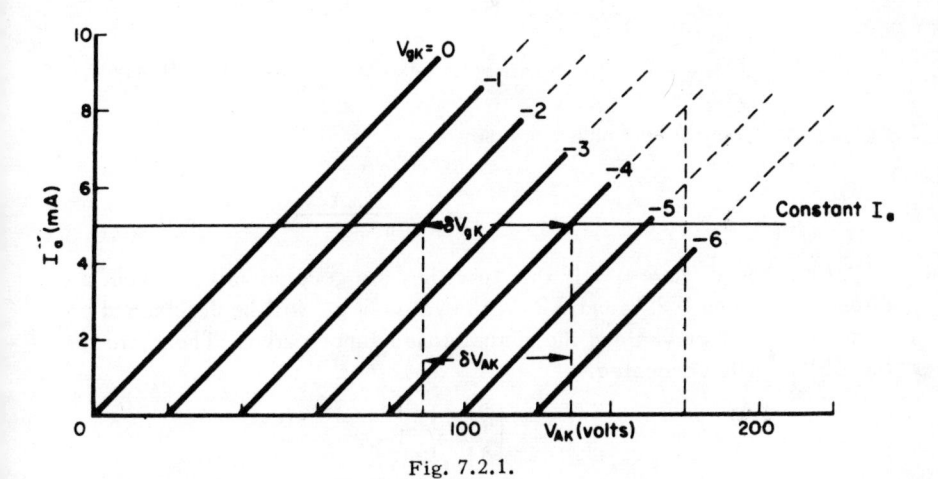

Fig. 7.2.1.

Let us now examine the characteristics and determine r_a.

$$r_a = \left| \frac{\delta V_a}{\delta I_a} \right| V_{gk} .$$

V_{gk} must be kept constant. Let us choose $V_{gk} = 4$ V. The line representing

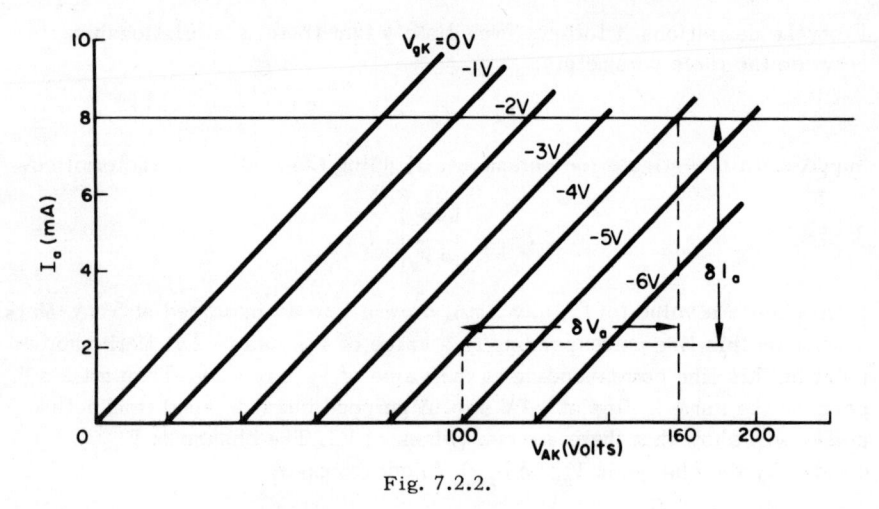

Fig. 7.2.2.

a -4 V bias is to be, in fact, the hypotenuse of a right-angled triangle of which the 'Y' axis will be I_a and the 'X' axis, V_{ak}. Draw a line connecting points (100 V, 2 mA) and (160 V, 8 mA); then complete the triangle as shown in figure 7.2.2. Then

$$\delta V_a = 160 - 100 = 60 \text{ V.}$$
$$\delta I_a = 8 - 2 = 6 \text{ mA.}$$
$$\therefore r_a = \frac{60 \text{ V}}{6 \text{ mA}} = 10 \text{ K}\Omega \text{ (as } I_a \text{ is in milliamperes).}$$

g_m may more easily be found because

$$g_m = \frac{\mu}{r_a} = \frac{20}{10 \text{ K}} = \underline{2 \text{ mA/V.}}$$

A final set of curves, only this time they are genuine valve curves, are offered in figures 7.2.3. and 7.2.4. The value of g_m will be determined both from the I_a/V_{ak} curves and the mutual conductance curves. These are for the EF86 triode connected.

$$g_m = \left| \frac{\delta I_a}{\delta V_{gk}} \right| V_{ak}.$$
$$= \left| \frac{6 - 3.2}{4 - 3} \right| 200$$
$$= \left| \frac{2.8}{1} \right| 200$$
$$= 2.8 \text{ mA/V, where the 200 V is a constant.}$$

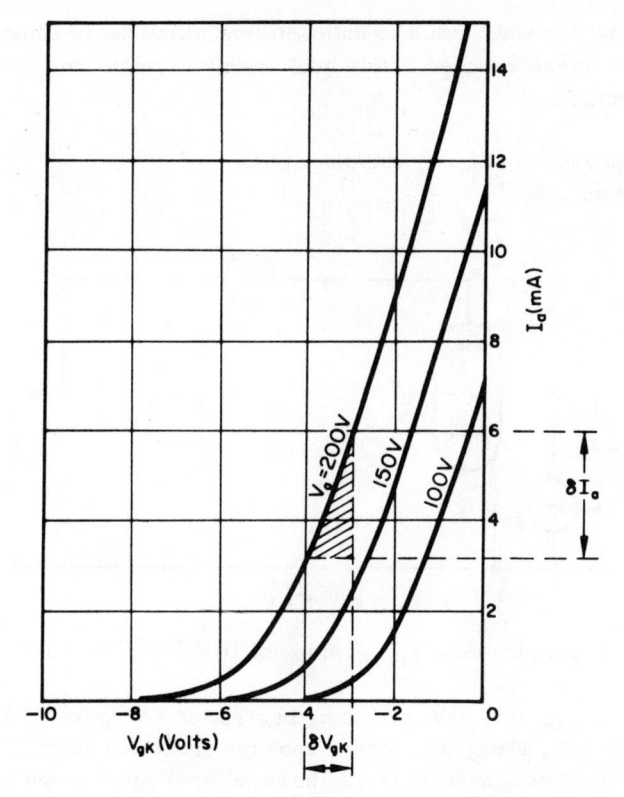

Fig. 7.2.3. Mutual conductance curves.

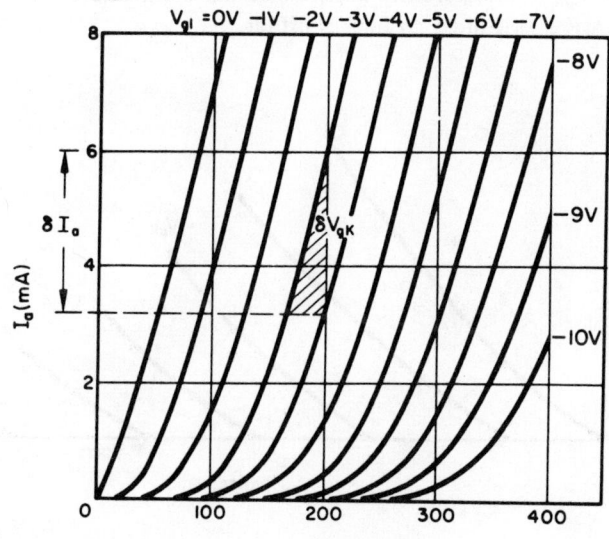

Fig. 7.2.4. I_a/V_{ak} curves.

It is seen that the same result is obtained from either set of characteristics. The I_a/V_{ak} curves are used in this book mainly because they provide so much information.

7.3. Measurements of I_a/V_a characteristics. Common cathode.

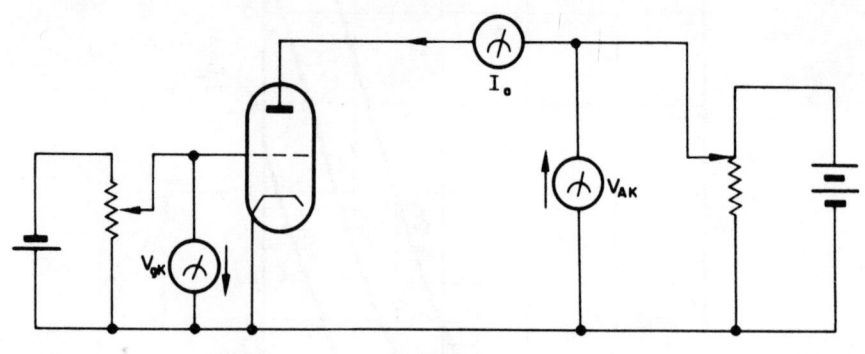

Fig. 7.3.1.

Set V_{gk} to 0 V. Increase V_{ak} in steps of 10 V from 0 to 350 V. Record I_a for each step.

Repeat for V_{gk} to -30 V increasing in steps of 1 V up to -10 V, then at greater intervals. These will depend upon the particular valve. The characteristics should be similar to those shown in the diagram given here. Note that once V_{gk} is sufficiently negative to cut off anode current, making V_{gk} more negative will have no further effect.

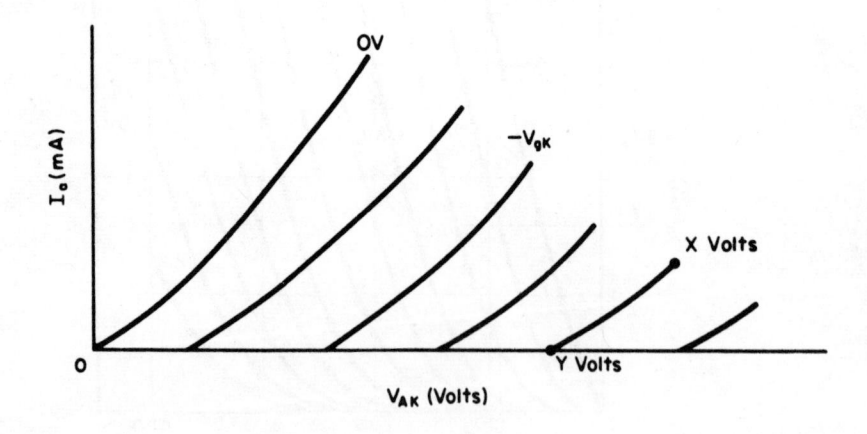

Fig. 7.3.2.

TRIODE VALVES, VOLTAGE REFERENCE TUBES AND THE THYRATRON

Do not exceed either the rated maximum I_a or power limit specified by the valve manufacturer. An important feature to note is that V_{gk} required to cut the valve off at different values of V_{ak}. In figure 7.3.2., x volts V_{gk} are required to cut the valve off with a V_{ak} of y volts. This applies to each of the family of grid curves.

Exercise 1.

From the characteristics of the EF86 Triode connected shown in figure 7.3.3. obtain the values of the parameters μ, g_m and r_a, at around a point on the graph corresponding to 200 V, 1 mA, shown as point P on the figure.

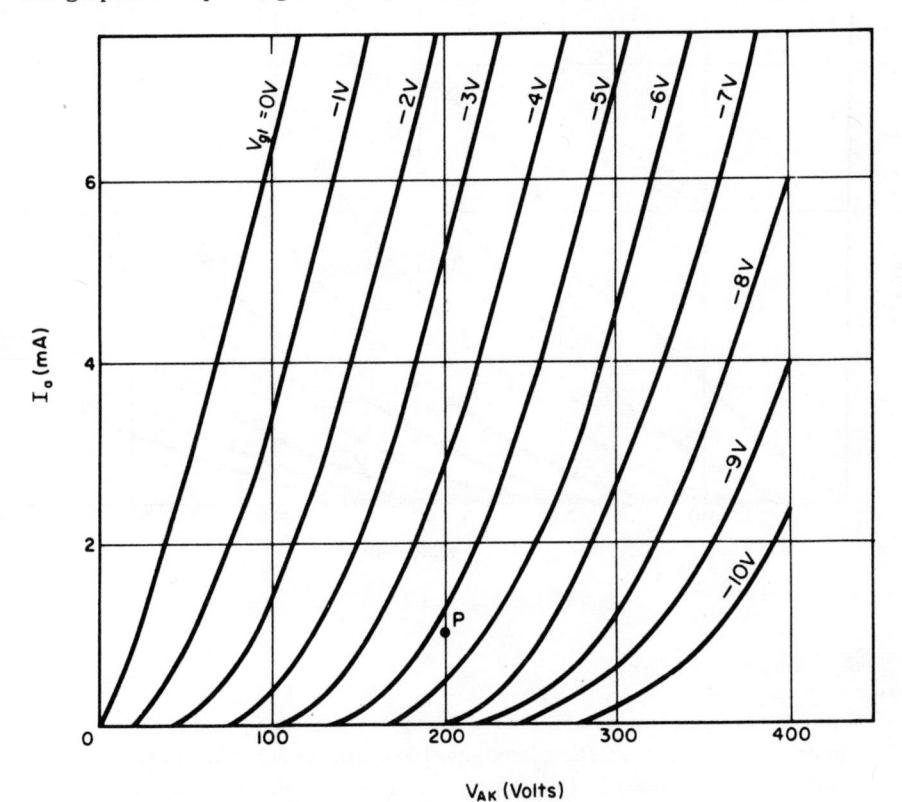

Fig. 7.3.3.

Note that when compared with figure 7.3.2., cut off values of V_{gk} for values of V_{ak} are shown in 7.3.3., e.g., -10 V is necessary to cut the valve off when it has a V_{ak} of 280 V. -4 V will cut the valve off when $V_{ak} = 100$ V, etc. Due to non-linearity of the curves, this ratio is not consistent, although later in the book, a consistent ratio will be assumed on occasions when dealing with 'small signal analysis'.

Exercise 2.

Repeat the above for the triode in figure 7.3.4. The reader will see that as these curves are less linear, quite different results will be obtained depending upon the area in which the parameters are taken. Two suggested points around which to work are shown as P and Q on the characteristics.

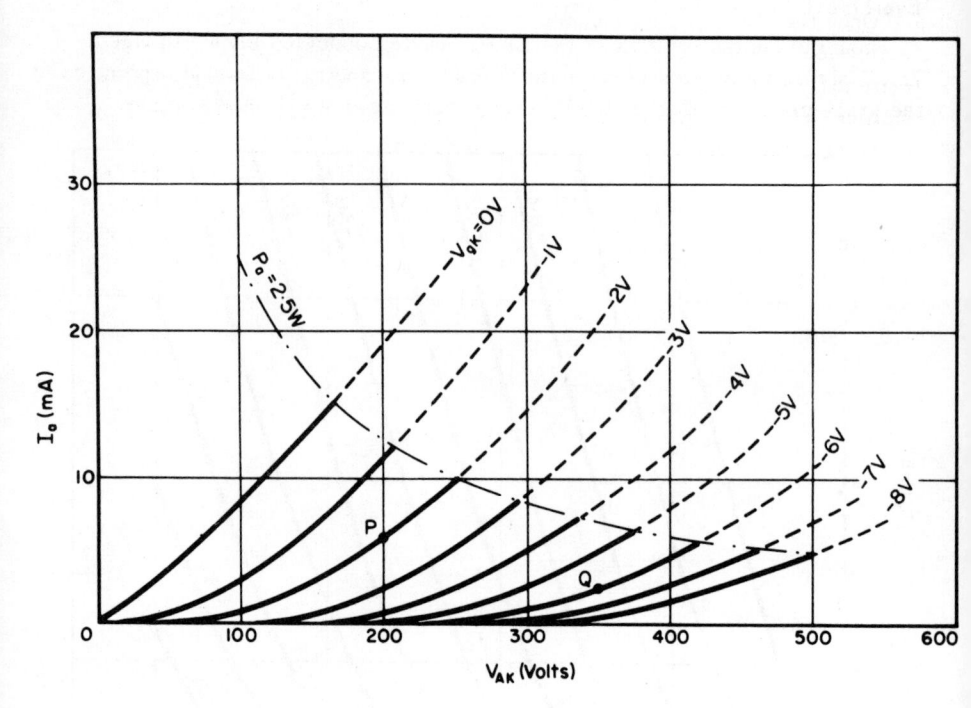

Fig. 7.3.4.

7.4. Gas-filled devices.

There are two types of gas-filled valve, the cold cathode and the hot cathode. In these valves the glass envelope after evacuation is filled with an inert gas (e.g. neon) at low pressure and the presence of the gas molecules drastically alters the characteristics compared with those of an evacuated valve.

With a vacuum diode the anode current is proportional to the anode voltage within the working range, but with a gas-filled diode the anode current remains at zero as the anode voltage is increased until a certain value, called the ionisation potential, is reached. Once this potential is reached anode current rapidly flows, the value of this current being determined mainly by the external resistance in the circuit.

TRIODE VALVES, VOLTAGE REFERENCE TUBES AND THE THYRATRON

In the cold cathode diode (voltage reference tube) if sufficient anode voltage is applied the gas ionises into electrons and positive ions. The electrons travel at high velocity towards the anode while the ions, which have a much greater mass, move much more slowly towards the cathode. The tube glows during conduction with a colour which is characteristic of the particular gas used in the tube.

Once the gas has been ionised by raising the anode voltage to the ionisation potential, a much lower voltage is needed in order to keep the current flowing. This lower voltage is called the maintaining potential or burning voltage. This voltage remains substantially constant over a limited range of current and can be used as a stabiliser.

In the case of the hot cathode type, the cathode and heater assembly is similar in principle to that of the vacuum tube, so that an initial supply of electrons is present as a space charge. The gas filled triode, or thyratron, will be mentioned shortly.

7.5. Simple stabiliser circuits.

A voltage stabiliser has the following typical characteristics.

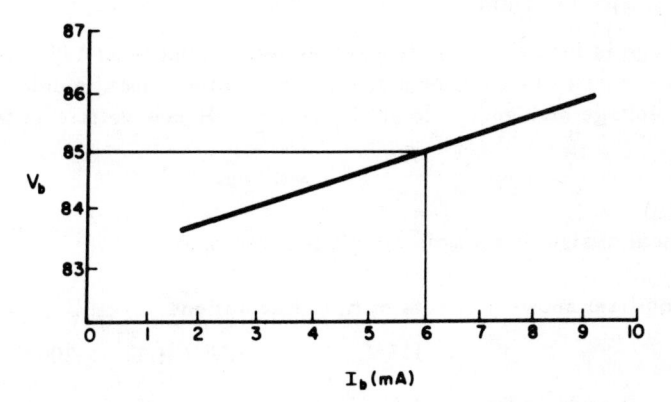

Fig. 7.5.1.

A simple circuit using a neon stabiliser is illustrated in figure 7.5.2. the object being to obtain from a d.c. supply marked H.T., which itself may have rather poor regulation, a stable d.c. across the load resistor R_L.

The neon needs 115 V (V_s) in order to 'strike'. Once struck, the p.d. across the neon will stabilise at approximately 85 V. This p.d. is subject to slight variation according to the current through it, with I_b at 6 mA, the p.d. = 85 V. This p.d. is known as the burning voltage (V_b) and is reasonably constant. This device therefore may be used as a stabiliser.

R_s would be chosen such that the load current in R_L plus the current through the neon causes a voltage drop across R_s, which, added to 85 V,

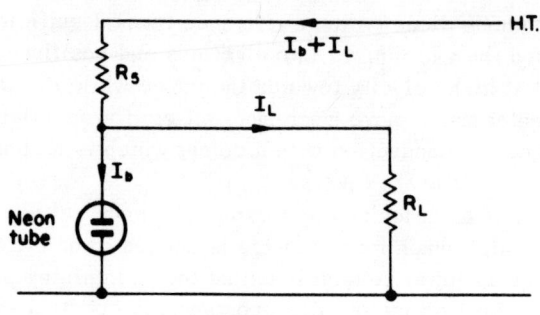

Fig. 7.5.2.

would equal the H.T.

$$R_s = \frac{V_s - V_b}{I_L + I_b} = \frac{V_s - V_b}{I_b + V_b/R_L}$$

Suppose the H.T. to be 200 V, $R_L = 10\,\text{K}\Omega$, $R_s = 10\,\text{K}\Omega$. Then the maximum voltage that could initially appear across the neon would be

$$\frac{200 \times \text{Load}}{\text{Total}} = \frac{200 \times 10\,\text{K}}{20\,\text{K}} = 100\,\text{V}.$$

This voltage is insufficient to 'strike' the neon, as at least 115 V is needed, hence care must be taken when determining resistor values in order to ensure a voltage high enough to strike the neon. Hence we may write

$$\frac{V_s \cdot R_L}{R_s + R_L} \not< V \text{ striking.}$$

A typical design of a simple circuit is given here.

7.6. Stabiliser showing effects of H.T. fluctuations.

$$V_s = 115\,\text{V}, \quad V_b = 85\,\text{V}, \quad \text{H.T.} = 300\,\text{V}.$$

Current in R_L = 4 mA.

$$R_L = \frac{85\,\text{V}}{4\,\text{mA}} = 21.25\,\text{K}\Omega \text{ and from } \frac{\text{H.T.} \times R_L}{R_L + R_s} = V_s$$

$$\frac{V_s}{\text{H.T.}} = \frac{R_L}{R_s + R_L}$$

$$R_s = \frac{(21.25)(300) - (115)(21.25)}{115}$$

$$\frac{6375 - 2440}{115} = \frac{3935}{115} = \underline{34.4\,\text{K}\Omega}$$

TRIODE VALVES, VOLTAGE REFERENCE TUBES AND THE THYRATRON

R_s must not have a value greater than 34.4 K as otherwise the neon will not strike. (This should be checked by using the load upon the total). And seeing that VR_L just = 115 V, any increase in R_s will mean that the striking voltage will never be reached.

Suppose the neon current (I_b) to be 6 mA, then

$$R_s = \frac{H.T. - V_b}{I_b + I_L} = \frac{300 - 85}{(6 + 4)mA} = \frac{215}{10} = \underline{21.5 \text{ K}\Omega}.$$

If the H.T. falls to, say, 270 V, the total current will be

$$\frac{270 - 85}{21.5 \text{ K}} = \frac{185 \text{ V}}{21.5 \text{ K}} = \underline{8.6 \text{ mA}}$$

therefore the neon current would be 8.6 mA $- I_L$ = 8.6 − 4 = 4.6 mA.

It is seen that any H.T. variation is cancelled by a change in current through the neon, while the load current and the voltage remain constant.

7.7. Stabiliser showing effect of load current variations.

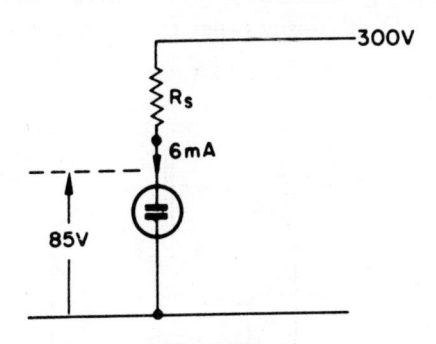

Fig. 7.7.1.

The next step is to connect the load (figure 7.7.2.).

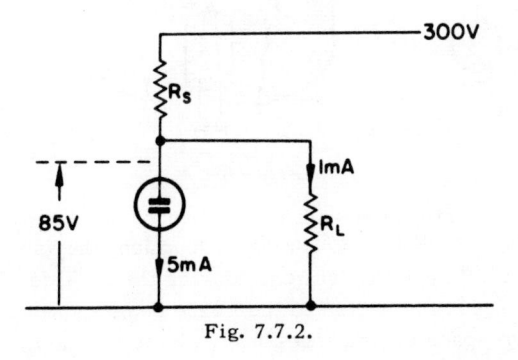

Fig. 7.7.2.

Assume the load current = 1 mA, therefore

$$R_L = \frac{85}{1\,\text{mA}} = 85\,\text{K}\Omega.$$

The neon (within its working range) will remain at 85 V.

The current in $R_s = I_b + I_L$, which, as before, must be 6 mA. The neon current will have dropped to 5 mA.

The neon stabiliser, therefore, has characteristics such that it will burn at 85 V; any change in load current will be taken up by the neon, so that the current through R_s will remain substantially constant. Neons vary a great deal, and some burning currents may be in the order of 5–60 mA, with burning voltages from (60–150) V. Some neons are meant only to provide a constant voltage, which will normally be used as a reference voltage. The 85A2 in its preferred operating conditions in regulated power supply units will be discussed in Chapter 18.

7.8. The gas-filled triode.

The Thyratron exhibits characteristics very different from those of the vacuum triode. The Thyratron may consist of an electrode structure as shown in figure 7.8.1.

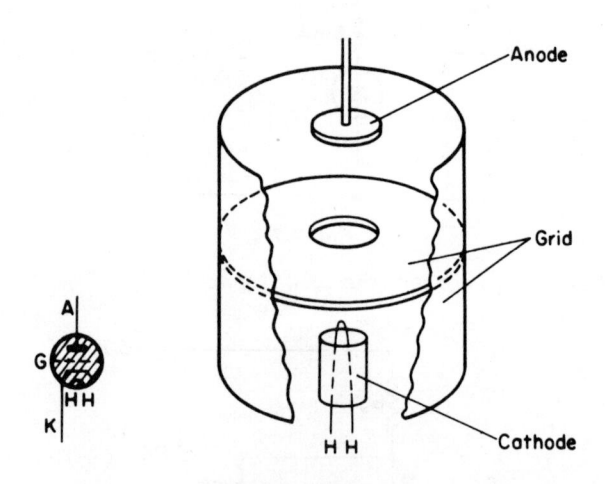

Fig. 7.8.1.

The anode and cathode are not unlike that of the normal triode. The grid however may be very different. During conduction, the gas atoms are ionised by the act of colliding with electrons leaving the cathode. The positively ionised gas atoms are attracted to the space charge surrounding the cathode. As the negative space charge loses many electrons due to combination with

TRIODE VALVES, VOLTAGE REFERENCE TUBES AND THE THYRATRON

the positive ions, it becomes smaller in size. A smaller space charge results in a lessening repelling force on the cathode emitted electrons, consequently a much larger electron flow to the anode results, and the larger current for a constant anode voltage indicates a lower internal resistance than that of the vacuum triode valve.

The electrons released by collisions are, as they have less mass, swept away more rapidly than the heavier positive ions. Much of the inter-electrode space consists of a region known as the 'plasma'. Here exists a combination of normal gas molecules with positive ions and electrons; this is the prime luminous source. It is in the cathode vicinity where a larger potential exists with it's subsequent electric field, and it is in this vicinity that positive ions leave the plasma and are attracted to the cathode. Cathode bombardment by the ions may cause even further electrons to be released, if this process continues, the ionisation may become self maintained.

The grid may be initially negatively biassed in order to repel the electron flow from the cathode, as with a vacuum triode. The anode voltage, if raised to a high enough potential, will create an electric field which will cancel the repelling effect of the grid. When current begins to flow, the electrons collide with the gas molecules and dislodge electrons from the gas atoms. Many of the gas atoms thus become ionised. The positive ions are attracted to the negative space charge and soon reduce it's size and effectiveness. The space charge now greatly inhibited, allows a very large anode current to flow. Some of the positive ions are attracted to the grid thus cancelling the repelling action and consequently the grid has no further control over the anode current flow, and the anode current will now be proportional to the anode voltage. In order to reduce undesirable effects due to electrostatic charges on the glass envelope, the grid may be constructed as a cylinder in two parts as shown. The outer part may be a cylinder whilst the second part may be a disc connected inside of the outer one. It is through the inner disc that the anode current will flow. More elaborate arrangements may be employed. The gas may be mercury vapour, inert gas or hydrogen. When the device strikes, at a little above the ionisation potential, the resultant anode current flow is sufficient to maintain the ionisation.

If a small negative potential is applied to the grid, a very much larger anode potential will be required to cause the device to strike. The grid is extremely efficient and will repel so many of the emitted electrons that the anode voltage will need to be very much larger if the device is to strike.

The time taken for the device to strike is known as the 'ionisation time', which may be influenced by the anode load particularly if it has an inductive component. With a vacuum triode the control grid whilst negative with respect to the cathode, will always control the anode current but with the gas-filled triode, the grid will, after ionisation, lose all further control, as during conduction the negative charge on the grid is neutralised as it collects positive ions. Once fired, the Thyratron will behave as a closed switch. De-ionisation may be accomplished by removing the anode voltage. The de-ionisation time may be in the order of $10-100\,\mu S$. This time may be reduced by applying a negative potential to the anode thus 'dragging out' many of the positive ions causing the device to switch off much sooner. As stated earlier, during conduction the grid attracts many positive ions and these cause a positive sheath to form in the vicinity of the grid. This sheath tends to neutralise the electrostatic attraction of the grid. The maximum anode current may be controlled by determining the value of the anode load resistor by applying simple ohms law.

If the thyratron anode potential $= V_a$ and with an H.T. supply, the anode current

$$I_a = \frac{\text{H.T.} - V_a}{R_L}$$

where R_L is the external load resistor.

The mutual characteristics of the Thyratron are given in figure 7.8.2.

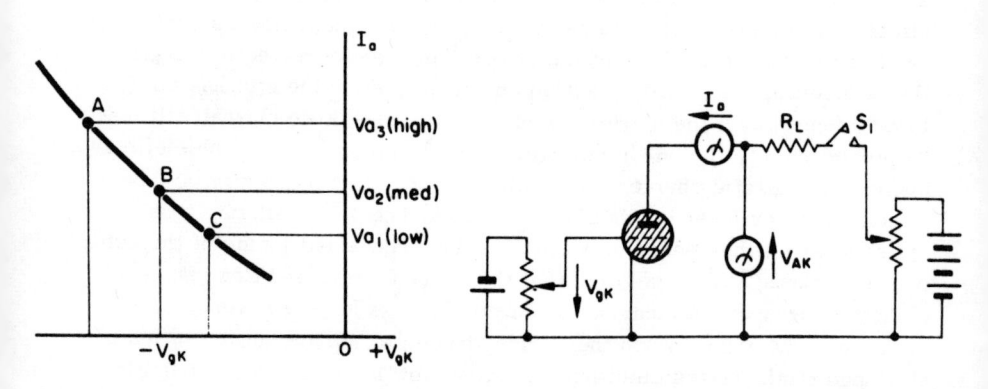

Fig. 7.8.2.

When determining the characteristics, the following measurements should be taken. The grid should be set to it's most negative value. S1 should be closed. The anode voltage should be set to a value V_{a_1}. The grid voltage should be adjusted to a less negative value until the device strikes. Whilst ionised, V_a will fall to about 10 volts. The value of the anode and critical grid voltage should be plotted. S1 should be opened in order to de-ionise the device. The grid should be set to its most negative position.

The anode voltage should be set to a higher potential, V_{a_2}. S1 should be closed. The grid should be taken less negative until the device strikes.

TRIODE VALVES, VOLTAGE REFERENCE TUBES AND THE THYRATRON

Both the anode and critical grid voltage should be plotted. The foregoing should be repeated until a family of characteristics have been plotted. The grid, once the device has fired, should be taken to its extreme negative potential in order to show quite clearly, that the grid has lost all control and that the anode current remains at a steady value. It will be seen that the anode current will be at a higher value for a higher anode voltage. This is comparable to that of a vacuum triode.

7.9. Control ratio.

The control ratio is defined as the change of anode voltage divided by the change in critical grid voltage. If the points A, B and C in figure 7.8.2. are connected, the resultant characteristic is known as the control characteristic. The slope of the characteristic defines the control ratio. The characteristic is reasonably straight and may be seen to be almost constant over the normal working range.

A control ratio characteristic is shown in figure 7.9.1. It may be seen that as the line $A-C$ is considered to be a straight line, the slope of the characteristic will give the control ratio.

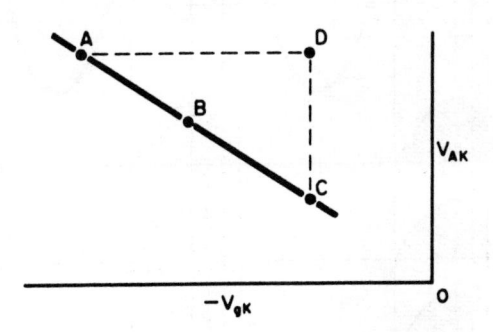

Fig. 7.9.1.

If the approximation of a straight line is acceptable, then the control ratio may be defined as the ratio of the anode voltage to the critical grid voltage.

In the idealised example shown in figure 7.9.1. the control ratio is given as CD/AD, around the point B.

7.10. Grid current.

When the Thyratron is 'off', a very small grid current will flow. This will be in the order of less than $1 \mu A$. If the grid is driven positive with respect to the cathode, the grid will collect electrons and current will effectively enter the grid, and this current may be considerable. If there is any likelihood

126 ELECTRONICS FOR TECHNICIAN ENGINEERS

that the grid may be driven positive up to the ionising potential, the grid would become an anode in effect and an arc would form between the grid and the cathode. A resistance should be inserted in the grid circuit in order to prevent damage to the device, the resistor not exceeding $100\,K\Omega$ to prevent erratic firing. The effective grid voltage would be lower due to the drop across the resistor, therefore the anode voltage would have to be correspondingly lower.

7.11. Firing points.

As stated earlier, varying the firing point by changing the bias will cause an advance or delay in time around an arbitrary time reference, as shown in figure 7.11.1.

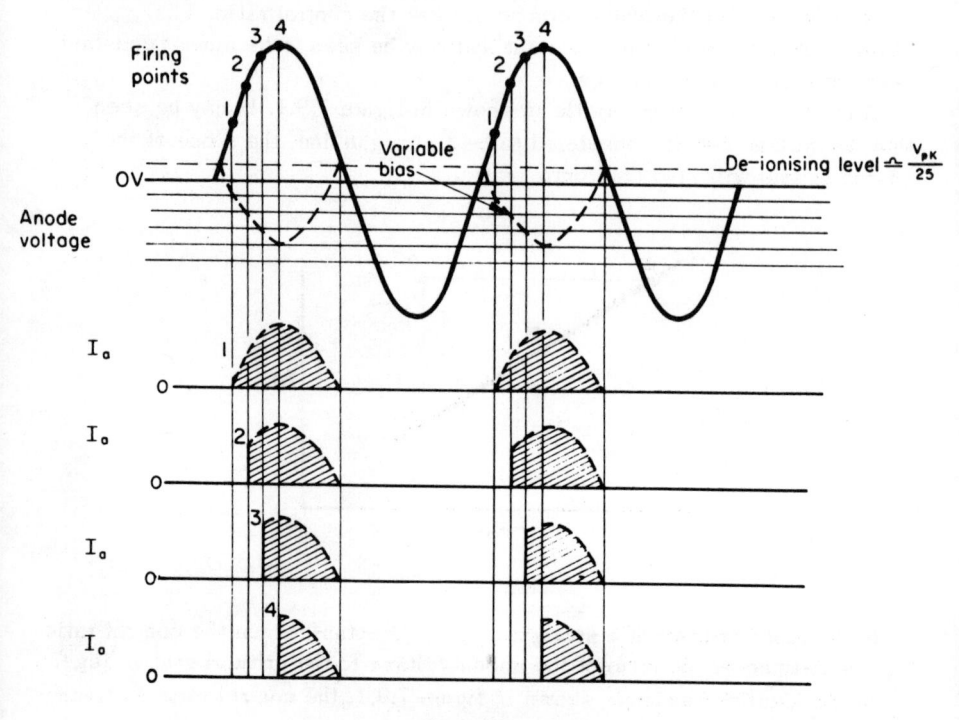

Fig. 7.11.1.

The maximum possible delay using variable bias is $90°$, beyond that point the bias line just touches the most negative peak of the critical bias curve. Increasing the bias further will prevent the device from conducting. The principle is adopted in grid controlled rectifier systems.

A simple illustration in figure 7.11.2. shows the effect of the anode voltage upon the anode current with a fixed grid bias. The illustration is not to scale because the de-ionisation voltage is $\ll$ the peak voltage shown.

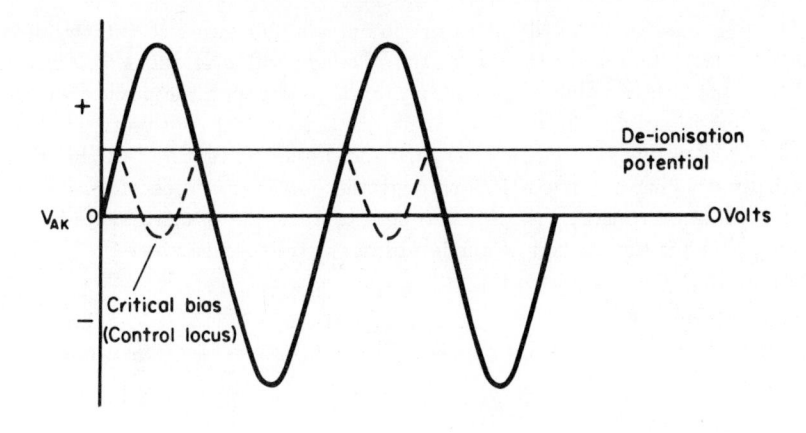

Fig. 7.11.2.

During the period $A-B$, the anode current follows the anode voltage as with a normal diode. At the point B, the anode voltage falls below the level at which ionisation can be maintained. This is the de-ionisation voltage point.

The valve will not conduct whilst the anode voltage is below, or negative to, the ionisation voltage. The anode current does not commence until point A is reached where the anode voltage reaches the ionisation potential. The actual potential at which the device will 'fire' is determined by the grid bias. If a larger anode potential is required to fire the device, a larger negative grid bias is necessary. If this is so, then the effect is for ionisation to commence at a later time. Consequently if it is required to fire the device earlier in time, the grid bias will have to be reduced.

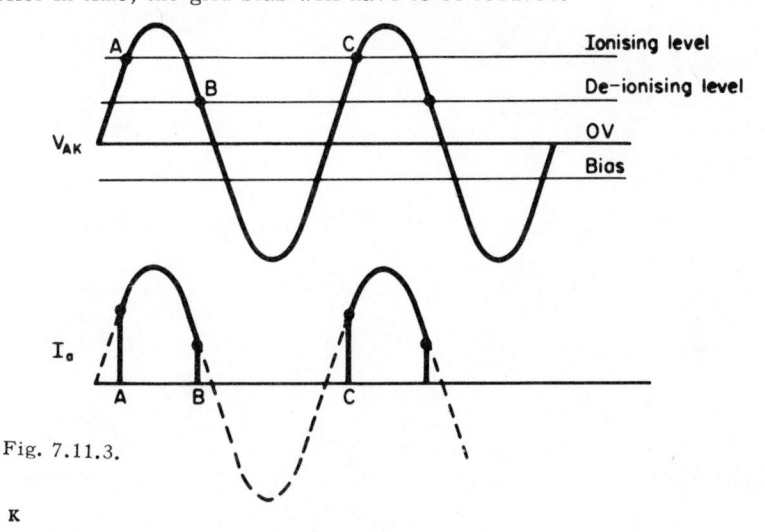

Fig. 7.11.3.

The combination of anode and grid potentials necessary to fire the device is an important feature of the Thyratron. For any value of anode voltage there will be a particular value of grid potential beyond which the device will not ionise. The diagram in figure 7.11.3. shows the critical grid voltage necessary for a chosen anode voltage if the device is required to fire.

Simply drawing a vertical line through the particular value of anode voltage required to fire, will give the necessary value of critical grid voltage.

Note that for low values of anode voltages, the grid may need to be positive, whilst for higher voltages,

$$V_g = \frac{-V \text{ striking}}{\text{control ratio}}.$$

CHAPTER 8

Amplifiers

The parameters μ, gm and ra were discussed in the previous chapter. These parameters are normally used for 'small signal' a.c. considerations. When we wish to investigate say the voltage gain of an amplifier, we may use the equivalent circuit technique or the large signal graphical approach. Both methods have their advantages on particular occasions.

When we use the parameters, we have to assume that all of the curves and spacings are linear. Further, we assume that the voltages and currents under consideration are quite small. If we are dealing with non-linear characteristics, we can assume so small a signal that it would be working on a very tiny part of the curve and consequently could be considered as linear over the very small range considered.

The picture for large signals however, is quite different. We use an actual characteristic, one that will probably contain definite non-linear curves. When we plot load lines we allow for these non linearities and quite accurate results are obtained.

This chapter will deal mainly with large signal tactics and although but a simple comparison of small signal to large signal results will be made, small signal techniques are discussed in detail at a later stage.

We will also be looking at power transference. We will show how an optimum amount of power may be transferred under certain circuit conditions and will also discuss the maximum anode power disipation allowed by valve manufacturers. The two should not be confused as although both related, they are dealt with quite separately.

8.1. The triode valve equivalent circuit.

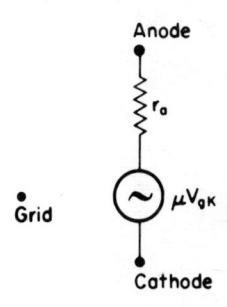

Fig. 8.1.1.

This is a 3-terminal device. The voltage generator (μV_{gk}) develops an e.m.f. of μ times the input V_{gk}. ra is the incremental internal resistance of the generator.

8.2. Voltage amplification using load lines. The operating point.

If we regard the valve as a voltage amplifier, we may use either a graphical or equivalent circuit technique in order to calculate the amplification. Let us consider the simple triode amplifier circuit diagram in figure 8.2.1.

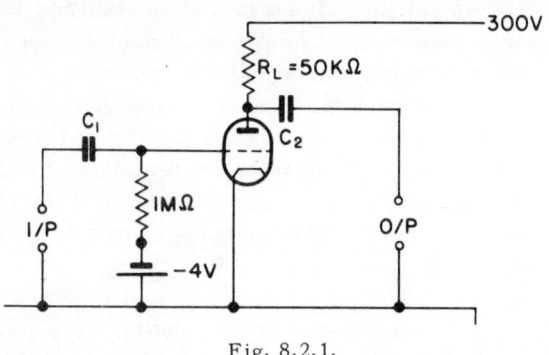

Fig. 8.2.1.

C_1 is the input coupling capacitor. C_2 is the output coupling capacitor (its other functions are dealt with later). Both capacitors will be considered open circuit at d.c. and short circuit to a.c.

Now consider figure 8.2.2.

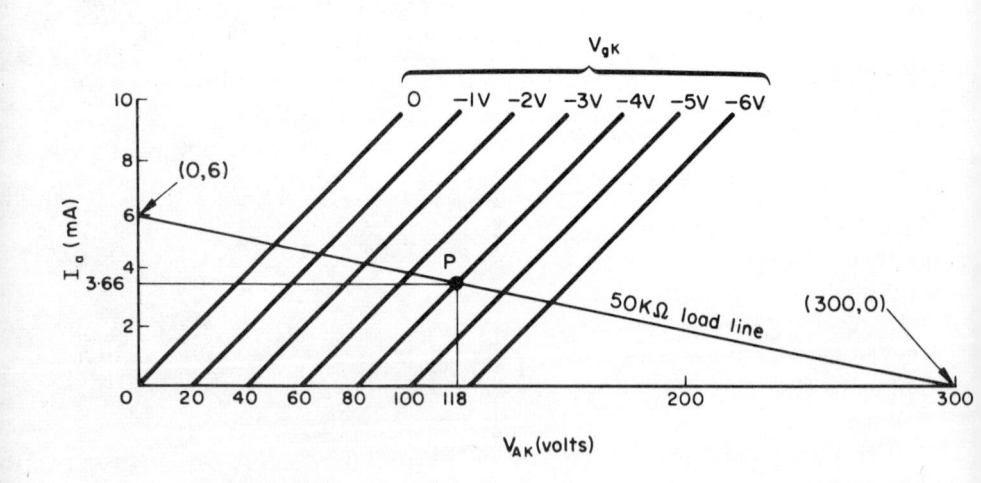

Fig. 8.2.2.

Establishing the d.c. conditions.

Figure 8.2.2. is that of the I_a/V_{ak} characteristics of an ideal valve. We can see from the circuit diagram that the valve is operating with an anode load of 50KΩ and an H.T. of 300 V. We must first establish the d.c. or quiescent states before we can examine the signal states of the amplifier.

We need to know the *operating point*, shown as point P, then we can vary the position of this point (apply a signal) and determine the gain. Assume that we have a meter connected between anode and cathode and that we can record V_{ak} for the two following tests. If the valve is assumed to be 'short circuit', V_{ak} will be zero, and the 'short circuit' anode current will be 300/50KΩ = 6mA. These values identify the point at which we should connect the top end of our 50KΩ load line. If now we assume the valve is 'open circuit', V_{ak} will be 300 V and the anode current will be zero. These values give us the point for the lower end of the load line. These points may be shown as (0, 6.) and (300, 0).

The load line is drawn as shown and properly identified as 50KΩ load line.

The valve must be operating on the load line but in order to say just where, we must refer to another factor. This factor is the grid bias.

The circuit diagram shows a bias of – 4V and at this stage, due regard should be made to the fact that the grid is negative with respect to the cathode. The operating point P occurs at the intersection of the load line and the – 4V grid bias curve.

The dotted lines show that the standing or steady anode current is 3.6 mA and that the steady V_{ak} is ≃118 V. The latter is that voltage that exists across the valve, between anode and cathode. It is not the anode voltage (that is the anode voltage with respect to earth) although in this simple circuit V_{ak} is the same voltage as V_a. This however is not the general case and we should try from the start to name these potentials correctly so as to avoid confusion later on.

8.3. Signal amplification.

Suppose we were to apply a signal of 2 V $P-P$ to the grid. As the cathode is earthed, the full 2V – $P-P$ would be applied between grid and cathode; the grid would traverse the d.c. load line up from – 4V to – 3 V and down from – 4 V to –5V as shown in figure 8.3.1.

The anode voltage (WRT cathode) would change from 118 to 102 and from 118 up to 134. This change in anode potential is available as an output voltage.

For a total grid 'swing' of 2 V $p-p$ the anode would 'swing' 32 V $p-p$. The gain of the circuit $\dfrac{O/p}{I/p} = \dfrac{32V}{2V} = \underline{16}$

132   ELECTRONICS FOR TECHNICIAN ENGINEERS

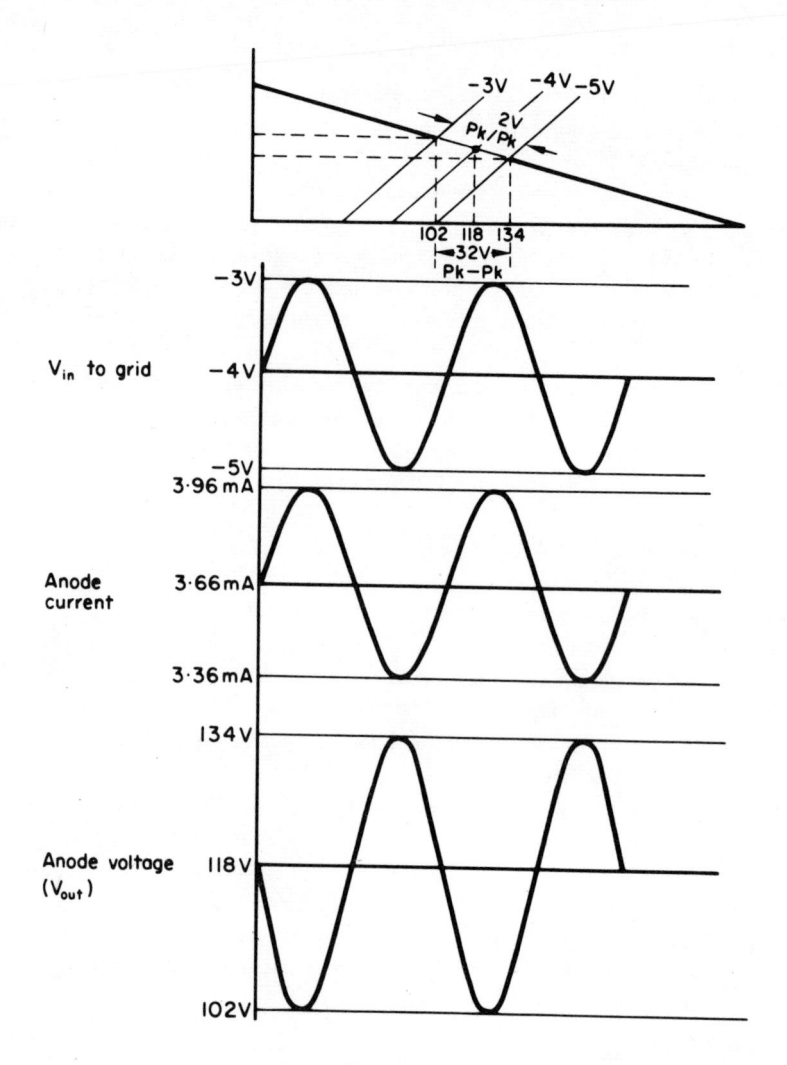

Fig. 8.3.1.

Note from figure 8.3.1. that as the grid voltage is positive going, the anode voltage is negative going. Let us further investigate the I_a/V_a characteristics and their use. Figure 8.3.2. shows a normal single stage amplifier using a small power pentode, triode connected, with an anode load of 48KΩ and a cathode bias resistor of 2KΩ. Assume once more that all capacitors are short circuit to the signal frequencies involved. The characteristics for this valve are shown in figure 8.3.3.

In most circuits, the first piece of information required relate to the d.c. conditions before a signal is applied.

AMPLIFIERS

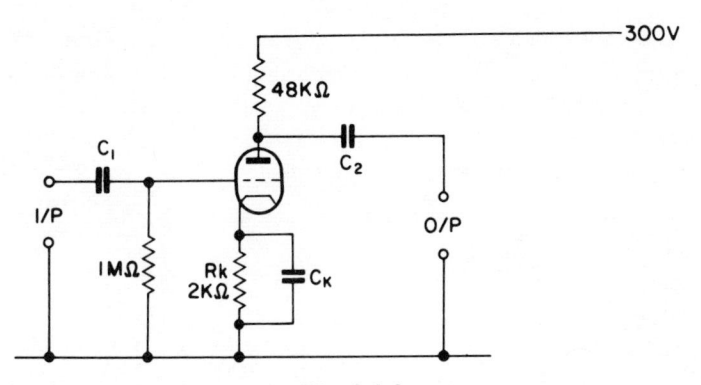

Fig. 8.3.2.

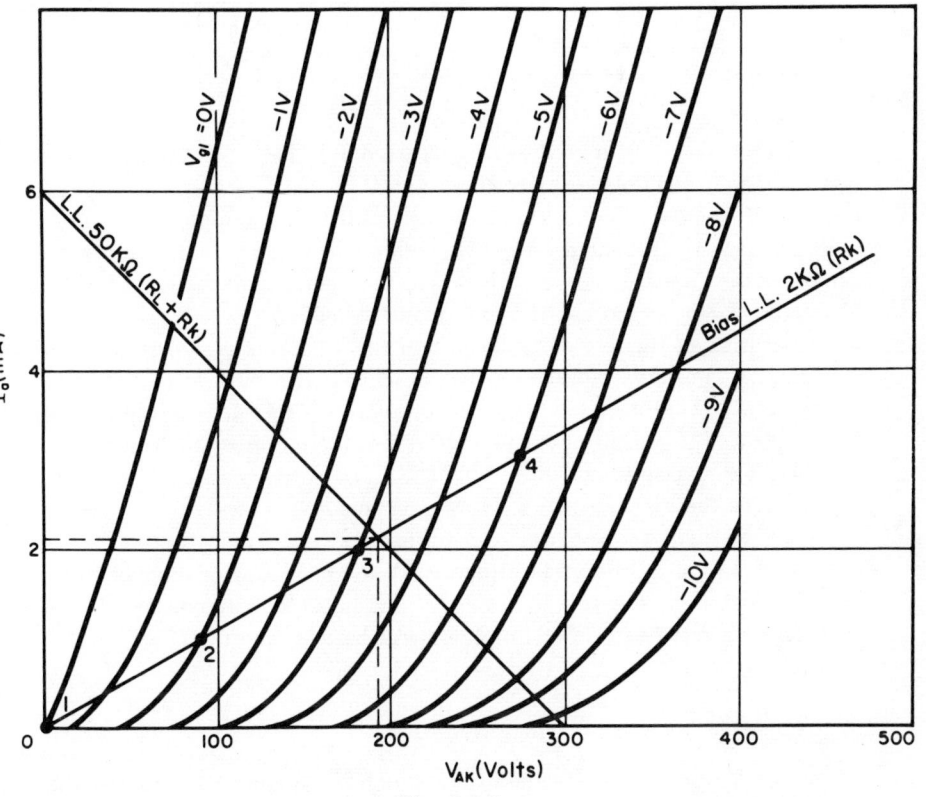

Fig. 8.3.3.

1st step.

Construct a d.c. load line. We need a point on the I_a axis. i.e., $V_{ak} = 0$ V.

Consider the valve short circuit, $\therefore$ maximum $I_a = \dfrac{300 \text{ V}}{(48 + 2)\text{K}\Omega} = \underline{6\,\text{mA}}$

Consider the valve open circuit, $V_{ak} = 300\,V$. This is the point at $I_a = 0$, the lower end of the load line. In fact with this type of straightforward load line technique, the upper load line point is nearly always at H.T. while the other point is at The full H.T./Total resistance in series with valve.

Figure 8.3.4. shows these two conditions.

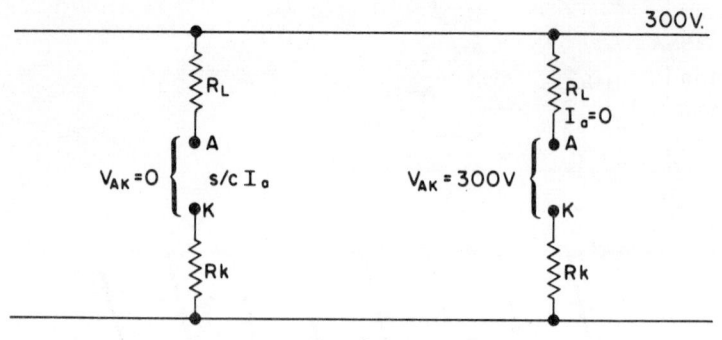

Fig. 8.3.4.

The d.c. Load line has been drawn in figure 8.3.3. The coordinates for the upper and lower ends of the local lines are (0,6) and (300,0) respectively.

We know that the valve operating point is sitting on the d.c. load line somewhere and if we consider the bias resistor in conjunction with the valve characteristics, we can say exactly where the operating point must be. Consider figure 8.3.3. which has the bias load line added.

Examination of figure 8.3.2. shows that the −ve bias produced is obtained by arranging a positive voltage on the cathode. If the cathode is positive, then the grid (which is at 0 V) must be negative with respect to the cathode; and using this principle, a positive cathode voltage corresponds to negative grid bias. The bias resistor $R_k = V_k/I_a\,\Omega$.

+ 1 V on the cathode gives − 1 V grid bias.

+ 2 V on the cathode gives − 2 V grid bias, and so on. Remember, to cause a change in I_a, the grid voltage must be changed with respect to the cathode; if we raise the grid by, say, 1 V and raise the cathode by 1 V, then the difference between grid and cathode will be 0, hence the I_a will not change.

8.4. Construction of a bias load line.

Assuming 0 V bias ($V_{gk} = 0$), then the current in R_k must be zero. (as the cathode d.c. voltage is the bias). Position a point where $V_{gk} = 0\,V$ curve, intersects with $I_a = 0$. Point 1 on the load line.

Assume − 2V bias. 2 V across 2 KΩ means that $I_a = 1\,mA$.

Plot a point at the intersection of − 2 V bias curve at $I_a = 1\,mA$. Point 2. Assume various bias values, plotting as in (1) (2) (3) until the points are on the right hand side of the d.c. load line. Connect these points with a line (which will be less straight the more non-linear the bias curve spacings). The operating point is located where the bias load line intersects the d.c. load line. The point should be reinforced that the bias load line is a straight line only when the grid curves are straight and equispaced.

In our graph, the operating point co-ordinates are $I_a = 2.1$ mA and $V_{ak} = 195$ V. the grid bias $= 4.2$ V. The d.c. conditions of the circuit are shown in figure 8.4.1. Note the difference between V_a and V_{ak}. V_a is the the anode d.c. voltage measured from earth.

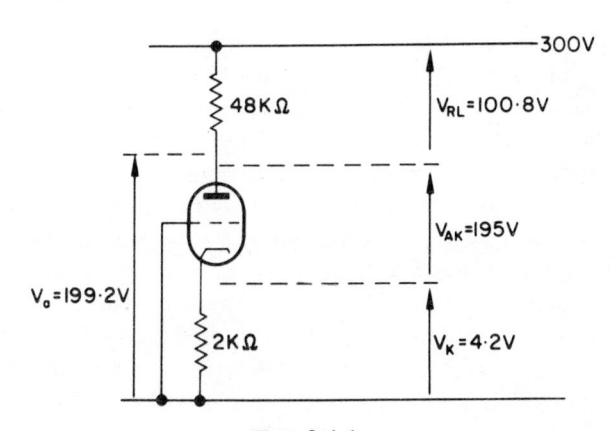

Fig. 8.4.1.

If we add $V_k + V_{ak} + V_{RL}$, they must be equal to the H.T. voltage.

$$V_K + V_{ak} + V_{RL} = \text{H.T.}$$
$$4.2 + 195 + 100.8 = 300 \text{ V}.$$

The capacitors have been omitted for clarity as they do not affect the d.c. condition in any way. Note particularly that V_a is in fact always $V_{ak} + V_a$, and only if the cathode has no resistor (R_K) will $V_{ak} = V_a$. i.e. $V_a = V_{ak} + V_k$. and if $V_k = 0$ then $V_a = V_{ak} + 0$ or $V_a = V_{ak}$.

If a valve is biassed at -4V, this may be obtained in one of two ways:

1. Earth the cathode and put -4 V on the grid, or
2. Earth the grid (to d.c.) and put $+4$ V on the cathode.

Remember, it is the voltage difference between grid and cathode that controls or determines anode current. Therefore, should a valve be biassed at -4V, a signal (change of grid to cathode voltage) may be applied of $+4$V. The signal will *add* to the existing steady bias voltage at any instant. The table shows this effect.

Input Signal	d.c. Bias	Actual Grid to Cathode Voltage
0	− 4	− 4 V (grid negative to cathode)
1	− 4	− 3 V (grid negative to cathode)
2	− 4	− 2 V (grid negative to cathode)
3	− 4	− 1 V (grid negative to cathode)
4	− 4	0 V (grid potential equals cathode potential)
− 1	− 4	− 5 V (grid negative to cathode)
− 2	− 4	− 6 V (grid negative to cathode)
− 3	− 4	− 7 V (grid negative to cathode)
− 4	− 4	− 8 V (grid negative to cathode)

If an input signal of + 5 V is applied, then the grid to cathode will tend towards + 1 V. The grid will become positive to the cathode, and electrons leaving the cathode will be attracted by the grid, so that electrons will flow out of the grid. A large I_a will also flow. When electrons leave the grid, conventional current *enters* the grid, and this current passes through the cathode circuit. The 'grid' current condition must never be allowed in ordinary circuits, and henceforth we will assume that grid current is inadmissible until we reach the chapter devoted to circuits where grid current is acceptable. If grid current does flow, the grid acts as an anode to the cathode, and the grid cathode input circuit behaves as a rectifier; hence a low impedance is presented to the input signal, and severe distortion or clipping results. As the grid is not capable of dissipating larger power, heavy grid current can damage the valve. The large anode current that also flows can also cause considerable damage unless very special care is taken

8.5. Maximum anode Dissipation.

The maximum power to be dissipated by the anode is slways given by the valve manufacturer, and it may be necessary to plot a curve to this effect upon existing I_a/V_a characteristics.

Figure 8.5.1. shows such a curve.

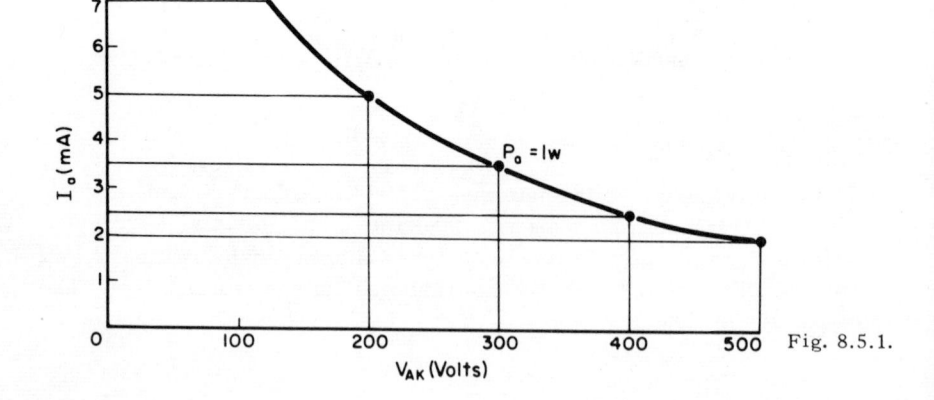

Fig. 8.5.1.

AMPLIFIERS

Assume the maximum anode dissipation to be 1 W.
Then as power (watts) = $I_a \times V_a$.
Then 1 watt = $I_a \times V_a$. Where I_a is in mA and V_a in volts.
All we do is select an arbitrary value for V_a, say and calculate the coresponding I_a to give 1 W.

Let us construct a table, and plot the co-ordinates on the characteristics.

P_a	V_a	×	I_a	Coordinates
1 W	100 V	×	10 mA	(100 V, 10 mA)
1 W	200 V	×	5 mA	(200 V, 5 mA)
1 W	300 V	×	3.33 mA	(300, 3.3 mA)
1 W	400 V	×	2.5 mA	(400, 2.5 mA)
1 W	500 V	×	2.0 mA	(500 V, 2 mA)

It is of course convenient to choose a value of V_a to keep the arithemetic easy. As an example,

If $P = 1.2$ W, say, then V_a could be 120 V, $I_a = 10$ mA.
and $V_a = 240$ V, $I_a = 5$ mA.

or If $P = 2.5$ W, then $V_a = 250$ V, $I_a = 10$ mA
$V_a = 500$ V, $I_a = 5$ mA

and so on; doubling the voltage would mean halving the current, and so on. A little practice will show just how easy it is.

From now on, when positioning load lines, care must be taken to ensure that the load line does not cut through the shaded area above the p.a. curve. Neither must the grid be made more positive than $V_{gk} = 0$. In figure 8.5.2.

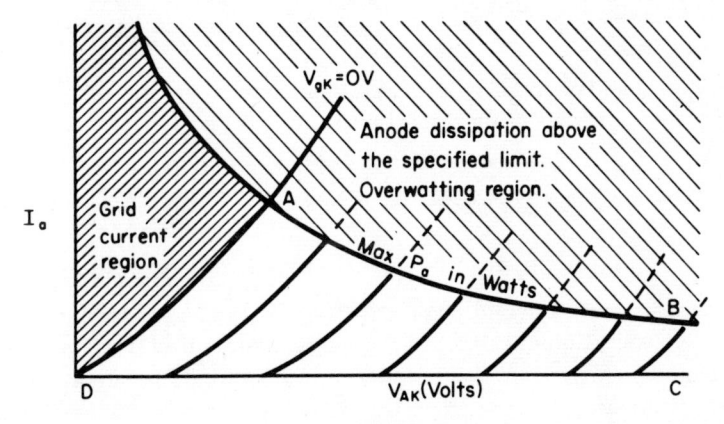

Fig. 8.5.2.

both Pa_{max} area and grid current areas are shaded in order to show clearly which part of the characteristics we must *not* use for normal amplifiers.

Only area *ABCD* may be used. In the design of an amplifier, an operating point would initially be positioned somewhere in the centre (or most linear portion) of this area. This is a class A amplifier.

8.6. Derivation of resistor values for an amplifier.

Let us consider the practical design of a single stage amplifier using this technique. We will confine our design to the d.c. state at this time.

Figure 8.6.1. gives the theoretical circuit diagram, and figure 8.6.2 gives the I_a/V_a characteristics.

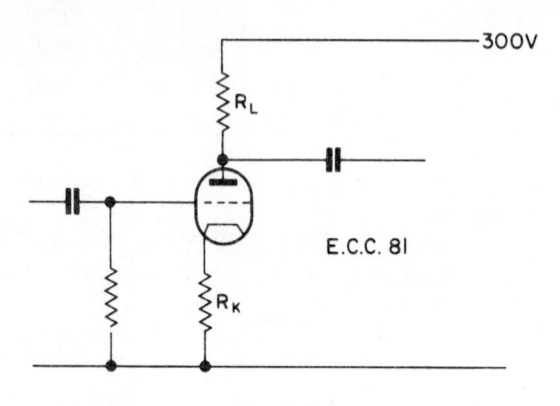

Fig. 8.6.1.

We have positioned an operating point such that there is a reasonably equal swing either side of the point — well clear of Pa area and approximately midway between 300 V and $V_{gk} = 0$. We know that one point for the load line is the H.T. (300 V). Drawing a line from 300 V — operating point — up to the I_a axis completes the load line. The current shown on the I_a axis is seen to be 30 mA.

This d.c. load line must represent a total resistance of 300 V/30 mA = 10 KΩ. It also represents, for this circuit, $(R_L + R_K)$.

∴ $R_L + R_K$ = 10KΩ. The bias point is seen to be 2.5 V.

Hence R_K = 2.5/6.5 mA = 390Ω (I_a is seen to be 6.5 mA from the operating point).

Thus $R_L + 390Ω = 10 KΩ$ ∴ $R_L = 10KΩ - 390Ω = 9.610 KΩ$

In practice, having established values for R_L and R_K, standard values would be chosen nearest to that of the calculated values. The load lines would then be repositioned, and slight variations would be recorded for final analysis. Also, the d.c. load line would be positioned close to the p.a. curve, but at this stage it is wise to allow for tolerances etc., and not position it too close.

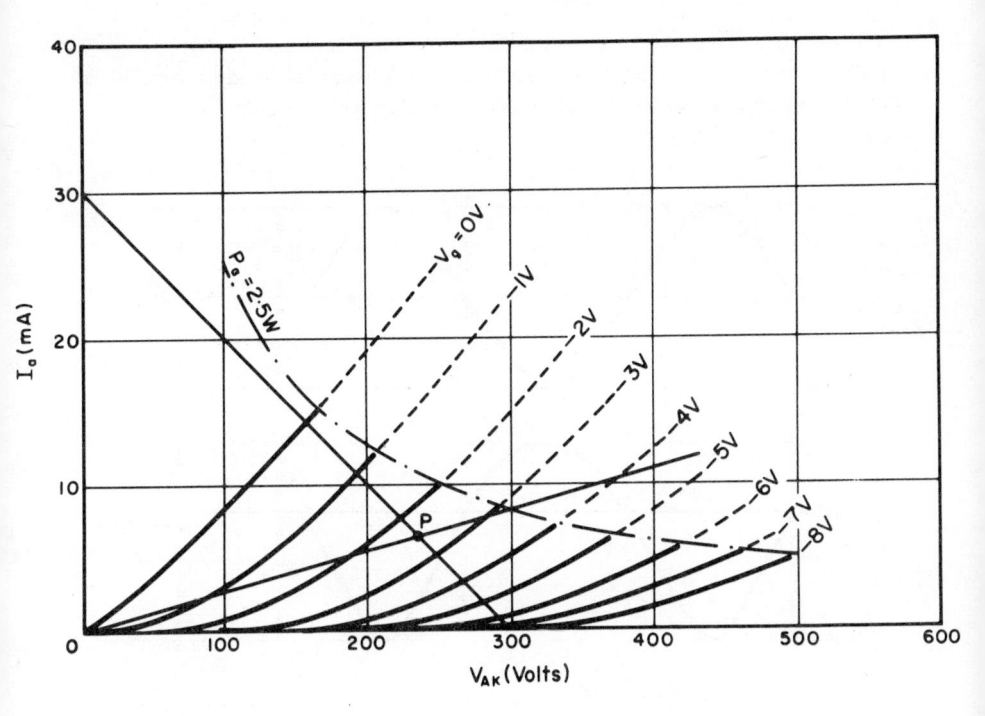

Fig. 8.6.2.

8.7. Voltage gain, (a.c. condition).

Once the d.c. conditions have been established, and the operating point is either known or determined, a signal voltage input will result in an amplified signal voltage at the amplifier output.

Figure 8.6.1. shows a simple amplifier and figure 6.6.2. shows the d.c. and bias load lines plotted on the I_a/V_{ak} characteristics.

With a bias of 2.5 V, the steady anode current is 6.5 mA and the anode is sitting at a potential of approximately 238 V.

If a signal of + 2.5 V is applied to the grid, the operating point will move along the d.c. load line from − 2.5 to 0 V.

The anode voltage will fall from 238 V to 158 V.

Therefore for a change in grid voltage of 2.5 V, positive going, an anode voltage change of 8 0 V, negative going, results.

If the + 2.5 V input is removed and a − 2.5 V input applied instead, the operating point will move down the d.c. load line to a point corresponding to − 5 V, V_{gk}.

The anode voltage is seen to be 282 V.

Hence for a -2.5 V input, a positive going change of anode potential of 124 V occurs. Figure 8.7.1. shows both grid and anode waveforms but of course, is not to scale.

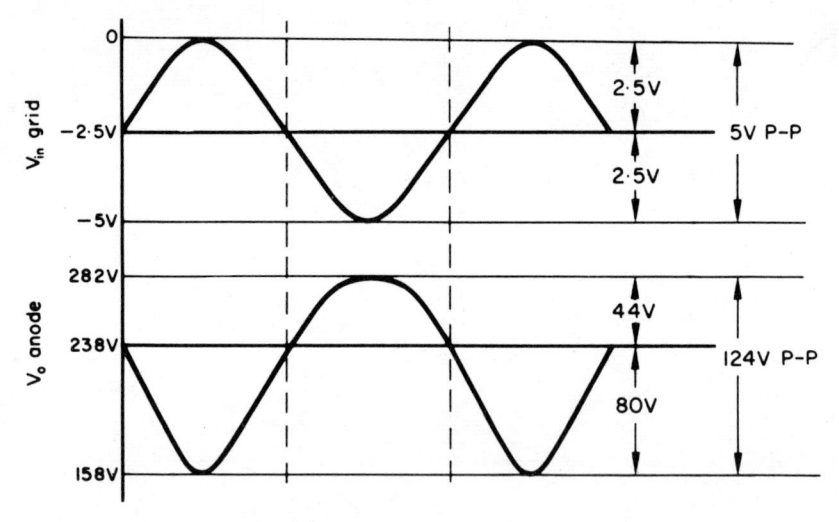

Fig. 8.7.1.

The gain of the amplifier is given as

$$\frac{v_o}{v_{in}} = \frac{282 - 158}{5} = \frac{124}{5} \doteq 25.$$

The output, for an assumed sinusoidal input, is seen to be far from symmetrical and is described as distorted.

We shall see later how we can overcome this problem by using negative feedback.

If we were to have removed the cathode bypass capacitor C_K, the gain would have been very much less than 25. Removing C_K would cause negative feedback to develope, but we will not concern ourselves with feedback systems at this stage, other than the following simple explanation.

We saw in the table (8.4.1.) that an input signal adds to the existing steady bias between grid and cathode. Let us take this a step further and see the effect of an unbypassed cathode.

Assume a simple amplifier is operating with -5 V bias. Further, the bias is obtained by producing $+5$ V across the cathode resistor R_K. If we were to apply an input of say $+2$ V, the anode current would increase. This increase would cause a larger voltage to develop across R_K. The effective input to this simple amplifier is that change of potential between grid and cathode. Hence if V_k was allowed to 'follow the grid', and this is known as a cathode-follower action, then the rise in cathode volts tends to cause the steady difference between grid and cathode to remain almost constant.

This change in V_{gk} will not remain absolutely constant but if for a + 2 V input, the cathode voltage increased by say 1.8 V, the effective input, or change in bias, would be $2 - 1.8 = 0.2$ V. Hence with an effective input of only 0.2 V, the output signal would be a signal resulting from amplification of 0.2 V only.

Hence the overall gain would be reduced.

This reduction in gain, caused by an un-bypassed cathode, is a form of negative feedback.

The bypass capacitor C_K, is chosen so that its reactance, at the signal input frequency, is about 1/10th of the Ohmic value of R_K. When a signal is applied the capacitor behaves as something approaching a short circuit and consequently holds the cathode potential at an almost constant level, i.e. no change in V_k.

The capacitor C_K performs this function at the signal frequency mentioned and all higher frequencies. Its effectiveness reduces however at frequencies below that discussed and as the input approaches zero frequency, so the capacitive reactance increases in value and does not bypass R_k so effectively.

When designing a simple amplifier, one decides upon the lowest frequency at which the amplifier is to work, and C_K is chosen to have a reactance of $R_K/10$ at that frequency. When this is the case, the amplifier gain will, if all other factors are ignored, fall to a value of its middle frequency gain divided by $\sqrt{2}$. This level, 70.7% down, is known as the -3dB point and is often quoted as the acceptable fall in gain when considering the amplifier frequency response as shown in figure 8.7.2.

The number of dB's $= 20 \log_{10} V_0/V_{in}$.

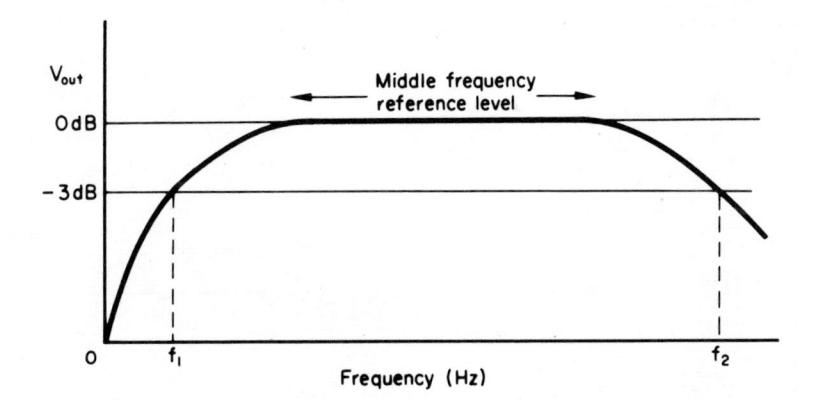

Fig. 8.7.2.

This amplifier gain falls to −3 dB at f_1 and f_2.

Hence the response of the amplifier may be said to be 'flat' between f_1 and f_2 to within 3 dB's. When deciding upon the value of C_K, it is the frequency f_1 to which the amplifier is to be used.

8.8. Maximum power transference.

Suppose that the circuit is as shown in figure 8.8.1. It is desired to find the exact value of R_L in order to obtain the maximum possible power across R_L for a given value of valve internal resistance.

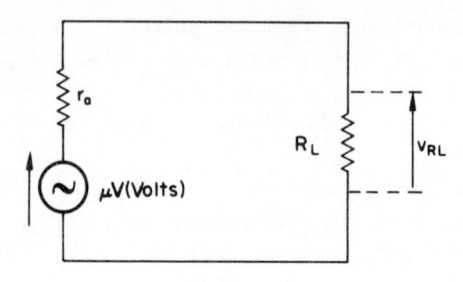

Fig. 8.8.1.

μv is a voltage source, r_a is the internal resistance and R_L the load resistance. vR_L is (by load over total technique)

$$\frac{v \times R_L}{ra + R_L}$$

For maximum power across R_L, R_L must $= r_a$. A graph of P against R_L shows this quite clearly. (figure 8.8.2.)

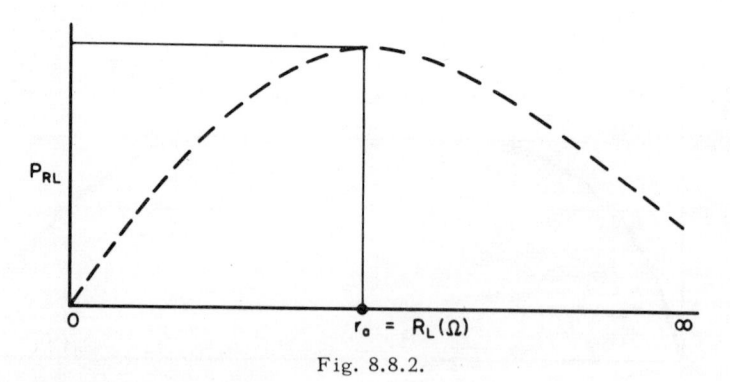

Fig. 8.8.2.

When R_L has a value equal to r_a, power transfer to R_L is at a maximum. With amplifiers, this may give rise to distortion, but this defect will be

AMPLIFIERS

ignored for the time being. If R_L is the anode load and r_a is the r_a of the valve, then the theory is just the same. Assume the generator to have an e.m.f. of say 10 V; then if $R_L = 0$, $V_{RL} = 0$, therefore power = 0, and as $R_L \to \infty$, $V_2/R_L \to 0$.

It is reasonable to assume, then, that the maximum power point must lie somewhere between $R_L = 0$ and $R_L = \infty$.

A simple circuit complete with all calculations may be studied in order to verify P max. when R_L has the same value as the internal resistance. Assume that the circuit is as shown in figure 8.8.3.

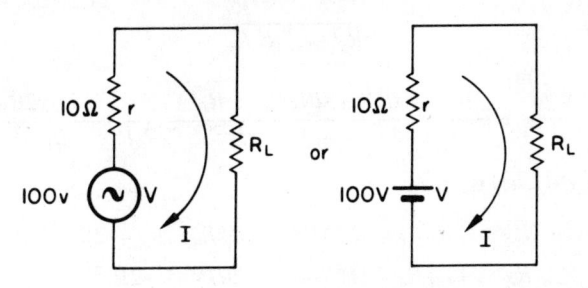

Fig. 8.8.3.

R_L	Circuit current	p.d. across R_L	Power in R_L
0	10 A	0	0
5	6.66 A	33.30 V	222 W
10	5 A	50 V	250 W
15	4 A	60 V	240 W
20	3.3 A	66.6 V	222 W
1,000	0.09 A	90 V	8.1 W
1,000,000	÷100 mA	÷100 V	÷0.0001 W
$\to \infty$	$\to 0$	$\to 100$	$\to 0$

Note that the maximum power in the load occurs when $R_L = r$. If R_L is either smaller or greater than r, the power across the load becomes smaller.

8.9. Maximum power theorem (d.c.)

Maximum power occurs in a load when the load impedance is equal to the source impedance. Let us discuss this in a different manner.

In a purely resistive circuit, R_L is the load and R_S is the source resistance. Figure 8.9.1.

$$\text{Power in the load} = 1^2 R_L \text{ but } I = \frac{V}{R_S + R_L}$$

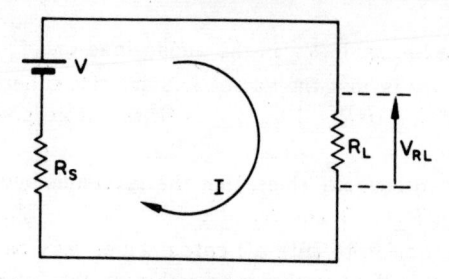

Fig. 8.9.1.

Therefore
$$P = \frac{V^2 R_L}{R_S^2 + 2R_S R_L + R_L^2}$$

and differentiating,
$$\frac{dP}{dR_L} = \frac{V\,du - u\,dV}{V^2} = \frac{(R_S^2 + 2R_S R_L + R_L^2)V^2 - V^2 R_L(2R_S + 2R_L)}{(R_S^2 + 2R_S R_L + R_L^2)^2}$$

and equating dP/dR_L to zero

$$(R_S^2 + 2R_S R_L + R_L^2)V^2 = V^2 R_L(2R_S + 2R_L)$$
$$R_S^2 + 2R_S R_L + R_L^2 = 2R_L R_S + 2R_L^2$$
$$\therefore (R_S)^2 = (R_L)^2 \qquad \therefore \underline{R_L = R_S} \text{ for maximum power}$$

There is, of course, no phase difference between the voltage across R_S and R_L with respect to the circuit current, as both the resistors are non-reactive.

8.10. Maximum power theorem, a.c.

We will now consider a circuit containing a reactive component figure 8.10.1.

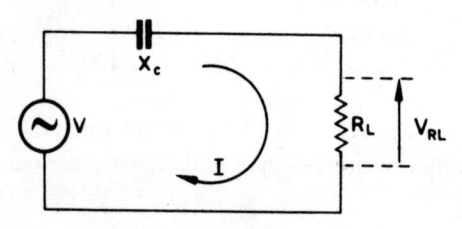

Fig. 8.10.1.

$$P = i^2 R_L \qquad \text{but} \qquad i = \frac{v}{\sqrt{X_c^2 + R_L^2}}$$

$$\therefore P = \frac{v^2 R_L}{X_c^2 + R_L^2}$$

$$\therefore \frac{dP}{dR_L} = \frac{(X_c^2 + R_L^2)v^2 - v^2 R_L(2R_L)}{(X_c^2 + R_L^2)^2}$$

AMPLIFIERS

and equating dP/dR_L to zero

$$(X_c^2 + R_L^2)v^2 = v^2 R_L/(2R_L)$$
$$\therefore \quad X_c^2 + R_L^2 = 2R_L^2$$
$$\therefore \quad X_c^2 = R_L^2$$
$$\text{hence} \quad X_c = R_L$$

$\therefore \quad R_L = X_c$ for maximum power in R_L

As $R_L = X_c$ there must be a 45° phase difference as shown.

This simple principle applies to valves; if R_L is made equal to ra (the internal resistance of the valve), then maximum power is developed across R_L. (This ignored any distortion that might be present; mention will be made of this later on.)

The phase difference would appear as in figure 8.10.2.

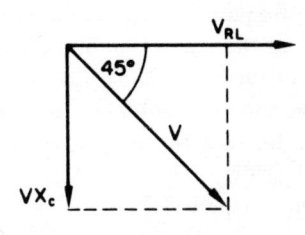

Fig. 8.10.2.

One must not confuse the 'maximum power curve' mentioned in figure 8.5.1. That curve merely indicates the maximum allowable dissipation of the anode itself, and must not be confused with the 'power transference' into a given load, R_L.

Examples.

A triode valve had an anode current of 6 mA with an anode voltage of 200 V and a grid bias of -2 V. During subsequent tests the anode potential was raised to 250 V, the anode current was found to have increased to 8 mA. Resetting the grid volts to -3 V restored the anode current to its original value. What are the valve parameters?

$$r_a = \left|\frac{\delta V_a}{\delta I_a}\right| \; V_g = \left|\frac{250 - 200}{8 - 6}\right| = \frac{50}{2} = 25 \text{ K}\Omega$$

$$g_m = \left|\frac{\delta I_a}{\delta V_g}\right| \; V_a = \left|\frac{8 - 6}{3 - 2}\right| = \frac{2}{1} = 2 \text{ mA/V}.$$

$$u = \left|\frac{\delta V_a}{\delta V_g}\right| \; I_a = \left|\frac{250 - 200}{3 - 2}\right| = \frac{50}{1} = 50.$$

The product of $gm \times ra = \mu$, therefore $2 \text{ mA/V} \times 25 \text{ K}\Omega = 50$.

A triode valve was found to have an anode current of 10 mA with a grid bias of −5V and an anode voltage of 250 V. When the bias was increased to −7.5 V, the anode current fell to 5 mA. Increasing the anode voltage to 300 V restored the anode current to 10 mA. An anode load of 10 K was connected and a 5 V signal applied. What then would the output voltage be?

Answer ... 50 V. It is left to the reader to verify this answer.

8.11. An inductive loaded amplifier.

In this concluding section we will deal with an amplifier to which has been connected an inductor as an anode load.

The inductor contains two distinct components, a pure resistance r and pure inductance, L.

We will see that this exercise will extend the previous work on load lines just a little further, and produce a particular complication.

A brief introduction to frequency response has been given but the reader is advised to refer to one of the many fine books on the subject, should he wish to pursue this subject further.

We will discuss the methods of analysing the circuit from both a d.c. and a.c. point of view. The latter will be an approximation only and the reason for this will be explained.

The amplifier is shown in figure 8.11.1.

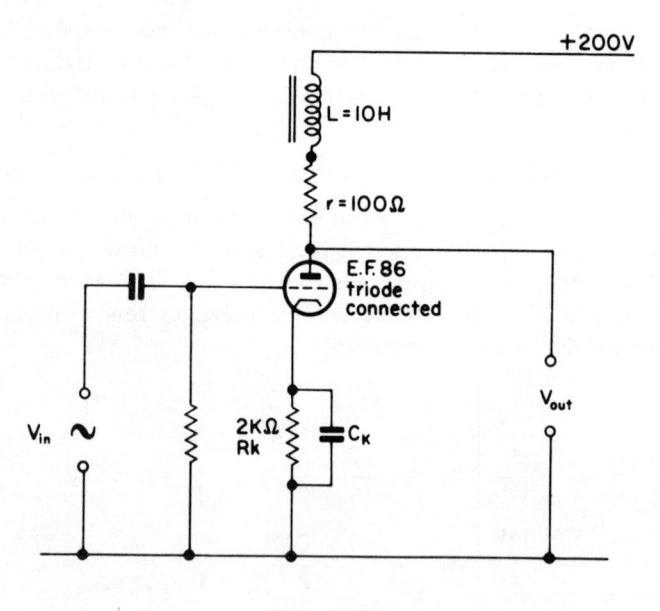

Fig. 8.11.1

Load Lines. 1. (d.c.)

We saw in an earlier section, how to deal with the plotting of load lines on a graph whose Y axis has a maximum value smaller than the *short-circuit* current point representing the top end of the load line. In this case, the maximum value is 8.0 mA on the graph.

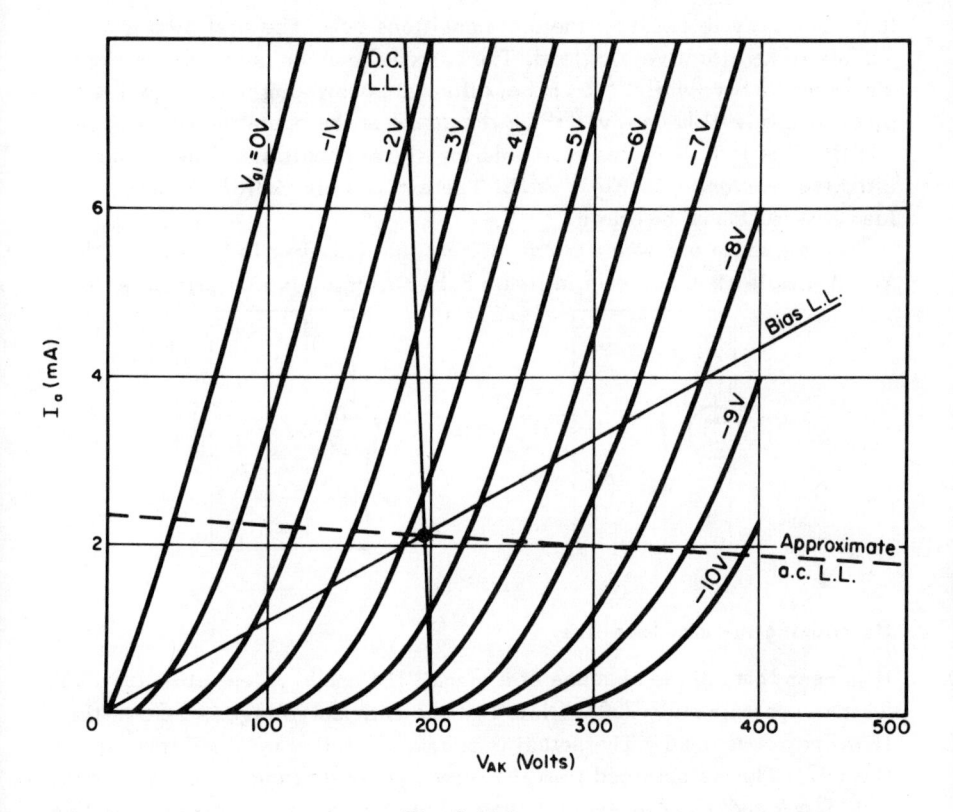

Fig. 8.11.2.

A voltage V of 16.8 V will be chosen as this will cause a *short-circuit* current of 8.0 mA, figure 8.11.2. The coordinates (183.2, 8) result. This then, is the point for one end of the load line. The other, as usual will be the H.T. line (in our case, 200 V and 0 mA). The second coordinate then is (200, 0). These points must be joined in order to draw the d.c. load line. Although this load line may look unusual, it is quite correct, remember it is the angle that matters.

Note that the 16.8 V were subtracted from the 200 V, giving the value of 183.2 V in the first point, (183.2, 8).

The valve may be operating anywhere on the load line, and without proper biassing, will be damaged. If a bias load line for the 2KΩ cathode resistor is plotted, the operating point P will be established. The steady conditions will be as follows. $V_{ak} = 195.7\text{V}$. $I_a = 2.15\text{mA}$. $V_{gk} = -4.3\text{V}$.

Load Lines. (2). a.c.

It is necessary to consider the a.c. conditions now. The first step is to calculate the effective a.c. load. The cathode resistor is *shorted* at the signal frequency, whilst the choke, although possessing a very low resistance at d.c. will have a very high *reactance* at the operating frequency of 10 KHz. The H.T. becomes *earthy* during signal conditions. The actual effective *reactance*, is $X_L = 2\pi fL$. The a.c. load is 628KΩ. An a.c. load line must now be drawn.

The maximum a.c. *short circuit* current that can flow is V_{ak} divided by X_L. A simple sketch shown in figure 8.11.3 illustrates this principle.

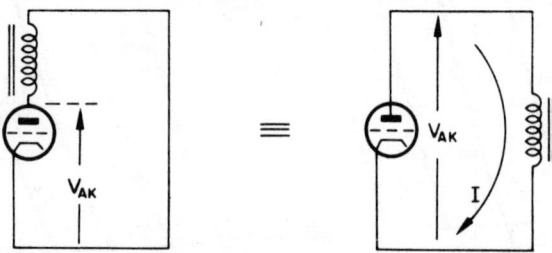

Fig. 8.11.3.

Positioning the a.c. load line.

It is seen that, in the absence of a signal, the anode potential is 199.75V. This is approximately 200 V (Note that the cathode voltage is zero at the frequency concerned). The actual coordinates for the a.c. load line are (0, 2.47). This is obtained from the maximum *short circuit* a.c. signal current, of 0.32 mA and added to the standing anode current of 2.15 mA. A load line is drawn from (0, 2.47) to the operating point P. The line is then carried on towards the V_{ak} axis. The point at which the line will cut the V_{ak} axis is seen to be greater than 600 V. This then, is the theoretical second point of the a.c. load line (1500, 0).

It is quite evident that the anode will swing to a potential much higher than the H.T. line.

This load line is an approximation and is not valid for inductive loaded amplifiers. The actual load line will be an ellipse. When the load has a power factor approaching unity, i.e., a resistance, the load line will be straight, but as the power factor approaches zero, the ellipse approaches almost a circle.

AMPLIFIERS

When an a.c. input is applied to the amplifier, the signal component of the anode current will alternate about the operating point.

This causes an alternating voltage to be developed across the inductor, hence an alternating component of anode voltage will be present.

The load line in the illustration does give an approximate idea as to the variations of anode current and voltage but must be regarded as a rough approximation only.

One can see that the voltage *swing* may easily be much greater than the H.T. supply. Figure 8.11.4. is an enlarged set of characteristics and does not represent the EF86.

It is seen that in figure 8.11.4, as the anode current increases from zero in a position direction (point 1), the back e.m.f. of the inductance is at its maximum negative value.

This is substantiated from Back e.m.f. $= -L(di/dt)$ and is seen to be negative for a positive increase in current.

At this point, the back e.m.f. is in series opposition to the H.T. so that V_{ak} = the quiescent V_{ak} minus the negative peak excursion of the anode at this instant.

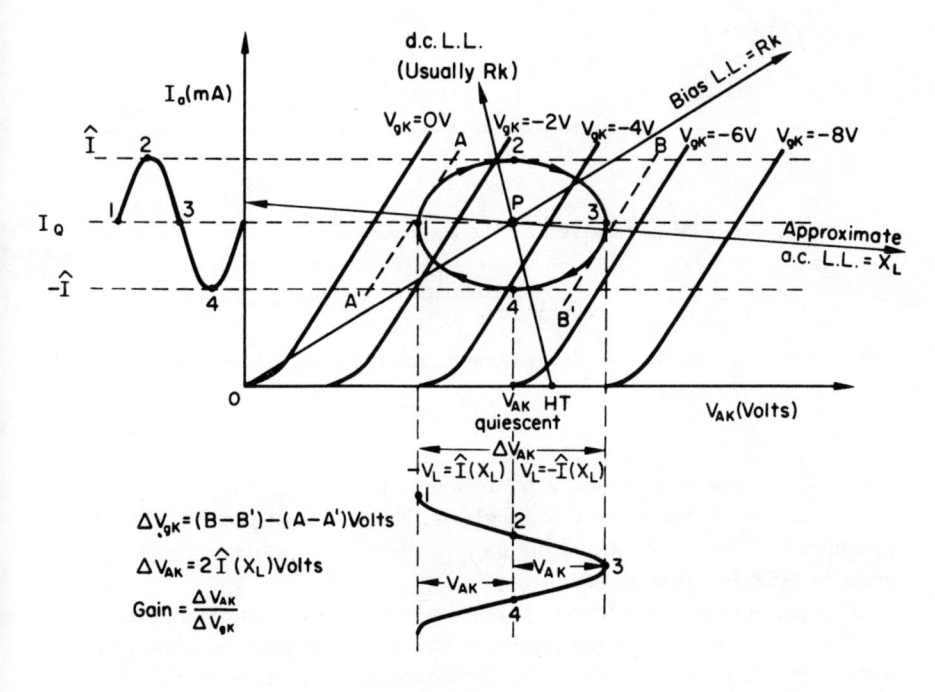

Fig. 8.11.4.

When the anode current reaches its maximum value and the rate of change (di/dt) is zero, there is no back e.m.f. induced in the inductance. Point 2.

When the anode current is going negative from its zero point, (at point 3) the V_{ak} will be the sum of the quiescent potential of V_{ak} plus the positive excursion of the anode due to the positive going maximum e.m.f. developed across the load. Hence, when di/dt is zero, the e.m.f. is zero. As i_a goes positive, through zero, the e.m.f. is at maximum negative. As i_a goes negative through zero, the e.m.f. is at its maximum positive.

An ellipse will result, as is shown in the illustration.

The process of plotting the ellipse for a given input voltage is tedious and consequently the reader is advised to use the equivalent circuit technique as shown in figure 8.11.5.

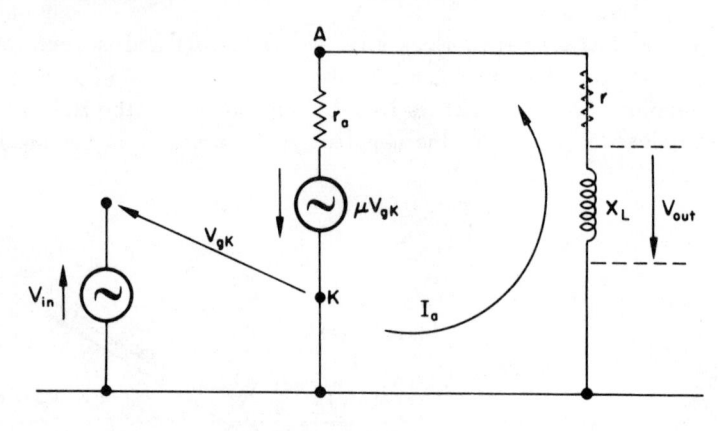

Fig. 8.11.5.

If we ignore r, V_{out} is given by the expression

$$\frac{\mu \; j\omega L}{ra + j\omega L} \qquad \text{load over total}$$

and if $r_a \ll \omega L$ as it is in this example, then the expression may be written as

$$\frac{\mu j \omega L}{j\omega L} \simeq \mu$$

and is in antiphase to the input as demonstrated in figure 8.11.5.

The a.c. load line is seen to be almost horizontal, if it were, then the current would be constant and the gain of the circuit — from a large signal point of view — would be equal to μ.

The gain of the circuit, using the elliptical load is approximately 35.5.

The μ of the valve in the region of the quiescent point is also approximately 35.5. Let us summarise the key points in this example;

An enlarged, but not to scale, drawing is shown of the ellipse in figure 8.11.4. It shows a d.c. load line representing R_k, as the inductor is assumed to have little resistance.

The bias load line is shown for R_k and gives the operating point P.

The approximate a.c. load line is shown representing X_L at a given frequency.

I_q and V_{ak} quiescent are shown.

At point 1, the current passes through zero and is positive going. The corresponding point 1 on the ellipse is $\hat{I}(X_L)$ V below V_{ak} quiescent of the I_q line. Points 3 on both the current waveform and the ellipse are reversed.

Points 2 and 4 show no change of current input, hence the ellipse, at points 2 and 4, are at V_{ak} quiescent on the corresponding I_a value line.

The actual input voltage, i.e., change of V_{gk}, is plotted by drawing two lines, $A - A^1$ and $B - B^1$ parallel to the printed V_{gk} curves, and at a tangent to the 'peaks' of the ellipse.

Therefore for a change of $5.5 - 1 = 4.5$ V V_{gk}, the anode current has changed $2\hat{I}$ and an output of $2\hat{I}(X_L)$ is obtained. The gain will, for large values of X_L, approach the value of μ.

CHAPTER 9

Simple transformer coupled output stage

We discussed a simple inductive loaded amplifier in 8.11. We saw that the load 'looked like' a 100Ω resistor in the absence of an alternating signal.

Once a signal was applied, the reactance of the choke became a most important factor as the effective anode load became 628KΩ at the signal frequency concerned. We saw that an inductive reactance caused the operating point to traverse an elliptical path under signal conditions. (Note that we ignored the d.c. resistance of 100Ω as it would have negligible effect upon 628KΩ).

In this chapter we will extend the previous discussion a little by considering an amplifier using a load that is transformer coupled. We will not concern ourself with frequencies or reactance, but simply to examine the 'transformer action' of the load under a.c. conditions.

9.1. Simple concept of 'transformer action' on a resistive load.

Figure 9.1.1. shows a simple transformer to which is connected a 10Ω resistor, R_L.

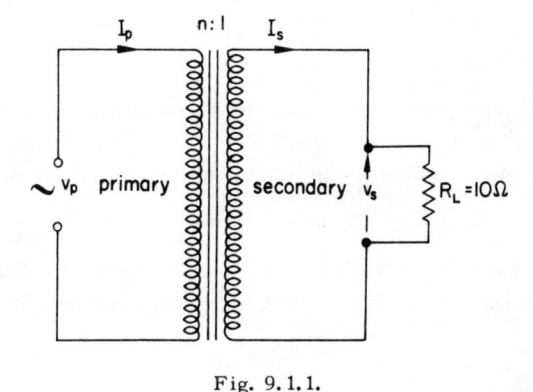

Fig. 9.1.1.

A number of factors need to be discussed before we examine a transformer coupled amplifier. A perfect loss free transformer must be assumed for our basic study.

9.2. Equating power in primary and secondary.

If the primary and secondary winding turns are equal in number then the primary and secondary volts are the same.

The input and load current must be the same also. Power in both input and output must equate even when the turns ratio is not unity.

Example:

Suppose the primary has 1000 turns and the secondary 200. Should 2V be applied to the input, and at a current of 5A, then the power in the primary circuit would be $P = 2 \times 5 = 10W$.

The secondary power must also be 10W, but with 1/5 of the primary turns, $2V/5 = 0.4V$ will be developed across the output. The output, or secondary current is given as $i_s = \dfrac{10W}{0.4Vs} = 25A$.

9.3. Equality of ampere-turns ($\bar{A}$).

The product of the number of turns and the current through those turns, written as ampere-turns, $\bar{A}$, must be the same for both primary and secondary.

From the foregoing example, the primary ampere-turns is seen to be $5 \times 1000 = 5000\bar{A}$. The secondary ampere-turns is also $25 \times 200 = 5000\bar{A}$.

The voltage ratio is proportional to the turns ratio whilst the current is inversely proportional to the turns ratio.

One can never get something for nothing and as the reader might imagine, all transformers contain losses of some kind.

A study of a practical design of a transformer is contained in a later chapter, but for now, we will assume an ideal transformer unless specifically stated otherwise.

9.4. Reflected load.

The load resistance connected across the secondary winding will be reflected into the primary circuit once alternating signals are applied.

If we equate input (primary) and output (secondary) powers,

$$Ip^2 Rp = Is^2 Rs \qquad (1)$$

where Rp and Rs are the effective primary and secondary resistances.

Let the turns ratio $= n$. Equating $\bar{A}$ for primary and secondary,

$$Ipn = I_s \qquad (2)$$

where the primary has n times the secondary turns.

Hence (1) $\quad Ip^2 Rp = I_s^2 Rs$ and substituting for ip
from (2) $\quad \dfrac{Is^2}{n^2} Rp = I_s^2 Rs$

thus $\quad Rp = n^2 Rs \times \dfrac{Is^2}{Is^2}$

SIMPLE TRANSFORMER – COUPLED OUTPUT STAGE 155

Hence the effective resistance reflected into the primary from the secondary, $Rp = n^2 Rs$ and will add to the primary resistance under a.c. conditions.

It is well worth mentioning that either winding of a transformer may be called the primary, it depends how the transformer is used. Some primaries have n times the secondary turns, where n may be greater or less than unity. A bench isolating transformer will often have a value of $n = 1$.

9.5. Simple transformer output stage.

Figure 9.5.1. shows part of a transformer coupled output stage.

The valve is a power amplifier and the transformer delivers the power output from the valve into the 3Ω load.

The maximum power theorem discussed in a previous chapter revealed that for maximum power to be delivered to the load, the load resistor R_L must be of the same numerical value as the source resistance. By suitably choosing a transformer of a required value of n, the load can be made to 'look' like the required source resistance (in this case, of the valve) from $Rp = n^2 Rs$.

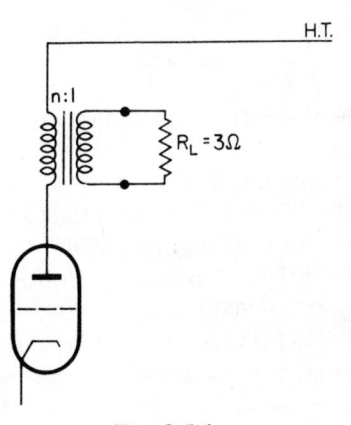

Fig. 9.5.1.

The valve has an output resistance of 3KΩ. The load resistance is 3Ω. From $Rp = n^2 Rs$, we can evaluate n.

$$n^2 = \frac{Rp}{Rs} = \frac{3000}{3} = 1000$$

$$n = \sqrt{1000} = \underline{31.6 : 1}$$

The problem however, is not quite this simple as in practice, the transformer primary contains its own d.c. resistance – even before the secondary resistance is reflected back into the primary.

Let us assume that the primary d.c. resistance is 300Ω.
Figure 9.5.2. shows both d.c. and a.c. conditions of the stage.

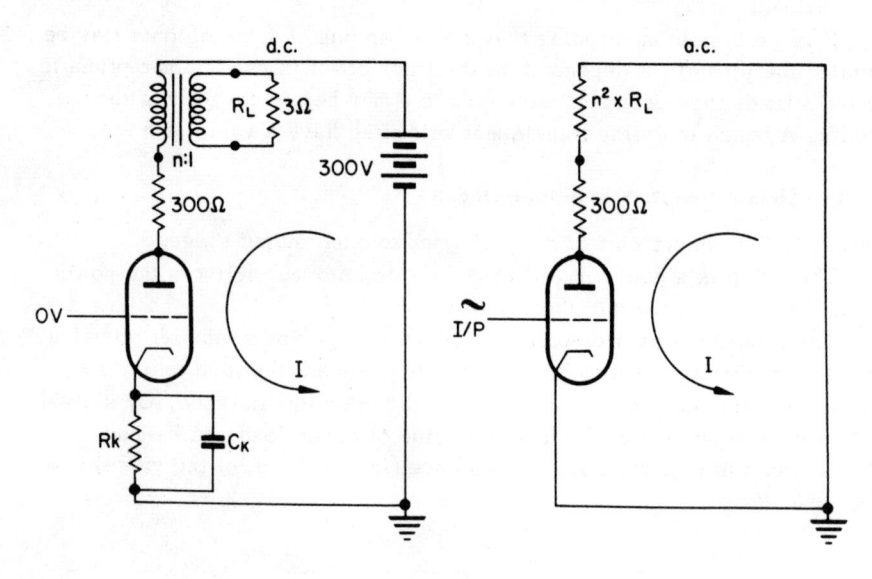

Fig. 9.5.2.

In the absence of an alternating signal, the valve will 'see' a d.c. resistance of 300Ω only.

Once alternating currents are flowing in the transformer windings, the valve will 'see' not only the 300Ω, but the reflected secondary resistance, $n^2 R_L$, in series with it. We need a further 2700Ω to give us a total of 3KΩ so as to match the 3KΩ ra of the valve. With $R_L = 3Ω$ and from $n^2 R_L = 2700$ ∴ $n^2 = 2700/3 = 900$ hence $n = \sqrt{900} = 30$.

With $n = 30$, the reflected resistance is $30^2 \times 3 = 2700Ω$.

The total resistance required is 3000Ω and this meets the maximum power transference requirements.

The reader should carefully study figure 9.5.2. and ensure that he can see the difference between d.c. and a.c. conditions before proceeding further. It will then be time to continue with an analysis of an output stage.

9.6. Plotting the d.c. load line.

Figure 9.6.2. shows the I_a/V_{ak} characteristics of a power triode. Figure 9.6.1. shows the output circuit we are to discuss. We have chosen a steady bias of $-9V$ in this example.

SIMPLE TRANSFORMER – COUPLED OUTPUT STAGE

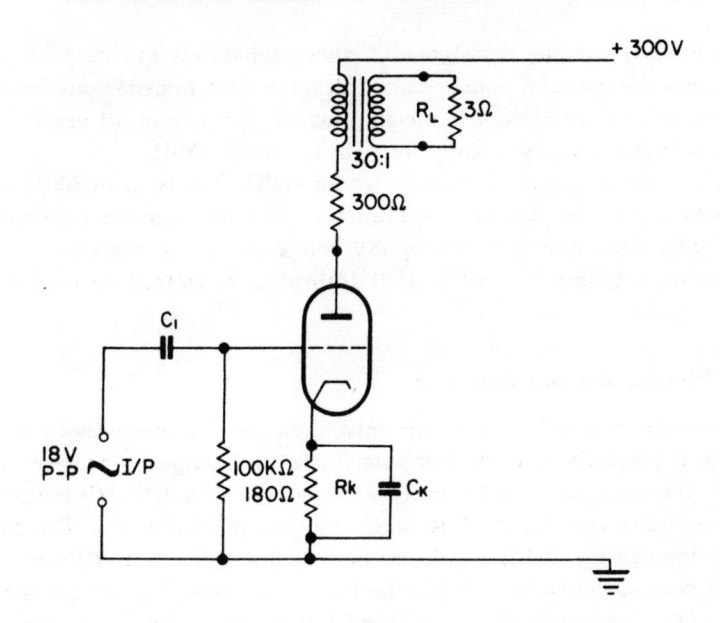

Fig. 9.6.1.

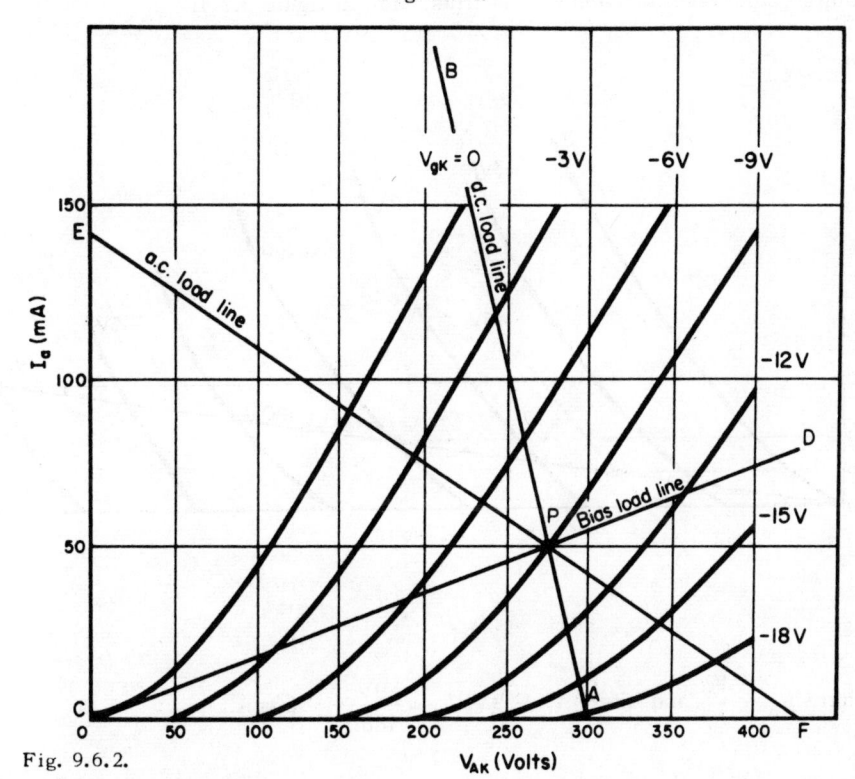

Fig. 9.6.2.

The primary winding resistance is shown separately in figure 9.6.1. This is not the case for normal circuit diagrams but is necessary here if we are to build an easy picture in simple stages. The sum of all resistances in series with the valve = $Rp + Rk$. These sum to 480Ω.

A d.c. load line must first be drawn for 480Ω. The H.T. is 300V, and is the starting point for the d.c. load line (A). In order to produce an upper point for the load line, let us take 48V below 300 so as to give us coordinates of (point B) = (252, 100) following the method described in 1.11

9.7. Plotting the bias load line.

The bias load line will be straight only when the characteristics are straight and equally spaced. For normal practical purposes however a straight line is usually sufficiently accurate. The line $(C - D)$ is the bias load line drawn from (0, 0). One single calculation was made. The current flowing through the 180Ω, R_K, to maintain a bias voltage of 9V = 50mA. A point corresponding to –9V (on the bias curve) and 50mA (I_a) was drawn and the bias load line was drawn from (0, 0) to this point. A number of points could have been chosen as illustrated in figure 9.7.1.

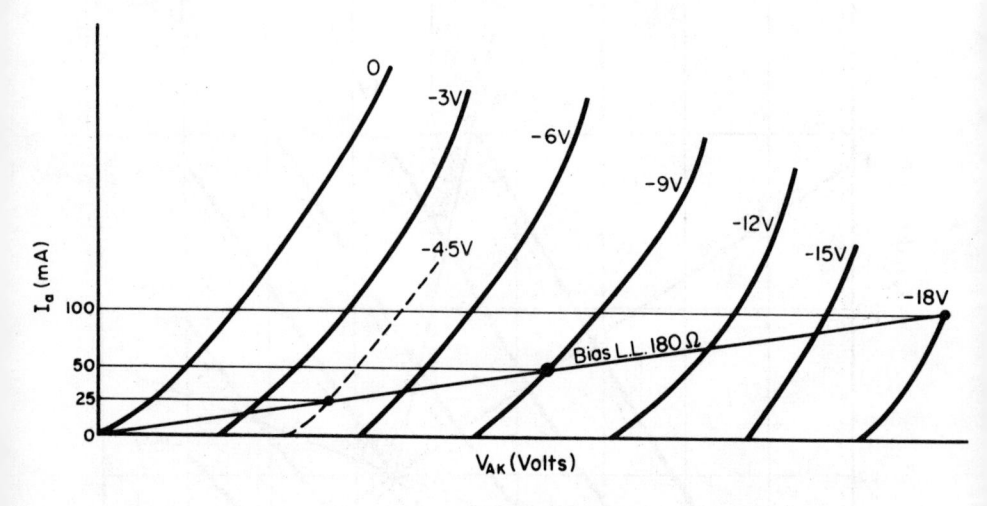

Fig. 9.7.1.

Where $R_K = \dfrac{V_K}{I_a}$ and where $V_K = 9V$, $I_a = \dfrac{9V}{180Ω} = 50mA$.

SIMPLE TRANSFORMER — COUPLED OUTPUT STAGE

9.8. Operating point.

The operating point, from which all d.c. values may be obtained, is seen to be the intersection of the d.c. and bias load lines. The coordinates are (276,50) and is marked as point P on **figure 9.6.2. The steady d.c. values are** $V_{ak} = 276V$, $I_a = 50mA$ and $V_{gk} = -9V$.

9.9. a.c. load line.

Once an alternating signal is applied, we assume that all capacitors become short circuit (to the frequency concerned) and transformer action occurs, thus the transformer reflects the load into the primary circuit.

Figure 9.9.1. shows the effective load in the primary.

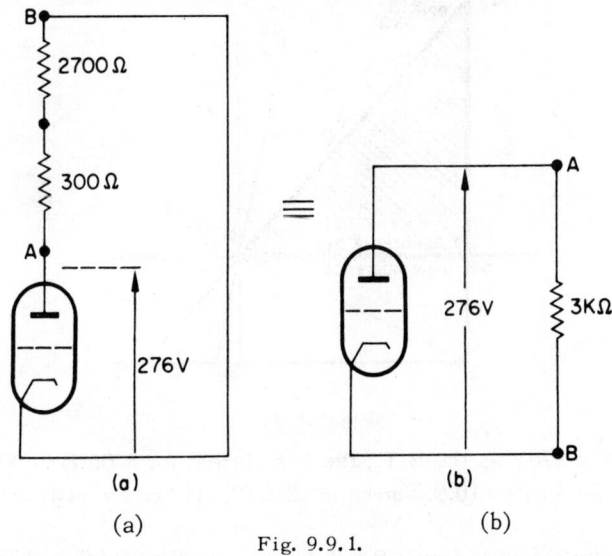

(a)   (b)

Fig. 9.9.1.

With no signal applied, we have 276V across the valve. If we apply an a.c. signal, C_K becomes short circuit and R_K becomes zero.

The 276V will be present across anode to cathode and as the cathode is now earthed, 276V exists from anode to earth. The effective resistance in the anode circuit is now $2700 + 300 = 3000\Omega$.

The H.T. is also at earth potential to a.c. and may be redrawn as in figure 9.9.1. (b).

We can see therefore that 276V exist across the 3KΩ also. The maximum a.c. signal current that can flow through the effective resistance of 3000Ω, is given from Ohm's law as

$$i_{max} = \frac{276V}{3k\Omega} = 92mA\ Pk.$$

M

It follows therefore, that as 50mA anode current is flowing with no signal, 50 + 92 = 142mA would flow if the maximum a.c. current were to flow.

The 142mA is the absolute maximum anode current that could theoretically flow and is the upper point for the a.c. load line.

This is shown as point E in the figure 9.6.2.

An alternative explanation may be useful. We have an operating point as shown in figure 9.9.2. (276, 50).

If we *ignore* the 50mA steady d.c. current and create an artificial X axis as shown, we need to draw an a.c. load line in the same manner as for previous d.c. load lines.

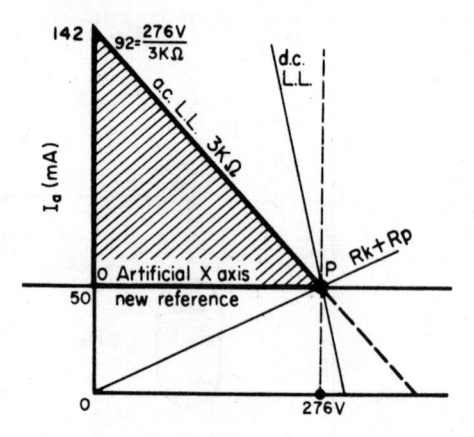

Fig. 9.9.2.

Viewing the 276V as the H.T., the s/c current for 3000Ω, is 92mA. A load line is drawn from (0.92) down to (276,0). These are still *artificial values*.

If the reader refers to figure 9.6.2. he will readily see that the load lines are identical.

All that remains is to allow for the case where the valve grid is driven in a negative direction.

We do this by simply continuing the a.c. load line from (0,142) through point P, and down to the V_{ak} axis.

The lower end of the load line is seen to terminate at approximately (425,0). This is marked as point F in figure 9.6.2.

The three load lines are now complete.

9.10. Applying a signal.

In the absence of a signal, the anode is at approximately 276V. As the valve is biassed at −9V, we are limited to a positive going input of +9V.

SIMPLE TRANSFORMER – COUPLED OUTPUT STAGE

This will take the grid to the threshold of grid current. Assuming a sinusoidal input of 9V peak, the grid will travel up the load line from −9V to 0V and for the other half cycle of input waveform will travel from −9V to −18V.

The effective input is therefore ±9V or 18V $P-P$.

The anode, sitting at 276V d.c. will change its potential accordingly. Figure 9.10.1. shows the grid and anode variations.

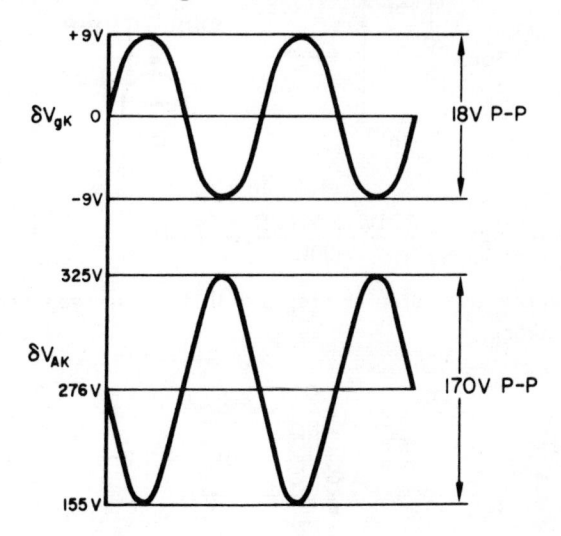

Fig. 9.10.1.

We can see that for an input of 18V $P-P$ a corresponding change in anode potential of 170V $P-P$.

As the anode 'swing' is 170V $P-P$, then we must re-examine the anode load, as the voltage distribution appears as shown in figure 9.10.2.

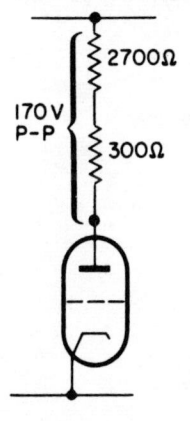

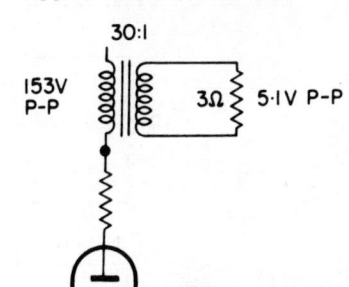

The actual 'useful' voltage is across the referred 3Ω (2700Ω), while the signal across the primary resistance of 300Ω is lost in useless dissipation.

Fig. 9.10.2.

Using load upon total once again, the useful peak to peak voltage is as in figure 9.10.3.

Fig. 9.10.3.

The useful voltage $\dfrac{170V \times 2700\Omega}{3000\Omega} = P{-}P$

Figure 9.10.4. shows, step by step how this voltage appears as power in the 3Ω (speaker).

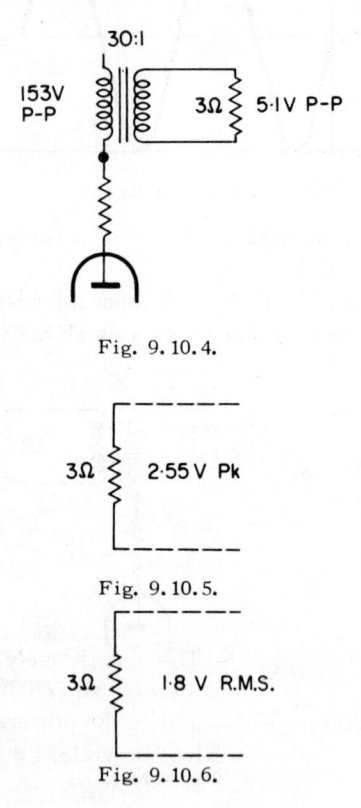

Fig. 9.10.4.

Fig. 9.10.5.

Fig. 9.10.6.

3Ω ⌇ 1·08 Watts

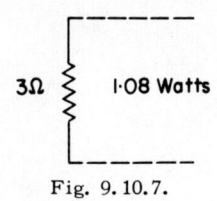

Fig. 9.10.7.

Hence, for an input of 18V $P-P$, an output of a little over 1W is obtained.

CHAPTER 10

Miller effect

10.1. Miller effect in resistance-loaded amplifiers.

In a triode valve, a capacity exists between anode and grid. This is due to the proximity of the grid to the anode within the glass envelope. The capacity C_{ag}, varies with the gain and is expressed as the effective capacity between grid and anode, $C_{ag}(1 + A)$ where A is the gain of the stage.

The anode voltage $V_a = -AV_g$ where V_g is the input to the grid. When the positive going input voltage is applied (V_g), the anode potential falls. V_a is therefore 180° out of phase with the input and is expressed as $-AV_g$.

The voltage across the capacitor C_{ag} is the difference between V_g and $V_a = V_{ga}$. (The difference is actually $V_g - V_a = V_g - (-AV_g) = V_g + AV_g = V_g(1 + A)$. The C_{ag} current leads V_{ag} by 90°. The current in the capacitor, C_{ag}, leads V_g by 90°. This has the effect of presenting a capacitance to the generator at the grid. The effective resistance in the grid circuit remains unaltered (normally infinity). This would not be so if there was a reactive component in the anode load. The resultant phase shift might present a negative resistance, into the grid circuit. With V_g applied, an input current taken from the input generator would be $(1 + A)$ times that value which would result if V_a were zero. The effective input capacitance may be given as $C_{ag}(1 + A)$.

There is a constant capacity between grid and cathode, C_{gk}, and as this is in shunt with $C_{ag}(1 + A)$, the total input capacitance would become $C_{in} = C_{gk} + C_{ag}(1 + A)$.

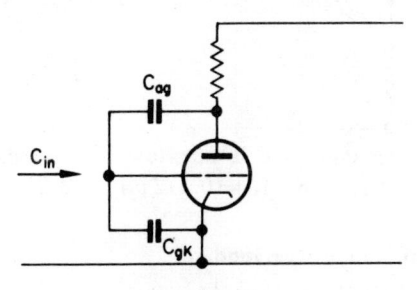

Fig. 10.1.1.

Example.

(a) Figure 10.1.2. What is the effective input capacitance in the following circuits?

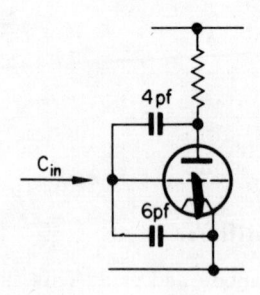

Gain = 20
C_{in} = 6 p.f. + (1 + 20) 4 p.f. = 90 p.f.

(b) Figure 10.2.3.

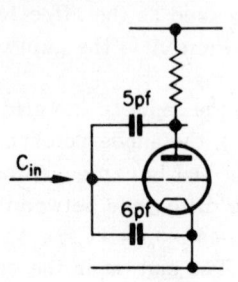

Gain = 30
C_{in} = 6 p.f. + (1 + 30) 5 p.f. = 161 p.f.

(c) Figure 10.1.4. This is a cathode follower, hence the gain is less than unity.

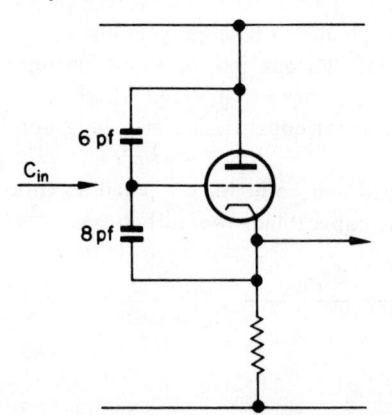

Gain = 0.9
As the output is in phase with the input, the voltage across C_{gK} =
$V_g - V_K = V_g - AV_g = V_g'(1 - A)$

C_{ag} is in shunt with the input and is constant. C_{in} = 6 p.f. + (1 − 0.9) C_{gk}
∴ C_{in} = 8 p.f. + (0.1) 2 p.f. = 8.2 p.f.

10.2. Amplifier with capacitive load.

The anode voltage (AV_g) lags the input (ignoring the valve phase shift). A capacitor current through C_{ag} leads V_{ag} by 90°.

The capacitor current leads V_g by less than 90° and subsequently has an active and reactive component. This causes negative feedback,

THE MILLER EFFECT

$$C_{in} = C_{ag}(1 + A\cos\phi) + C_{gk}.$$
$$R_{in} = \frac{XC_{ag}}{|A|\sin\phi}$$

Where ϕ is the phase difference between the voltage across the anode load, and the voltage μV_{gk} generated within the valve.

10.3. Amplifier with inductive load.

The anode voltage V_a, leads the voltage V by an angle ϕ. V_{ag} leads the input voltage V_g and subsequently the capacitive current has two components, active and reactive.

The active component is now in antiphase with V_g, **consequently** the input resistance becomes negative.

$$-R = \frac{1}{X_c|A|\sin\phi}$$

A detailed discussion on these circuits is given in 19.15.

10.4. Miller timebase.

A saw tooth generating circuit, the Miller timebase generator, exploits the Miller effect to the full. A large external capacitor is connected between anode and grid. This capacitor becomes $(1 + A)$ times its value. Further consideration of the Miller circuit is given in Chapter 31.

10.5. Cathode follower input impedance.

A triode has capacity between anode and grid (C_{ag}) and capacity between grid and cathode (C_{gk}). The following example illustrates the loading effect of these capacitances upon the previous stage or voltage input source.

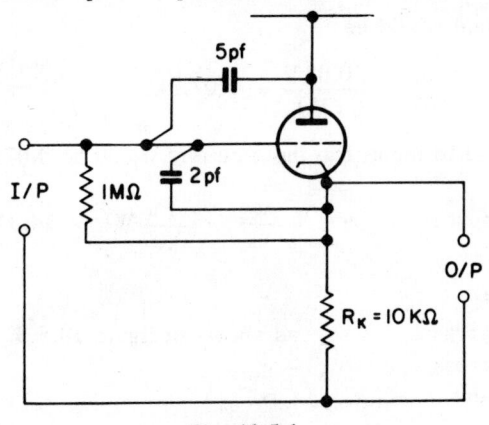

Fig. 10.5.1.

Consider the cathode follower shown in figure 10.5.1. The valve has a μ of 30 and an ra of $10\,\text{K}\Omega$. $C_{ag} = 5\,\text{p.f.}$ whilst $C_{gk} = 2\,\text{p.f.}$

We require to find the input impedance. The input impedance, in practice, is much better quoted as a value of R in shunt with a value of C. This eliminates the need to calculate the impedance for a given frequency unless it is specifically required. As the anode is at earth potential to a.c. the circuit becomes as shown in figure 10.5.2.

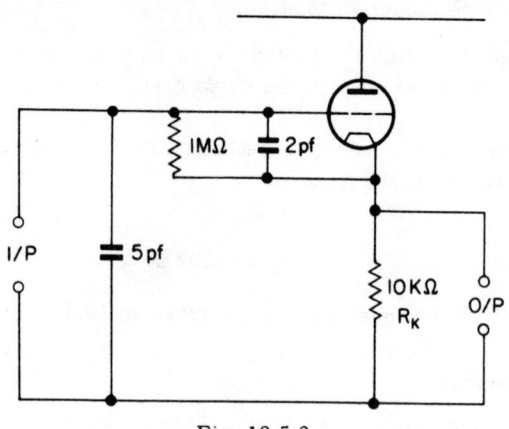

Fig. 10.5.2.

The stage gain

$$\frac{\mu R_K}{ra + (1 + \mu)R_K} = \frac{30 \times 10}{10 + 31 + 10} = \frac{300}{320} = 0.93.$$

Input resistance.

The effective potential across the $1\,\text{M}\Omega$ with an input of $1\,\text{V} = 1 - 0.93 = 0.07\,\text{V}$.

The input current would be

$$\frac{0.07\,\text{V}}{1\,\text{M}\Omega} = 0.07\,\mu\text{A}.$$

With $1\,\text{V}$ applied to input, the input current would be $0.07\,\mu\text{A}$,

$$\text{the input resistance} = \frac{v_{in}}{v_{in}} = \frac{1\,\text{V}}{0.07}\,\text{M}\Omega = 14.3\,\text{M}\Omega.$$

Input capacitance.

The voltage across $C_{ag} = v_{in}$ as shown in figure 10.5.3.
The voltage across $C_{gk} = v_{in} - v_0$.
The input current $i_1 = i(C_{ag}) + i(C_{gk})$

THE MILLER EFFECT

$$\therefore \quad i_1 = \frac{v_{in}}{XC_{ag}} + \frac{v_1 - v_0}{XC_{gk}}$$

$$\therefore \quad i_1 = v_{in} j\omega C_{ag} + (v_1 - v_0) jw C_{gk}$$

hence

$$i_1 = j\omega v_{in} \left(C_{ag} + 1 - \frac{v_0}{v_{in}} C_{gk}\right)$$

but $\frac{v_0}{v_{in}}$ is the gain of the stage, A,

$$\therefore \quad i_1 = j\omega v_{in} [C_{ag} + (1 - A) C_{gk}].$$

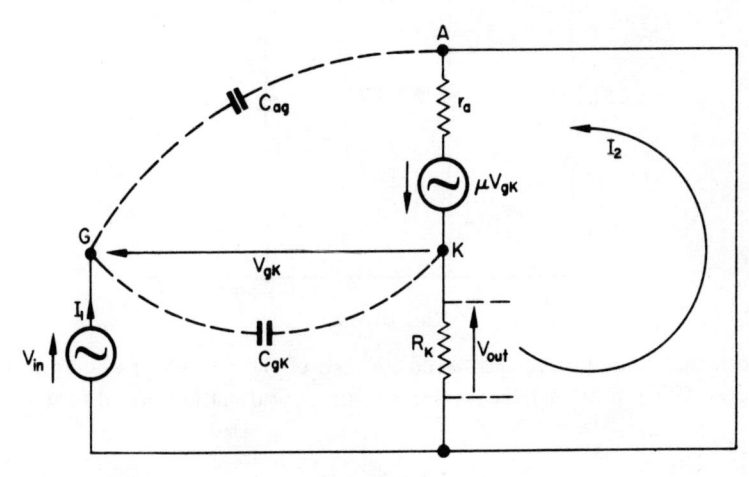

Fig. 10.5.3.

The effective reactance
at the input $= \dfrac{v_{in}}{i_1} = X_{in} = \dfrac{v_{in}}{j\omega v_{in} (C_{ag} + (1-A)C_{gk})}$

$$= \frac{1}{j\omega [C_{ag} + (1-A)C_{gk}]}$$

hence

$$C_{in} = C_{ag} + (1-A)C_{gk}.$$

The grid cathode capacitance is seen to be reduced by a factor $(1-A)$ where A is the cathode follower gain. The anode grid capacitance will be unaffected as this is now across the input source.

The total input capacitance is the sum of C_{ag} plus the modified value of C_{gk} (depending upon the gain).

$$C_{in} = C_{ag} + (1-A) C_{gk}.$$
$$= 5 \text{ p.f.} + (1 - 0.93) \, 2 \text{ p.f.}$$

$$= 5\,\text{p.f.} + (0.07 \times 2)\,\text{p.f.}$$
$$= 5\,\text{p.f.} + 0.14\,\text{p.f.}$$
$$= 5.14\,\text{p.f.}$$

Input impedance.

The input impedance, expressed as two components, therefore will be 14.3 MΩ in shunt with 5.14 p.f. as illustrated in figure 10.5.4.

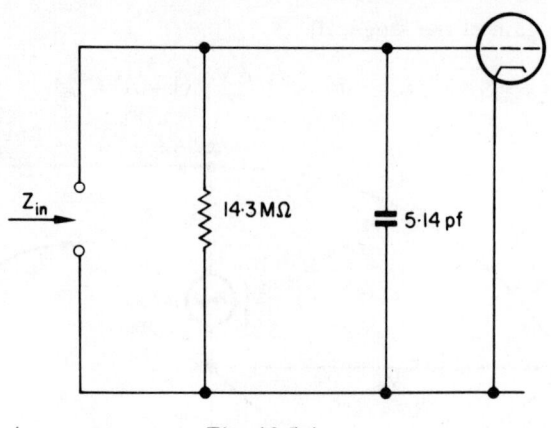

Fig. 10.5.4.

(Readers that have not yet mastered the use of the operator j should refer to chapter 19 for a brief introduction to this useful mathematical tool.)

CHAPTER 11

The pentode valve

11.1. The tetrode and pentode.

A capacitance exists between the anode and grid of a triode valve, figure 11.1.1. This capacitance, C_{ag}, is an interelectrode capacitance and proves very troublesome on many occasions. The capacity C_{ag}, increases with the gain of the stage and at high frequencies presents a capacitance into the grid circuit which is of a magnitude often comparable with externally connected capacitors.

Tetrode characteristics.

The screen grid valve, or tetrode, was developed in order to reduce C_{ag}. A second grid, the screen, was inserted between the grid and anode. Capacity exists between anode and screen and a further capacity is present between screen and grid. (Figure 11.1.1a.). The screen grid is normally connected to a high voltage potential (if it were at the anode potential, no effective capacity could exist) but draws very little current due to the fact that it often consists of a spiral. Each turn of the spiral has a relatively large space between it and its neighbour consequently most of the electrons are attracted, through these gaps, to the anode.

Varying the anode voltage of a tetrode does little to vary the anode current compared to that of a triode. The tetrode however, has a certain undesirable property. Electrons reaching the anode at a high velocity may cause other electrons to be freed from the anode and would be attracted to the screen grid. This increases the screen current and subsequently reduces the net anode current. This is known as secondary emission. If the screen is at a lower potential than the anode (as is the grid in a triode), these freed electrons return to the anode and the original anode current is maintained.

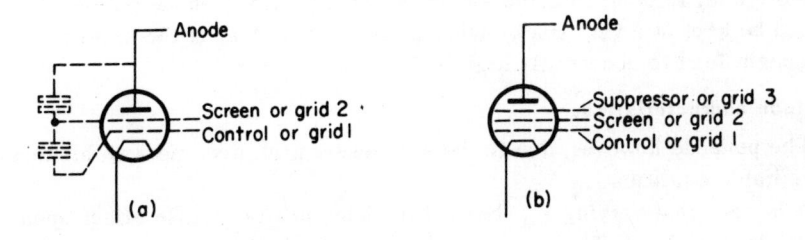

Fig. 11.1.1.

A method of reducing the secondary emission was to insert a third grid between the screen and the anode. Figure 11.1.1b. The potential of the third grid is usually at 0 V, or at most, a few volts positive. The electrons that are freed from the anode are no longer attracted by the grid nearest the anode, as its attraction is very small compared with the anode. The third grid is known as the suppressor grid. In a tetrode, with the screen at a constant potential, and a fixed grid bias, the following typical curve will be obtained showing I_a/V_a.

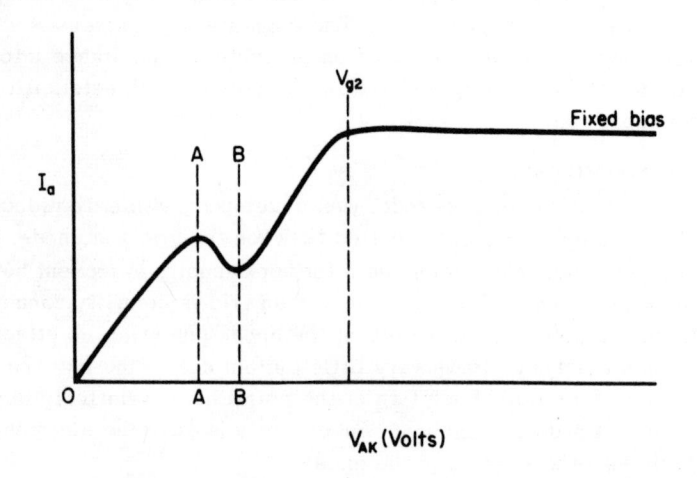

Fig. 11.1.2. Anode characteristic of a tetrode.

The portion of the curve $A-B$, represents a negative resistance. The tetrode is used on occasions as an oscillator which utilizes this negative resistance region.

Increasing the V_{ak} results in an increase in I_a from 0–A.
Increasing V_{ak} further results in a decrease of I_a at points $A-B$.
This is due to the screen grid attracting the freed anode electrons.

Increasing V_{ak} further still, results in a normal increase in I_a, as the anode is much more positive than the screen, therefore secondary emission is negligible. In order to avoid secondary emission, the anode potential should be kept at a very much higher potential than that of the screen and this might lead to abnormally high H.T. lines.

Pentode characteristics.

The pendode however, due to the suppressor grid, does not exhibit these undesirable qualities.

It is seen that varying V_{ak} beyond the knee has very little effect upon the I_a, figure 11.1.3. The *ra* of a pentode is very high and is not normally

THE PENTODE VALVE

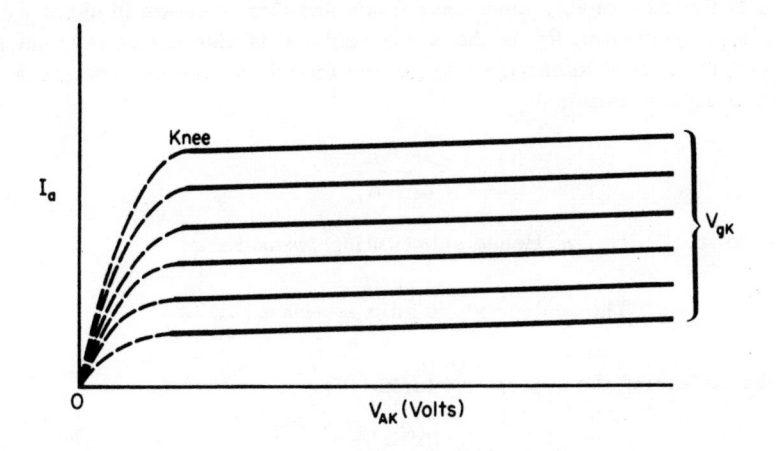

Fig. 11.1.3. Anode characteristics for a pentode

derived from the characteristics as the change in I_a due to the change in V_{ak}, is so small. The d.c. load line is often plotted so as to pass through the upper knee region. There is no secondary emission in the pentode, as when the electron stream has passed through the suppressor grid and arrived at the anode, the anode is positive with respect to the suppressor and any electrons displaced from the anode return to the anode. The suppressor is usually at a potential of a few volts and in many cases, at zero potential.

The stage gain of a triode is given as

$$\frac{\mu \times R_L}{ra + R_L}$$

and is derived from voltage generator series equivalent circuit.

The pentode, as its ra is very very high, is often shown as follows.

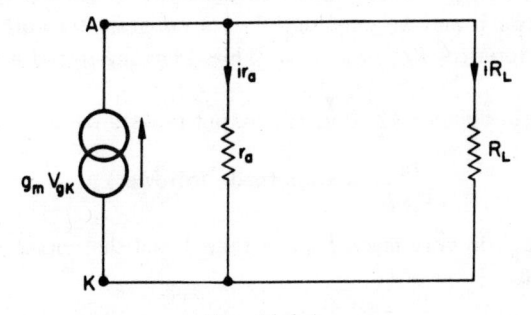

Fig. 11.1.4.

Where the current generator, $gm\ V_{gk}$ represents the constant current characteristics in the I_a/V_a graph. Figure 11.1.3.

ra is the incremental anode resistance and this is shown in shunt with the current generator. R_L is the anode load and is also shown in shunt with the generator. It is intended to derive the formula for the pentode stage gain, using the formula

$$\frac{\mu R_L}{ra + R_L}$$

as a basis. $\mu = gm \cdot ra$. Hence substituting $gm\,ra$ for μ.

$$\text{The pentode stage gain} = gm\,ra \cdot \frac{R_L}{ra + R_L}$$

and by collecting the appropriate terms

$$(gm)\frac{(R_L\,ra)}{(ra + R_L)}$$

This may be seen to be a current (gm) entering a shunt resistor network whose effective resistance is

$$\frac{ra\,R_L}{ra + R_L}$$

If ra is very much higher than R_L, the approximate formula

$$gm \times ra \times \frac{R_L}{ra}$$

may be employed ignoring R_L in the denominator as $R_L \ll ra$. This then resolves to $\simeq gmR_L$. The ra of a pentode is extremely high. The gm is comparable to that of a triode.

From the identity $\mu = gmra$ it may be seen that if gm has a value similar to that of a triode it may be considered as a constant for both triode and pentode. Therefore $\mu = Kra$ or $\mu \propto ra$. Thus showing μ had a very high value also.

We will see in chapter 12, that the output resistance for a triode is

$$\frac{ra}{1 + \mu} \text{ (as a cathode follower)}$$

In a pentode μ is very much greater than 1 and the output resistance may be expressed as

$$\simeq \frac{ra}{\mu} \simeq \frac{1}{gm}.$$

It is important to appreciate that the *cathode current* of a pentode consists of the *anode current plus* the *screen current*. The following circuit diagrams in figure 11.1.5. should illustrate the point.

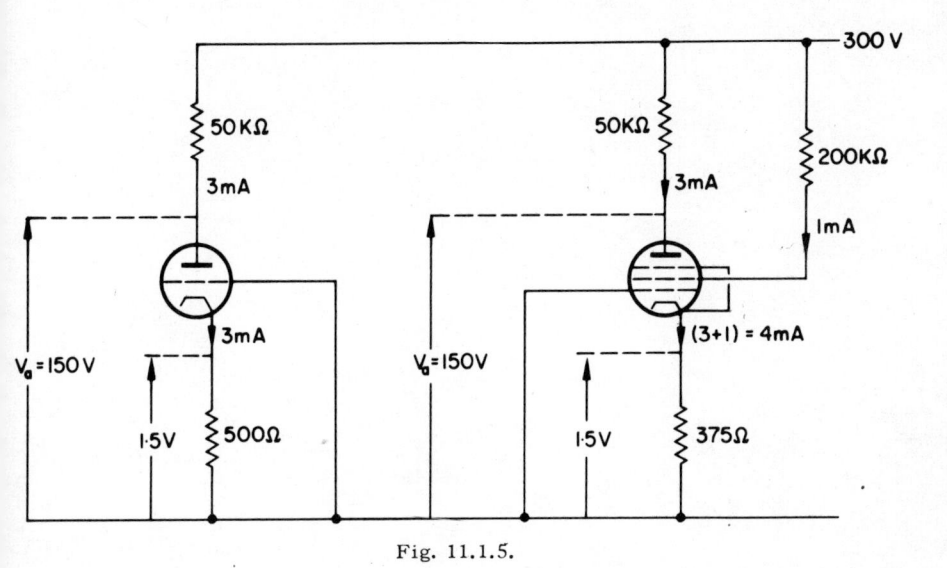

Fig. 11.1.5.

Both valves have an anode current of 3 mA. Both are running at $V_{ak} = 148.5\,\text{V}$. Both have 1.5 V bias. R_k in the pentode is smaller by the ratio of

$$\frac{I_a}{I_a + I_g} \quad \text{or} \quad \frac{3}{4} \times 500 = 375\,\Omega$$

This allows for the constant screen current of 1 mA also flowing through the cathode resistor.

CHAPTER 12

Equivalent circuits and large signal considerations

We have already discussed a number of basic network theorems including Kirchhoff's laws. We 'look into' a number of 'black boxes' and calculate their input and output resistances.

We saw that, when determining say, the output resistance of a box, we were told to quote the conditions at the input. Similarly, we learnt that if we were to apply a simple test to, say, the output in order to calculate the output resistance, we could get different answers depending upon whether we had the input terminals open or short circuit.

With valves, it is usual to short circuit the input when establishing the output resistance.

When dealing with transistors however, we sometimes leave the output either open, short circuit or connected across a load, when 'looking into' the input. We will not concern ourselves with these devices in this chapter as the amount of coverage warrants a more detailed discussion later on.

There is nothing mysterious about these techniques, they are quite straightforward and should be mastered. We intend to discuss a large number of different examples on valves in this chapter because of the importance of the subject. Transistors will be similarly discussed in another (chapter).

A positive value of μ has been adopted for all of these equivalent circuit examples, any 180° phase shift will be clearly seen by the direction of current flow and associated voltages. Small signal equivalent circuits are valid only when the changes are so small that the parameters μ, ra and gm remain constant.

12.1. Simple triode valve.

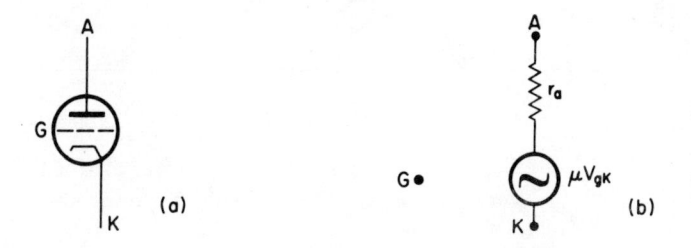

Fig. 12.1.1.

The equivalent circuit for the triode valve is shown in figure 12.1.1.(b) where ra is the incremental or a.c. resistance of the valve and μV_{gk} is the internal voltage generator. We must remember that an input to the grid can be effective only when it causes a *change in level between the grid and cathode*; This difference is the true effective input to the valve and has the symbol v_{gk}. The generator develops an e.m.f. of an amplitude μ times the true valve input v_{gk}. This e.m.f. must be shown antiphase to v_{gk} so as to allow for the amplifier phase shift.

The generator in the equivalent circuit is therefore identified as μv_{gk}. The generated e.m.f. appears across the terminals $A - K$ in figure 12.1.1.(b) and is the anode to cathode e.m.f.

The reader will recall that the p.d. developed across the load resistor say, in an amplifier, will be less than the e.m.f. generated within the amplifier.

Equivalent circuits discussed in this chapter are concerned with instantaneous values of a.c. only; we assume that for complete circuits, the d.c. conditions have been correctly chosen.

Hence we further assume that all capacitors may be considered to be short circuited for all practical purposes; this will be evident as we proceed. All d.c. supplies are assumed to be short-circuited to a.c. also, hence the normal H.T. rail to a stage is considered to be at earth potential in equivalent circuits.

12.2. Common cathode amplifier.

We will consider a very simple amplifier that employs cathode bias. The bias resistor is adequately bypassed with a suitable capacitor.

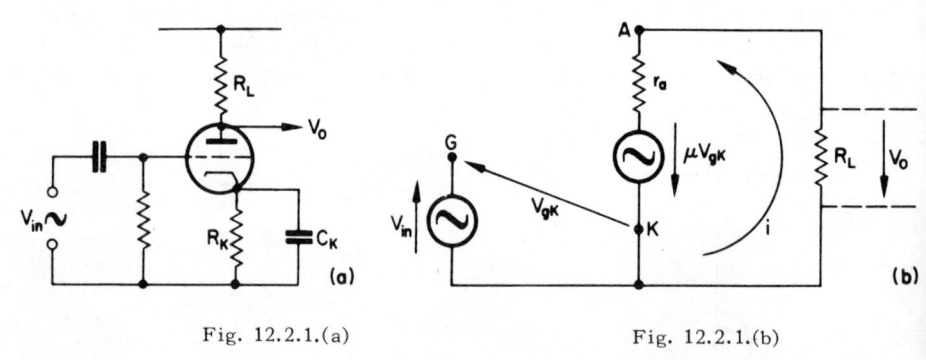

Fig. 12.2.1.(a)   Fig. 12.2.1.(b)

Figure 12.2.1.(b) shows the complete equivalent circuit for the simple amplifier shown in figure 12.2.1.(a). The equivalent circuit should be drawn according to the following steps:

1. Draw the equivalent circuit for the triode valve as in figure 12.1.1.(b).
2. All external resistors should be added to the circuit. (H.T. is earthy)
3. Short circuit all resistors that are shunted by capacitors.
4. Draw the input signal, v_{in}, showing it to be, say, positive going.

EQUIVALENT CIRCUITS AND LARGE SIGNAL CONSIDERATIONS 179

5. Draw an arrow representing v_{gk} assuming say, that the grid is rising.
6. Draw an arrow antiphase to (5) against the generator μv_{gk}.
7. Draw the anode signal current in direction indicated by arrow in (6).
8. Draw an arrow to denote the polarity of the output voltage observing direction of anode current.

We can now derive the formula for the stage gain of the amplifier. Considering v_{gk}, this is seen to be equal to v_{in}. Substituting this in r.h.s. loop, we equate e.m.f. and p.d.'s.

$$\mu v_{gk} = \mu v_{in} \quad \therefore \quad \mu v_{in} = i(ra + R_L) \quad \text{hence } i = \frac{\mu v_{in}}{ra + R_L}$$

but as $v_0 = i R_L \quad \therefore \quad v_0 = \dfrac{\mu R_L \cdot v_{in}}{ra + R_L}$

hence stage gain $\dfrac{v_0}{v_{in}} = \dfrac{-\mu R_L}{ra + R_L}$

and is negative as indicated by the arrow depicting v_0.

If $\mu = 20$, $ra = 10\,\text{K}\Omega$ and $R_L = 50\,\text{K}\Omega$, determin the gain. The reader is invited to carry out this simple calculation and compare the result with the example in 8.3. where R_L and the valve parameters were as shown here.

12.3. Unbypassed cathode – common cathode amplifier.

We will now discuss the equivalent circuit and formula for the stage gain of an amplifier with the cathode unbypassed as shown in figure 12.3.1. and 12.3.2.

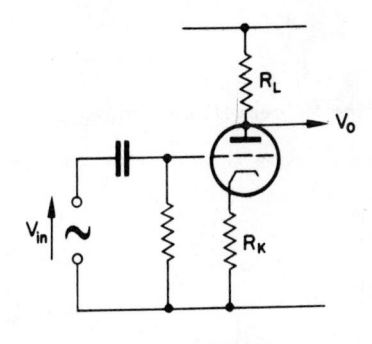

Fig. 12.3.1.

The first step is to express v_{gk} in terms of v_g and v_k. $v_g = v_{in}$ and $v_k = i R_K$. We can see that v_k is positive with respect to earth, the input v_{in} is also positive.

The two voltages are in series opposition, the input will be the larger of the two. Hence $v_{gk} = v_{in} - v_k = v_{in} - i R_k$. Substituting for v_{gk} in the

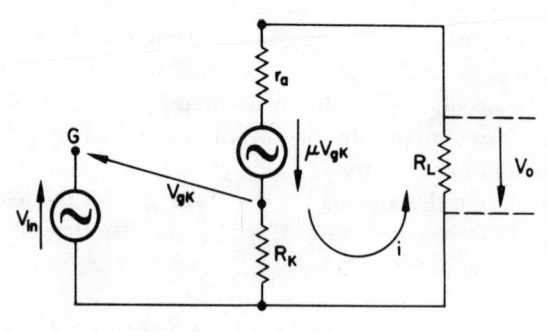

Fig. 12.3.2.

r.h.s. loop, and equating e.m.f.'s and p.d.'s.

where $\quad \mu v_{gk} = \mu(v_{in} - iR_k)$

hence $\quad \mu(v_{in} - iR_k) = i(ra + R_L + R_k)$

$\quad\quad\quad \mu V_{in} = i[ra + R_L + (1+\mu)R_k]$ but $v_0 = iR_L$,

$\quad\quad\quad \mu v_{in} R_L = v_0 [ra + R_L + R_k(1+\mu)]$

Stage gain $\quad \dfrac{v_0}{v_{in}} = \dfrac{-\mu R_L}{ra + R_L + (1+\mu)R_k}$

and is again negative as can be seen from the equivalent circuit. The gain is obviously smaller than the previous example due to the $(1 + \mu)$ term in the denominator. The expression for the stage gain may be simplified to

$$\dfrac{\left(\dfrac{\mu}{1+\mu}\right)R_L}{\left(\dfrac{ra + R_L}{1+\mu}\right) + R_k}$$

12.4. The phase splitter or 'concertina' stage.

Figures 12.4.1. and 12.4.2. show the circuit diagram and equivalent circuits.

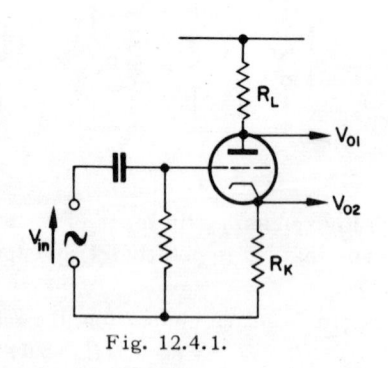

Fig. 12.4.1.

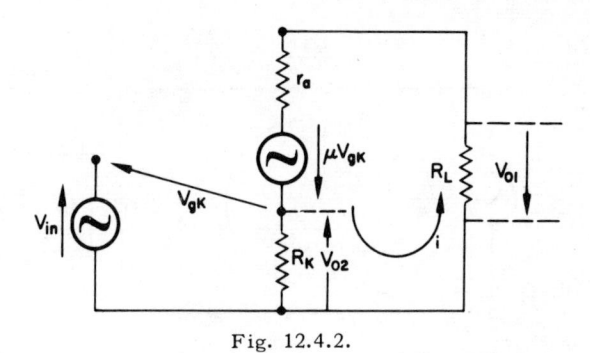

EQUIVALENT CIRCUITS AND LARGE SIGNAL CONSIDERATIONS 181

Fig. 12.4.2.

This circuit will provide two output signals in antiphase for one input. The relative amplitudes of the output are determined by the value of R_L and R_k. $v_{o_1} = i R_L$ and $v_{o_2} = i R_k$. The current is common to both R_L and R_k. We will derive the formula for i and for v_{o_1}, multiply i by R_L and for v_{o_2}, by R_k.

$$v_{gk} = v_{in} - i R_k. \text{ Hence } \mu v_{gk} = \mu[v_{in} - i R_k]$$

Considering the r.h.s. loop once again, and equating e.m.f.'s and p.d.'s;

$$\mu[v_{in} - i R_k] = i[ra + R_L + R_k]$$

hence
$$\mu v_{in} = i[ra + R_L + R_k + \mu R_k]$$

therefore
$$i = \frac{\mu v_{in}}{ra + R_L + (1 + \mu) R_k}$$

and for v_{o_1}, multiply both sides by R_L.

hence
$$v_{o_1} = \frac{-\mu R_L v_{in}}{ra + R_L + (1 + \mu) R_k} \quad \text{(out of phase)}$$

similarly for v_{o_2}, we multiply both sides by R_k in order to obtain

$$v_{o_2} = \frac{\mu R_k v_{in}}{ra + R_L + (1 + \mu) R_k} \quad \text{(in phase)}$$

These expressions may be simplified by dividing top and bottom by $(1 + \mu)$.

This gives
$$v_{o_2} = \frac{\mu v_{in}}{1 + \mu} \frac{R_k}{\frac{ra + R_L}{1 + \mu} + R_k}$$

and a simplified equivalent circuit may be drawn as in figure 12.4.3. v_{o_2} may be seen to be easily derived from $\frac{\text{load}}{\text{Total}} \times \text{input}$.

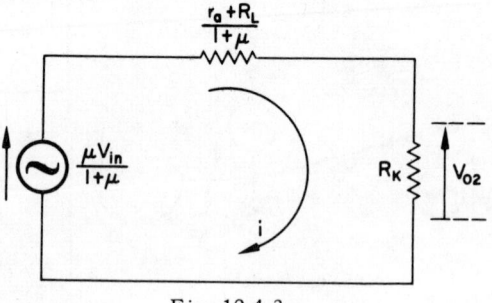

Fig. 12.4.3.

12.5. Grounded (or Common) grid amplifier.

Figures 12.5.1. and 12.5.2. show both the circuit diagrams and equivalent circuits.

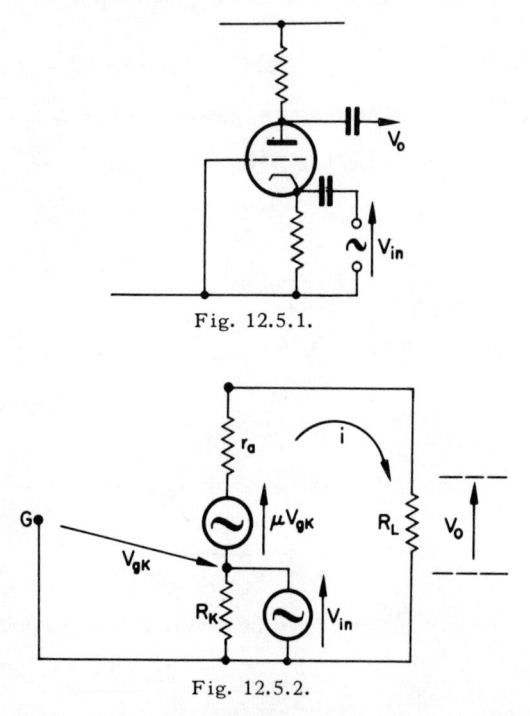

Fig. 12.5.1.

Fig. 12.5.2.

This amplifier has the input signal v_{in} applied to its cathode whilst the output signal is taken from the anode. It is essential that no signal appears at the grid and the grid therefore must be effectively earthed to a.c. (this can be arranged by connecting a capacitor in shunt with the grid resistor or by taking the grid direct to earth as shown in figure 12.5.2.)

EQUIVALENT CIRCUITS AND LARGE SIGNAL CONSIDERATIONS 183

This circuit is often used to load a low impedance output (we will deal with this later in detail). The transistor equivalent of the common grid amplifier is also covered in a later chapter.

When we apply v_{in} to the cathode, the signal source is connected across R_k. If we retain R_k in our discussion, it may tend to confuse, and as our generator is assumed to have zero resistance, R_k will be effectively short circuited. We will therefore 'remove' R_k for the purpose of deriving our formula for gain.

We can see from the equivalent circuit, the grid is earthed. The grid is negative with respect to the cathode which is being dragged up by the input signal.

$v_{gk} = v_{in}$, (note that the polarity for v_{gk} is allowed for by the v_{gk} arrow shows the grid to be negative. Note also that the generator μv_{gk} is seen to be antiphase to V_{gk} and is moving positively).

$$\text{as } v_{gk} = v_{in}, \mu v_{gk} = \mu v_{in}.$$

Hence as we now have two e.m.f.'s in series aiding in the r.h.s. loop and equating these to the p.d.'s,

$$\mu v_{gk} + v_{in} = i(ra + R_L) \text{ but } v_{gk} = v_{in}$$

therefore $\mu v_{in} + v_{in} = i[ra + R_L]$

Thus $V_{in}[1 + \mu] = i[ra + R_L]$

hence $\dfrac{\mu v_{in}}{v_{in}} = \dfrac{1 + \mu}{ra + R_L}$ but $v_o = iR_L$, therefore $\dfrac{v_o}{v_{in}} = \dfrac{(1 + \mu)R_L}{ra + R_L}$

The output signal voltage is in phase with the input. This is depicted by the arrow identified as v_o.

12.6. Input resistance, common grid amplifier.

The input resistance of a common cathode amplifier, assuming that no grid current flows, is simply R_g. This will be so whether the cathode is bypassed or not. It will remain R_g for variations in value of the anode load resistance, R_L.

The input resistance of the common grid amplifier shown in figure 12.5.1. will now be dealt with. The equivalent circuit will be the same as that shown in figure 12.5.2.

An expression for i was obtained earlier and is given as $i = \dfrac{(1 + \mu)v_{in}}{ra + R_L}$.

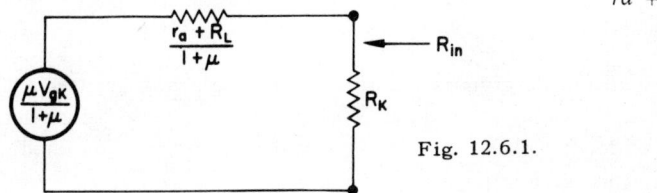

Fig. 12.6.1.

The input resistance is given by Ohm's law, as $R_{in} = v_{in}/i_{in}$ where i_{in} is the current taken from the input generator. The reader will see that the generator v_{in}, is in series with the anode signal current.

Therefore
$$R_{in} = \frac{v_{in}}{i_{in}} = \frac{ra + R_L}{(1 + \mu)}$$

We might need to know not only the input resistance of the amplifier alone but of the complete circuit including the cathode bias resistor R_k. It is simply a matter of shunting, the expression with R_K using the product/sum rule.

Hence, R_{in} with R_k in circuit $= \dfrac{(ra + R_L) R_k}{(ra + R_L) + (1 + \mu) R_k}$

This simplifies to
$$\frac{\left(\dfrac{ra + R_L}{1 + \mu}\right) \cdot R_k}{\left(\dfrac{ra + R_L}{1 + \mu}\right) + R_k}$$

which gives R in shunt with $\dfrac{ra + R_L}{1 + \mu}$

and a simplified equivalent circuit is given in 12.6.1.

12.7 Cathode follower (Common anode).

Figures 12.7.1. and 12.7.2. show both the circuit diagram and the equivalent diagram to be discussed.

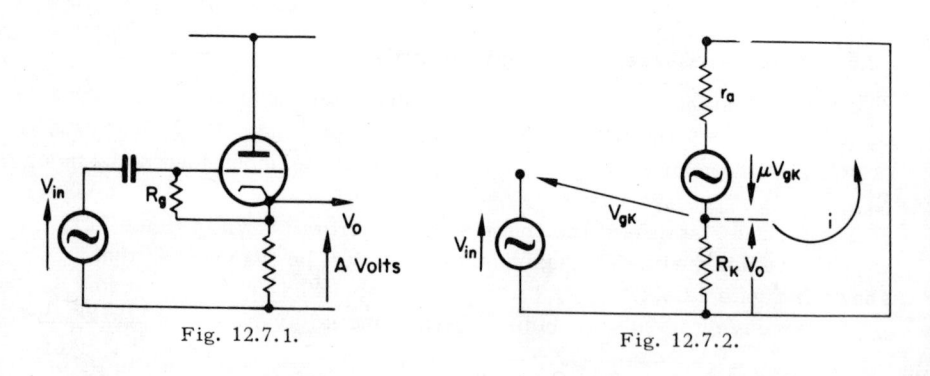

Fig. 12.7.1.   Fig. 12.7.2.

The circuit diagram is similar to that of a phase splitter with the exception that it has no anode load. The gain of a cathode follower may be derived as for the phase splitter except that as there is no anode, R_L will be zero in the expression.

EQUIVALENT CIRCUITS AND LARGER SIGNAL CONSIDERATIONS

The gain of the phase splitter (with the output taken from the cathode) was given as

$$\frac{v_0}{v_{in}} = \frac{\mu R_k}{ra + R_L(1+\mu)R_k}$$

but as there is no anode load resistor, R_L becomes zero, hence a cathode follower gain

$$A = \frac{v_0}{v_{in}} \quad \frac{\mu R_k}{ra + (1+\mu)R_k}$$

There is no phase reversal. Sometimes a resistor is connected between the anode and the H.T. but is bypassed by a suitable capacitor. The anode is therefore still at earth potential and the result given is unchanged.

12.8. Output resistance of a cathode follower.

The output resistance of a cathode follower may be derived by means of a simple test.

A voltage is applied to the output terminals (this is the cathode in this case) and an expression derived for the current that flows into the cathode from the generator, v_{in}. R out (output resistance) is given once more by Ohm's law as $R\text{ out} = v_{in}/i_{in}$.

The grid must be earthed during this test, Hence $v_{gk} = v_{in}$. The two generators in the r.h.s. loop add, giving $v_{in} + \mu v_{in}$ as the total e.m.f. in the loop. Equating them to the iR drops,

$$v_{in} + \mu v_{in} = i(ra)$$
$$v_{in}(1+\mu) = i(ra)$$

$$\text{Rout} = \frac{v_{in}}{i} = \frac{ra}{1+\mu}$$

simply allowing for R_k in shunt with this result gives the effective output resistance of the circuit as a whole.

Thus
$$\text{Rout} = \frac{ra}{1+\mu} \bigg/\!\!\bigg/ R_k.$$

The reader is invited to draw the equivalent circuit and check this result.

12.9. Input resistance of a cathode follower.

The input resistance of a cathode follower can be very high.

This circuit is of immense value when connected as an *impedance transformer*, i.e., a high impedance input, a gain approaching unity and a low output impedance.

Figure 12.7.1. shows a cathode follower circuit.

The gain of a cathode follower is given as $\dfrac{\mu R_k}{ra + R_k(1+\mu)}$

This will always be less than unity. Let A be the gain of a cathode follower.

We can see from figure 12.7.1. that we have v_{in} applied and A volts available at the cathode output terminal.

The potential across $R_g = (v_{in} - A)$ V.

The input current, i, flowing from the generator through

$$R_g = \dfrac{(V_{in} - A)}{R_g} \; \mu A,$$

where R_g is in MΩ.

$$R_{in} = \dfrac{v_{in}}{i} = \dfrac{v_{in} \cdot R_g}{(v_{in} - A)} \; M\Omega$$

and if $A = 0.9$, $R_g = 1M$ and $v_{in} = 1V$, then the effective input resistance

$$R_{in} = \dfrac{1}{1 - 0.9} \; M\Omega = \underline{10 \, M\Omega}.$$

A useful expression for R_{in} of a cathode follower is $R_{in} = R_g/1 - A$.

12.10. Output resistance of the anode.

Figure 12.10.1. shows both the circuit diagram and the equivalent circuit of a common cathode amplifier.

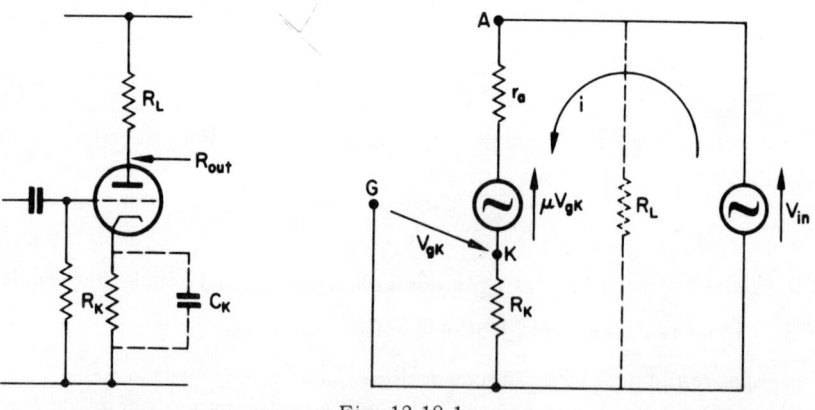

Fig. 12.10.1.

In order to derive an expression for the output resistance, Rout, we must begin by earthing the grid. This ensures that the input is short-circuit.

EQUIVALENT CIRCUITS AND LARGER SIGNAL CONSIDERATION 187

We then apply a voltage to the anode, calculate the current flowing from the external generator into the anode and apply Ohms law. We should particularly note that the grid is negative with respect to the cathode in this instance. $V_{gk} = iR_k$.

Note: C_K is not connected in this example.

Therefore we have in the right hand loop, two generators in series opposition, and equating e.m.f.'s to iR drops, we have;

$$v_{in} - \mu v_{gk} = i[ra + R_k] \qquad \therefore v_{in} - \mu i R_k = i[ra + R_k]$$

Hence $v_{in} = i[ra + R_k + \mu R_k] \qquad \therefore v_{in}/i = ra + R_k(1 + \mu)$

$\therefore$ Rout $= ra + R_k(1 + \mu)$. R will shunt this output resistance.

If C_k is connected, R_k will be zero due to the *short-circuited* capacitor across it. As R_k would be zero, the bracketed term would vanish, hence the output resistance from the anode would therefore be ra alone. This of course we would expect as the valve's internal resistance is simply the ra.

A simplified equivalent circuit for Rout, with an un-bypassed cathode is given in figure 12.10.2.

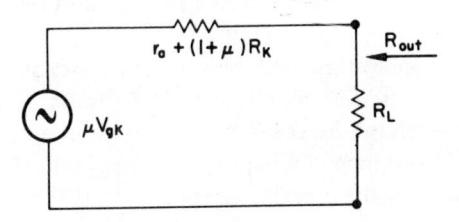

Fig. 12.10.2.

12.11. Cathode coupled amplifiers – Long Tailed Pairs (L.T.P.s)

A larger number of circuits fall into the category of the 'Long Tailed Pair' family. Figure 12.11.1. shows a basic circuit of this kind.

The L.T.P. was originally a circuit that contained a very high value of resistance for R_k which in turn was returned to high negative voltage rail. A true long tailed pair has $R_k \to \infty$.

The circuit function is as follows;

A positive going input to the grid of V_1 causes V_1 anode current, I_1, to increase. This increase has a twofold effect. It causes a larger voltage drop across R_{L_1} thus causing V_1 anode to fall in a negative direction. This fall is in fact an amplified and inverted input signal. The other effect of the increase in I_1 is that it will cause V_k to increase in a positive direction.

The increase in cathode potential is transmitted of course to the cathode of V_2. As V_2 cathode is raised, and as its grid is held at earth potential,

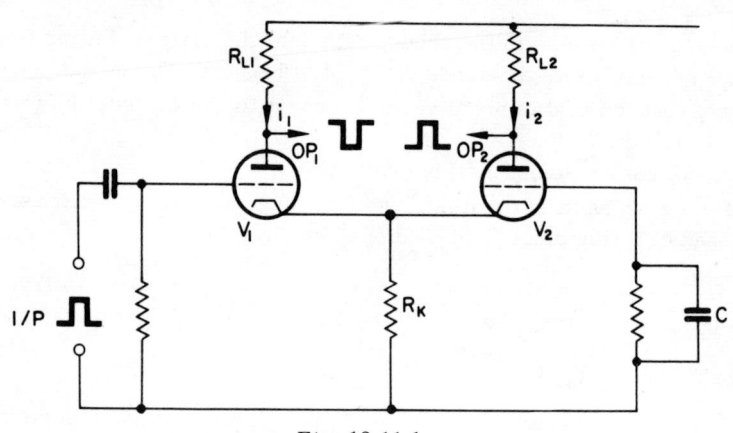

Fig. 12.11.1.

via C, the effect is to cause a reduction in I_2. As I_2 falls in value, the voltage across R_{L_2} reduces likewise. The anode of V_2 therefore rises in a positive direction.

This rise is in effect, the second output signal and is seen to be in phase with the input signal. This circuit then will provide two outputs for a single input.

We will see throughout this book that we can discuss tests in terms of either voltage or current. The results obtained should be the same, as voltage and current are related by Ohms law. We have recently concerned ourselves with input resistance. We applied a voltage in order to arrive at our answers, we could have discussed current inputs if it would have been more convenient.

The long tailed pair function showed that the signal voltage to V_2 was applied via the cathode. We would have shown the signal current flowing from V_1 into V_2 and got the same answer. If we consider the circuit diagram in figure 12.11.2 we will recognise a long tailed pair complete with relevant circuit component values. We intend to analyse this circuit and obtain expressions for the voltage gains from each output. R_k has been chosen as 1500Ω in order to emphasise its effects upon other factors in the circuit. Normally $R_k \gg \dfrac{ra + R_L}{1 + \mu}$ for a true L.T.P.

It is often most convenient to 'remove' one of the valves for a while and examine the remaining valve on its own. Once a certain stage in the analysis is reached, we replace the valve and complete the analysis. We will adopt this approach for this example. Figure 12.11.3. shows one of the valves, V_2, replaced in the circuit by its input resistance. (The reader will begin to appreciate the importance of the previous examples on input resistance).

Fig. 12.11.2.

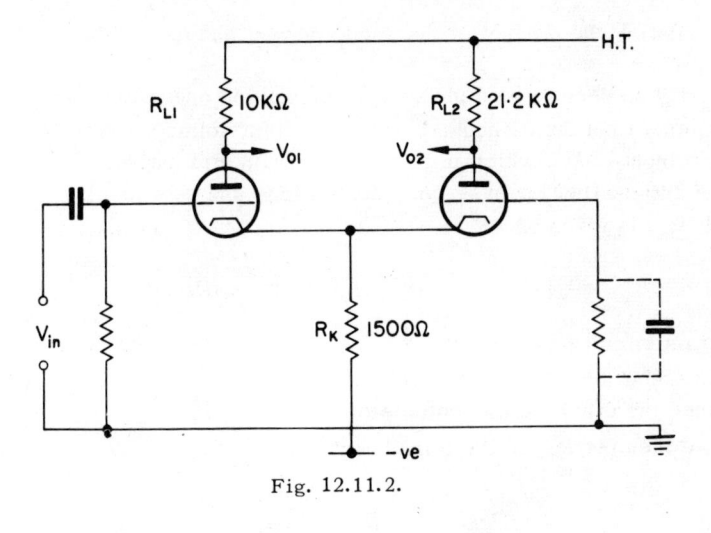

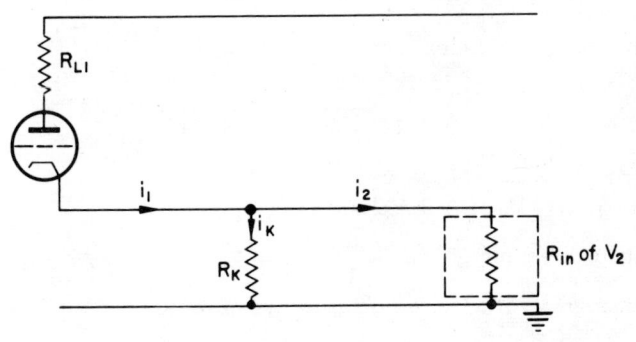

Fig. 12.11.3.

The input resistance 'looking into' the cathode of V_2, represented by the 'black box' in figure 12.11.3, is given as

$$\frac{ra + R_L}{1 + \mu} = \frac{35.3}{21} = 1.675 \, K\Omega.$$

The input resistance to V_2, shown as R_{in}, is in shunt with the cathode resistor R_k. The signal current in V_1, due to the input signal, flows through both R_k and R_{in} in the following proportions;

$$i_2 = \frac{i_1 \times R_k}{R_k + R_{in}}$$

The signal current i_2 enters V_2 cathode and constitutes the signal (anode) current in V_2. The output voltages may be established for each valve by calculating the product of the anode current and the anode load resistor in each case.

V_1 may be seen to be a phase splitter, having one output from the anode and another from the cathode. The latter output voltage is transmitted to the cathode input of V_2, which in turn is a common grid amplifier.

The current flowing in V_1 considered alone and as a result of the input signal v_{in}, is given as

$$i_1 = \frac{\mu \, v_{in}}{ra + R_{L_1} + (1+\mu)R_k}$$

In this circuit however, R_k consists of two components, i.e. R_k in shunt with the input resistance of V_2.

Hence the effective cathode resistance $R_k^1 = R_k /\!/ R_{in}$.

If we then put R_k^1 in the formula for i_1, we get

$$i_1 = \frac{\mu \, v_{in}}{ra + R_{L_1} + (1+\mu)R_k{'}}$$

and putting in known values,

$$i_1 = \frac{20 \, v_{in}}{14 + 10 + (21)0.792}$$

where $R_k^1 = 0.792$.

Hence for $v_{in} = 1\,\text{V}$, $i_1 = 0.492\,\text{mA}$.

i_2, from the expression given $= \dfrac{0.492 \times 1.5}{1.5 + 1.675} = 0.232\,\text{mA}$.

The output signal voltages are obtained as follows;

$$v_{o_1} = i_1 R_{L_1} = 0.492 \times 10 = 4.92\,\text{V (negative going)}$$
$$v_{o_2} = i_2 R_{L_2} = 0.232 \times 21.2 = 4.92\,\text{V (positive going)}$$

The outputs are seen to be equal in amplitude. They are the same because R_{L_2}, was chosen to be $21.2\,\text{K}\Omega$ for that very reason.

In order to obtain equal outputs from V_1 and V_2, the value of R_{L_2}, must be carefully chosen to have a certain ratio to R_{L_1}.

An expression which enables this choice to be made is derived as follows:
W require both outputs to be equal in magnitude, i.e. $|v_{o_1}| = |v_{o_2}|$.

Hence $\dfrac{\mu R_{L_1}}{ra + R_{L_1} + (1+\mu)R_k{'}} = \dfrac{\mu R_{L_1}}{ra + R_{L_1} + (1+\mu)R_k{'}} \times \dfrac{(1+\mu)R_{L_2}}{ra + R_{L_2}}$

$$R_{L_1} = \frac{R_k{'} \times (1+\mu)R_{L_2}}{ra + R_{L_2}}$$

EQUIVALENT CIRCUITS AND LARGER SIGNAL CONSIDERATIONS 191

$$\therefore R_{L_1} = \frac{R_k \left(\frac{ra + R_{L_2}}{1 + \mu}\right)}{R_k \left(\frac{ra + R_{L_2}}{1 + \mu}\right)} \times \frac{(1 + \mu) R_{L_2}}{ra + R_{L_2}}$$

$$\therefore R_{L_1} = \frac{(1 + \mu) R_k \cdot R_{L_2}}{(1 + \mu) R_k + ra + R_{L_2}}$$

thus $\quad R_{L_1} = [(1 + \mu) R_k + ra + R_{L_2}] = (1 + \mu) R_k \cdot R_{L_2}$

and $\quad R_{L_1} = [(1 + \mu) R_k + ra] \quad\quad = R_{L_2}[(1 + \mu) R_k - R_{L_1}]$

therefore $\quad R_{L_1} = \dfrac{R_{L_1}[ra + (1 + \mu) R_k]}{(1 + \mu) R_k - R_{L_1}}$

Hence, provided $ra_1 = ra_2$ and $\mu_1 = \mu_2$ for equal amplitude outputs, R_{L_2} is determined by

$$\frac{R_{L_1}[ra + (1 + \mu) R_k]}{(1 + \mu) R_k - R_{L_1}}$$

once R_{L_1} is known or chosen.

The reason that the outputs are unbalanced when $R_{L_1} = R_{L_2}$ is that some of the V_1 anode current flows through R_k, and only a part of this useful signal current enters V_2 as its input signal current.

If R_K is made high enough in value, as in the case of a true long tailed pair, it then becomes possible to obtain almost equal outputs.

Figure 12.11.4. demonstrates this well.

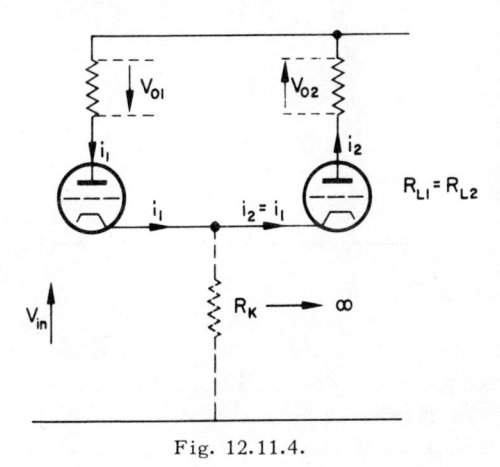

Fig. 12.11.4.

As R_k tends to infinity, there can be one current path only and therefore i_1 must equal i_2.

This is not possible in this circuit when a normal value cathode resistor is used, but if made

$$\gg \frac{ra + R_L}{1 + \mu}$$

output voltages within 1% – 2% of each other can be achieved. An approximate expression for the output, where $R_{L_1} = R_{L_2}$ is given as

$$\frac{1}{2}\left(\frac{\mu R_L}{ra + R_L}\right) \text{ provided } R_k \gg R_{in} \text{ of } V_2.$$

12.12. Long tailed pair approximations.

Let us now consider a practical down-to-earth design of a long tailed pair. We will confine our discussions to the design of d.c. conditions, as if these are not right, the circuit will not function from an a.c. point of view. We will make one basic assumption and by using Ohm's law, derive all of the component values. The circuit is shown in figure 12.12.1.

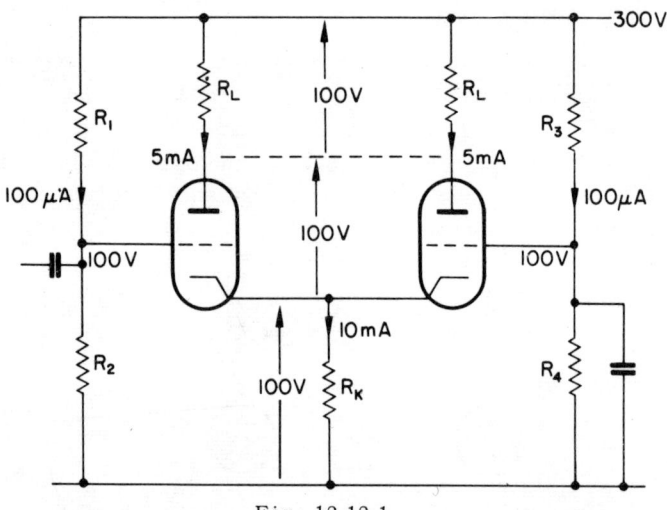

Fig. 12.12.1.

We would refer to the valve characteristics, decide upon a suitable operating point, note the V_{ak} and I_a, mark these and all other potentials on the circuit and complete the picture by applying Ohm's law.

Each valve is to have 100 V across anode to cathode. i.e. $V_{ak} = 100$ V. 5 mA is to flow through each valve. The operating point will have shown the bias we require, and if small, can be ignored. Larger valves requiring say

10 – 20 V bias should not be ignored, but let us assume that our valve in this example needs say 3 V. We will ignore it for the moment.

As we have 300 V H.T., and will use 100 V across each valve, there remains 200 V to be shared between the anode and cathode resistors.

Suppose we let these components have equal voltages, then 100 V will exist across the anode load, the cathode resistor and the valve itself.

The current flowing through R_k will be $2 \times 5\,\text{mA} = 10\,\text{mA}$. R_k will have a value of $100\,\text{V}/10\,\text{mA} = 10\,\text{K}\Omega$. Each R_k will have 100 V across them, and with 5 mA anode current, will need to be 20 KΩ. We will choose say 100μA. to flow down through the potential dividers so as to swamp any very tiny grid current. Each grid will be at the same potential as the cathode because in this example, we are assuming zero bias. The grids will have therefore 100 V at the junction of R_1 and R_2 and R_3 and R_4.

If we need 100 V across R_4 and a current through it of 100μA, then R_4 will be 1 MΩ. As there must be 200 V across R_3, its value must be 2 MΩ. This applies also to R_1 and R_2.

Let us now consider an analysis of a similar circuit. We will approach this in a similar manner, assuming zero bias, as before. We need to know all d.c. values. The circuit is shown in figure 12.12.2.

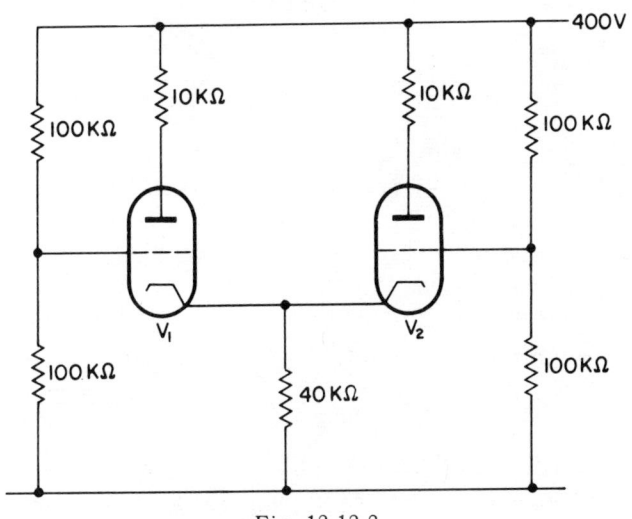

Fig. 12.12.2.

The grid voltage for both valves is given by

$$V_g = \frac{400 \times 100\,K\Omega}{200\,K\Omega} = 200\,\text{V}.$$

The cathode will be at 200 V also (when $V_{gk} = 0$). If we have 200 V across a 40 KΩ, R_k then the cathode currents in $R_1 = 200\,\text{V}/40\,\text{K}\Omega = 5\,\text{mA}$.

This 5 mA is shared between V_1 and V_2, hence there is 2.5 mA anode current per valve. 2.5 mA through the 10 KΩ anode load will produce a 25 V drop. The anode potential will therefore be 400 V − 25 V = 375 V. V_{ak} = 375 − 200 = 175 V. To sum up, $V_{g_1} = V_{g_2} = 200$ V. $V_{a_1} = V_{a_2} = 375$ V. $I_1 = I_2 = 2.5$ mA. $V_k = 200$ V.

Compare this with the next example, the circuit of which is shown in figure 12.12.3.

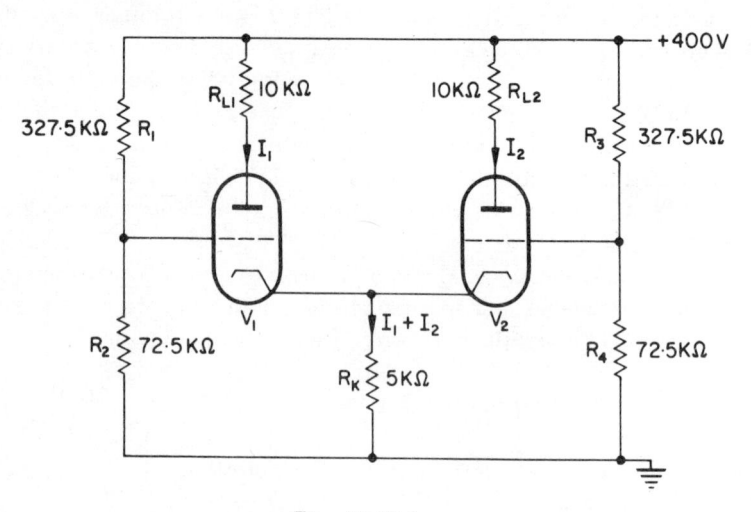

Fig. 12.12.3.

We will discuss two completely different methods of analysis on this occasion; one will be similar to that discussed during the previous example whilst the other method will be much more detailed and quite accurate as no assumptions are made for the bias.

We will compare the results of both analyses and give the reader an indication as to when he should use either method to his advantage.

12.13. Graphical analysis of a long tailed pair.

We will simplify the circuit in figure 12.12.3 a little first by removing V_2 and doubling the value of R_k so as to retain the same d.c. conditions for V_1. The simplified circuit is shown in figure 12.13.1.

The voltage at the grid is

$$\frac{400 \times 72.5 \text{ K}\Omega}{400 \text{ K}\Omega} = 72.5 \text{ V}.$$

Whatever we might do during this analysis, this grid voltage will remain constant. It will depend entirely upon the ratio of R_1 and R_2.

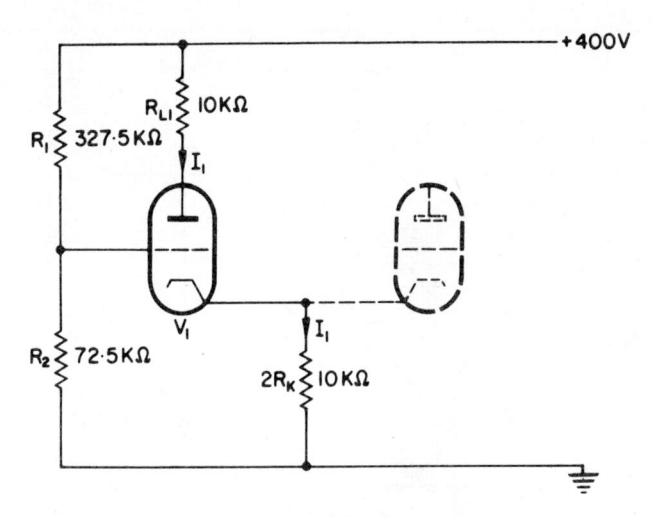

EQUIVALENT CIRCUITS AND LARGER SIGNAL CONSIDERATIONS

Fig. 12.13.1. Simplified version of figure 12.12.3.

The next step is to plot a d.c. load line for R_L and R_k, i.e., for 20 KΩ. on the I_a/V_{ak} graph shown in figure 12.13.2. for the valve we are using.

Having drawn the d.c. load line (points $A - B$) we need to establish the operating point, P. Before we can decide this, we need to draw a bias load line.

We cannot do this as we have done before becasue even at $V_{ak} = 0V$, there will still be some anode current flowing. A different technique is required. In other words, we cannot draw a line from the point $(0,0)$, as in previous examples.

Perhaps the most simple way in which to construct the line is to draw up a table as shown below. We will record in column 1, the fixed V_g of 72.5 V. In column 2, we will write in a number of assumed bias voltages, taken from the graph (although this is not essential, but it simplifies things).

The third column will contain the cathode voltages that must result from adding figures in the previous two columns. Finally the fourth column will contain the anode (cathode) current necessary to maintain V_k for the cathode resistor in this particular circuit.

V_g (Constant). (V)	Assumed bias. (V)	V_k . (V)	I_a (mA)
72.5	0 V	72.5 V	7.25 mA
72.5	2.5 V	75 V	7.5 mA
72.5	5.0 V	77.5 V	7.75 mA
72.5	7.5 V	80 V	8.0 mA

We need to plot now, a number of points having co-ordinates (I_a and V_{gk}). These points are shown on the I_a/V_{ak} graph. The points are then joined by

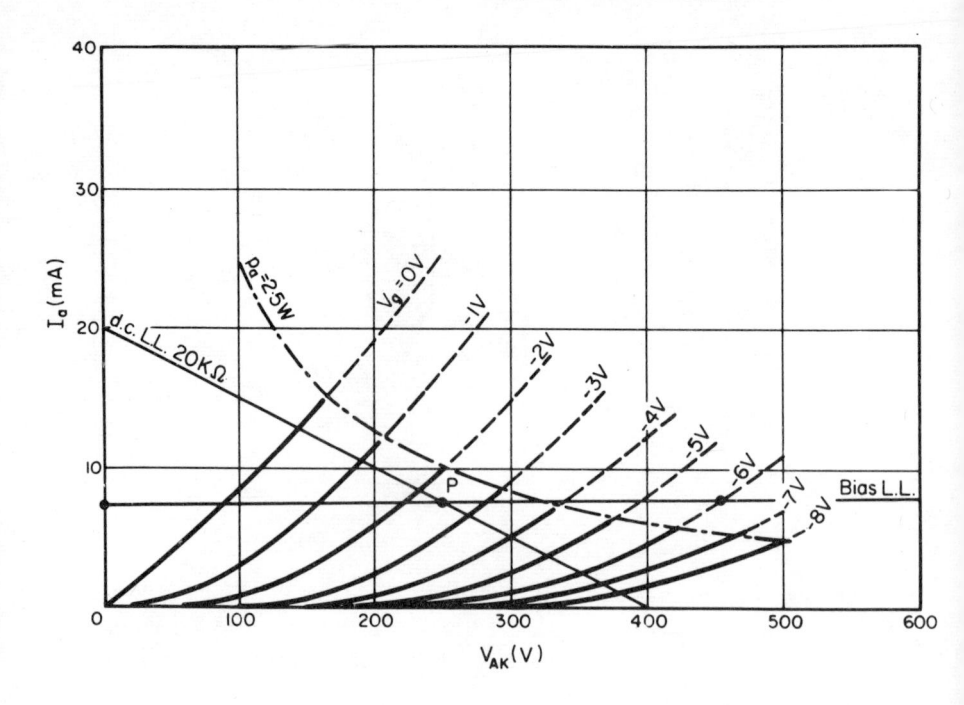

Fig. 12.13.2.

a continuous line; this will be pretty straight and as the reader will see for himself, is much more horizontal than previous bias load lines drawn, and will *not* start at (0,0).

The intersection of the bias and d.c. load lines establishes the operating point. The steady anode current is seen to be 7.75 mA. V_{ak} is 250 V. V_{ak} is 7.75 mA × 10 KΩ = 77.5 V. The bias, V_{gk} = 2.5 V. V_{RL} = 7.75 mA × 10 K = 77.5 V. V_a = 400 − 77.5 = 322.5 V.

V_k + V_{ak} + V_{RL} must sum to the full H.T. Putting in the values obtained, 77.5 + 250 + 77.5 = 400 V. V_{ak} = 250 V.

We must now return to the original circuit and replace V_2.

As the circuit is symmetrical, the voltages and currents will be *balanced* i.e., both anode currents will be identical, etc., and is shown in figure 12.13.3.

We should now compare the results in this example with the following values using the *zero bias* approach. Referring to figure 12.12.3. and by adopting the method previously discussed; V_g = 72.5 V. V_{gk} assumed to be zero. V_k = 72.5 V. Total **cathode current** = 72.5 V/5 K = 14.5mA. I_1 = I_2 = 7.25mA. V_{RL} for both valves = 7.25 × 10 = 72.5 V. V_{ak} = V_a − V_k = 327.5 − 72.5 = 255

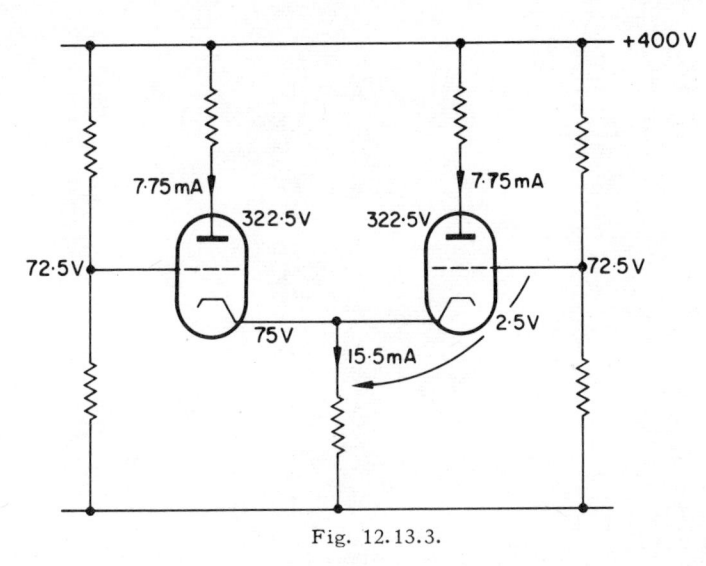

EQUIVALENT CIRCUITS AND LARGER SIGNAL CONSIDERATIONS 197

Fig. 12.13.3.

This error is approximately 2% and in practice may not matter overmuch, particularly if the voltmeter used has a 2% error at f.s.d. This 'quick' method is very useful for rapid fault-finding and enables a very good approximation to be obtained very rapidly. The graphical approach can only be as accurate as the actual valve characteristics and for really close work, the valve curves should be drawn individually for the valve in question when designing a given circuit.

CHAPTER 13

Linear analysis

13.1. Elementary concept of 'flow diagrams'.

The subject of flow diagrams can become very complicated. This chapter is included so as to give the technician some insight into the subject.

Only the most elementary circuits will be considered as it is not thought necessary for the technician to investigate the subject too deeply at this stage, although of course, the depth to which the individual reader will study, will be a matter for him to decide.

These diagrams provide a means of displaying by a simple drawing, the functions of many devices ranging from valve networks to servomechanism systems, etc.

They also allow the reader to derive formulae in an alternative manner than that shown in the previous chapter. The signal variations are also assumed to be very small in this chapter, thus the valve parameters are assumed to remain constant.

A circuit diagram for which a flow diagram is drawn is shown in figure 13.1.1. This is of course, a d.c. case by means of introduction.

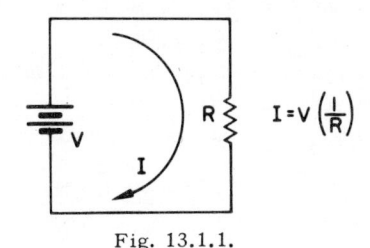

Fig. 13.1.1.

The signal flow diagram for this simple circuit is shown in figure 13.1.2.

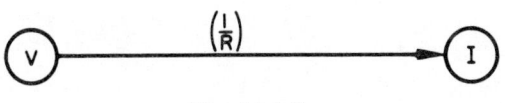

Fig. 13.1.2.

This tells us that the current (in the right hand circle) is due to the product of the voltage (in the left hand circle) and the term $1/R$ on the line representing the flow from left to right. The high potentials are on the left, and the end result on the right. Higher potentials go even further to the left. Signal flows from left to right and are positive going when doing so.

The direction of flow must be observed now that we have decided upon a convention.

If this is taken a stage further, a Triode valve may be introduced, figure 13.1.3. There is no anode or cathode load. It is intended to derive an expression for the anode current i_a. The following steps have been carefully chosen so as to build up the final signal flow diagram in easy stages.

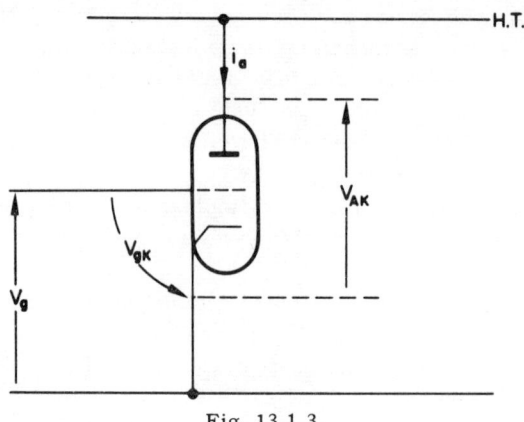

Fig. 13.1.3.

Consider the circuit shown. It is seen that the two external factors which determine the anode current, i_a, are (1) The H.T. (v_{ak} in this case) and (2) v_{gk} (v_g in this case). These then are the two factors available externally; the parameters of the valve must of course be considered as they too play an important role.

In the absence of a signal to the grid the anode current will be

$$i_a = \frac{v_{ak}}{ra}.$$

If however, a signal v_{gk} is applied the anode current will be modified. The anode current will be due to the sum of

$$\frac{v_{ak}}{ra} \text{ and } v_{gk} gm.$$

An expression for the anode current is

$$i_a = \frac{v_{ak}}{ra} + gm\, v_{gk} \quad \ldots \ldots (1)$$

The first step, then, is to draw a cricle and write inside the circle the term under investigation, which, in this case is i_a, figure 13.1.4.

LINEAR ANALYSIS USING FLOW DIAGRAMS

Fig. 13.1.4.

The next step is to draw a second circle and write inside, one of the external factors upon which, the i_a depends. Figure 13.1.5.

Fig. 13.1.5.

The next step is to connect both circles with an arrow, taking care to point the arrow in the correct direction. Figure 13.1.6.

Fig. 13.1.6.

The picture is not yet complete, however, as it must include the ra of the valve. Any term positioned on the line, is multiplied by the term in the left hand circle, i.e.

$$v_{ak} \times \frac{1}{ra} = i_a.$$

$1/ra$ must, therefore, be positioned on the line; this completes the anode circuit.

The story is not yet complete, allowance must be made for the signal in the grid circuit. This is allowed for by drawing a third circle as shown in figure 13.1.7.

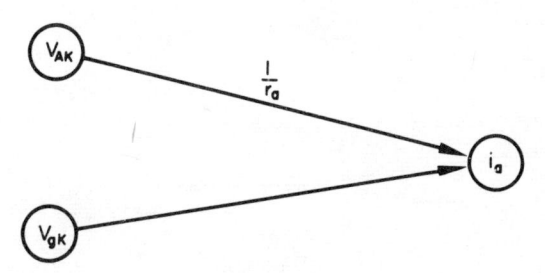

Fig. 13.1.7.

The effective grid cathode voltage, v_{gk}, times the gm, will cause further anode current to flow, consequently the parameter gm must be positioned on the second arrow. Figure 13.1.8.

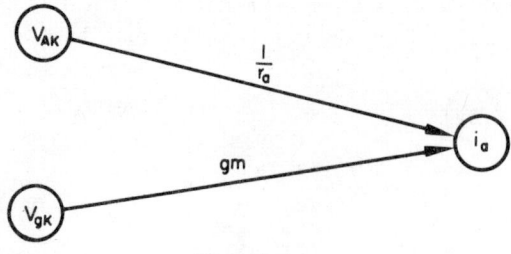

Fig. 13.1.8.

For this simple circuit, the flow diagram is complete. Figure 13.1.9.

$$i_a = \frac{V_{ak}}{r_a} + V_{gk} \cdot gm$$

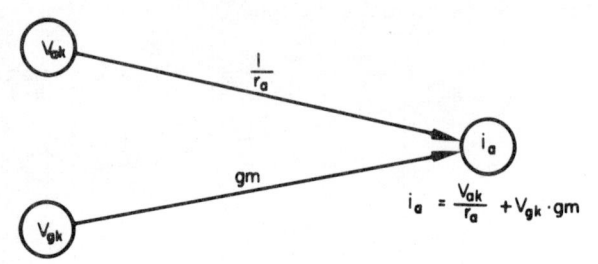

Fig. 13.1.9.

having obtained the expression for the anode current, a term may be taken out of the bracket. If gm is taken outside, the expression becomes,

$$i_a = gm\,\frac{v_{ak}}{\mu} + v_{gk}\ .$$

We might have decided to take the term $(1/ra)$ out of the bracket which would have given a third expression, $i_a = 1/ra\,(v_{ak} + \mu v_{gk})$.

The flow diagrams for both these expressions are given in figure 13.1.10.

$$i_a = gm\left(\frac{V_{ak}}{\mu} + V_{gk}\right)$$

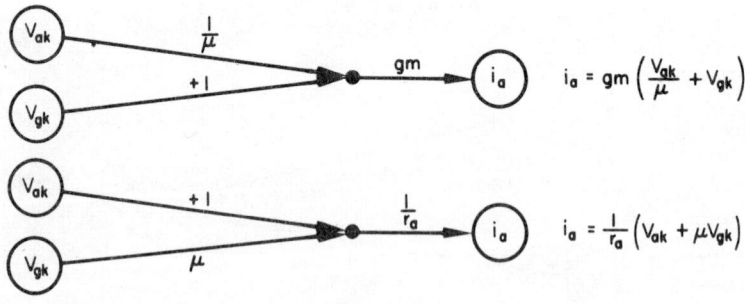

$$i_a = \frac{1}{r_a}\left(V_{ak} + \mu V_{gk}\right)$$

Fig. 13.1.10.

LINEAR ANALYSIS USING FLOW DIAGRAMS   203

Any of these expressions may be used although the first one perhaps, is the most common. The first expression will be used in the following pages. Note that the parameters gm and $1/ra$ are outside of the bracket. All terms within the bracket are to be multiplied by those outside of course. When a common multiplier is taken out of the bracket, the term will have a line to itself on the signal flow diagram. The appropriate arrow for the term is common to any expression to its left.

13.2. Simple amplifier with resistive anode load.

Let us now consider a Triode valve with a resistive anode load. Figure 13.2.1.

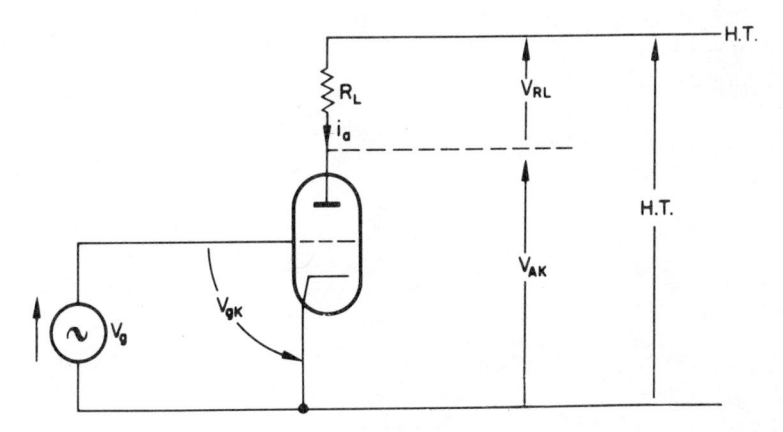

Fig. 13.2.1.

It is seen that the voltage v_{ak} is the H.T. less the voltage drop across the anode load. The voltage across the anode load is $i_a(R_L)$. ....(1)

Therefore the voltage v_{ak} = H.T. $- i_a R_L$ ..... (2). The anode current due to the anode circuit, is, therefore,

$$i_a = \frac{\text{H.T.} - i_a R_L}{ra}$$

This simplifies to $i_a = 1/ra \,(\text{H.T.} - i_a R_L)$ ..... (3).

The grid circuit remains as before. Note that there is a common term outside of the bracket, $1/ra$ and, following the same procedure as before, this term has an arrow to itself.

It is intended to begin to build the signal diagram, in stages, as before.

First, draw a circle representing the unknown, which in this instance, is the anode current, i_a. Figure 13.2.2.

(i_a)

Fig. 13.2.2.

The next step is to draw a circle for v_{ak}. Figure 13.2.3.

(V_AK)   (i_a)

Fig. 13.2.3.

Obviously i_a is the product of v_{ak} and $1/ra$. Therefore position $1/ra$ on the arrow. Figure 13.2.4.

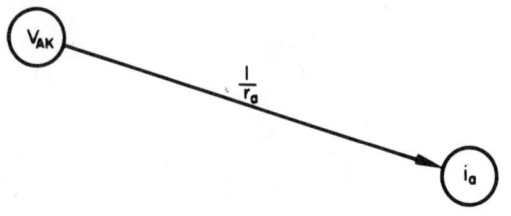

Fig. 13.2.4.

But the H.T. must be allowed for, and as this is at a higher potential than the circle, v_{ak}, draw another circle to the left of the one previously drawn for the potential, v_{ak}. Figure 13.2.5.

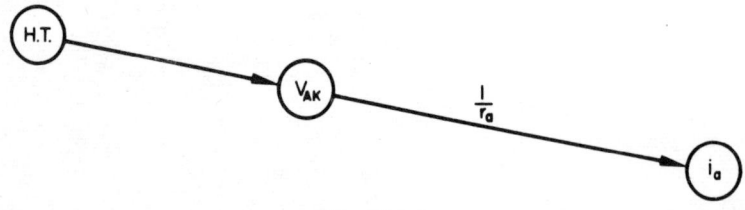

Fig. 13.2.5.

The product of a circle and the 'arrow' parameter must be added to the potential of the next circle. It is obvious that one must subtract $i_a R_L$ at the point v_{ak}, if the value of v_{ak} is to be valid. (From 2.)

This can be done quite easily by drawing an arrow from the circle i_a, back to the circle v_{ak}. Then position a term-R_L on the arrow, as a *voltage* is required to be fed back, *not a current*. Figure 13.2.6.

LINEAR ANALYSIS USING FLOW DIAGRAMS 205

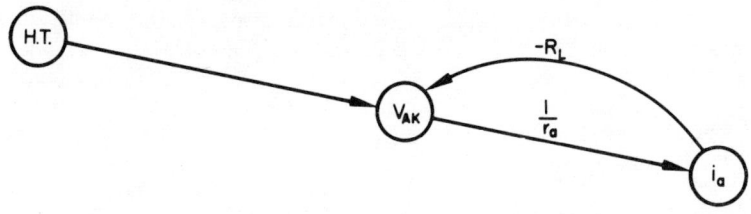

Fig. 13.2.6.

Although the arrow is in the reverse direction (away from the i_a circle) it is important to add the product of $i_a R_L$ to the v_{ak} circle, giving $i_a = \text{H.T.} - i_a R_L$. (The term $-R_L$ gives a negative voltage).

There is a need to identify the arrow leaving the H.T. circle; reference to the formula in (3) shows that the coefficient of H.T. is $1/ra$. Consequently we must position a 1 on the H.T. arrow, as the term $1/ra$ already exists between the two right hand circles. Figure 13.2.7.

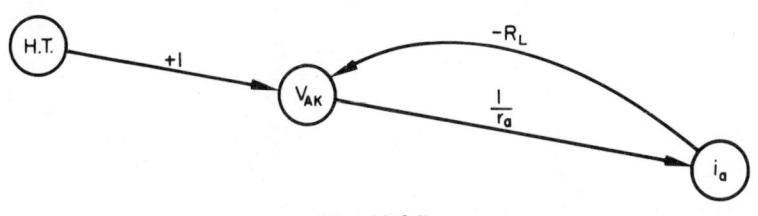

Fig. 13.2.7.

It now remains to draw a circle for the potential v_{gk} as before. Figure 13.2.8.

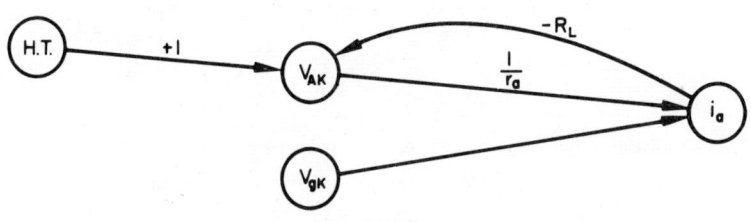

Fig. 13.2.8.

The arrow must be marked gm, as with the previous case. Figure 13.2.9.

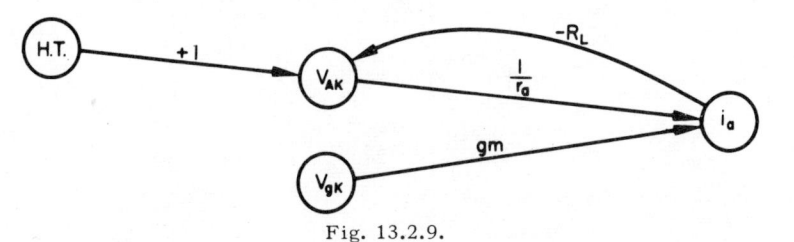

Fig. 13.2.9.

The last step for this circuit, is to draw an arrow for the input, v_g. As this is a voltage and is not modified in any way, the arrow is marked unity. Figure 13.2.10.

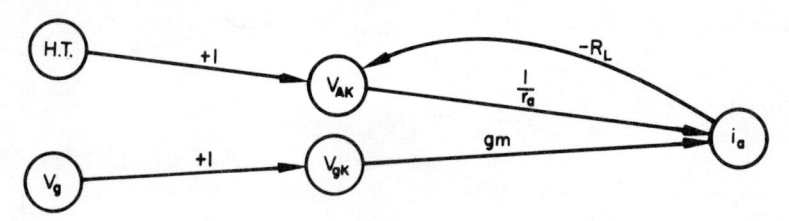

Fig. 13.2.10.

This, then, is the final signal flow diagram for the circuit given. Let us now examine each circle, one at a time.

v_{ak} $v_{ak} = $ H.T. $- i_a R_L$.

v_{gk} $v_{gk} = v_g$ (in this circuit only.)

i_a $i_a = \dfrac{v_{ak}}{ra} + gm V_{gk}$ and $i_a = \dfrac{\text{H.T.} - i_a R_L}{ra} + gm v_{gk}$

as the term, v_{ak} no longer appears in the expression, re-draw the signal flow diagram, as in figure 13.2.11.

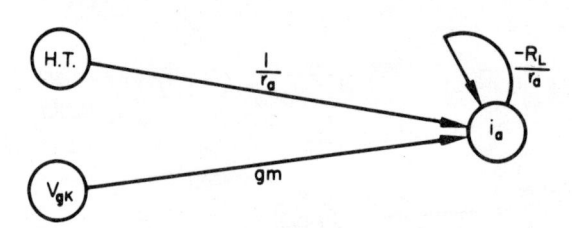

Fig. 13.2.11.

Upon examination of the diagram it can be seen that the expression for i_a is given by

$$i_a = \frac{\text{H.T.}}{ra} - \frac{R_L i_a}{ra} + gm V_{gk}.$$

By combining the terms containing i_a,

$$i_a \left(1 + \frac{R_L}{ra}\right) = \frac{\text{H.T.}}{ra} + gm V_{gk}$$

Hence, $i_a = \left(\dfrac{\text{H.T.}}{ra} + gm V_{gk}\right)\left(\dfrac{1}{1 + \dfrac{R_L}{ra}}\right)$

LINEAR ANALYSIS USING FLOW DIAGRAMS 207

The term in the second bracket is non dimensional, this is a very useful way of separating terms, as the term in the first bracket, will, for this circuit be constant, and the value of the anode current will be determined by the external component contained in the second bracket.

The signal flow diagram may be re-drawn as follows, in figure 13.2.12.

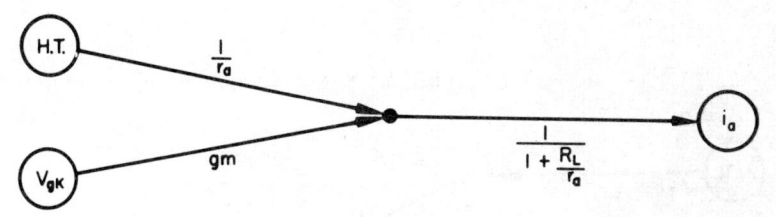

Fig. 13.2.12.

The dimensionless term

$$\frac{1}{1 + (R_L/ra)}$$

may be written as

$$\frac{1}{1 - (-R_L/ra)}$$

if desired as in this form, it is often useful and convenient.

Now consider the following circuit. This is a Triode cathode follower. It has an unbypassed cathode resistor, R_K. Figure 13.2.13.

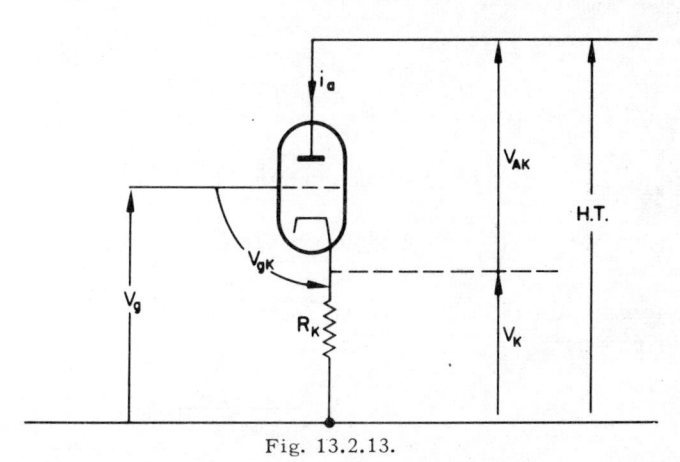

Fig. 13.2.13.

Part of the flow diagram is shown, in figure 13.2.14. No allowance has yet been made for the unbypassed cathode resistor, R_K.

The next step is to allow for the cathode potential v_k.

$$v_k = i_a R_K.$$

P

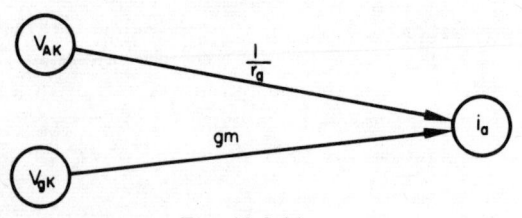

Fig. 13.2.14

Figure 13.2.15. shows the full signal flow diagram including the cathode potential circle.

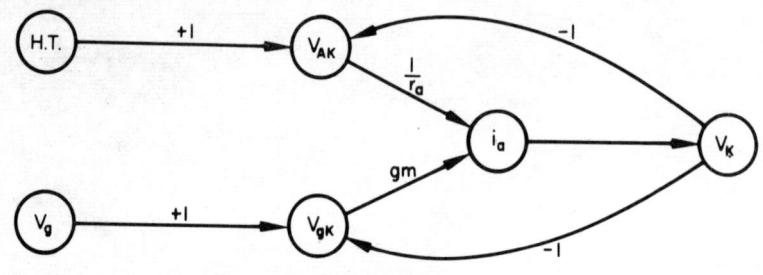

Fig. 13.2.15.

$v_k = i_a R_K$. $v_{ak} =$ H.T. $- v_k$. $\therefore v_{gk} = v_g - v_k = v_g - i_a R_K$.
Substituting these expressions in formula (1), then from

$$i_a = \frac{v_{ak}}{ra} + gm\, v_{gk} = \frac{\text{H.T.}}{ra} - R_K \frac{i_a}{ra} + gm\, v_g - i_a R_K gm$$

once more, collecting i_a terms,

$$i_a \left(1 + \frac{R_K}{ra} + gm R_K\right) = \frac{\text{H.T.}}{ra} + gm\, v_g.$$

Therefore the expression for the anode current is

$$i_a = \left(\frac{\text{H.T.}}{ra} + gm\, v_g\right) \left(\frac{1}{1 + R_K(1/ra + gm)}\right)$$

If the terms in the right hand bracket are multiplied, by ra, the more familiar expression,

$$i_a = \left(\frac{\text{H.T.}}{ra} + gm\, V_g\right) \left(\frac{ra}{ra + R_K(1 + \mu)}\right)$$

is obtained (where $gm \times ra = \mu$).

Upon expanding the expression

$$i_a = \frac{\text{H.T.} + \mu V_g}{ra + R_K(1 + \mu)}$$

LINEAR ANALYSIS USING FLOW DIAGRAMS

The final circuit to be examined is that of a phase splitter, figure 13.2.16.

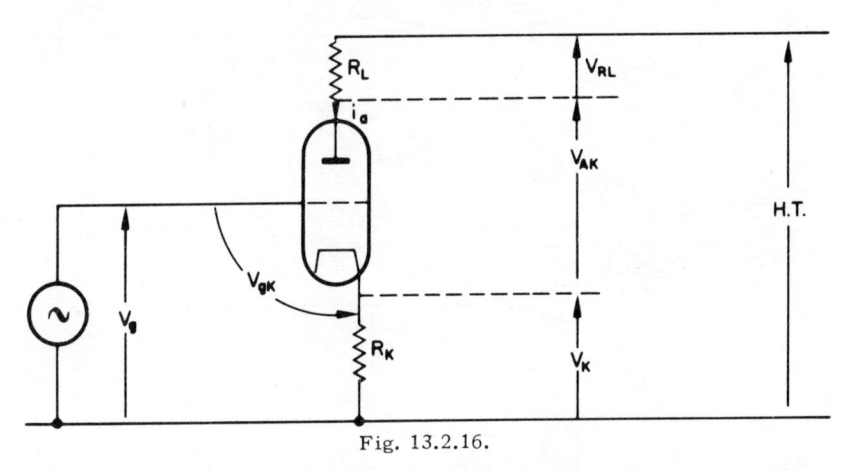

Fig. 13.2.16.

The signal flow diagram is very similar to the one considered previously, except that allowance must be made for the voltage drop across both the anode and cathode loads. This is shown in figure 13.2.17.

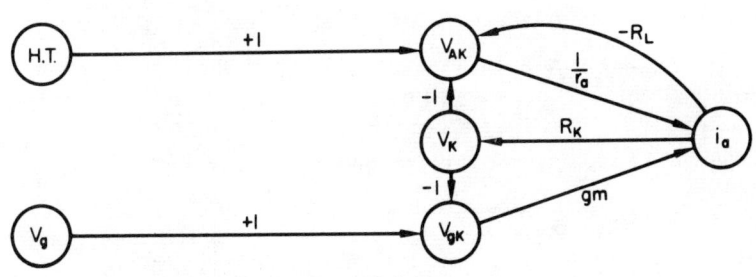

Fig. 13.2.17.

$v_{gk} = v_g - v_k$ $\therefore = v_g - i_a R_K$ and $v_{ak} = $ H.T. $- i_a(R_K + R_L) = $ H.T. $- v_k - v_o$

(where v_o is the output taken from the anode).

from

$$i_a = \frac{v_{ak}}{ra} + v_{gk}, \quad i_a = \frac{\text{H.T.}}{ra} - i_a\left(\frac{R_K + R_L}{ra}\right) gm v_g - i_a gm R_K$$

hence

$$i_a = \left(\frac{\text{H.T.}}{ra} + gm v_g\right)\left(\frac{1}{1 + R_L/ra + R_K/ra + gm R_K}\right)$$

The simplified signal flow diagram is shown in figure 13.2.18.
Which further simplifies to the diagram shown in figure 13.2.19.

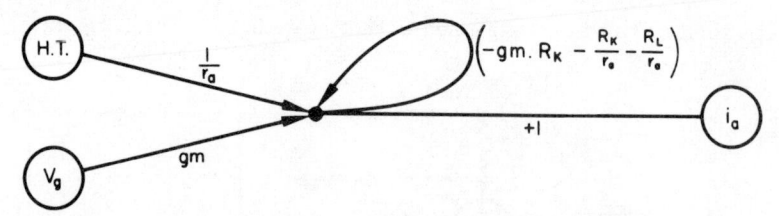

Fig. 13.2.18.

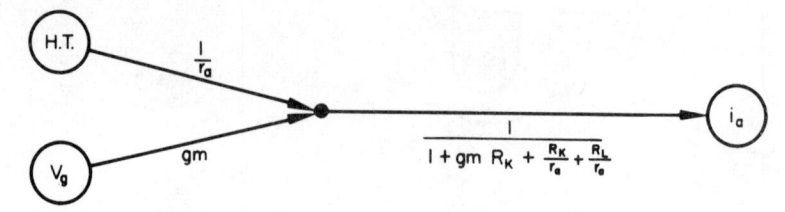

Fig. 13.2.19.

This has been a simple introduction to 'linear analysis' techniques.
Using these techniques, investigation of one or two circuits will be undertaken in order to attempt to consolidate the position reached, so far.

13.3. A linear analysis of a clipper stage.

The following analysis assumes that all valve parameters are linear over the working range. A formula will be derived and subsequently used during the analysis.

Consider a simple triode valve as shown in figure 13.3.1.

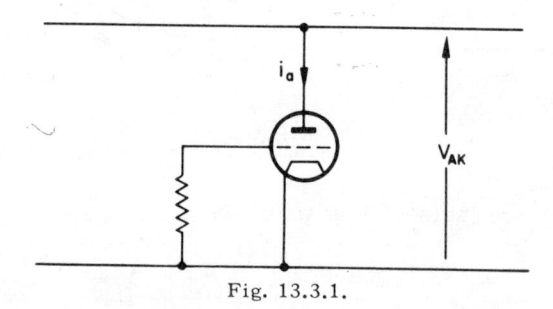

Fig. 13.3.1.

The anode current due to the anode voltage only in the absence of an input signal, may be expressed as

$$i_a = \frac{v_{ak}}{r_a}$$

When a signal input is applied to the grid, as in figure 13.3.2, the total anode current is given as

$$i_a = \frac{v_{ak}}{ra} + gm\, V_{gk}.$$

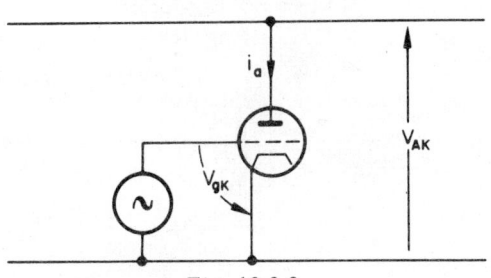

Fig. 13.3.2.

Finally, figure 13.3.3. shows a further development of the circuit which now includes a cathode resistor R_K.

Fig. 13.3.3.

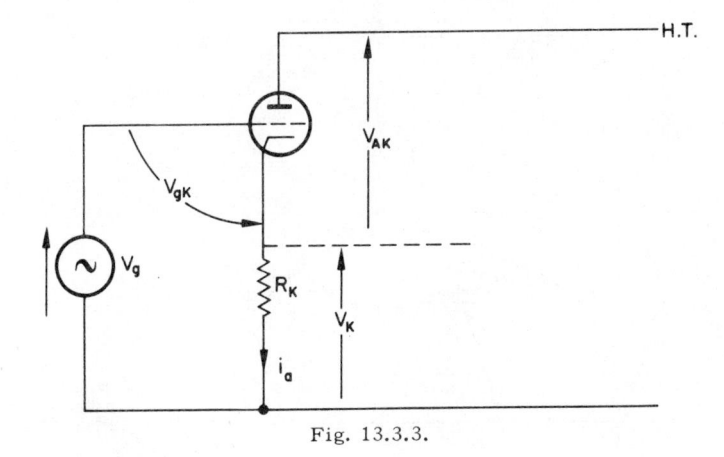

The current may be expressed as

$$i_a = \frac{v_{ak}}{ra} + gm\, v_{gk}$$

as before, but $v_{gk} = v_g - i_R{}_K$. Also $v_k = i_a R_k$ and $V_{ak} = $ H.T. $- v_k$. Hence, substituting these terms into the expression for the anode current,

$$i_a = \frac{\text{H.T.} - i_a R_k}{ra} + gm(v_g - i_a R_k)$$

This is the formula we shall use during our analysis of the clipper stage. Let us consider the clipper circuit shown in figure 13.3.4. The input signal is represented by the generator, e. This applies a sinusoidal input to the grid of V_1 and, if the amplitude is correct, will cause an output signal to appear that will be a 'squared off' version of the input waveform. The generator is isolated from a d.c. point of view, from the 100 V potential existing at the grid of V_1, by the input capacitor.

The signal e, generated will be algebraically added to the steady 100 V and this sum will become the effective input to V_1 grid.

The effective input signal to V_1 grid is shown as V_g in the figure.

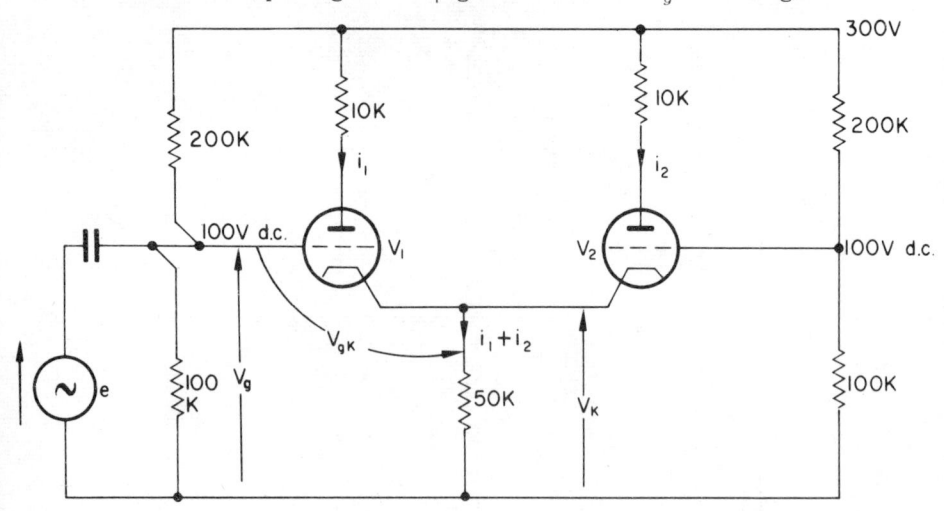

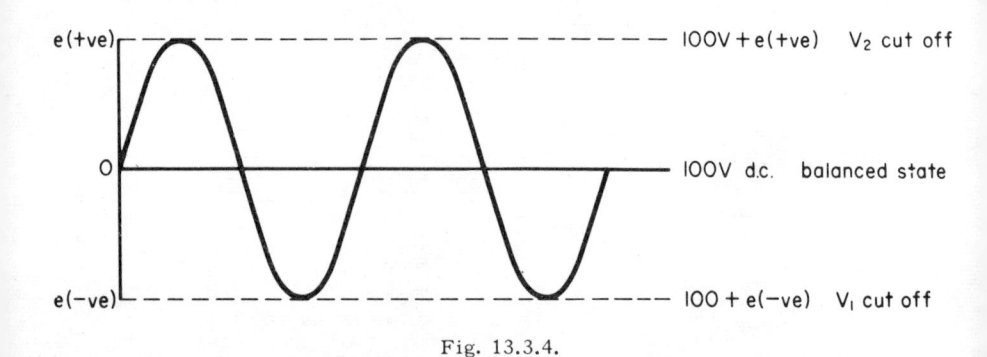

Fig. 13.3.4.

Each valve has a gm of 2 mA/V, an ra of 10 KΩ and a μ of 20. We will assume that for this discussion, these parameters remain constant throughout.

The grid of V_1 will experience a change in level of 100 ± the input e from the generator. When the generator passes through its zero point, the grid will experience a 100 V input at that instant.

The circuit will at that point, be balanced, and both anode currents will be of the same magnitude. When the generator is at its most positive peak, V_1 will be 'hard on' and V_2 will be cut off.

When the generator is at its most negative peak, V_1 will be cut off and V_2 will be conducting.

There are then, three levels of V_g we need to discuss, they are;
  (1) The balanced state where $i_1 = i_2$.
  (2) The level of input e to cause cut off of V_1.
  (3) The level of input e to cause V_2 to be cut off.

We will assume that the circuit is symmetrical and that all components are accurate and have no tolerances, i.e., the complete circuit is balanced at the instant V_1 grid has an instantaneous value equal to the steady d.c. potential at V_2 grid, i.e., 100 V.

Establishing i_1 and i_2 when the circuit is balanced.

We have previously discussed a method of analysing a balanced L.T.P. circuit. We 'remove' one valve, double the cathode resistor, R_k to allow for the current that would flow due to the valve we have 'removed', and examine the remaining valve.

Suppose that we 'remove' V_2 and change the value of R_k to $2R_k$. We can then establish i_1. Using the derived formula,

$$i_1 = \frac{\text{H.T.} - i_a R_k}{ra} + gm(v_g - i_a R_k)$$

but, to allow for V_2 current, R_k must be doubled, hence the formula becomes

$$i_1 = \frac{(\text{H.T.} - \text{drops across } R_{L1} \text{ and } 2R_k)}{ra} + gm(V_g - \text{drop across } 2R_k)$$

$$i_1 = \frac{300 - 110 i_1}{10} + 2(100 - 100 i_1)$$

$$10 i_1 = 300 - 100 i_1 + 20(100 - 100 i_1)$$

$$10 i_1 = 2300 - 2100 i_1$$

$$i_1 = 230 - 210 i_1$$

$$i_1 = i_2 = \frac{230}{211} = 1.09 \text{ mA in the balanced state.}$$

Having established i_1 and i_2 we can 'replace' V_2 and restore R_k to its original value.

Establishing i_2 when V_1 is cut off.

With V_1 cut off, V_2 would pass an anode current approximately equal to the sum of both anode currents when both valves are conducting. The next step is to determine the actual value of i_2, with V_1 cut off.

$$i_2 = \frac{300 - 60i_2}{10} + 2(100 - 50i_2)$$

$$i_2 = 30 - 6i_2 + 200 - 100i_2$$

$$i_2 = \frac{230}{107} = 2.15 \text{ mA}.$$

Input required to cut off V_1 considered alone.

The input required to cut off V_1 is given as

$$0 = 2\left(v_{gk} + \frac{v_{ak}}{ra}\right) \text{ as } i_1 = 0.$$

$$0 = 2v_{gk} + \frac{300}{10} - 2v_{gk} = 30 \text{ hence}$$

the required V_{gk} to cut off $V_1 = -15$ V. When considered alone, V_1 is a cathode follower.

Input required to cut off V_1 with V_2 conducting.

This -15 V however, does not allow for other circuit components. A cathode voltage will exist due to the anode current of V_2.
The anode potential of V_1 will be at 300 V, as $i_1 = 0$, hence the input, v_g, required to cut off V_1 is given as

$$0 = 2v_{gk} + \frac{(300 - v_k)}{ra}$$

$$0 = 2v_{gk} + (30 - 10.75) \quad v_{gk} = -9.63 \text{ V}$$

We have v_{gk}, but of course need v_g, only. Hence the effective input signal

$$v_g = v_k - v_{gk} = 107.5 - 9.63 = \underline{97.87 \text{ V}.}$$

Therefore $v_g = 97.87$ V to cut off V_1 anode current in order to cause clipping in one direction. We need now to determine the value of v_g that will drive V_1 hard on, cutting off V_2, thus clipping in the reverse direction.

Input required to cause V_1 to be hard on, cutting off V_2.

If V_1 is hard on, then V_2 will be cut off. i_2 will be zero. When V_1 is on, i_1 will be flowing and as i_2 will be zero, $i_1 = i_k$.

With V_2 off, $i_2 = 0$. Hence

$$0 = 2(100 - v_k) + \frac{(300 - v_k)}{10}$$

and multiply both sides by 10,

$$0 = 20(100 - v_k) + (300 - v_k)$$

$$0 = 2000 - 20v_k + 300 - v_k$$

$$\therefore \quad 21v_k = 2300$$

and

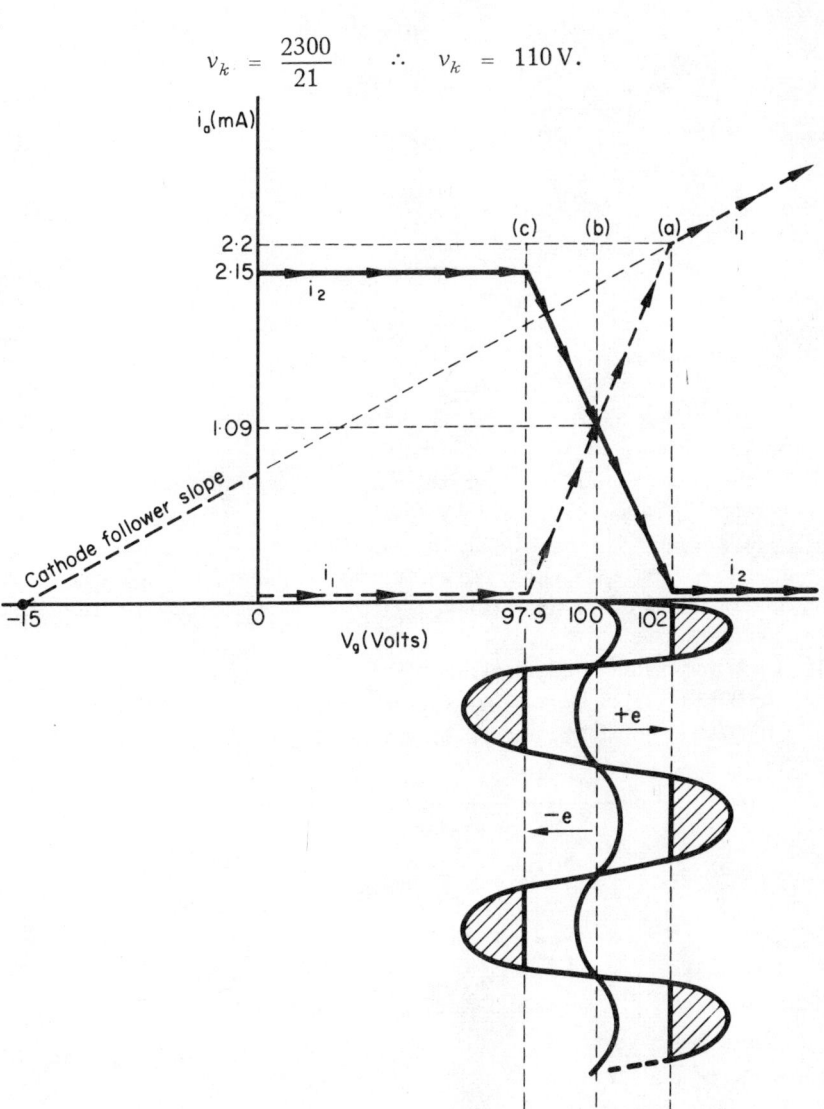

$$v_k = \frac{2300}{21} \quad \therefore \quad v_k = 110\,\text{V}.$$

Fig. 13.3.5.

$$i_k = \frac{v_k}{R_k} = \frac{110\,\text{V}}{50\,\text{K}} = 2.2\,\text{mA}.$$

But $i_k = i_1$ as V_1 only is conducting and we now require the value of v_g (for V_1) to cause V_1 to conduct heavily enough to cut off V_2.

Hence,
$$i_1 = \frac{300 - 60\,i_1}{10} + 2(v_g - 50\,i_1)$$

$$\therefore\ 10\,i_1 = 300 - 60\,i_1 + 20(v_g - 50\,i_1)$$

$$\therefore\ 10\,i_1 = 300 - 60\,i_1 + 20\,v_g - 1000\,i_1$$

$$\therefore\ v_g = \frac{1070\,i_1 - 300}{20} \text{ and as } i_1 = 2.2\,\text{mA},$$

$$v_g = \frac{1070\,(2.2) - 300}{20} = 102\,\text{V}.$$

Hence a minimum v_g of 102 V is necessary to drive V_1 on sufficiently to cut V_2 off.

A graph showing the values, including those in the transitional period, is given in figure 13.3.5.

Hence, for clipping to occur, the external sinusoidal input from the generator, e, is approximately 4 V P–P.

The greater the signal input, the more square the output signal. With a signal less than ± 2 V peak, the circuit will behave as an amplifier.

The changeover, or transitional period, shows the state of both valves in figure 13.3.5.

(a) With an input causing V_1 to conduct heavily cutting off V_2.
(b) With no input from the generator, leaving the circuit balanced.
(c) With an input causing V_1 to be cut off, switching V_2 on.

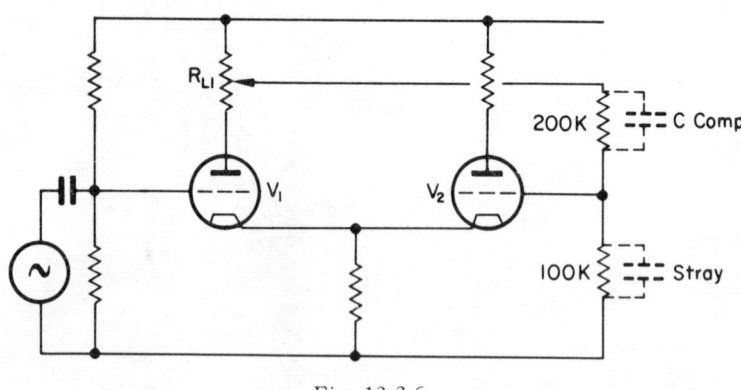

Fig. 13.3.6.

The steeper the characteristics, during the transition, the squarer the output. The Clipper can be converted into a Schmitt trigger circuit by causing positive feedback as shown in figure 13.3.6.

Positive feedback is achieved by connecting the 200 K in V_2 grid circuit to the anode of V_1. If we replace R_{L_1} by a potentiometer as shown, and connecting the 200 K to the slider of the potentiometer, the desired amount of positive feedback can be obtained.

The cathode remains substantially constant and is therefore 'earthy' and due to the Miller effect in V_2, a stray capacity will exist across the 100 K grid leak of V_2. This capacity is detrimental to the output pulse and the effects can be reduced by compensating the divider by connecting a capacitor C comp across the 200 K as shown in the figure. The method of calculating the value of this capacitor is given in 18.12. The anode of V_1 will 'see' this capacitance and V_1 anode circuit output impedance should be as low as possible. R_{L_1} should therefore be as low value as possible.

The transitional characteristics become, not only vertical, as in figure 13.3.7., but in fact, become slightly negative as in figure 13.3.8.

The dotted line shown during period x, is theoretical only. The line will be curved due to the valve.

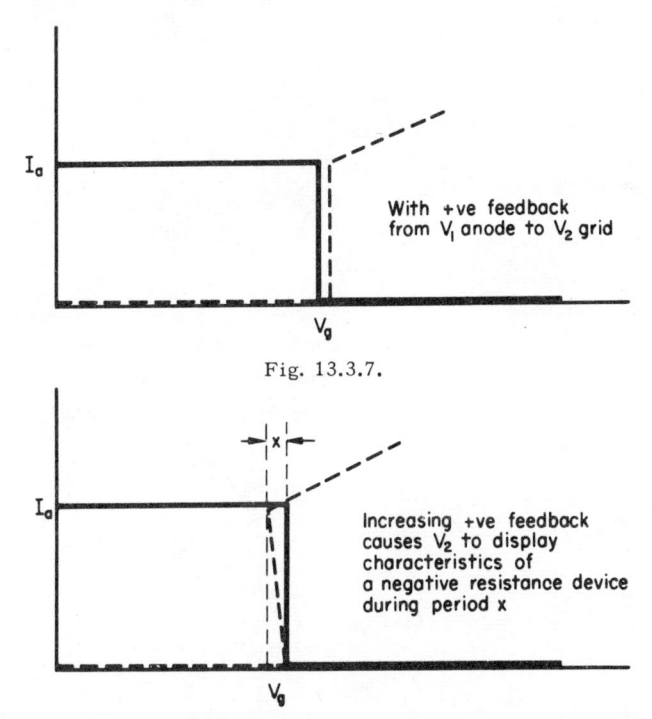

Fig. 13.3.7.

Fig. 13.3.8.

Figure 13.3.9. shows the inherent 'backlash' in the Schmitt Trigger circuit.

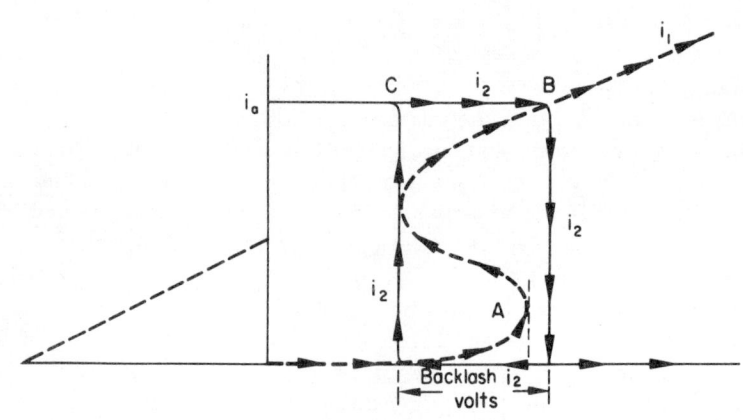

Fig. 13.3.9.

When reaching point A, i_1 jumps to point B, which is the other stable state. There is no control during the unstable period and examination of unstable point is not possible. The current i_1 returns on the same path as shown dotted in the figure.

The current path for i_2 is shown in solid lines and may be seen to traverse two distinctly separate paths for its forward and return journey. The distance between points $B-C$ is the measure of backlash.

Backlash of 5–7 V is a common value.

An alternative method.

An alternative method for determining the peak input signals to cause the clipper to produce a squared off version of the sinusoidal input is as follows. It assumes that R_k is many many times greater than the input resistance to the cathode of V_2.

Assume $I_1 = I_2$ and assume zero bias. With $R_k = 50\,\text{K}\Omega$, $V_k = 100\,\text{V}$.
Hence $I_k = 2\,\text{mA}$ $I_1 = I_2 = 1\,\text{mA}$.
With 1 mA flowing through R_L, $V_a = 300 - 10 = 290\,\text{V}$.
Hence $V_{ak} = 290 - 100 = 190\,\text{V}$.
From

$$I = \frac{V_{ak}}{ra} + gm\,V_{gk}, \qquad I = \frac{190}{10} + 2V_{gk}.$$

$$I = 19 + 2V_{gk} \text{ hence } V_{gk} = -\frac{18}{2} = -9\,\text{V}.$$

as $V_g = 100\,\text{V}$, and $V_{gk} = -9$, then $V_k = 109\,\text{V}$.

Thus $\quad I_1 + I_2 = \dfrac{109\,V}{50\,K} = 2.18\,mA.$

Therefore $\quad I_1 = \dfrac{2.18}{2} = 1.09\,mA.$

Hence the assumption of zero bias, and the initial assumed current of 1 mA, is taken care of during the subsequent working. The current is seen to be 1.09 mA as with the previous example. When

$$R_k \gg \dfrac{ra + R_L}{1 + \mu}$$

the gain of each anode

$$\doteq \dfrac{1}{2}\dfrac{\mu R_L}{ra + R_L} \doteq \dfrac{1}{2}\dfrac{20}{20}\dfrac{10}{} \doteq 5.$$

the maximum positive voltage excursion of each anode = 1.09 mA × 10 KΩ = 10.9 V.

Hence the input e, required

$$= \dfrac{10.9\,V}{5} = 2.18\,V\,pk.$$

Therefore the solution becomes,

$$V_g = \underline{100 \pm 2.18\,V} = \underline{97.82\,V \text{ and } 102.18\,V.}$$

This is a most useful approach when approximate answers suffice.

CHAPTER 14

Pulse techniques

Sinewaves have been dealt with in the preceding chapters. In pulse technique, sinewaves are often replaced by rectangular pulses. Some common waveforms are shown in figure 14.1.1.

Square wave

Rectangular pulse

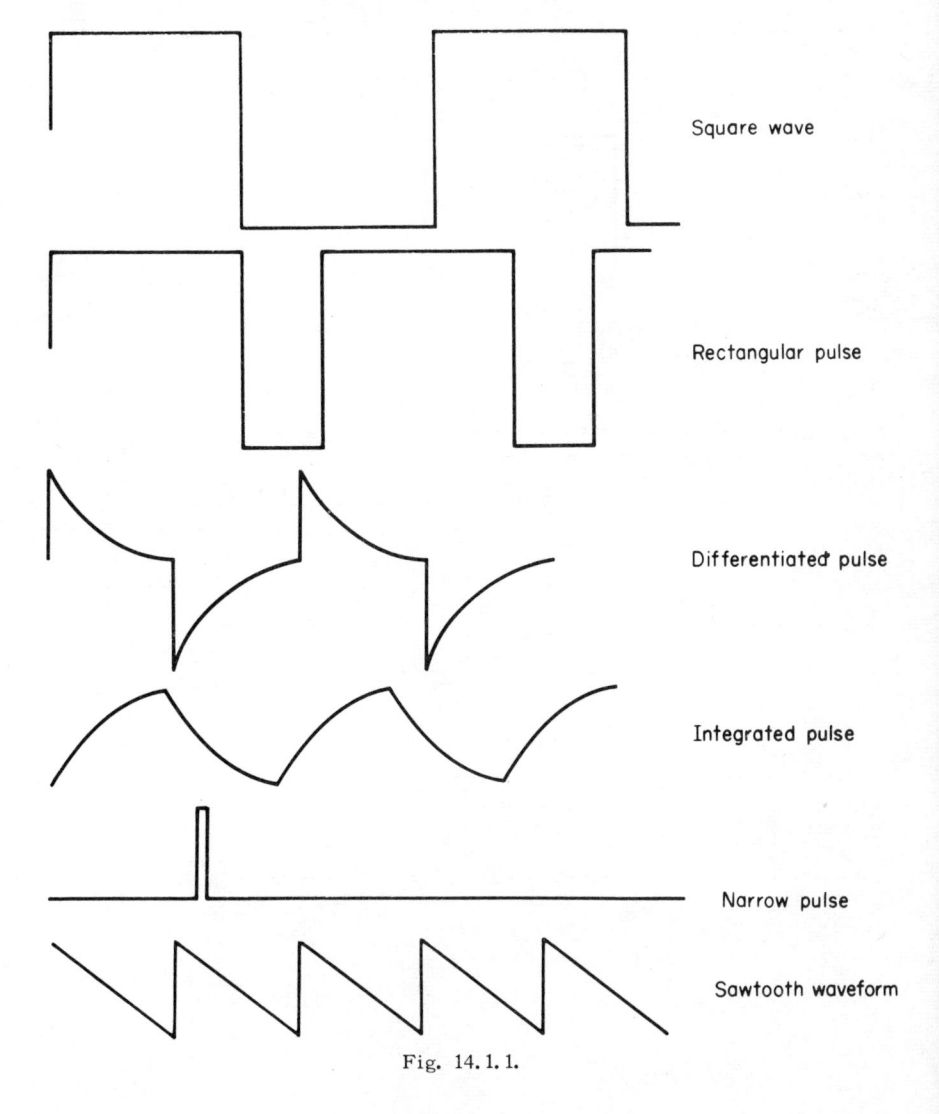

Differentiated pulse

Integrated pulse

Narrow pulse

Sawtooth waveform

Fig. 14.1.1.

ELECTRONICS FOR TECHNICIAN ENGINEERS

Step function.
(a rapid change of level.)

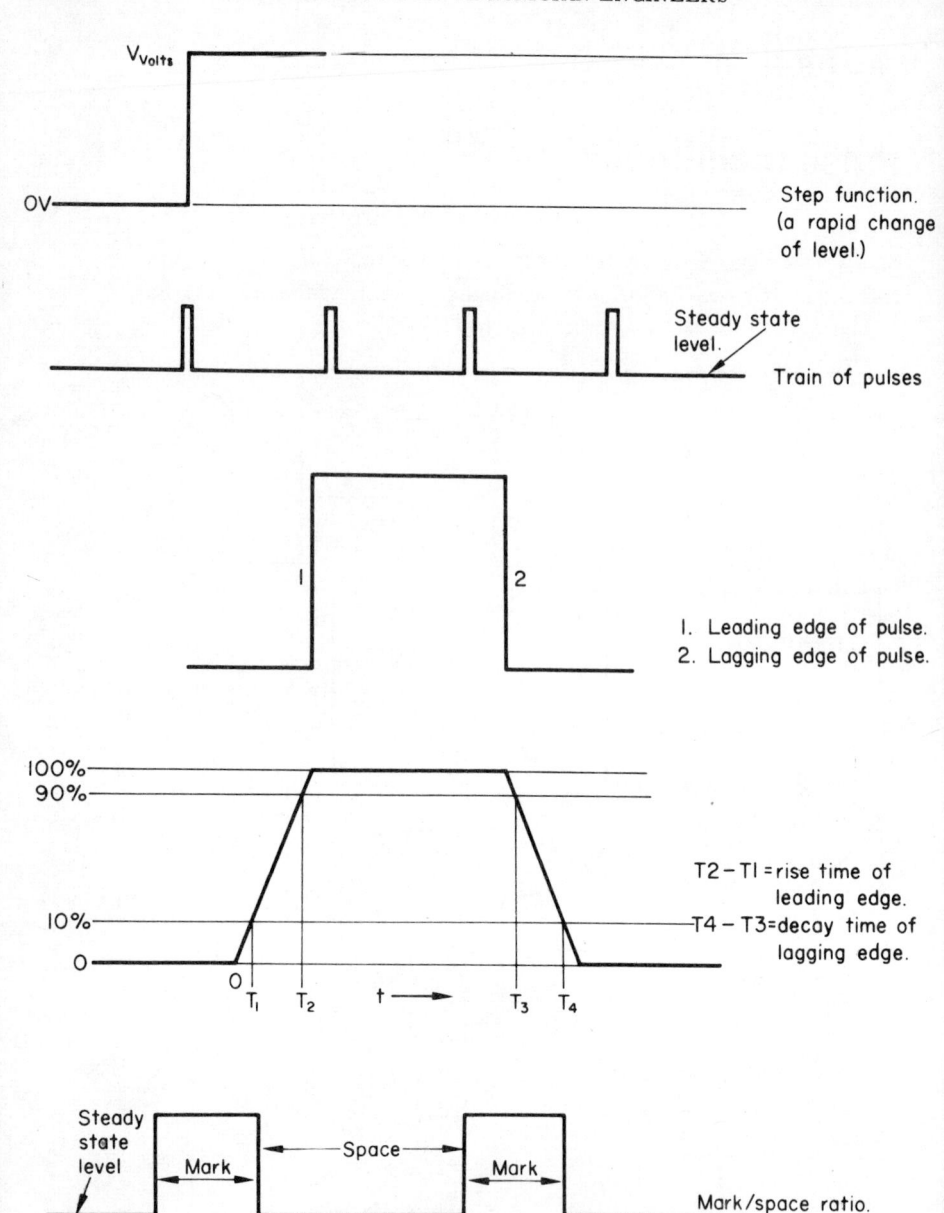

Steady state level.
Train of pulses

1. Leading edge of pulse.
2. Lagging edge of pulse.

$T_2 - T_1$ = rise time of leading edge.
$T_4 - T_3$ = decay time of lagging edge.

Mark/space ratio.

Fig. 14.1.1.

PULSE TECHNIQUES, MULTIVIBRATORS AND SAWTOOTH GENERATORS 223

14.2. Step function inputs applied to C.R. networks.

Figure 14.2.1. shows a step function voltage input. This input is assumed to rise from 0 to V volts instantaneously at $t = 0$.

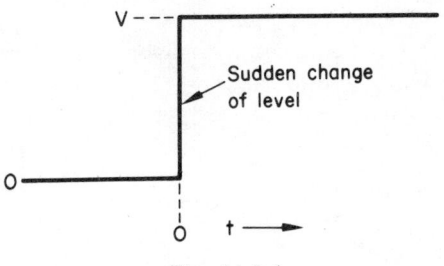

Fig. 14.2.1.

A capacitor will try to resist any voltage changes across itself and, in figure 14.2.2., when V volts are applied, and with the capacitor initially uncharged, current will flow at an amplitude given by $I = V/R$. This current flows at $t = 0$ and a potential begins to build up across C.

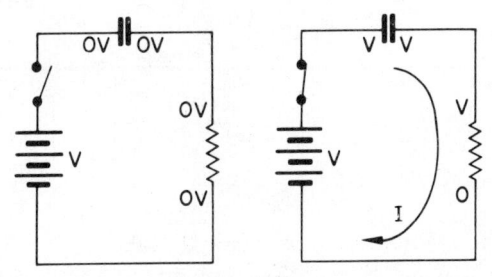

Fig. 14.2.2.

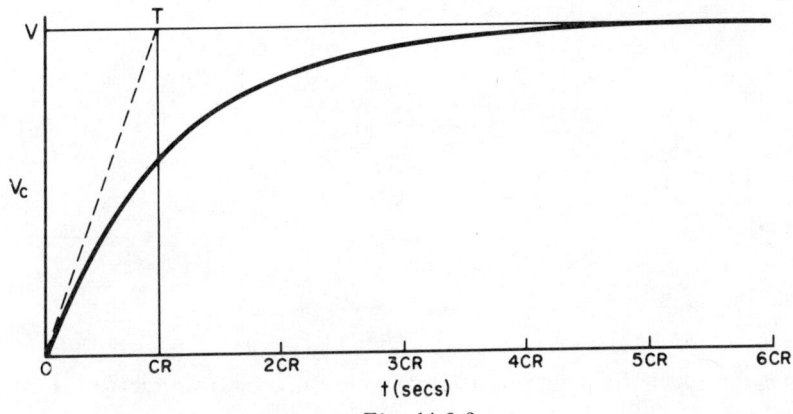

Fig. 14.2.3.

Q

The potential across C builds up as shown in figure 14.2.3. If the initial slope of the curve had remained constant, the potential across C would have reached maximum at point T in time. This point is known as the time constant of the circuit, $C.R.$, seconds.

The curve is that of the expression $V_c = V(1 - e^{-t/C_R})$ and maximum voltage V would be reached at $t = \infty$.

The potential across the capacitor changes little after $t = 5\,C.R.$ and we are justified in assuming that for most practical purposes, $V_c = V$ at $5\,C_R$ seconds after closing the switch at $t = 0$.

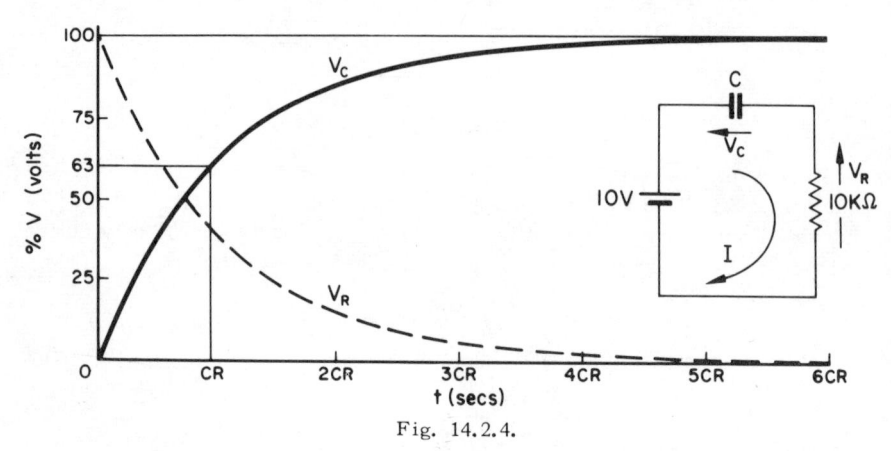

Fig. 14.2.4.

Therefore at $t = 0$, $V_c = 0$, $V_R = 10V$, $I = 1mA$. At $t \doteq 5\,C.R.$, $V_c = 10V$, $V_R = 0V$, $I = 0$.

A general case is illustrated in figure 14.2.4.

Note that at $t = C.R.$, $V_c = 63\%$ of the voltage existing across R at $t = 0$. (Figure 14.2.4.).

Examine the circuit in figure 14.2.5.

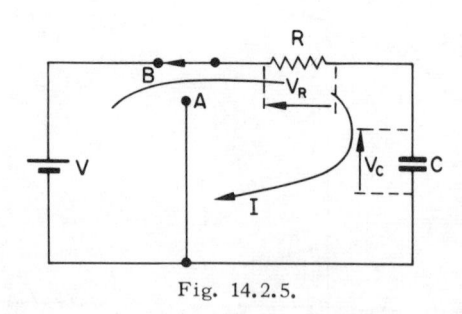

Fig. 14.2.5.

With an input changing from $0V$ to V volts at $t = 0$, (switch from A to B) $V_R = V$, $V_c = 0$.

Using Kirchoff's laws,

$$V = V_R + V_C.$$
$$V = V + 0$$

and this satisfies the equation. If however after a period 5 $C.R.$, the switch is changed from B to A then Figure 14.2.6. results.

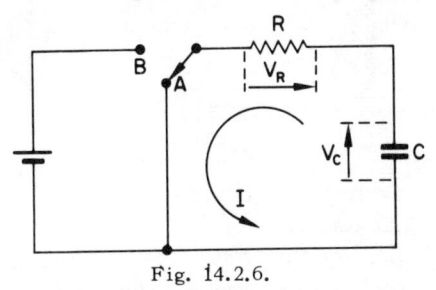

Fig. 14.2.6.

C, now fully charged to V volts, discharges current through R as shown. Using Kirchoff's laws:

$$0 = V_R + V_c$$
$$V_c = -V_R$$

It is clear, then, that $V_R = V_c$, but is negative. Suppose we apply a square wave as shown in figure 14.2.7. equivalent to the battery and switch in figure 14.2.5. except that we switch from A to B several times at regular intervals.

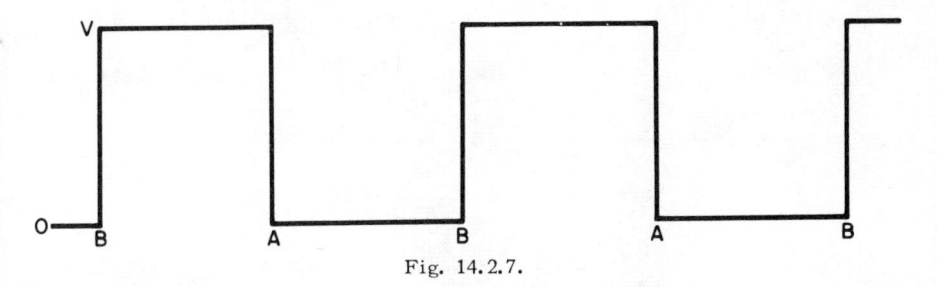

Fig. 14.2.7.

The shape of the waveform V_c is known as an integrated waveform; the shape of the waveform V_R is known as a **differentiated waveform**. The input waveform has a mean level given by the dotted lines. Figure 14.2.8. The position of the dotted lines is such that the area above the line is equal to the area below the line (area 1 = area 2). The V_c waveform also has a mean level for the same reason. The V_R (Shaded) waveform has no mean level, because area 2 is negative and exactly equals area 1, which is positive.

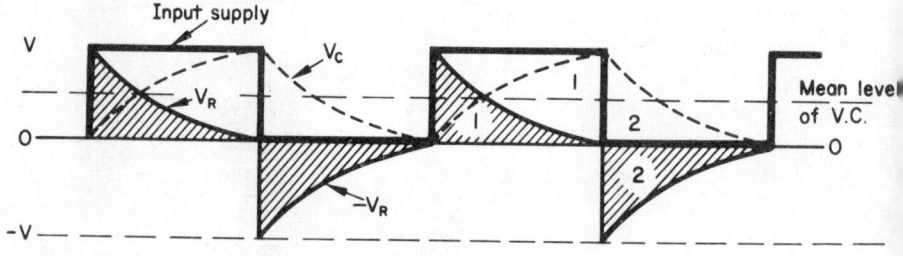

Fig. 14.2.8.

If the mean level is measured from any arbitrary reference level, earth, for instance then the difference between the mean level and the arbitrary reference level is known as the d.c. level.

The input has a d.c. level, so has V_C but V_R has neither a mean or d.c. level, hence no d.c. voltage will appear across R.

If figure 14.2.9. is considered, then it is evident that there is no d.c. level on the grid of V_2; (if there were, distortion would inevitably occur).

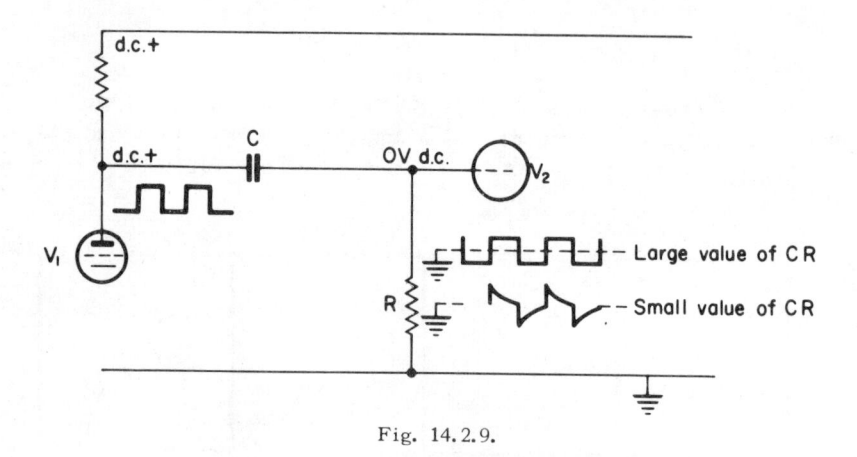

Fig. 14.2.9.

where V_R is now the voltage across R, the grid resistor.

It is in this fashion that the d.c. voltage (usually from the anode of a preceding valve) is across the capacitor but never across the grid leak. Hence the bias arrangement for V_2 is unaffected. The capacitor C blocks the d.c. level at V_1 anode from reaching V_2 grid. Further, any d.c. component or d.c. level due to the V_1 anode waveform is never transmitted via C to V_2 grid.

Now take a few examples of V_C at various times after $t = 0$ (Figure 14.2.10.). Figure 14.2.4. is shown but with the curve V_c only.

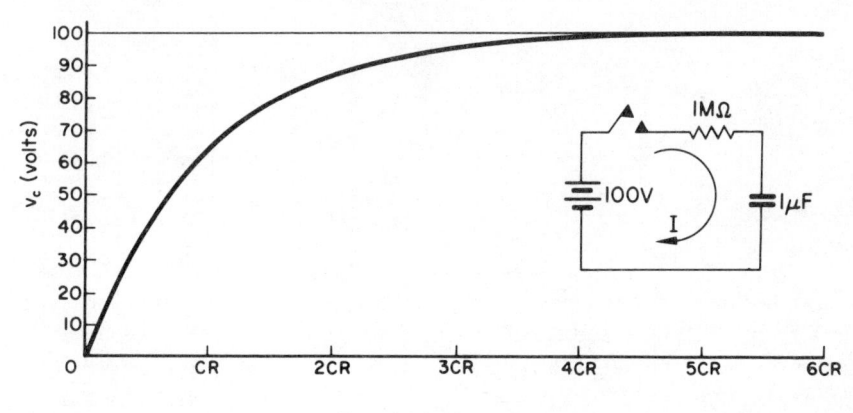

Fig. 14.2.10.

Example 1.

What value would V_C become at $t = 1$ secs.? (Figure 14.2.10.).
The time constant $C.R. = 1M \times 1\text{ufd} = 1\text{sec}$.
From the curve, V_C is seen to be 63% of 100V at 1 C_R.
$V_C = 63$V at 1 second after closing the switch.

Example 2.

What is the approximate value of V_C at 5 secs. After $t = 0$? Answer: 100V.
What is the value of V_C at 4 secs. after closing the switch?
Answer: 98.2V.

This curve is a natural growth and has many applications.
The mathematical expression for the potential across the capacitor C is given by

$$V_c = V[1 - e^{-t/C_R}]$$

Where V, in this example, is 100V,

$$C = 1\mu F \text{ and } R = 1M\Omega.$$

When $t = 0$, $V_c = 0$ and when $t = \infty$, $V_c = V$.

14.3. Pulse response of Linear circuit components.

Some time will elapse after a train of pulses is applied to a $C.R.$ network, before a steady state is reached.

Consider the following circuit diagram, figure 14.3.1.

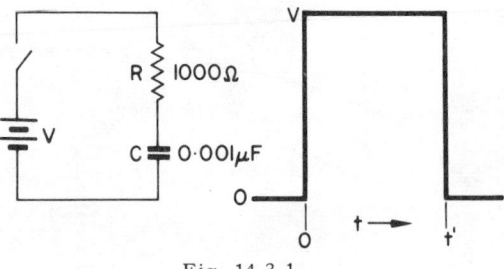

Fig. 14.3.1.

When the switch is closed, the input is applied immediately. The input is a step function, theoretically rising from 0V to V volts instantaneously. After a time t, the pulse falls to zero. If the capacitor is initially uncharged, an expression for V_R is $V_R = V(e^{-t/CR})$, whilst the expression for the capacitor voltage, $V_C = V(1 - e^{-t/CR})$. The time constant is $C.R$. Let the time constant be T.

After a time $5\,C.R.$, V_C is approximately V. After a time $5\,C.R.$, V_R is approximately 0V. When these conditions prevail, the circuit is said to have reached it's steady state. When a train of pulses are applied to a circuit, several pulses will need to be applied before the circuit has 'settled down'. Figure 14.3.2. shows a few pulses from a train of pulses, it may be seen that the amplitudes are varying; it is this amplitude that will be considered. The following will be accurate to two figures only.

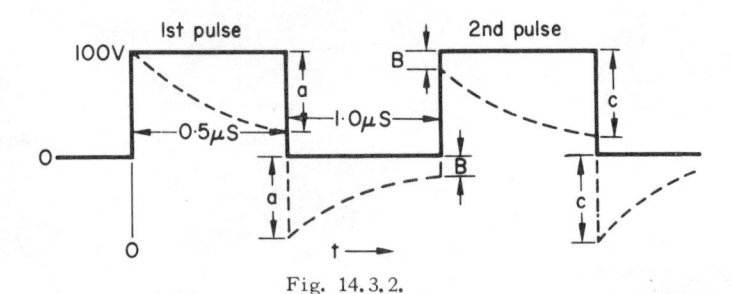

Fig. 14.3.2.

The pulse duration will be assumed to be of $0.5\mu S$. The space between the pulses will be assumed to be $1.0\mu S$. Let the input voltage be 100V. It is intended to compute how many pulses will need to be applied before the steady state is reached.

The actual circuit to be investigated is shown in figure 14.3.3.

The resistor is a 1000Ω, and the capacitor is a $0.001\mu F$. The time constant is $1.0\mu S$. From the theory previously covered, it may be seen that at time zero, ($t = 0$), the voltage across the resistor would be 100V.

The pulse will fall to zero, however, at a time $0.5\mu S$ after the leading edge of the input, but the short time available does not allow the capacitor

PULSE TECHNIQUES, MULTIVIBRATORS AND SAWTOOTH GENERATORS 229

to acquire the full voltage. The first pulse is applied at time zero, $t = 0$. The resistor has, at this instant, the full 100V across itself. This voltage immediately begins to fall, and will after 5μS, be zero. Before this can occur, the input pulse 'falls' to zero. The voltage across the resistor falls to the amplitude of the input, but, this time, is negative. During the space between the lagging edge of the pulse and the leading edge of the second pulse, the voltage across the resistor 'falls' from −100V towards zero.

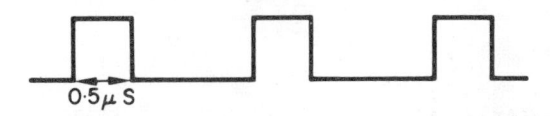

Fig. 14.3.3.

It cannot reach zero however, as the leading edge of the second pulse arrives and once more, the voltage across the resistor becomes 100V less its value at that instant.

The train of pulses is shown in figure 14.3.4. This also shows the pulse amplitudes at time intervals after time zero.

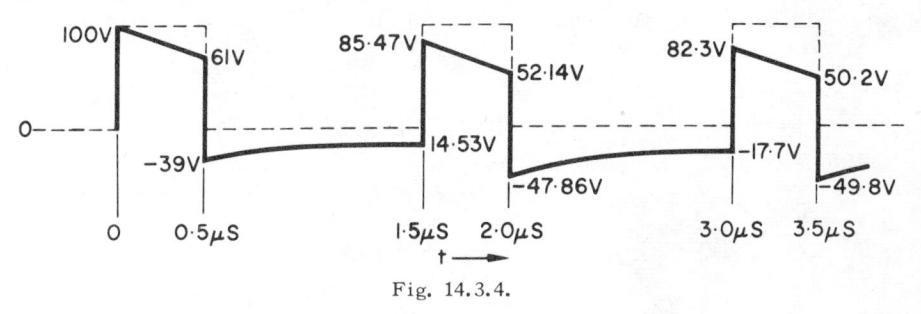

Fig. 14.3.4.

At $t = 0$, $V_R = 100$V. The time duration of the pulse, T is 0.5μS. The time constant $T = CR$ is 1.0μS. During the pulse, the index of e, (in $e^{-t/CR}$) becomes $t/T = 0.5$.

During the space, the index becomes $t/T = 1.0$. A few calculations will initially be made and from there on, the remainder will be placed in the table.

At $t = 0$, $V_R = 100$V. At $t = 0.5\mu S$, $V_R = 100\ (e^{-t/CR}) = 100\ (e^{-0.5}) = 61$V.
At $t = 0;5$ $V_R = -100 + 61 = -39$V.
At $t = 1;5$ $V_R = -39\ (e^{-t/CR}) = -39\ (e^{-1}) = -39\ (0.37) = -14.53$V.
At $t = 1;5$ $V_R = 100 - 14.53 = 85.47$V. 2nd Pulse

At $t = 2$ $V_R = 85.47\ (0.61) = 52.14\text{V}$
At $t = 2$ $V_R = -100\ 52.14 = -47.86\text{V}$
At $t = 3$ $V_R = -47\ (0.37) = -17.7\text{V}$
At $t = 3$ $V_R = 100 - 17.7 = 82.3\text{V}$ 3rd Pulse
At $t = 3;5$ $V_R = 82.3\ (0.61) = 50.2\text{V}$
At $t = 3;5$ $V_R = -100\ 50.2 = -49.8\text{V}$
At $t = 4;5$ $V_R = -49.8\ (0.37) = -18.4\text{V}$
At $t = 4;5$ $V_R = 100 - 18.4 = 81.57\text{V}$ 4th Pulse
At $t = 5$ $V_R = 81.57\ (0.67) = 50\text{V}.$
At $t = 5$ $V_R = 50 - 100 = -50\text{V}.$
At $t = 6$ $V_R = -50\ (0.37) = -18.5\text{V}.$
At $t = 6$ $V_R = 100 - 18.5 = 81.5\text{V}$ 5th Pulse
At $t = 6;5$ $V_R = 81.5\ (0.61) = 50\text{V}.$

The steady state is reached at the time the fifth pulse is applied. The amplitude is seen to be a maximum of approximately 81.5V. This value will, from now on, remain constant. The amplitude of the lagging edge is seen to be 50V.

The measure of these amplitudes are often used when analysing simple amplifiers. Knowing the frequency, the 'droop', and one or two constants, the c.w. frequency response is determined without the lengthy procedure of plotting the amplitude of output signal at predetermined frequency intervals.

With a very low frequency input, and a long pulse duration, the period before the steady state is reached, may be important.

14.4. Simple relaxation oscillator.

Reference to 7.5.1. shows that 115V is necessary to strike the neon, but once struck, the neon potential falls to 85V.

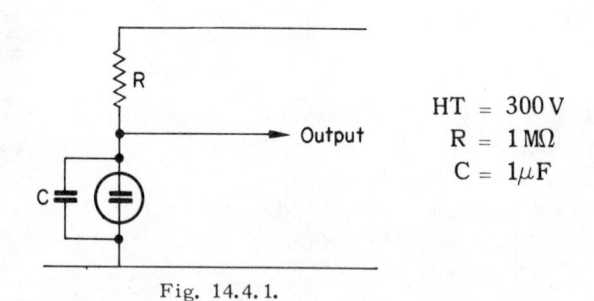

$$\text{HT} = 300\,\text{V}$$
$$R = 1\,\text{M}\Omega$$
$$C = 1\mu\text{F}$$

Fig. 14.4.1.

If switch is closed, $V_C = 0$ (at $t = 0$) therefore the neon is extinguished: $V_R = 300\text{V}.$ The output $\therefore$ is 0V.

After a time $5CR$ (5 secs) V_C would become 300V (as this was the voltage across R at $t = 0$). But as V_C climbs up to 300V, it reaches 115V;

PULSE TECHNIQUES, MULTIVIBRATORS AND SAWTOOTH GENERATORS 231

the neon 'strikes' and immediately falls to 85V, dragging the capacitor down with it. The capacitor then starts climbing again to 300V, but when it reaches 115, it is once again dragged down to 85V by the neon. R is chosen such that it is high enough to prevent the neon from remaining in the 'struck' condition once it has fired. The neon would need 1mA burning current or so, but the maximum burning current can only be $300 - 85\text{V}/1\text{M}\Omega = 215\mu\text{A}$.

Figure 14.4.2. shows the curve of the expression $(1 - e^{-t/CR})$. The output waveform is sawtooth in shape although there is a slight curve as can be seen from the illustration.

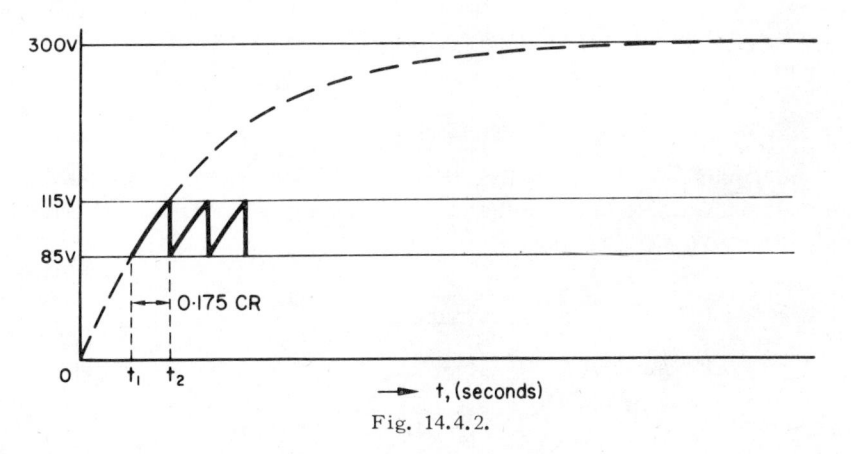

Fig. 14.4.2.

The duration of the pulse, or pulse width, is from $0.35\,CR$ to $0.535\,CR = 0.175\,CR = 0.175$ secs or $\underline{175\,\text{m/secs.}}$

The amplitude of the output $115 - 85 = 30\text{V}$.

This output waveform could be used in a limited manner as a simple deflection voltage for a cathode ray oscilloscope. The amplitudes are proven and the pulse duration is obtained as follows:

From

$$V_c = V[1 - e^{-t/CR}],$$

For t_1. $85 = 300[1 - e^{-t_1/CR}]$, $300 - 85 = 300e^{-t_1/CR}$ (1)

For t_2 $115 = 300[1 - e^{-t_2/CR}]$, $300 - 115 = 300e^{-t_2/CR}$ (2)

and dividing (1) by (2)

$$\frac{300 - 85}{300 - 115} = e^{(t_2 - t_1)/CR}$$

and taking $\log_e$s of both sides,

$$\therefore \qquad \log_e\left[\frac{300-85}{300-115}\right] = \frac{t_2-t_1}{CR}$$

hence

$$T = t_2 - t_1 = C.R.\log_e\left[\frac{300-85}{300-115}\right]$$

where T is the pulse width.

Generally, the expression becomes

$$T = t_2 - t_1 = C.R.\log_e\left[\frac{V-V_b}{V-V_s}\right]$$

where V_s is the striking potential and V_b the burning, or extinguishing potential.

14.5. Simple free running multivibrator.

As previously stated a *basic* long tailed pair is a balanced circuit, both valves conducting equally in the absence of a signal. With a multivibrator, only *one* valve at a time is conducting. Consider figure 14.5.1.

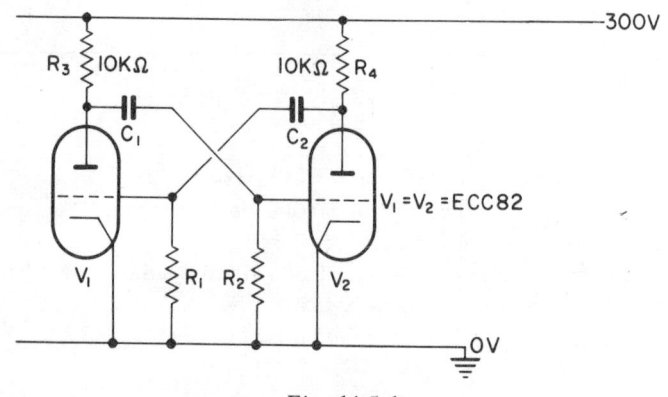

Fig. 14.5.1.

Assume that the H.T. is applied. Although V_1 and V_2 both start conducting slightly, V_1, say, due to unbalance because of component tolerances, conducts a little heavier than V_2. V_1 anode starts falling due to the drop in R_3 caused by the increasing anode current. This negative excursion is transmitted via C_1 to V_2 grid. This negative 'pulse' on the grid of V_2 causes V_2 anode current to reduce; V_2 anode climbs in the direction of 300V. This positive change of level is transmitted via C_2 to the grid of V_1, causing V_1 to conduct even more heavily. This heavy

PULSE TECHNIQUES, MULTIVIBRATORS AND SAWTOOTH GENERATORS

conduction causes V_1 anode to fall further still; taking V_2 grid down even more until V_2 is completely off and V_1 completely on. V_2 grid is now sitting at a large negative level. (This is $t = 0$). C_1 now begins to discharge via R_2, until the grid of V_2 climbs up to a point where the bias is such that V_2 commences to conduct. Immediately V_2 conducts, its anode falls and takes the grid of V_1 down beyond cut off, so that the position is reversed. The point at which either valve starts to conduct (when cut off) is given on the valve characteristics at the intersection of the load line and the H.T. on the V_{ak} axis. It will be helpful to analyse a circuit of this type, step by step.

We need to know the shape of the waveform at either grid, together with width of the pulse.

Figure 14.5.1. shows the circuit to be analysed.

Step 1.

Assume that V_1 is 'ON' and V_2 'OFF'.
We have therefore just one valve to consider, as V_2 anode current is zero.
For the moment consider V_1 as an ordinary amplifier.

Step 2.

Construct a d.c. load line for the 10KΩ anode load. Figure 14.5.2.

Note carefully that V_1 is ON (hard on, $V_{gk} = 0$); if the grid should go any further along the load line in a positive direction grid current would flow, but for the time being we will assume no appreciable grid current flow. It is seen that $V_{ak} = 130$V volts when $V_{gk} = 0$.

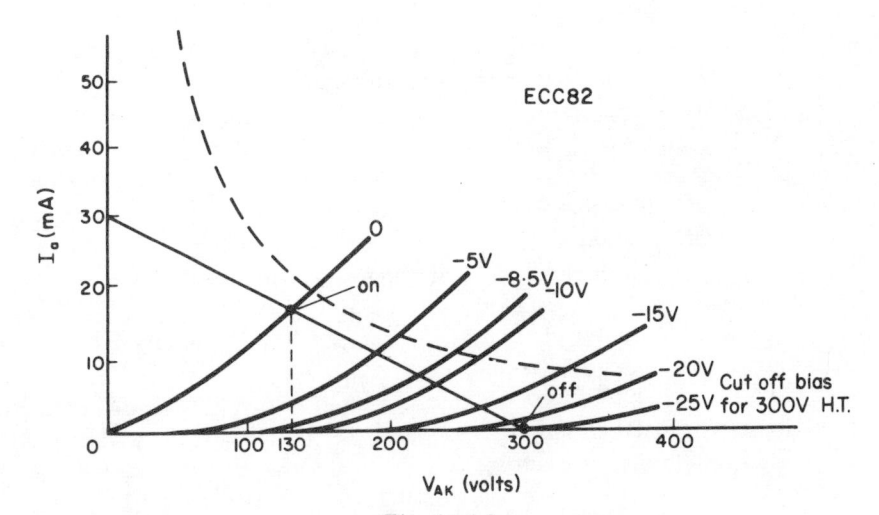

Fig. 14.5.2.

Step 3.

V_1 can only be in one of two states, ON or OFF. It cannot sit anywhere else on the load line, but must be 'ON' or 'OFF' due to the feedback effect from V_2.

Now assume that V_1 is 'OFF'. It is seen that the value of V_{gk} to *just* cause the valve to be 'OFF' is $-25V$ volts (where the d.c. load line corresponds to (H.T., 0).

Step 4.

Consider now that V_2 is 'ON'.
Since V_1 is 'OFF', its anode is at 300V (Figure 14.5.3.)

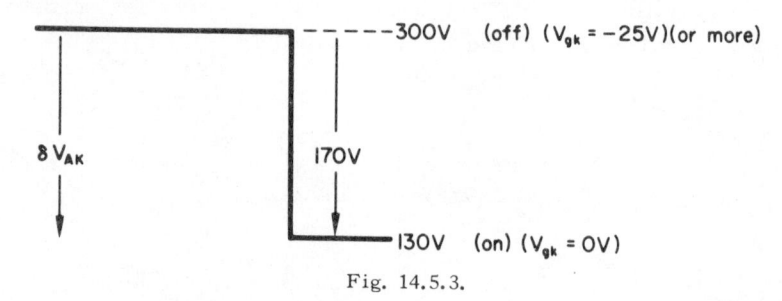

```
----------------- - - - - - 300V    (off) (V_gk = -25V)(or more)
|                |
|                |
8 V_AK           170V
|                |
|                |
                 ---------- 130V    (on) (V_gk = 0V)
```

Fig. 14.5.3.

If V_1 is now hard 'ON', its anode is at 130V. The total 'fall' of the anode is $300 - 130 = 170V$.

This 170V is falling, and represents a negative going rectangular pulse of 170V amplitude. All of this pulse arrives at V_2 grid at $t = 0$, and drags the grid from $V_{gk} = 0$ to $V_{gk} = -170V$; and V_2 is well and truly 'OFF' (only $-25V$ is needed for cut off.). The voltage at this instant across R_2 is $-170V$; remember that the series capacitor, C_1, was short circuit at this instant, but will in $5\,CR$ acquire the voltage across R_2, and V_{R_2} (V_{gk} of V_2) will become zero. But reference to figure 14.5.4. shows that before $5\,CR$ seconds, the grid climbs from $-170V$ to $-25V$; at this instant, V_2 begins to conduct, as the grid voltage (V_{R_2}) is insufficiently negative to keep it 'OFF'. Immediately V_2 starts conducting, the cycle is complete, and V_2 turns ON, switching V_1 off. At this stage an analysis of V_1 could be made, but in this circuit is quite unnecessary, because the circuit is symmetrical and the above analysis applies to both valves.

The pulse width (point of changeover from V_2 to V_1) is easily given by using the graph shown in figure 14.7.1.

V_g (V_{R_2}) is at $-170V$ at $t = 0$. This climbs towards 0V, and would reach this value at $5\,CR$. When the grid reaches 'cut-on' at $-25V$, the pulse duration is complete. Note that the voltage 'climb' is always the amplitude of voltage across the appropriate resistor at $t = 0$; in this case 170V.

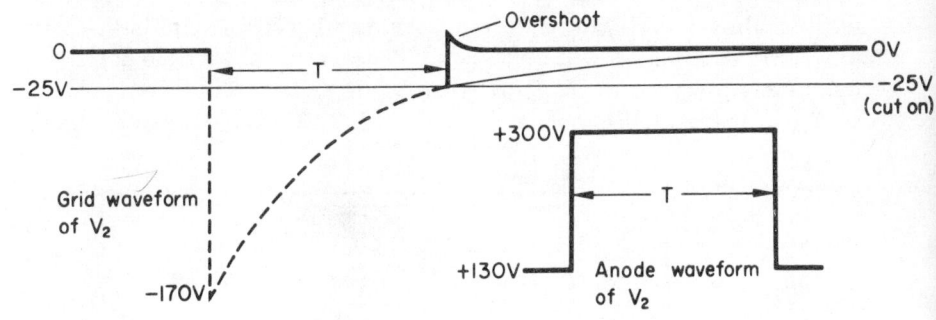

PULSE TECHNIQUES, MULTIVIBRATORS AND SAWTOOTH GENERATORS   235

Fig. 14.5.4.

Clearly the maximum possible climb = 170V, from −170V to 0V. The 'actual' climb = 145V, as it stops 25V short of 170V.

The maximum possible 'climb' is 170V. This is V.
The actual 'climb' is seen to be 145V. This is V_c.
Then from $V_c = V[1 - e^{-t/CR}]$, $t = 1.91\, CR$ seconds.
If we chose $C = 1\mu F$ and $R = 1M\Omega$, then the pulse duration, i.e., $t - 1.91$ seconds.

The overshoot occurs due to the grid being momentarily dragged into grid current, i.e., V_{gk} is taken positive. This overshoot is arrested by grid current action and C rapidly discharges through R_3 and the grid-cathode diode. The latter having a resistance of about 500Ω.

The overshoot 'decay' is seen to be of a similar shape, although inverted, to the pulse formation between $0 - T$, and decays rapidly due to the small time constant $C(R_L + R_D)$.

Another example is given in figure 14.5.5.

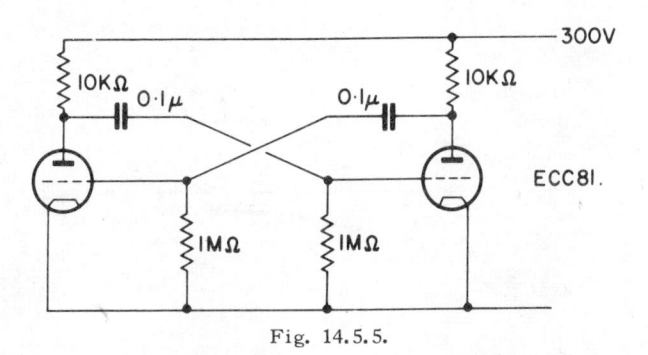

Fig. 14.5.5.

Examine the waveform on one grid, say V_2 grid. Assume that V_1 is conducting (ON). Draw a load line for 10KΩ for V_2 on the ECC81 characteristics in figure 14.5.11. At the two extremes of the grid swing, ($V_{gk} = 0$ and $V_{gk} = -8V$) gives $V_{ak} = 158V$ and $V_{ak} = 300V$.

236   ELECTRONICS FOR TECHNICIAN ENGINEERS

The anode then swings 300 − 158 = 142V. V_1 grid is at 0V. The fall in anode voltage (−142V) drags V_1 grid down to −142V. The grid is therefore at −142V, and in 5 CR will climb up to 0V. However at −8V the pulse is completely formed, and 'T' established.

T (our pulse width) = $2.85 \times 0.1 \mu\text{fd} \times 1\text{M}\Omega$ = 285 mS as in figure 14.5.6.

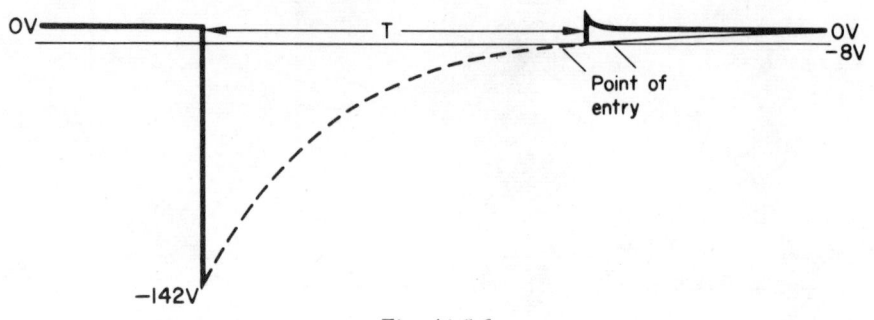

Fig. 14.5.6.

As the grid approaches 'cut on' (−8V in the above example) any noise or random fluctuations in the H.T., sudden impulses, etc, occasionally reach the grid, which may cause the valve to 'cut on' a little sooner than −8V. This is because the curve cuts −8V almost horizontally; in other words it enters the 'cut on' point so 'flat' that the point of entry is somewhat indeterminate, and is very susceptible to slight changes in potential (see Figure 14.5.6.); but if we arrange the point of entry such that it enters more vertically, then the point at which it enters is much more clearly defined and is less subject to random impulses.

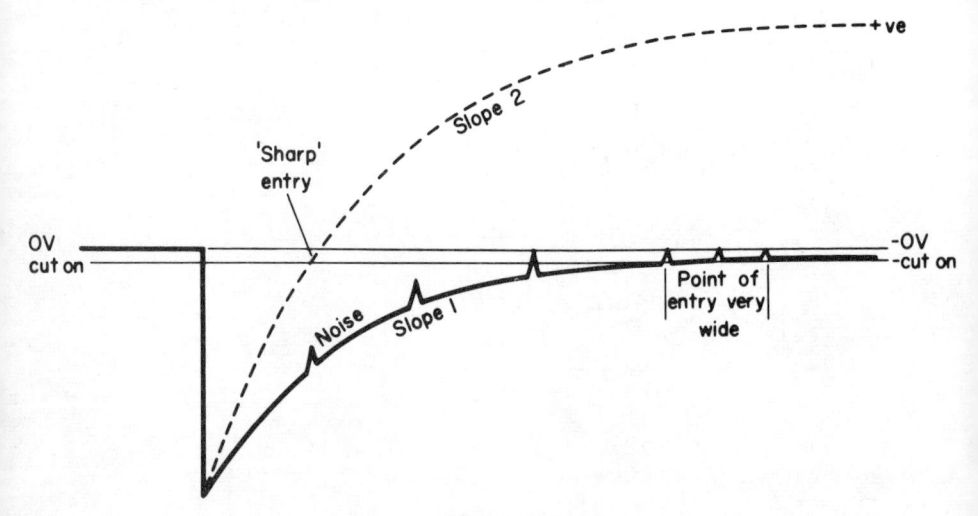

Fig. 14.5.7.

Slope 1 is more horizontal, and the point of entry quite 'wide' and indeterminate. **Intermittent noise can cause a random cut-on.**

Slope 2 enters much more sharply and is clearly defined.

We can determine component values for (or analyse) a circuit using the slope No. 2. as in figure 14.5.7. In order to accomplish this, we simply take the grid leak to a positive potential as in figure 14.5.8.

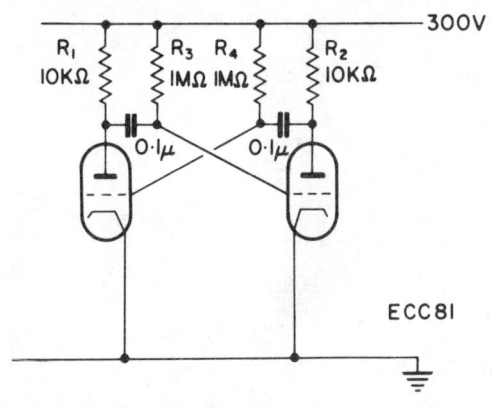

Fig. 14.5.8.

The reader should note that any predetermined positive potential will suffice. It does not have to be the H.T. rail although this is often the case.

The grid leaks (R_3 and R_4) are returned to +300V as an example. Assume that one valve is ON and the other OFF.

A load line for 10K shows that the anode swing is, as previously, 142V. 'Cut on' is −8V, as before.

The voltage across R_4 at $t = 0 = 300 - (-142) = 442V$. (i.e. the grid is at −142 whilst the top end of R_4 is at +300V).

The waveform will now look as in figure 14.5.9.

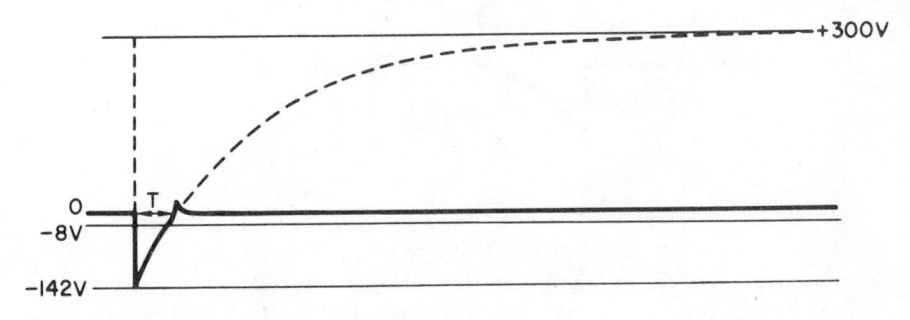

Fig. 14.5.9.

Note that the capacitor (V_{g_2}) *climbs towards the voltage across R at* $t = 0$ (442V). i.e. from -142V to $+300$V and will reach $+300$V in $5CR$ seconds. At -8V, however, as before, the pulse is complete and T is easily seen. The total possible climb = 442V. The actual climb = $142 - 8 = 134$V, as before.

The point of entry is sharply defined, and the circuit will be very much more stable. The pulse width is narrower, but is of little consequence as C or R may be chosen once T is established.

The expression
$$V_c = V[1 - e^{-t/CR}]$$
is simplified to
$$1 - \frac{V_c}{V} = e^{-t/CR} \quad \therefore \quad \frac{V - V_c}{V} = e^{-t/CR}$$

$$\therefore \frac{V}{V - V_c} = e^{+t/CR} \quad \therefore \quad t = CR \log_e \frac{V}{V - V_c} = CR \log_e \frac{442}{308} = CR \log_e 1.435$$

$$\therefore \quad t = 0.36\,CR.$$
Hence $\quad t = 36\,\text{mS}.$

With this practical idea as to the analysis of simple multivibrators, it is time to take a deeper look at the various component functions.

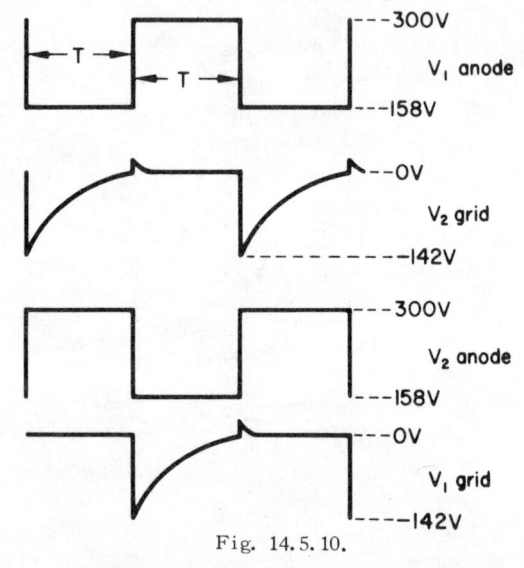

Fig. 14.5.10.

As V_1 anode falls by 142V down to 158V d.c. the 'upper' side of capacitor falls with it, and remains at 158V whilst V_1 is on. The 'lower' side of the capacitor during the anode fall, falls by 142V also and takes V_2 grid down to from zero to -142V. This -142V is present at the grid end of R_4 while the top end of R_4 is connected to $+300$V. The total voltage

PULSE TECHNIQUES, MULTIVIBRATORS AND SAWTOOTH GENERATORS 239

across R_4 is the difference between $+300V$ and $-142V = 442V$. The capacitor cannot maintain the current through R_4 (causing 442V across it), and begins to discharge through R_4; the grid starts climbing from $-142V$ up to the voltage at the top end of R_4. While climbing, however, the grid reaches $-8V$, and V_2 conducts; V_1 is OFF and the pulse is complete. Immediately V_2 conducts, V_2 anode falls and a pulse of $-142V$ appears at V_1 grid; V_1 is cut off, and its grid then starts its climb so that the cycle repeats itself. See figure 14.5.10.

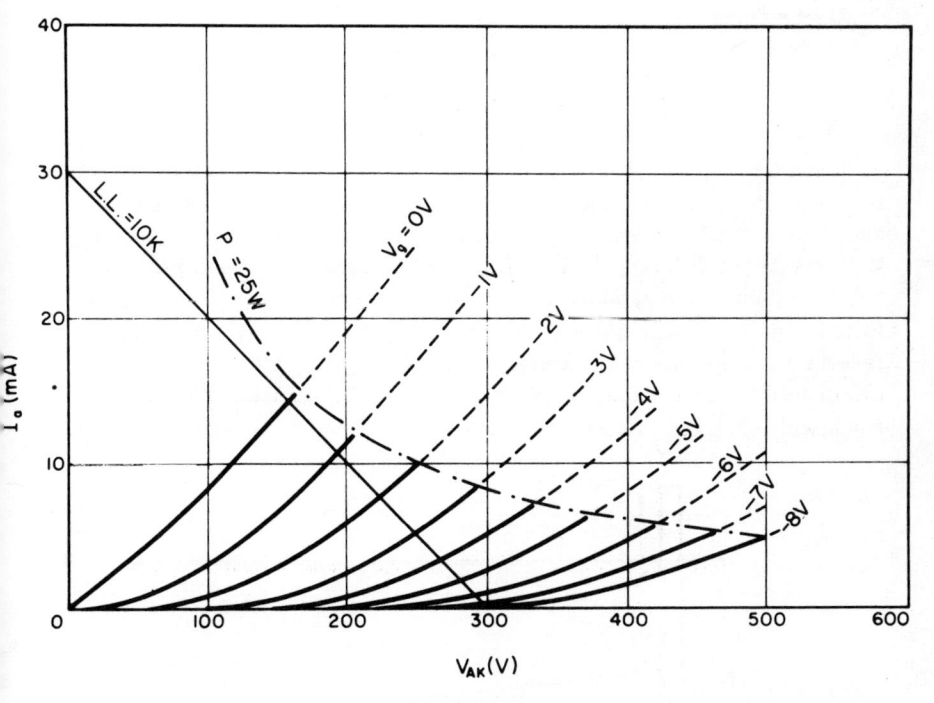

Fig. 14.5.11.

If taken from either anode, the output would be a reasonable square wave, with this symmetrical circuit, having a pulse width as shown and an amplitude of 142V.

If the time constant differs for each valve, then each valve must be analysed separately; rectangular pulses will still be obtained at the anode but will have a different pulse width according to the values of CR for each valve. The output would have a mark space ratio of something other than 1:1. **Multivibrators are one of a family of circuits that are pulse lengtheners when triggered (or set off) by an external pulse, which is usually very narrow.**

R

14.6. A basic pulse lengthening circuit.

A simple example of a pulse lengthener is the following circuit in figure 14.6.1.

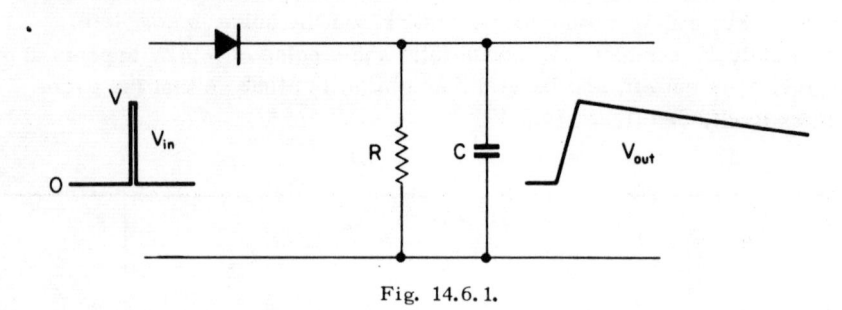

Fig. 14.6.1.

Consider the 'C' initially uncharged; therefore the cathode of the diode is at 0V. If V volts were applied to the input, the diode would conduct, and current would flow through R, causing a p.d. to be set up across C. If R is very much larger than R_d (the diode forward resistance) then the time constant $C \times R_d$, which is very short, and the output leading edge follows the input very closely. When the input falls to 0V, the diode is reverse biassed due to the charged capacitor, and is effectively open circuit. 'C' then discharges via R (by a time constant $C \times R$) and a very much wider pulse subsequently appears at the output. See figure 14.6.2.

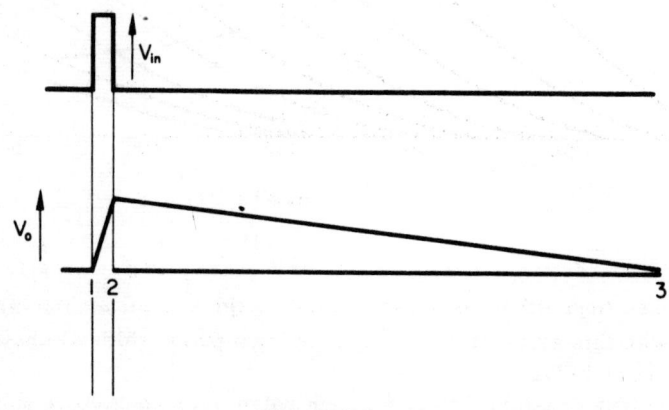

Fig. 14.6.2.

A second input pulse must not be applied until the output pulse has reached the required pulse width; this determines the maximum p.r.f. A pulse width 'amplification' of a thousand or so is easily obtained.

Note that if R_d is almost 'short circuit', then the output rises to 'V' immediately with no time lag whatsoever. The 'pulse width amplification

PULSE TECHNIQUES, MULTIBIBRATORS AND SAWTOOTH GENERATORS 241

factor' is given as the ratio of R/R_d, as C is a constant.

If $R = 1\text{M}\Omega$ and $R_d = 100\Omega$, then the output pulse width is $R/R_d \times \text{input} = 1\,000\,000/100 = 10\,000 \times \text{input pulse width}$.

The input pulse width should be much less than $5\,CR_d$.

Should $C = 0.1\mu\text{F}$. $R = 1\text{M}\Omega$. $R_D = 100\Omega$, and the input pulse width $100\mu/\text{s}$ then the output could be as great as $5\,CR = 5 \times 0.1\mu\text{F} \times 1\text{M}\Omega = 500\,\text{m/S}$. The charging time (points 1–2) = $100\mu\text{S}$ and point (2–3) 500 mS. Pulse width amplification 5000 : 1 (ignoring amplitude fall off).

This circuit is not very practical, as the output pulse falls to a very small amplitude and unless this is fed into a circuit that can discriminate between small changes of a low level of amplitude, the resultant circuits could become erratic or unreliable.

14.7. A charging curve.

Very rapid answers to any problem involving the term $V_c = V(1 - e^{-t/CR})$ may be obtained by using a 'charging curve'.

The law of the curve.

Figure 14.7.1. shows a typical series CR network across which is connected an e.m.f.

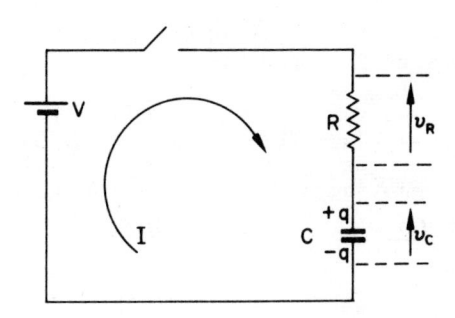

Fig. 14.7.1.

C is initially uncharged, the switch is open and $i = 0$.

The switch is closed at $t = 0$. Direct current cannot flow around the circuit due to the inclusion of the capacitor, C.

Current must be taken from the lower 'plate' of the capacitor and stored in the upper 'plate'.

Hence a negative charge will exist at the lower plate and a positive charge on the upper plate.

An expression for the voltages in the circuit is

$$V = v_c + v_R$$

$$\therefore \quad V = \frac{q}{c} + iR \quad \text{(where from } Q = cv,\ v_c = q/c\text{)}$$

and

$$V = \frac{q}{c} + \frac{dq}{dt} \cdot R \quad \left(\text{where } i = \frac{dq}{dt}\right)$$

and dividing both sides by R

$$\frac{V}{R} = \frac{q}{CR} + \frac{dq}{dt} \quad \text{and by rearranging terms,}$$

$$\frac{dq}{dt} + \frac{1}{CR} \cdot q = \frac{V}{R}$$

$$\therefore \quad \frac{dq}{dt} = \frac{V}{R} - \frac{1}{CR} \cdot q$$

and

$$\frac{dq}{dt} = \frac{CV - q}{CR} \quad \text{and integrating both sides,}$$

$$\int \frac{CR}{CV - q}\, dq = \int 1\, dt$$

$$CR \int \frac{dq}{CV - q} = \int 1\, dt$$

$$\therefore \quad CR \log_e (CV - q) = t + A \tag{1}$$

where A is the constant of integration.

But at $t = 0$, $q = 0$.

$$\therefore \quad -CR \log_e CV = A \quad \text{and substituting for } A \text{ in (1)}$$

$$-CR \log_e (CV - q) = t - CR \log_e CV$$

$$\therefore \quad -\frac{t}{CR} = \log_e \frac{CV - q}{CV}$$

thus $\quad -\dfrac{t}{CR} = \log_e \left(1 - \dfrac{q}{CV}\right)$ and taking antilogs,

$$e^{-t/CR} = 1 - \frac{q}{CV}$$

$$\therefore \quad \frac{q}{CV} = 1 - e^{-t/CR}$$

$$\therefore \quad \frac{q}{C} = V(1 - e^{-t/CR})$$

and

$$\frac{q}{C} = V_c$$

PULSE TECHNIQUES, MULTIVIBRATORS AND SAWTOOTH GENERATORS 243

$\therefore \quad V_c = V(1 - e^{-t/CR})$ and is the law of the charging curve.

The charging curve was plotted by assuming values for (t/CR) from 0–5 and the curve drawn as shown in figure 14.7.2.

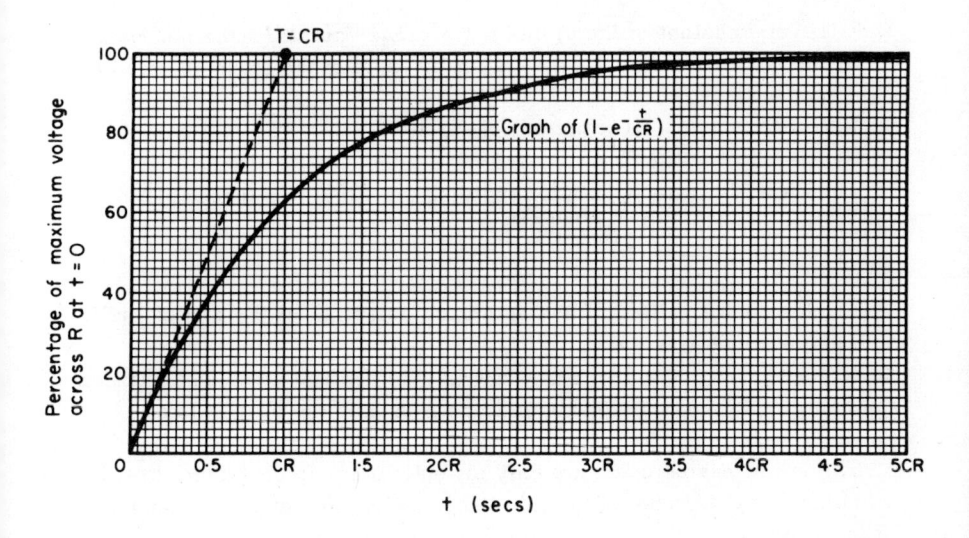

Fig. 14.7.2.

Using the charging curve.

The curve provides not only a very rapid means of obtaining an answer to a problem containing the term $V_c = \hat{V}(1 - e^{-t/CR})$, it also provides a method of obtaining accurate answers for the reader whose mathematical ability is not yet at the level where he can successfully deal with the term in question.

(This method was devised by the author for electronic technician apprentices who, at the age of 16, might have had some difficulty with the mathematics).

Example 1.

A 200V d.c. is suddenly applied to a series CR circuit. After what duration does the capacitor potential become 74V?

The voltage across R at $t = 0$ must be the full 200V. Hence 100% on the graph corresponds to 200V, i.e., each 1% corresponds to 2V.

74V is seen to be 37% on the graph. Using the graph in the normal manner shows that 37% corresponds to $0.5\,CR$ seconds.

If C and R are known, t can be calculated easily.

Example 2.

A 300V step function is applied to a series CR network. $C = 1\mu F$ and $R = 0.5M\Omega$.

What is the capacitor potential 0.75 seconds after the step function is applied?

The time constant of the circuit = $CR = 1\mu F \times 0.5M\Omega = 0.5$ seconds.

Hence 0.75 seconds = $1.5\,CR$ on the 'X' axis of the graph and $1.5\,CR$ corresponds to 77.5%.

100% on the graph must correspond to 300V.

Hence 77.5% = 232.5V.

Therefore, 0.75 seconds after applying the 300V step function, the capacitor potential will be 232.5V.

Example 3.

The multivibrator in 14.5.5. had an aiming potential for the grid of 142V. The actual climb was 134V.

Expressing $V_c/V = 134/142 \times 100$ as a percentage gives 94%. 94% on the charging curve corresponds to $2.85\,CR$ as was the case in that particular example.

The curve may be used, not only for electronic circuit problems, it is a 'natural' law and applies to many other topics. It can be used instead of formulae in the chapter on delay line pulse generator.

It is important to note that 100% on the graph must correspond to the total voltage across R at $t = 0$, i.e., the instant a step function is applied.

The curve can be measured and plotted in practice.

Unless one has a very very high impedance voltmeter, it might be advisable to measure, not V_c, but the current from $t = 0$ to $t = 5\,CR$.

Select a time constant of about 10–15 seconds, ensuring that the meter can cope with the initial current, $I = V/R$.

Take measurements at regular intervals and repeat the measurements, with C discharged initially, and plot the average results.

The curve is plotted as follows: One has measured I. $V_R = I \times R$ and as R and I are known, V_R is found easily. $V = V_R + V_C \therefore V - V_R = V_C$. and V_C can be plotted.

CHAPTER 15

Further large signal considerations, a binary counter

We will concern ourselves in this chapter with further large signal considerations. We will develop the discussion in such a manner that it will lead us to a detailed examination of a complete binary counter with meter readout.

We will then have a look at the design of a second binary stage, different from the first, and discuss several important design features.

The chapter will be confined to thermionic valves, as with these devices, no allowances need generally be made; the results in practice will normally be quite accurate. We will see in later chapters, the allowances we need to make for Transistors of the junction type.

The most important objective of this book is to discuss electronic circuit principles; the use of devices that need special considerations at this stage has been avoided; the reader therefore has but one unknown at a time to master.

15.1. A basic long tailed pair

Figure 15.1.1. shows a basic long tailed pair. This circuit was discussed in detail during previous chapters.

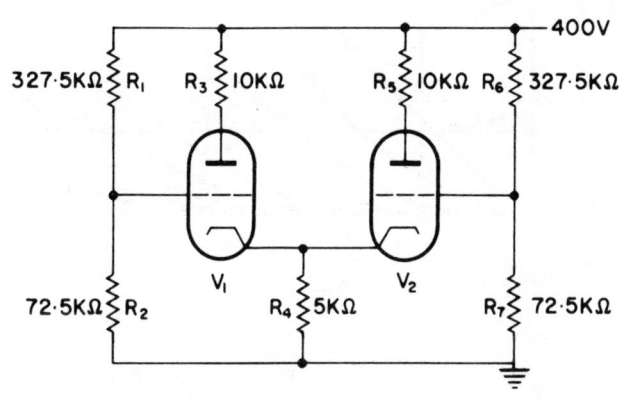

Fig. 15.1.1.

15.2. A basic Schmitt Trigger circuit

Figure 15.2.1. shows a basic Schmitt trigger circuit. We have converted the basic long tailed pair in figure 15.1.1. by simply disconnecting R_6 from the H.T. and reconnected it to the anode of V_1. We will assume that V_1 is on and

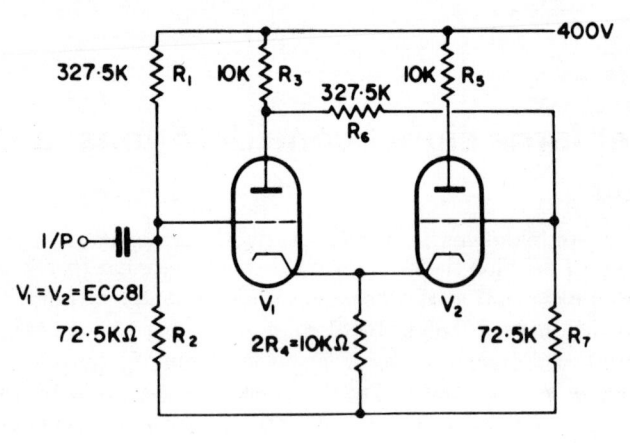

Fig. 15.2.1.

that V_2 is off. These conditions are essential in this circuit. Let us examine the d.c. conditions of the circuit and see whether our assumption that V_2 is off, is correct. The 20 K load line is drawn in figure 15.2.2. A bias line is plotted also.

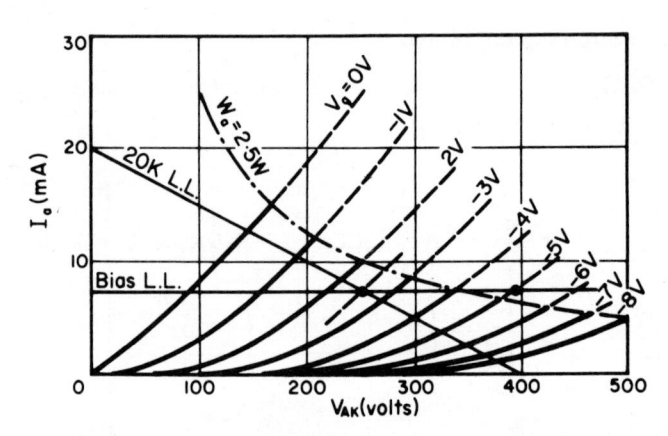

Fig. 15.2.2.

V_1.
$V_g = 72.5\,\text{V}$ if the bias was zero, $V_k = 72.5\,\text{V}$. Hence $I_a = 7.25\,\text{mA}$.
$V_g = 72.5\,\text{V}$ if the bias was $-2.5\,\text{V}$ $V_k = 75\,\text{V}$. Hence $I_a = 7.5\,\text{mA}$.
$V_g = 72.5\,\text{V}$ if the bias was $-5\,\text{V}$, $V_k = 77.5\,\text{V}$. Hence $I_a = 7.75\,\text{mA}$.

The points (V_{gk}, I_a) were plotted and a load line drawn. The intersection of the load line gives the following conditions for V_1. $V_{gk} = -2.5\,\text{V}$
$V_k = 75\,\text{V}$ $I_a = 7.5\,\text{mA}$ $V_a = 400 - 75 = 325\,\text{V}$ $V_{ak} = 325 - 75 = 250\,\text{V}$.
This ignores any current through $R_6 + R_7$.

V_2:

V_2 (if cut off) would have an anode to cathode potential of $400 - 75 = 325$ V. A bias of about -10 V is required to ensure cut-off of V_2. V_2 is sitting at a bias of $V_g - V_k$. $V_k = 75$ V.

$$V_g = \frac{325 \times 72.5}{400} \simeq 60 \text{ V}.$$

Hence V_2 is cut off with a bias of approximately -15 V and justifies the assumption. As V_2 is 'off', there will be no appreciable voltage drop across R_5 and thus the grid of V_1 will be virtually as shown in the graphical analysis of a long tailed pair in the previous chapter.

A negative going signal of sufficient amplitude fed into the grid of V_1 will cause V_1 to conduct less. The reduction of i_1 will cause V_1 anode to rise. This rise will be transmitted to V_2 grid and drag the grid up to a new level. As this occurs, i_2 will flow dragging the V_2 anode potential down. As V_2 anode falls, the negative going signal is transmitted to V_1 grid and takes this grid down even further than that due to the input signal. The effect is cumulative and V_1 is cut off. The top end of R_6 will be at H.T. potential, this will take V_2 grid up to the previous value of V_1 grid and V_2 will be 'on'. V_2 anode will have fallen and thus an output signal will have been available for the next stage. Once the input signal is removed, the circuit will revert to its former state.

15.3. A simple Bi-stable circuit

Figure 15.3.1. shows a circuit similar to the Schmitt trigger circuit except that in this instance, R_1 has been disconnected from the H.T. and connected to the anode of V_2. The circuit is now symmetrical and is a basic Bi-stable circuit.

If we were to add a few more components, we would have a Binary stage.

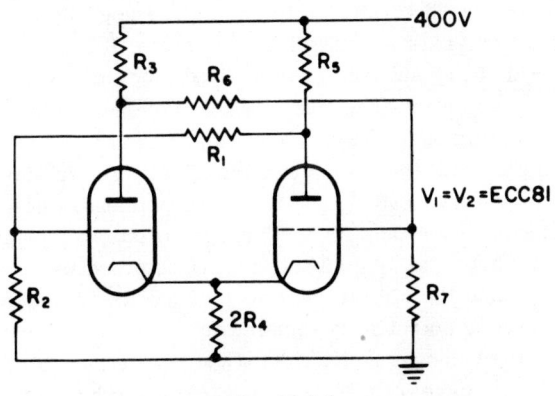

Fig. 15.3.1.

15.4. Introduction to Binary circuits – An analysis of the Eccles Jordan

The Eccles Jordan is one of the long tailed pair family. It has two stable conditions and is consequently a Bi-stable device.

A refinement often met is an electronic 'steering circuit' which ensures that the input signal always arrives at a predetermined grid at the right instant. The cathode voltage remains substantially constant, providing both valves are alike. The basic Eccles Jordan is shown in figure 15.4.1.

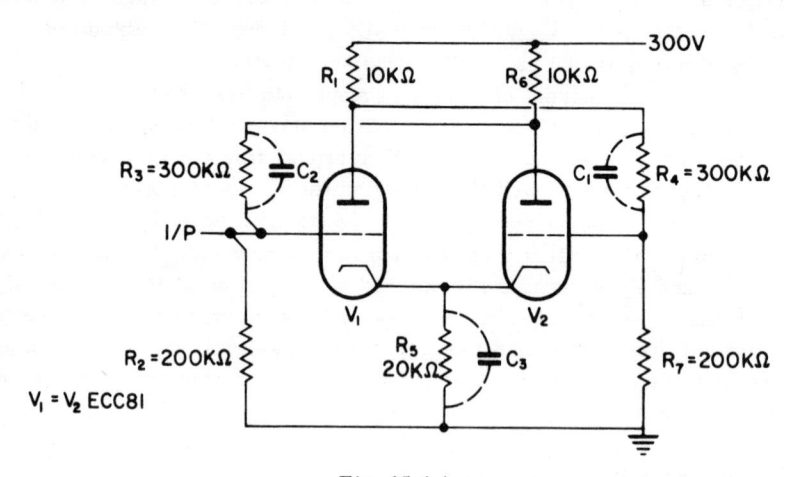

Fig. 15.4.1.

With this circuit, one valve only may conduct at any instant. Assume that V_1 is on, i.e. conducting. Its anode voltage will be lower than that of the H.T. consequently V_2 grid is lower than the cathode due to the potential divider action of R_4, R_7. V_2 anode current is zero as the low grid potential is sufficient to cut the valve off. V_1 anode is therefore almost at H.T. potential. V_1 grid is held within its grid base by the divider action of R_3, R_2 and V_2 anode p.d. This is the first stable state. If a negative signal is applied to V_1 grid, V_1 anode current immediately begins to fall, thereby raising the anode potential. The rise in anode potential is an inverted and amplified reproduction of the input waveform.

This rise in potential (positive going) taken V_2 grid up above cut-off. V_2 anode current is now flowing, and its anode potential falls, dragging V_1 grid even further negative than its previous level due to the input signal. V_1 is now off; its anode is approximately at H.T. potential, V_2 grid is within its working range and the circuit has now reached its second stable state. No change of cathode potential is required to assist with the change from one state to another; consequently the cathode is by-passed with a capacitor, which with R_K will have a time constant large enough to retain its charge

FURTHER LARGE SIGNAL CONSIDERATIONS. A BINARY COUNTER

and be unaffected by differences in V_1 and V_2 cathode currents.

Reference to the section on attenuator compensation will show quite clearly that due to grid cathode capacitance plus stray wiring capacitance, the anode to grid coupling resistor should have a small capacitor, connected in shunt so as to improve the edge of the pulse from anode to grid. It also has a memory function. This is described at the end of this chapter.

The circuit, may now be analysed, step by step, using the ECC81 valve characteristics shown in figure 15.4.3. It will be convenient to 'take a little liberty' occasionally by using very slight approximations should this appear desirable in the interests of simplicity.

It has been established that one valve only is conducting. Let us assume that V_1 is ON and V_2 OFF.

Consider the circuit in figure 15.4.2. as follows:

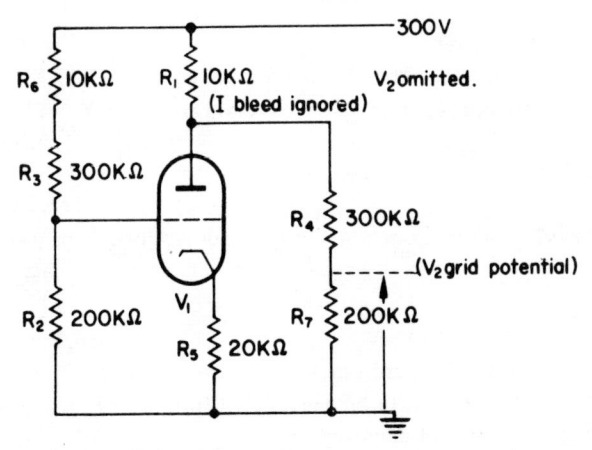

Fig. 15.4.2.

As V_2 does nothing to modify the circuit when off, it is omitted for clarity. We will also ignore any current flowing through $R_4 - R_7$, as this is small compared with the anode current. The first step is, of course to plot a d.c. load line as shown in figure 15.4.3. There is a slight difference between this circuit and the others previously described. If we consider the valve short circuit, the 'anode current' will be $300\,V/(20 + 10)\,K\Omega = 10\,mA$ and is one point required for our load line in figure 15.4.3. If we now consider the valve open circuit, V_{ak} will be

$$\frac{(200 + 300)\,K\Omega \times 300\,V}{(200 + 300 + 10)\,K\Omega} = 294\,V$$

(load over total once more). We have taken this to (300,0) as a close approximation. (In practice, the anode potential in this circuit could never rise above 294 V) V_1 grid is at a fixed potential of $300\,V \times 200\,K\Omega/510\,K\Omega = 117.5\,V$.

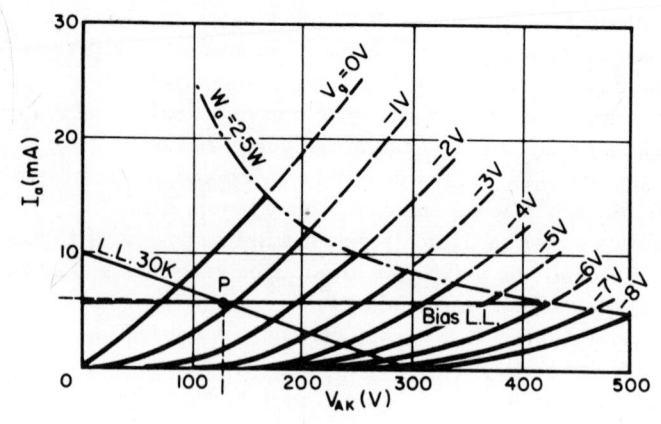

Fig. 15.4.3.

Bias load line

If $V_g = 117.5\,\text{V}$ and we assume zero bias, then $V_k = 117.5\,\text{V}$.
The current through R_k necessary to maintain $117.5\,\text{V} = 117.5\,\text{V}/20\,\text{K}\Omega$
$$= 5.84\,\text{mA}.$$
If $V_g = 117.5\,\text{V}$ and we assume $-5\,\text{V}$ bias, then $V_k = (117.5)\,\text{V}$
$$= 122.5\,\text{V}.$$
The cathode, or anode, current required to maintain this p.d.
$$= \frac{122.5\,\text{V}}{20\,\text{K}\Omega}$$
$$= 6.1\,\text{mA}.$$

These values will be sufficient for us to draw a bias load line. Remember that the line will theoretically be perfectly straight only when the spacing between the grid curves is constant.

If we mark a point where 5.84 mA intersects with $V_{gk} = 0$, and a second point where 6.1 mA intersects with $V_{gk} = -5\,\text{V}$, we have the two points we intend to join with a straight line. Identify the line as a bias load line. The intersection of the load lines drawn on the ECC81 characteristics show that the operating point P is at $V_{ak} = 122.6\,\text{V}$. $I_a = 5.91\,\text{mA}$, $V_{gk} = 0.8\,\text{V}$ and $V_k = 118.3\,\text{V}$. (This is shown in figure 15.4.3).

As V_1 is considered ON and V_2 considered OFF, we can reconstruct the circuit diagram in figure 15.4.4 and show the d.c. levels at each electrode.

$$V_g(v_2) = \frac{240\,\text{V} \times 200\,\text{K}\Omega}{300\,\text{K} + 200\,\text{K}\Omega} = \frac{240\,\text{V} \times 200}{500} = 96\,\text{V}.$$

V_2 is therefore cut off due to a bias of $118 - 96 = -22\,\text{V}$. Reference to the characteristics show that $-5\,\text{V}$ is sufficient to cut the valve off.

If we want both valves to change states, we must have to apply a negative pulse to V_1. The pulse will drag V_1 grid down thus reducing V_1 anode current.

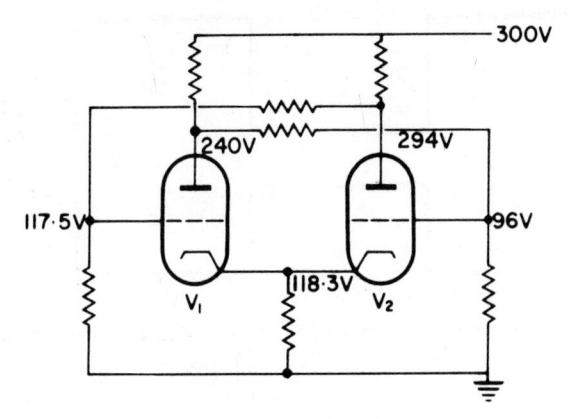

FURTHER LARGE SIGNAL CONSIDERATIONS. A BINARY COUNTER 251

Fig. 15.4.4

Consequently V_1 anode will rise, taking V_2 grid up with it into the region above -8 V thus causing V_2 anode current to flow. The resultant fall in V_2 anode voltage will drag V_1 grid farther down thus reducing V_1 anode current even more. V_1 anode voltage will rise further still, taking V_2 grid along the load line towards the $V_{gk} = 0$ point. V_2 anode has now fallen to 240 V, taking V_1 grid down to 96 V thereby cutting V_1 completely off; V_1 anode is then at 294 V. Both valves have changed states and the circuit is stable once more.

It is important to realise that the foregoing cumulative actions take place in a very short time. The actual time taken for the change over of states is easily measured by measuring the rise and decay time of the anode pulses. The sudden change in anode potential is once more a step function, and we require the majority of this signal to appear at the appropriate grid. The function of the Bi-stable circuit in figure 15.4.4. will now be examined. It has been shown that one negative input to V_1 (when on) will turn V_1 off. If a similar input is applied to V_2 (which is then on) it will turn V_2 off. The circuit will now be in its original state. It must be emphasised that both inputs must normally pass through some isolating circuit if the grids are not to follow both the leading and lagging edges of the input pulses. This refinement is discussed later, as we build the circuit step by step.

The waveform marked 'N' are negative going outputs taken *from V_1 anode only* (assuming V_1 on, initially), then for four input pulses we will have two output pulses as shown in figure 15.4.5.. The circuit divides by two. The method shown of applying an input alternately to each grid is not practicable. We need a device that, when input pulses are applied to a single input terminal, will automatically 'steer' the pulses to the appropriate grid of the valve that is ON at the instant each input pulse is applied.

The input might normally be a square wave or rectangular pulse. This is differentiated by $C_4.R_8$ and the positive going component is removed by D_3. This differential network may not be required with this steering circuit but

252 ELECTRONICS FOR TECHNICIAN ENGINEERS

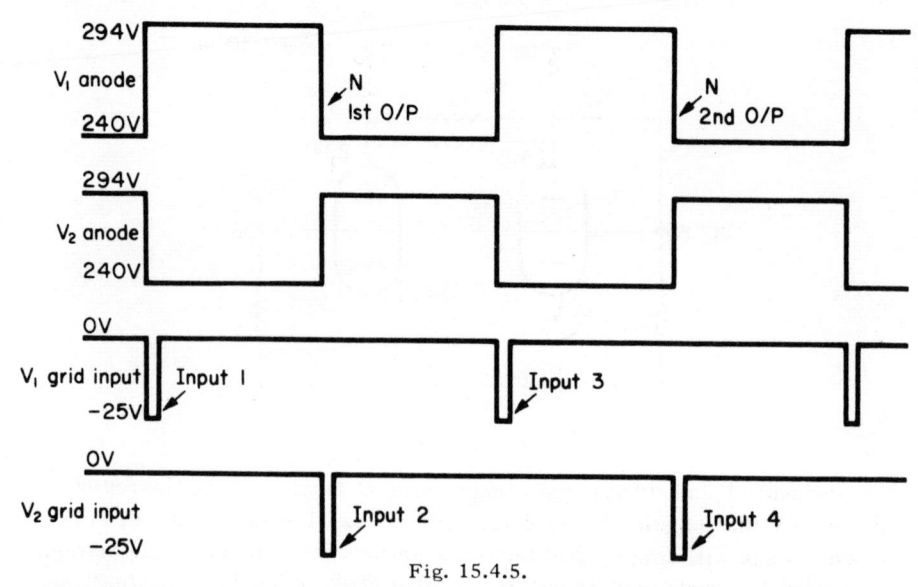

Fig. 15.4.5.

is included to emphasise the need for a negative going input. Hence a series of negative going pulses or pips is applied to *both* cathodes of the double diode, D_1 and D_2. Suppose that V_1 were ON, and that the negative input pulse is to be applied to V_1 grid. If V_1 is ON, V_1 anode is at 240 V. As D_1 anode is connected to V_1 anode, D_1 anode is at 240 V also. D_1 cathode is at 300 and the diode is therefore not conducting. D_1 is therefore 'OFF' by 60 V. The D_2 anode, which is connected to V_2 anode, is at 294 V. Although D_2 is non-conducting, it is 'off' by only 4 V. If the input signal is greater than 4 V, and negative going, D_2 will conduct, as its cathode is dragged down below its anode potential. When D_2 conducts, the input pulse passes through and arrives at the grid of V_1 as required. The circuit will then switch into its

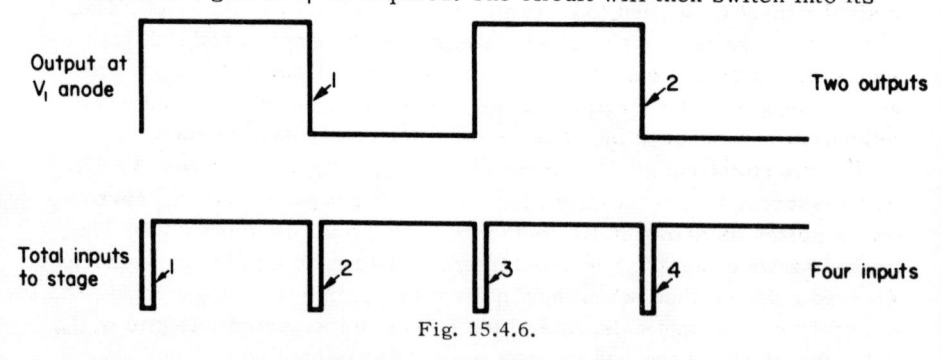

Fig. 15.4.6.

second stable state, and the next pulse is required to arrive at V_2 grid. D_1 will now be reverse biassed by 4 V, and will conduct upon receiving the next input pulse. The pulse will arrive at V_2 grid, causing the circuit to ultimately revert to its former stable state. The diodes D_1 and D_2 function as an electronic S.P.D.T. switch in this particular circuit. The capacitors C_1 and C_2 have a 'memory' function and are discussed later in this chapter. The input must be greater than 4 V in order to cause conduction of the appropriate diode but less than 54 V; a 54 V input might cause both diodes to conduct simultaneously. In our case -25 V input should suffice.

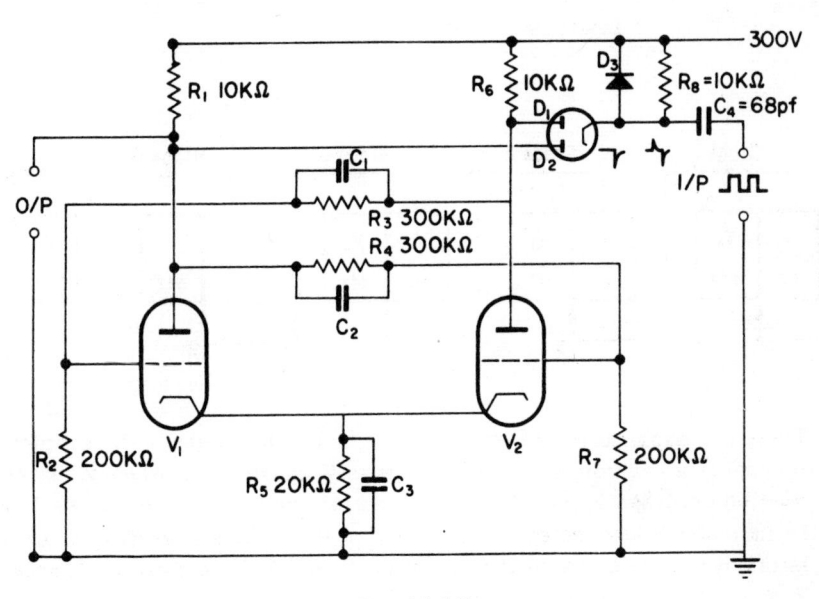

Fig. 15.4.7.

A great deal depends upon the negative going edge of the waveform in the circuit. R_2, R_3, R_4 and R_7 might be lower in value but then the 'bleed' current has to be allowed for in the calculations. For the moment it may be best to avoid any further complications. On occasions, the 'bleed' resistors might be very much higher than a megohm, in which case the anode of V_1 when cut off, would be almost the full H.T. voltage.

15.5. A simple binary counter

If the binary circuit is represented as a box, we can show how to connect each stage so as to form a simple counter. Figure 15.5.1.

It has been shown that for one negative pulse input, a positive output would result from V_1. A second negative input would cause V_1 to conduct, and a negative output would result. Therefore for every two negative inputs, one negative going output is obtained.

Suppose that four stages are connected as shown in figure 15.5.2.. Some indication is needed to show that V_1 is OFF.

Fig. 15.5.1.

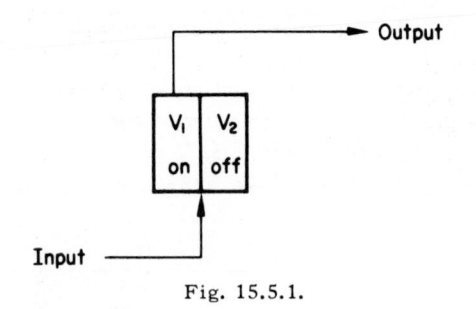

Fig. 15.5.2.

There are many methods from which to choose, but perhaps the simplest is to connect a low voltage indicator across V_2 anode load. When V_1 is off, V_2 must be on. When V_2 is on, a 50 V potential exists across its anode load. If the indicator is connected as described, it will indicate when V_2 is on and V_1 is off. It will be convenient to identify the stage 1 indicator as 1, stage 2 as 2, stage 3 as 4 and stage 4 as 8.

Let us show V_1 in the off state by placing a '1' in the following table and a '0' when V_1 is on. We must ensure that V_1 in each stage, is ON before commencing to count. This is known as the reset state.

Summary

Before any input is applied, all stages are reset to the '0' state, i.e. V_1 is ON. The first pulse causes V_1, stage 1, to stop conducting. A positive output results, and is fed to stage 2. A positive input to any stage does nothing, as D_3 removes all positive going pulse components. Stage 2 remains in the '0' state. Clearly the indicator marked '1' will be indicating one count.

When a second input is applied, stage 1 returns to the '0' state, but in doing so transmits a negative going pulse to stage 2, which also changes its state to a '1'. A positive going output leaves stage 2, but does not affect stage 3.

The third pulse changes stage 1 to a '1' state, but the resultant positive going pulse to stage 2 leaves stage 2 unaffected in the '1' state.

FURTHER LARGE SIGNAL CONSIDERATIONS. A BINARY COUNTER 255

	STAGE 1 INDICATOR MARKED	STAGE 2 INDICATOR MARKED	STAGE 3 INDICATOR MARKED	STAGE 4 INDICATOR MARKED
Input pulses.	①	②	④	⑧
0	0	0	0	0
1	1	0	0	0
2	0	1	0	0
3	1	1	0	0
4	0	0	1	0
5	1	0	1	0
6	0	1	1	0
7	1	1	1	0
8	0	0	0	1
9	1	0	0	1
10	0	1	0	1
11	1	1	0	1
12	0	0	1	1
13	1	0	1	1
14	0	1	1	1
15	1	1	1	1
16	0	0	0	0

The indicator now would show that (1 + 2) or 3 pulses had been counted. In each case one just adds the readings indicated by each indicator. At the 16th input, all stages revert to the '0' state, and a negative pulse leaves stage 4 and enters stage 5. Stage 5 would be labelled '16'.

15.6. Feedback in a simple counter

This simple counter might be required to count to 10 before resetting rather than 16. With the use of feedback, it is possible to arrange many different types of count. In all cases, the final number is the reset condition.

Reference to the tables shows that a following stage changes its state only when the preceeding stage changes from '1' to '0'. Changing from '1' to '0' is a negative output. Changing from '0' to '1' is a positive output. The table is reproduced, but this time the feedback paths are included which change a binary counter to one which completes its counting cycle at the 10th count.

It is known that the reset (16th count) state is '0' for all stages. This must now be the 10th count. Consequently the 9th count must be the previous 15th and the 8th count the previous 14th. Therefore the change due to feedback must take place after 7 (which is 7 for both systems) so that the new count number 8 is the old 14th.

S

Table showing feedback — Count of ten

The indicators are numbered as shown.

Pulses	1	2	4	2	
0	0	0	0	0	
1	1	0	0	0	
2	0	1	0	0	
3	1	1	0	0	
4	0	0	1	0	
5	1	0	1	0	
6	0	1	1	0	
7	1	1	1 Feed–	0	
8	0	0	0 back	1	We intend to
9	1	0	0	1	'jump' these
10	0	1	0	1	lines as we do
11	1	1	0	1	not need them
12	0	0	1	1	
13	1	0	1	1	
14	0	1	1	1	
15	1	1	1	1	
16	0	0	0	0	

This task is made easier because after 7 counts, stage 4 provides its first output pulse. This is a convenient point to change over. The feedback occurs at the eighth pulse. This changes columns 2 and 3 as shown below.

A final table shows the effect of the feedback. Note that counts 8–13 in the previous table have now been eliminated.

Final table — count of ten

Pulses	①	②	④	②	
0	0	0	0	0	
1	1	0	0	0	
2	0	1	0	0	
3	1	1	0	0	
4	0	0	1	0	
5	1	0	1	0	
6	0	1	1	0	
7	1	1	1	0	
8	0	1	1	1	(old 14)
9	1	1	1	1	(old 15)
10	0	0	0	0	(old 16)

(N.B. Stage 4 indicator is now marked 2 for this particular system).

FURTHER LARGE SIGNAL CONSIDERATIONS. A BINARY COUNTER

Consider count 8. No 8 pulse arrives, stage 1 (1 to 0) negative output goes to stage 2, which goes from 1 to 0: negative output causes stage 3 to go from 1 to 0; negative output to stage 4 which goes from 0 to 1, (and would in practice be 0001). But when stage 4 goes to 1, v_2 output is negative and changes stage 2 from 0 to 1 and stage 3 from 0 to 1, (both result in positive outputs which do nothing to following stages).

We have therefore, for the 8th count 0111. If we label the indicators 1, 2, 4, 2, this adds up to 8, corresponding to the 'old' 14th count. The circuit will, for the 9th count take up the old 15th state, and for the 10th count will take up the old 16.

The reader is recommended to try various feedback systems for different counts. Of course the indicators may have to be re-numbered, but this is of no consequence.

The feedback components may be as shown in figure 15.6.1.

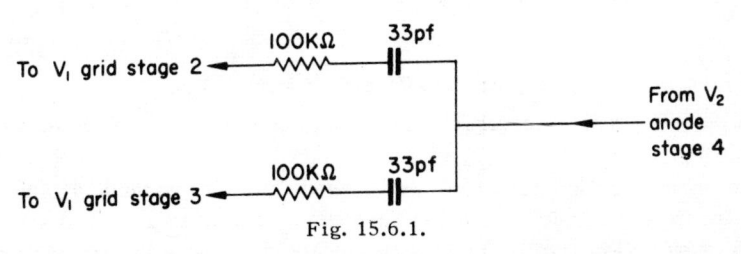

Fig. 15.6.1.

To reset all stages before applying a pulse, we might temporarily open circuit R_2 in each stage to chassis. This has the effect of raising V_1 grid potential thus ensuring V_1 is ON. A ganged switch contact permanently connected in the earthy ends of R_2 for each stage would be sufficient. This could be a spring loaded return push button as shown in figure 15.6.2.

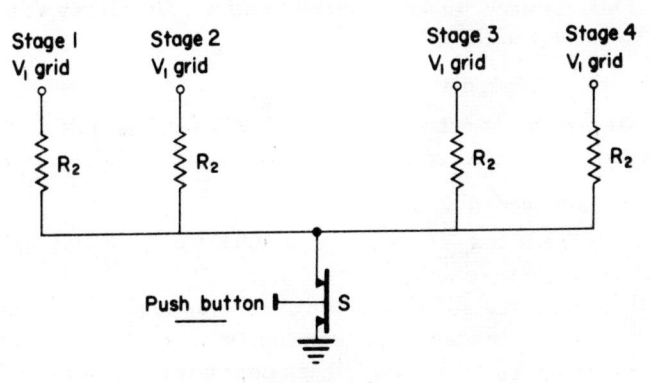

Fig. 15.6.2.

The grids are normally earthed via switch S and R_2.

15.7. Meter readout for a scale of 10

A rather more refined method of 'readout' is to use a meter, of the kind which is mechanically centre zero.

The potential across each anode to anode during conduction is ± 54 V. V. For the sake of simplicity, assume that this potential is 50 V. The exact potential is unimportant, as will be seen later.

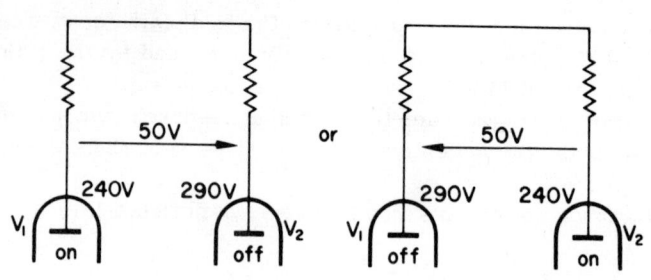

Fig. 15.7.1.

Consider stage 1. (Indicator marked 1)

A potential difference of 50 V has been assumed between the anodes of V_1 and V_2.

We connect a 1 MΩ resistor from V_1 anode to a rail marked 'A' and a 1 MΩ resistor from V_2 anode to a rail marked 'B' as shown in 15.7.2. A current of 25 μA will flow through each resistor provided that a low resistance is connected at the far end between rails A and B.

Stage 2 (Indicator marked 2)

A similar 50 V potential exists across the anodes of V_1 and V_2. As the indicator is 2, halve the resistors used in stage 1. Therefore a 500 KΩ resistor is taken from V_1 anode to rail A, whilst a 500 KΩ resistor is connected between V_2 anode and rail B.

Stage 3 (Indicator marked 4)

A 250 KΩ resistor is connected between V_1 anode and rail A. Similarly, a 250 KΩ resistor is connected between V_2 anode and rail B.

Stage 4 (Indicator marked 2)

This is similar to Stage 2, and uses a 500 KΩ resistor from v_1 anode to rail A and V_2 anode to rail B.

The 500Ω preset resistor is low in value so as to prevent a high impedance common load causing interaction between stages. It will also justify our assumption for low resistance connected across rails A and B.

Considering the currents in the 1 MΩ resistors 'stage 1', it is seen that the value of current flowing 50 V/2 MΩ = 25 μA. This figure ignores the 500 Ω as it is small compared with the 2 MΩ. Other stage currents may be similarly calculated.

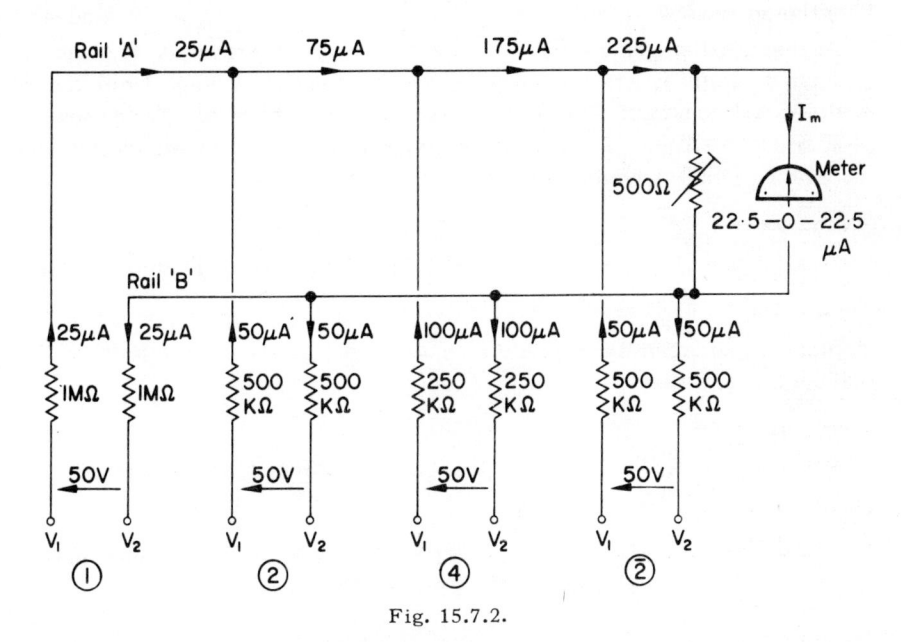

Fig. 15.7.2.

Assuming a meter resistance (R_m) of 4 KΩ, the current flowing through the meter (I_m) for a rail current of 225 μA will be

$$\frac{225 \,\mu A \times 0.5 \,K\Omega}{4 \,K\Omega + 0.5 \,K\Omega} = 25 \,\mu A.$$

This is the maximum current that could flow, and would occur only when *all* stages were either ON, or OFF at one time. When all states are ON, (V_1 ON) $I_m = -25 \,\mu A$, as shown in the circuit above. When all stages are OFF (V_1 OFF) as shown in figure 15.7.2. $I_m = +25 \,\mu A$. The amplitude of current is the same, but due to V_1 and V_2 in each stage changing states, the rail current in each stage would be reversed.

It is required to calibrate the scale from 0 to 9. There will be 9 divisions or spaces. The total F.S.D. (in both directions) $2 \times 22.5 \,\mu A = 45 \,\mu A$. At zero meter current, the pointer will indicate 4½ divisions (this is the mechanical centre zero). This cannot occur in practice, as some meter current must flow at all times. Remember that a negative current will cause a deflection to the left of the centre zero and a positive current a deflection to the right of the centre zero.

It is necessary to adjust the nominal 500 Ω preset resistor in order to adjust the meter current to 22.5 μA with all stages ON. (pointer indicating 0 counts). This will give $45/9 \,\mu A = 5.0 \,\mu A$ per division. This adjustment will 'take up' any tolerances in the assumed 50 V between anode-to-anode.

Function of readout circuit

At reset, all stages are 'ON', $I_m = -22.5\,\mu\text{A}$ pointer reads 0. At count of 1, Stage 1 of off. V_1 and V_2 change states. Subsequently, Stage 1 rail current reverses and substracts from other stage currents. The meter current now is $(-22.5 + 5.0)\,\mu\text{A}, = -17.5\,\text{A}$. This corresponds to a one division change and the pointer would indicate a count of 1.

Second pulse

Stage 2 now off. $I_m = -22.5 + 10 = -12.5\,\mu\text{A}$. Pointer indicates count of 2.

Third pulse

Stage 1 and Stage 2 are both off. $I_m = -22.5 + 15 = -7.5\,\mu\text{A}$. Pointer indicates count of 3.

Fourth pulse

Stage 3 now off. $I_m = -22.5 + 20 = -2.5\,\mu\text{A}$. Pointer indicates count of 4.

Fifth pulse

Stage 1 and 3 are both off. $I_m = -22.5 + 25 = 2.5\,\mu\text{A}$. Pointer now on the right hand side of centre zero, indicating the 5th count.

Sixth pulse

Stage 2 and 3 are both off. $I_m = -22.5 + 30 = +7.5\,\mu\text{A}$. Pointer now indicates the sixth count.

Seventh pulse

Stage 1, 2 and 4 are off. $I_m = -22.5 + 35 = +12.5\,\mu\text{A}$.

Eighth pulse

Stage 2, 3 and 4 now off. $I_m = -22.5 + 40 = +17.5\,\mu\text{A}$. Pointer now indicates the eighth count.

Ninth pulse

All stages now off. $I_m = +22.5\,\text{A}$. Pointer now indicates the ninth count. This is the final division.

Tenth count

All stages now ON. $I_m = -22.5\,\text{A}$. Pointer returns to 0 indicating no counts; the tenth count could be indicated on a second meter connected to a similar four stage counter. In practice however, should the meter read backwards, i.e. read 0 for 9, etc., it is a simple matter to reverse the meter terminals. In this example the rail current would be slightly higher than $25\,\mu\text{A}$ maximum because the assumed 50 V is really 54 V. This presents no problem however, as a slight re-adjustment of the 500Ω preset resistor

FURTHER LARGE SIGNAL CONSIDERATIONS. A BINARY COUNTER 261

would be sufficient to 'set up' the 22.5 μA needed for the F.S.D. of the meter.

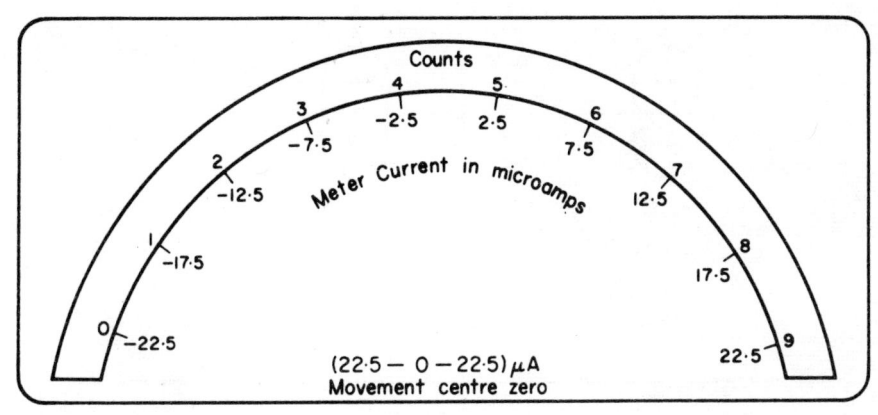

Fig. 15.7.3.

Time constants

The capacitor across R_K was connected to ensure that, if the valves passed different currents, there would be no variation of cathode potential during the changeover between one valve and the other. A later example in this book shows a similar circuit where the cathodes are taken direct to an earthy rail, hence this problem does not exist.

The capacitors C_1 and C_2 are called the 'speed up' capacitors. They have a multi-function. They help to keep the edge of the grid pulses sharp and by virtue of their respective charges, in relation to each other, they perform a 'memory' function and ensure that, when the circuit is momentarily in the unstable state during switchover, that the circuit continues to switch over and does not revert to its former state. Their values are influenced by many factors including the input p.r.f. and the resistors across which they are connected.

15.8. Design considerations of a simple bi-stable circuit

Many alternative approaches to the problems set in this book are offered. Consider the following circuit diagram in figure 15.8.1. Cathode variations are unwanted, therefore both cathodes are earthed. The grids obtain their bias via R_2 and R_6 and the negative rail. Let us consider the design of this circuit.

We are going to use a different set of I_a/V_{ak} characteristics in this circuit so as to provide wider experience in the use of valve characteristics.

The I_a/V_a characteristics are given in figure 15.8.2. It is necessary to establish the component values which will satisfy the circuit requirements.

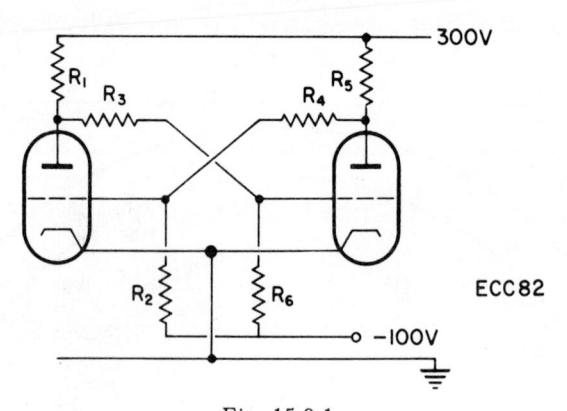

Fig. 15.8.1.

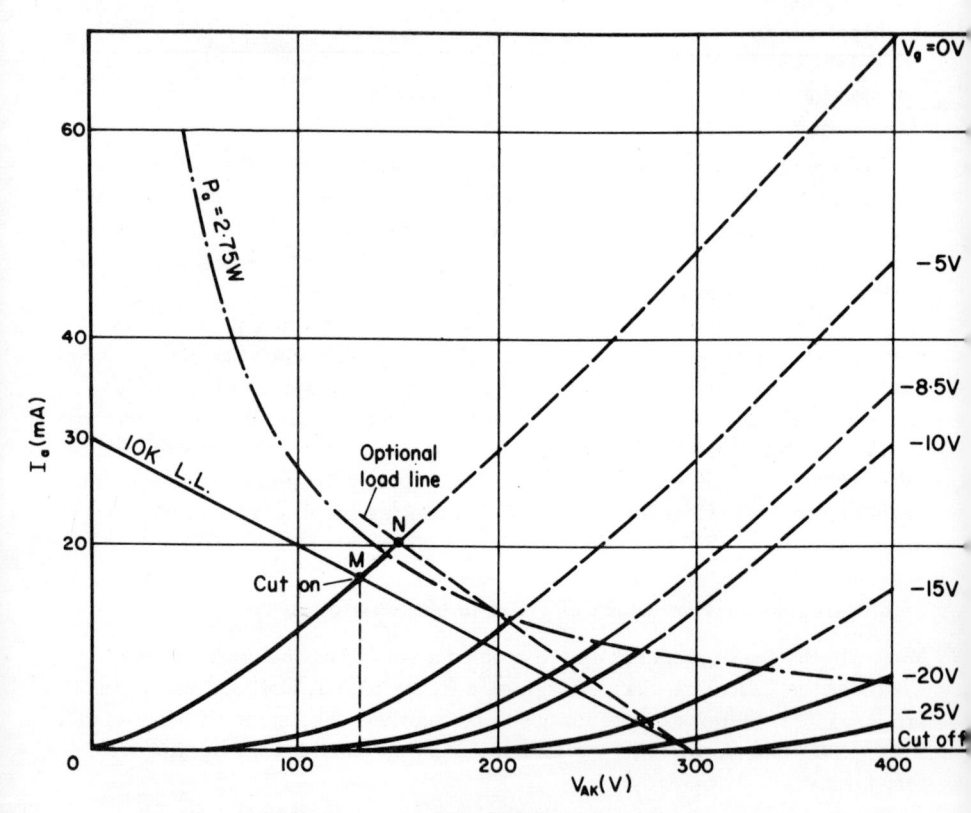

Fig. 15.8.2.

Supplies of $+300\,V$ and $-100\,V$ have been given.

Suppose we decide to 'aim' for an output signal of approximately half the h.t. i.e., 150 V. This will determine the anode load resistor and hence

FURTHER LARGE SIGNAL CONSIDERATIONS. A BINARY COUNTER 263

the slope of the load line. We have 300 V h.t. available, and therefore at $V_{gk} = 0$, we must ensure that the input V_{ak} below $V_{gk} = 0$, is approximately 150 V below our h.t.

The d.c. load line should be drawn now. This must normally, remain below the p.a. curve although we can cut it for one half of a double triode provided that only one valve is on at a time. The more vertical the load line, the lower the RL and the shorter the rise time of the output pulse. The load line has been chosen to connect points (300,0) and (0,30) corresponding to an RL of 10 KΩ. This allows some variation of the load line with respect to the p.a. curve due to the possible tolerance of the 10 KΩ. With V_1 'on' say, V_1 grid is at 0 V, reference to point M shows $V_{ak} = 132$ V. Consider now calculating the values of the coupling resistors R_3 and R_4. These should be at least 10 times R_1 therefore as $R_1 = 10$ KΩ, let $R_3 = R_4 = 150$ KΩ. The unwanted bleeder current may now be ignored.

Applying load over total and using -100 V as a reference, it is necessary to establish the minimum value of R_2 which would cause V_1 grid to be 0 V or more positive but *not* negative, to ensure V_1 is on.

(from figure 15.8.3.) $\quad 100\,V = \dfrac{400 \times R_2}{160 + R_2}$

$$\therefore \quad 16000 + 100 R_2 = 400 R_2$$

$$\text{and} \quad 16000 = 300 R_2$$

$$\text{thus } R_2 = \dfrac{16000}{300} = 53 \text{ KΩ (minimum)}$$

This is the minimum value for R_2.

Lowering the value would cause the grid to become more negative than 0 V, hence the valve would not be hard on. It is necessary to establish R_2 when V_1 is cut off by -25 V bias.

Fig. 15.8.3.

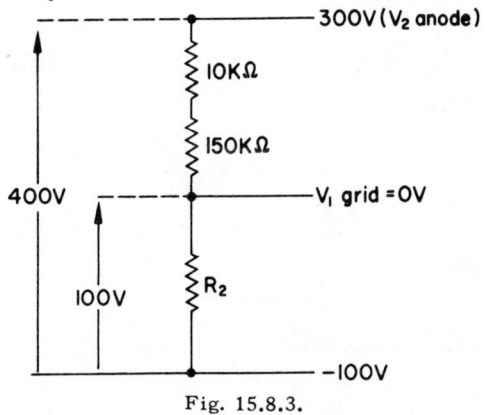

Fig. 15.8.4.

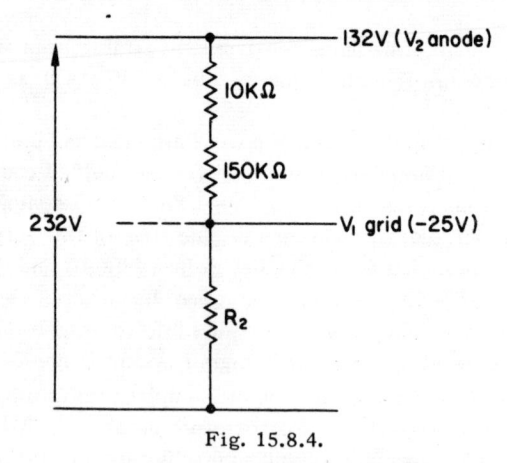

Using the load over total technique once more, (from figure 15.8.4.)

$$75\,V = \frac{232\,V \times R_2}{160 + R_2}$$

$$\therefore\quad 12000 + 75R_2 = 232R_2$$

$$\therefore\quad 12000 = 157R_2$$

$$\text{hence } R_2 = \frac{12000}{157} = 77.5\,K\Omega \text{ (maximum)}$$

A reasonable value will be 68 KΩ as it lies between these values.

A coupling resistor and the anode load resistor of the valve which is cut off, are in series with the h.t. and 0 V (at the other grid).

The anode voltage of the valve in the cut off state

$$V_A = \frac{300\,V \times 150\,K\Omega}{160\,K\Omega} = 280\,V. \text{ (By load over total)}.$$

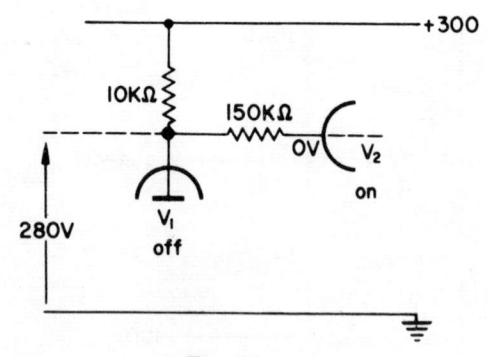

Fig. 15.8.5.

FURTHER LARGE SIGNAL CONSIDERATIONS. A BINARY COUNTER

V_2 grid is held at 0 V by grid current which clamps the grid to the cathode potential. This is shown in figure 15.8.5. The output pulse therefore would be as shown in figure 15.8.6.

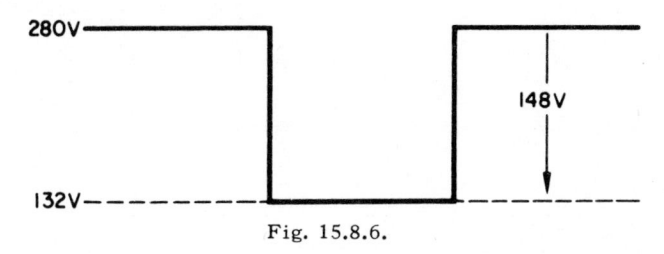

Fig. 15.8.6.

To complete the binary, a steering circuit would need to be added.

An optional load line is drawn on the characteristics in figure 15.8.2. It is drawn from (300, 0) to (150, 20) and will provide a 150 V anode swing from the point where the valve is in grid current (point N) to the cut off point. This represents $R_1 = 7.5\,\text{K}\Omega$.

This line would be a little more realistic in practice and does cut the p.a. curve a little. The reader should use this load line, assume $R_3 = 1.0\,\text{M}\Omega$ and evaluate R_2. The author has persistently erred on the safe side in his examples, for reasons which must be obvious. In practice however, once experience is gained, one might find that is is common practice to run components, including valves, much nearer to their maximum rating than has been shown throughout this book so far. This is something that each reader will have to decide for himself as he gains experience. When using a double triode within one envelope, and providing that one valve only is 'on', it is permissible to slightly overrate the maximum p.a. as that slight overheating within the envelope, for one valve, would be very much less than the stated maximum, which is for both valves 'on' at the same time.

The golden rule is however, to consult the manufacturer's data if in doubt on rating, etc.

CHAPTER 16

Further considerations of pulse and switching circuits

This chapter deals with a number of variations to the symmetrical multivibrators dealt with in chapter 15.

A number of further techniques are discussed and in particular, we will be looking at cathode follower circuits with a view to establishing maximum input voltages that may be applied before distortion occurs.

We will complete our discussion on Binary circuits, containing valves, with a final look at one technique enabling us to derive component values for given conditions.

This chapter will also complete our basic discussions on large signal analysis, and although in future chapters we will employ large signal analysis, we will not detail the basic steps as we have hitherto.

16.1. A cathode coupled binary stage

This circuit is a basic binary stage. It offers a further example in the application of Ohm's law. The circuit diagram is shown in figure 16.1.1.

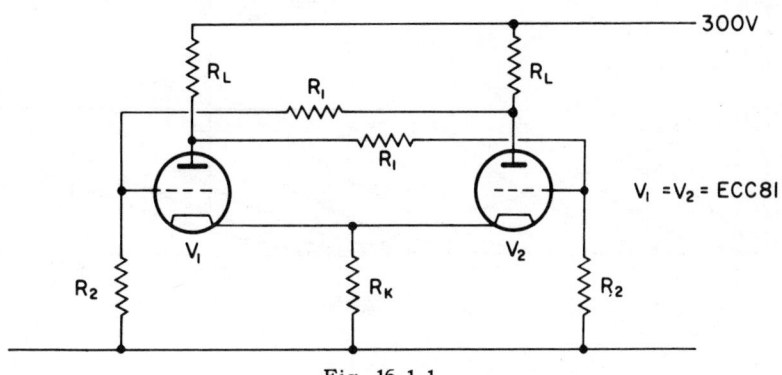

Fig. 16.1.1.

We shall determine the values of all of the resistors in the circuit but will not concern ourselves with the reasons for the choice of required circuit conditions. Let us assume that the conditions have been laid down and that we have to satisfy given requirements.

We have an ECC81 valve, an h.t. of 300 V, we have to ensure an output from the stage of between 60 and 70 V. We are not to let the valve current exceed 8 mA, and to ensure that the voltage across the valve, whilst 'on', does not exceed 100 V.

When the circuit is operational, one valve only will conduct at a time. Let us therefore consider one valve only. The circuit is symmetrical and hence it will be a simple matter to draw the whole circuit once we have established the values for the one valve alone.

We will also confine the discussion to d.c. considerations, and not concern ourselves with switching frequencies. This will allow us to examine changes in d.c. levels during both stable states.

These conditions allow an easy example to be shown and thus demonstrate the ease of evaluating the resistor values.

We require a maximum of 70 V output. A reasonable operating point for the valve when 'on' and held in grid current would be (89, 7.0). This will ensure that we do not exceed the 100 V across the valve, we will run it below 8 mA, and between conduction in grid current and complete cut off, we will have a change of 7 mA.

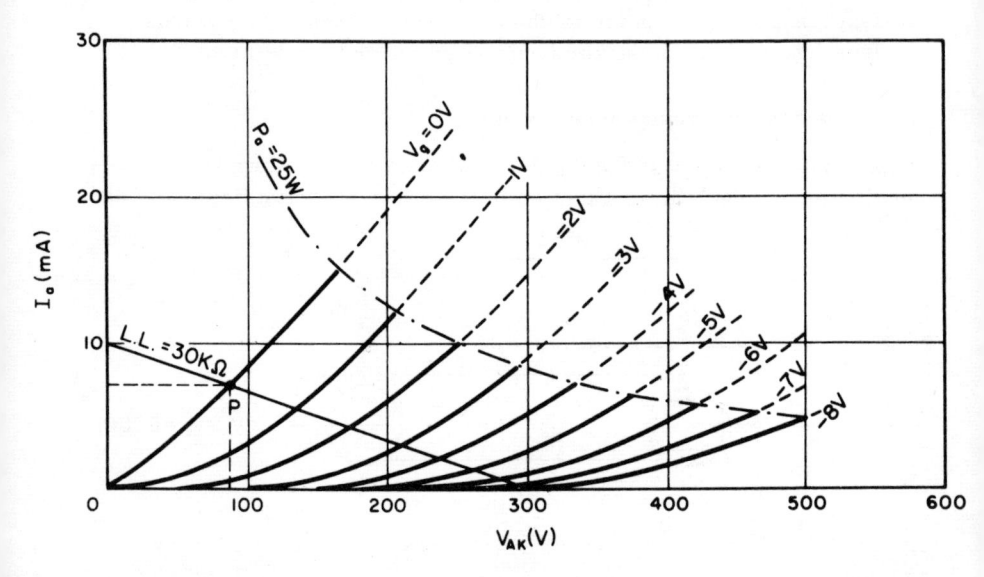

Fig. 16.1.2.

If we draw a load line from the point (89, 7.0), down to the 300 V point on the V_{ak} axis, this will determine the value of the sum of R_L and R_K. The load line has been drawn in figure 16.1.2. The load line represents a total resistance of $300/10 = 30$ K. As we will have a change in anode current of 7.0 mA, and we need, say, 70 V output, the value of $R_L = 10$ KΩ.

The cathode resistor must therefore be $30 - 10 = 20$ K.

When the valve is conducting, it will have a cathode voltage of 7×20 K $= 140$ V. The voltage across the valve will be $300 - 70 - 140 = 90$ V,

FURTHER CONSIDERATIONS OF PULSE AND SWITCHING CIRCUITS   269

and this is seen to be so on the characteristics. (The actual value is 89 V but we can ignore this slight discrepancy).

When the valve is to be cut off, its anode will be at approximately 300 V, if we make the resistors R_1 and R_2 large enough in value.

Its cathode however, will be at 140 V due to the other valve being in a conducting state. The voltage across the valve when 'off', will be 160 V. The characteristics show that, for a V_{ak} of 160 V, a bias of -4 V will cut it off.

We have established the values for the anode and cathode resistors. It now remains to determine values for R_1 and R_2.

Let R_1 be 1M. This is an arbitrary choice and is made high, so as not to drain an appreciable current through R_1 and interfere with our previous calculations.

It now remains to evaluate R_2, which has a particular ratio to R_1.

We intend to determine that ratio and hence, the value of R_2.

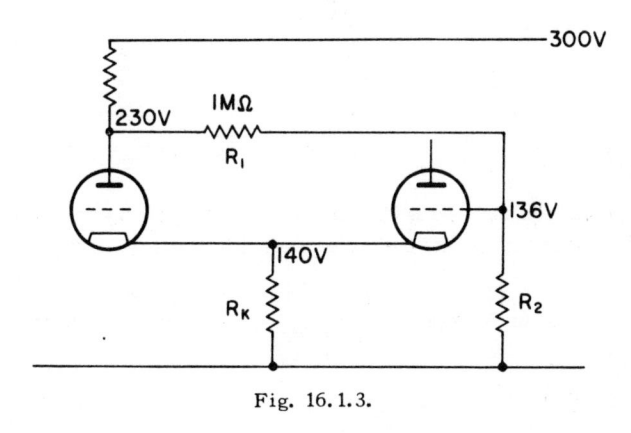

Fig. 16.1.3.

Figure 16.1.3. shows a simplified circuit. V_1 is in the 'on' state, and of course, V_2 is 'off'.

Under these conditions, V_1 grid must be at least -4 V with respect to the cathode voltage due to the current of V_1 flowing through R_K. As V_k is 140 V, the grid of V_2 must be at a potential of 136 V, thus ensuring a V_{gk} of -4 V.

The maximum value for R_2 to ensure that the grid of V_2 is at 136 V is determined as follows.

The potential at V_2 grid $= \dfrac{V_{a_1} \times R_2}{R_1 + R_2} = V_k - (V_{gk_0})$ where V_{gk_0} is the cut off bias.

$$\therefore (V_k - V_{gk_0})(R_1 + R_2) = V_{a_1} \times R_2$$

$$\therefore \quad R_2 = \frac{(V_k - V_{gk_0})R_1}{V_{a_1} - V_k + V_{gk_0}} = \frac{(140 - 4)(1\,M\Omega)}{230 - 140 + 4} = \frac{136\,M}{94} = 1.3\,M\Omega.$$

If the value of R_2 was to be increased above this value, there would be a danger that V_2 grid might not be sufficiently negative to cut the valve off.

Figure 16.1.4. shows the circuit, only this time, V_1 is off and V_2 is conducting, held in grid current at $V_{gk} = 0\,V$.

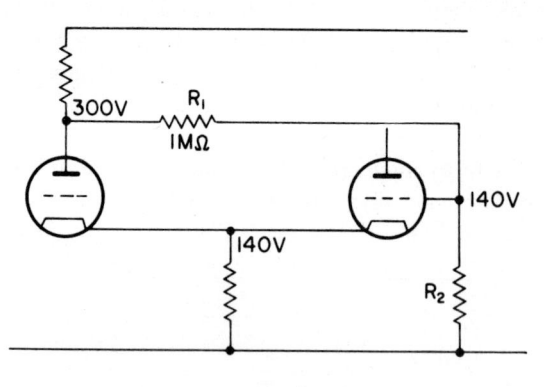

Fig. 16.1.4.

The value of R_2 must be such that, when V_1 anode has risen to 300 V, the grid of V_2 must be at the cathode potential, i.e., $V_{gk} = 0$.

Hence, $\quad V_2$ grid voltage $= \dfrac{V_{a_1} \times R_2}{R_1 \times R_2} = V_k$, for zero bias

$$\therefore \quad V_k (R_1 + R_2) = V_{a_1} \times R_2$$

$$\therefore \quad V_k R_1 = R_2 (V_{a_1} - V_k)$$

$$\therefore \quad R_2 = \frac{V_k R_1}{V_{a_1} - V_k}$$

$$\therefore \quad R_2 = \frac{140 \times 1\,M\Omega}{300 - 140} = \frac{140\,M\Omega}{160} = 875\,K\Omega.$$

This is the minimum value for R_2 to ensure that $V_{g_2} = V_k$ when $V_a = 300\,V$.

The value of R_2 therefore must lie between 1.3 MΩ and 875 KΩ. A 1 MΩ resistor seems a suitable choice.

The final d.c. circuit is shown in figure 16.1.5.

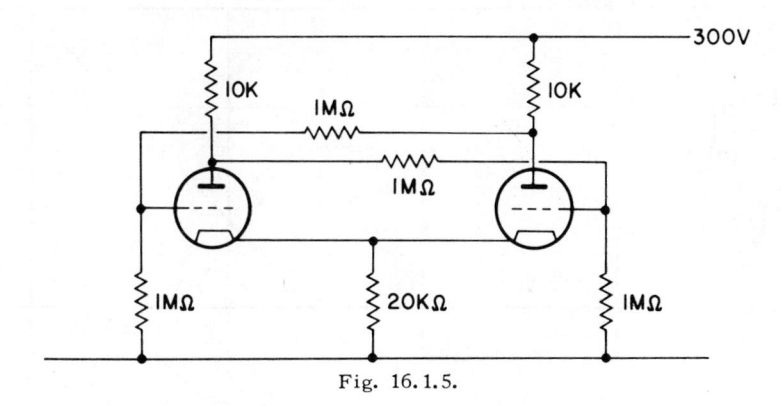

FURTHER CONSIDERATIONS OF PULSE AND SWITCHING CIRCUITS 271

Fig. 16.1.5.

16.2. A biassed multivibrator analysis

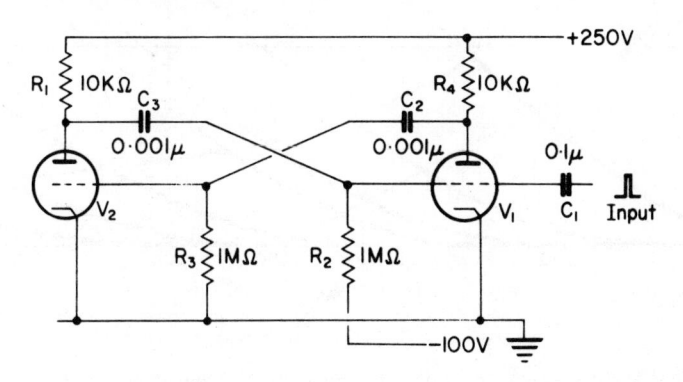

Fig. 16.2.1.

A monostable circuit is shown in figure 16.2.1. V_1 is held below cut off due to the bias obtained via R_2 and the -100 V rail. A glance at the anode characteristics will show that with 250 V h.t. and an anode load of 10 KΩ, a bias voltage of -20 V is sufficient to cause cut off (figure 16.2.2).

As V_2 grid is at 0 V, C_2 is fully charged to 250 V, as V_1 anode current is zero. V_2 grid is at 0 V and consequently V_2 anode current is 13.5 mA (VGK = 0 V).

This then is the only stable (hence monostable) state that the circuit has. A positive pulse applied to V_1 grid via C_1 causes V_1 grid to be lifted above cut off (any level positive to -20 V will do), which causes V_1 to conduct. V_1 anode falls rapidly, taking V_2 grid down below cut off. The actual fall is shown by the characteristics to be -138 V. As V_2 is cut off, V_2 anode rapidly rises to 250 V, taking V_1 grid up to 0 V. V_1 is now conducting heavily, and V_2 is cut off.

At this point, we will assume that the input pulse has been removed.
T

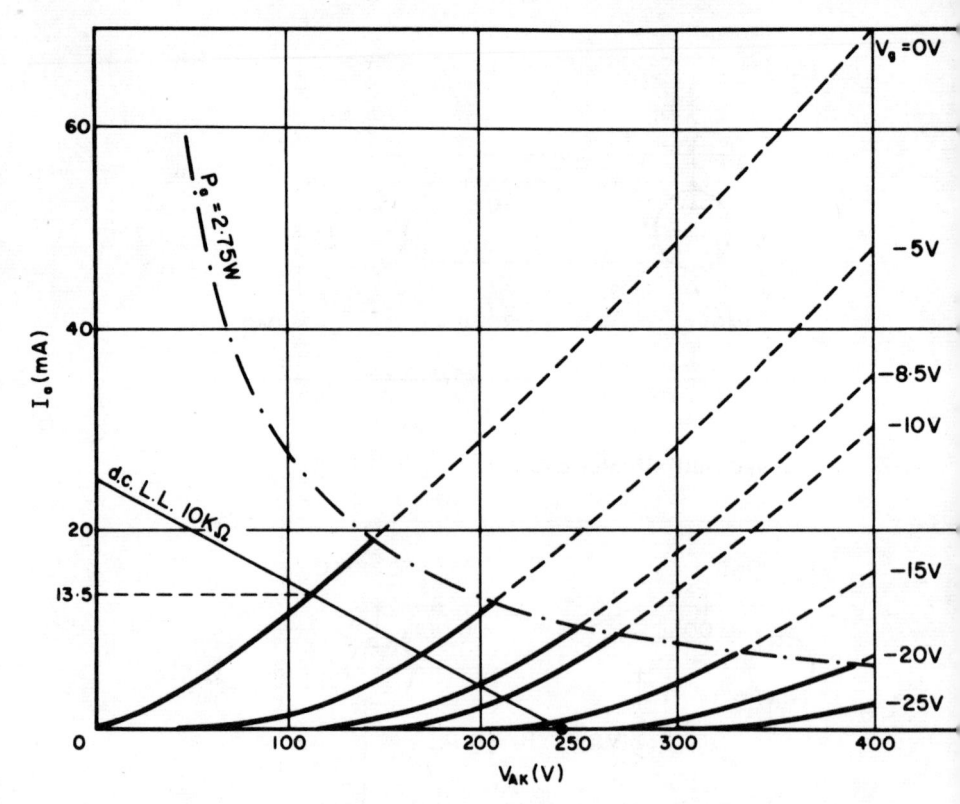

Fig. 16.2.2.

As this state is now unstable, the circuit will begin to revert to its previous stable state. V_2 grid is now sitting at -138 V. Therefore 138 V exist across R_3, and using the same arguments put earlier on, it is this voltage to which C_2 must 'charge'. This -138 V is due to the current in R_3 supplied by C_2. The capacitor cannot maintain this current and the potential across R_3 gradually diminishes in an exponential manner.

At the instant V_2 is off, C_2 must discharge through R_3 and the negative plate will attempt to rise exponentially to 0 V. This climb is arrested when the grid reaches -20 V as the circuit quickly reverts to its previous state.

The 'aiming' voltage is 138 V. The actual climb is 118 V.

An expression for the capacitor voltage is $V_c = V(1 - e^{-t/CR})$

Therefore $\dfrac{V_c}{V} = (1 - e^{-t/C.R})$ and $1 - \dfrac{V_c}{V} = e^{-t/C.R}$

thus $\dfrac{V - V_c}{V} = e^{-t/C.R}$

then $\dfrac{V}{V - V_c} = e^{-t/C.R}$ and taking $\log_e$ of both sides

FURTHER CONSIDERATIONS OF PULSE AND SWITCHING CIRCUITS 273

$$\text{Log}_e \frac{V}{V - V_c} = t/C.R.$$

Hence $\quad t = C.R. \log_e \dfrac{V}{V - V_c} = C.R. \log_e \dfrac{138}{138 - 118}$

and $\quad t = C.R. \dfrac{138}{20} = C.R. \log_e 6.9$

Thus $\quad T = C.R. \times 1.93$ hence $T = [(0.001 \mu F \times 1 M\Omega) \times 1.93]$ S

Therefore $\quad T = 1.93 \, mS.$

The complete pulse with its associated d.c. levels is shown in figure 16.2.3.

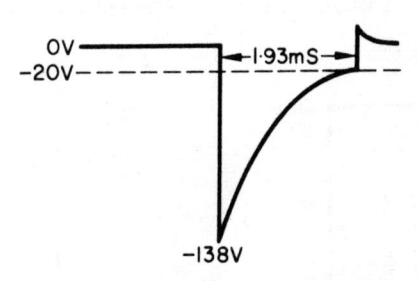

Fig. 16.2.3.

As V_2 grid passes above -20 V, the valve conducts, causing its anode to fall — thus dragging V_1 grid down below cut off. As the input pulse is of a very short duration and is no longer present, V_1 remains cut off due to the -100 V rail. The circuit will now remain in its stable state awaiting a further input pulse.

Note: The 'home study' reader with limited mathematical ability could derive the answer by using the 'charging curve' at the end of chapter 14.

16.3. A direct coupled monostable multivibrator — a simple analysis

Circuit diagram — Figure 16.3.1.

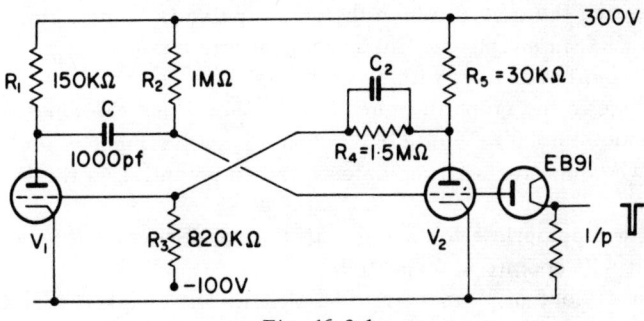

Fig. 16.3.1.

This circuit is not symmetrical. Each valve has a different value anode load. This necessitates considering each one quite separately. Following the principles established earlier, the first step is to plot a separate load line for V_1 and V_2. R_1 has a value of 150 KΩ, this will give a 'short circuit' current of 2 mA. The coordinates for the load line are (0, 2.) and (300, 0.).

As V_1 grid is returned to the -100 V rail, it may for the moment, be assumed that V_1 is not conducting which infers that V_2 is conducting.

The next step is to plot a load line for V_2 anode load. This has a value of 30 KΩ. The 'short circuit' current will be 10 mA. The coordinates for this load line are (0, 10.) and (300, 0).

The load lines should be identified at this stage in the normal way. (see figure 16.3.3). The anode voltage 'changes' are seen to be as shown in figure 16.3.2.

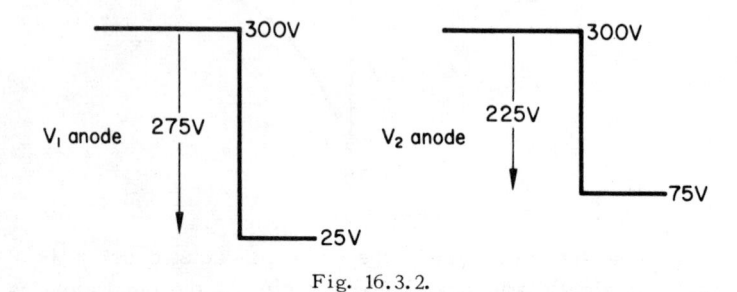

Fig. 16.3.2.

A negative input to V_2, via the isolating diode, causes V_2 to be cut off. V_2 anode voltage rises towards the h.t. line. This rapid rise drags V_1 grid up from cut off into its operating region. V_1 is now on and its anode potential has fallen. The fall in anode potential takes V_2 grid even further down and ensure complete cut off of V_2. The negative potential at V_2 grid causes the diode anode to become negative with respect to its cathode; the diode is now cut off also. The diode, whilst off, cannot effect the timing circuit consisting of R_2 C_1, thus the input circuit is isolated from the multivibrator during the time in which the output pulse is being formed. V_2 grid begins to rise, in an exponential manner, at a rate governed by R_2 C_1., until V_2 is on, once more, with V_2 grid at 0 V.

Reference to the graph in figure 13.3.3, shows that the *change* in output from V_2 anode is $300 - 75 = 225$ V, when a negative input is applied to V_2 grid. The negative sign indicates a fall in potential as the valve V_2 conducts.

It will be appropriate to consider all the components and ascertain that the circuit will operate as expected.

The anode load of V_2 is seen to be 30 KΩ. The coupling resistor R_4 should be at least 10 times that of R_5. R_4 is seen to be 1.5 MΩ, and that is quite satisfactory. The resistor R_3 forms the lower half of a potential divider, and its value in relation to R_4 is very critical. It must have a value such that, when V_1 is meant to be off, its grid is cut below cutoff, yet, when V_1 is meant to be on, its grid is at 0 V, at least. With an h.t. of 300 V, the grid potential necessary to ensure cutoff, is seen from the anode characteristics to be -25V.

It is desirable to simplify the circuit under investigation. Hence it will be easy to determine the correct value of R_3 in relation to the known value of R_4.

The fall at V_2 anode is seen to be 225 V. This occurs when V_2 is on. V_1, then, must be off. The -225 V from V_2 anode must drag V_1 down to -25 V at least. Knowing R_4, it is an easy matter to establish the minimum value of R_3. Using the load over total technique, once more, consider this divider and apply ohms law.

If the two conditions of V_2 anode are shown, also the required potentials of V_1 grid, the problem is greatly simplified.

Note that V_2 anode is either at 300 V or 75 V.

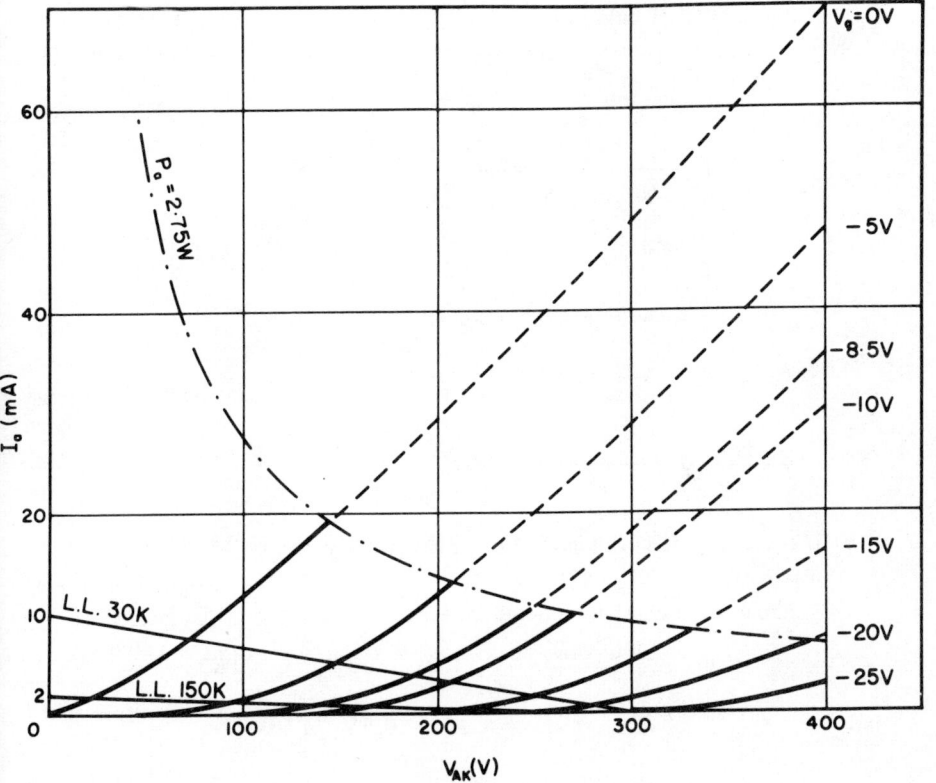

Fig. 16.3.3.

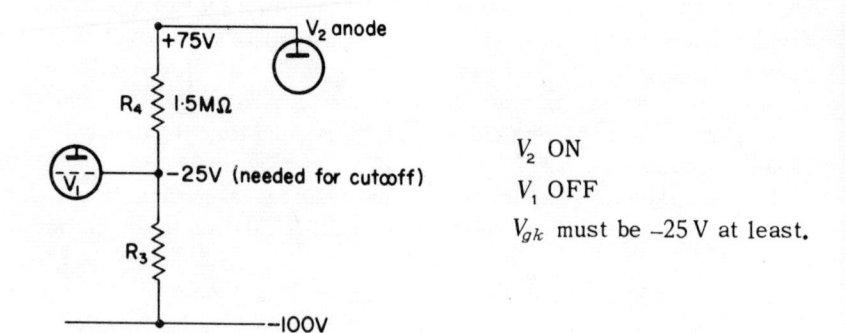

Fig. 16.3.4.

V_2 ON
V_1 OFF
V_{gk} must be -25 V at least.

Referring all voltages to the -100 V rail, V_{gk} must be 75 V, or less.

$$75 = \frac{175\, R_3}{1.5 + R_3} \quad \therefore \quad 112.5 + 75 R_3 = 175 \, R_3 \quad \therefore \quad 100 R_3 = 112.5$$

$$\therefore \quad R_3 = 1.125\,\text{M}\Omega$$

This is the MINIMUM value of R_3.

When V_2 is off, v_1 must be on. V_{gk} must be at 0 V, at least to ensure that the valve is in grid current.

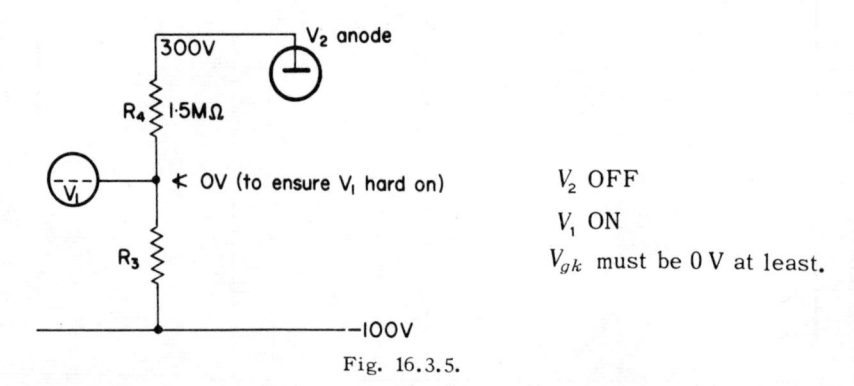

Fig. 16.3.5.

V_2 OFF
V_1 ON
V_{gk} must be 0 V at least.

Still referring all voltages to the -100 V rail, V_{gk} must be 100 V. (This is identical to $V_{gk} = 0$ V, w.r.t. earth).

Using load over total technique once more,

$$100 = \frac{400 \times R_3}{1.5 + R_3} \quad \therefore \quad 150 + 100 R_3 = 400 R_3$$

and $\quad\quad\quad 300 R_3 = 150 \quad \therefore \quad R_3 = 0.5\,\text{M (or less)}.$

This is the MAXIMUM value of R_3.

FURTHER CONSIDERATIONS OF PULSE AND SWITCHING CIRCUITS 277

R_3 then, must have a value that lies between $0.5\,\text{M}\Omega$ and $1.125\,\text{M}\Omega$. The value originally chosen of 820 K, appears to meet the circuit requirements.

The capacitor C_2, is a speed up capacitor as with the circuits previously considered. From the characteristics, it is seen that the change in anode voltage for V_1 is 275 V. This appears at V_2 grid, as a negative signal.

The V_2 grid waveform is as shown in figure 16.3.6.

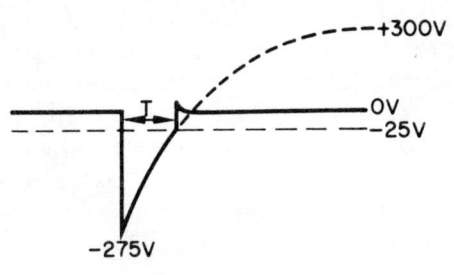

Fig. 16.3.6.

The pulse width T may be calculated, as follows.

$$T = C.R. \log_e \frac{V}{V - V_c} \quad \text{where } V = 575\,\text{V}.\ V_c = 250\,\text{V}.$$

The output waveform from V_2 anode is as shown in figure 16.3.7.

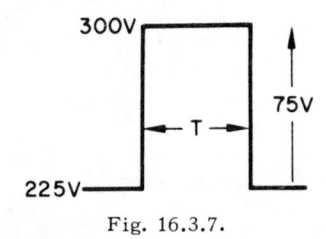

Fig. 16.3.7.

16.4. Cathode followers. Maximum input and grid current threshold

It has been previously demonstrated that the effective input to a valve amplifier is the change in potential between grid and cathode, V_{gk}.

For a valve with a bypassed cathode resistor, and operating at say, $-10\,\text{V}$ bias, the input to the grid could rise to 10 V only, if the valve is not to run into grid current. Beyond this value, V_{gk} would be zero and the valve would be operating on the threshold of, or in, grid current. The grid could if driven harder, rise a little above this value and would be limited by the forward drop of the resulting grid-cathode diode.

If the cathode resistor is bypassed, the cathode will, during the period of time that the input signal is applied, be effectively short circuited to chassis and the input V_g will be equal to the grid cathode voltage, V_{gk}. Therefore the full input will be the effective input, and, as stated, 10 V only may be applied.

If the cathode resistor is unbypassed, the cathode will follow the grid. As the grid rises, so will the cathode. Had the valve been operating at -10 V, the maximum input may be many times the 10 V before the threshold of grid current is reached. V_{gk} will, under these conditions, remain almost constant. If the gain of the cathode follower was 0.9, then with an input to the grid of 10 V, the effective input, V_{gk}, would become 1 V.

It follows therefore, that many times the normal input may be applied to a cathode follower before grid current will flow. With some circuits, the input may approach almost half of the h.t. before grid current flows.

Figure 16.4.1. shows a cathode follower, the bias is obtained by connecting the grid leak to a tapping point in the cathode resistor. The circuit will be examined from a d.c. point of view followed by consideration of the maximum input signal.

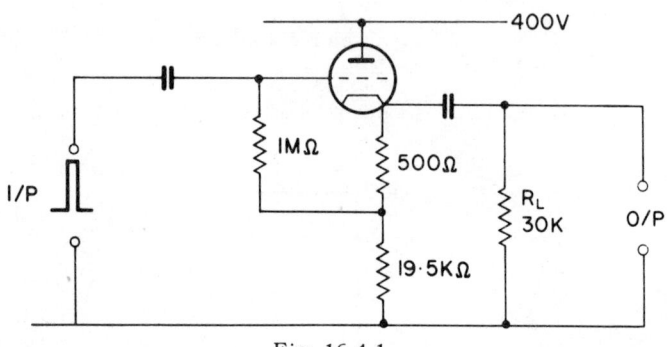

Fig. 16.4.1.

Reference to figure 16.4.2. shows that plotting a d.c. load line for the 20 KΩ and a bias load line for the bias resistor of 500 Ω, shows that the quiescent conditions are as follows. V_{ak}, 274 V. I_a, 6.3 mA. Bias, 3.1 V. V_k, 126 V.

As there is no grid current, and no appreciable input signal current, there can be no drop across the grid leak, therefore the potential V_g is the same as the potential V_k plus V_{gk} ∴ V_g = 128 − 3.1 = 122.9 V say 123 V and of course, V_k = 126 V.

a.c. conditions

An a.c. load line drawn next, is constructed in the usual manner. The uppermost point of the load line is obtained by dividing the V_{ak} by the a.c. load. The a.c. load is 20 K/30 K = 12 KΩ. The point corresponding to the upper end of the load line is

FURTHER CONSIDERATIONS OF PULSE AND SWITCHING CIRCUITS 279

$$I_q = \frac{V_{ak}}{\text{a.c. load}}$$

therefore the point becomes $6.3 + 274/12\,\text{K} = 6.3 + 22.8 = 29.1\,\text{mA}$. The coordinates are $(0, 29.1)$. A second point is of course, point P.

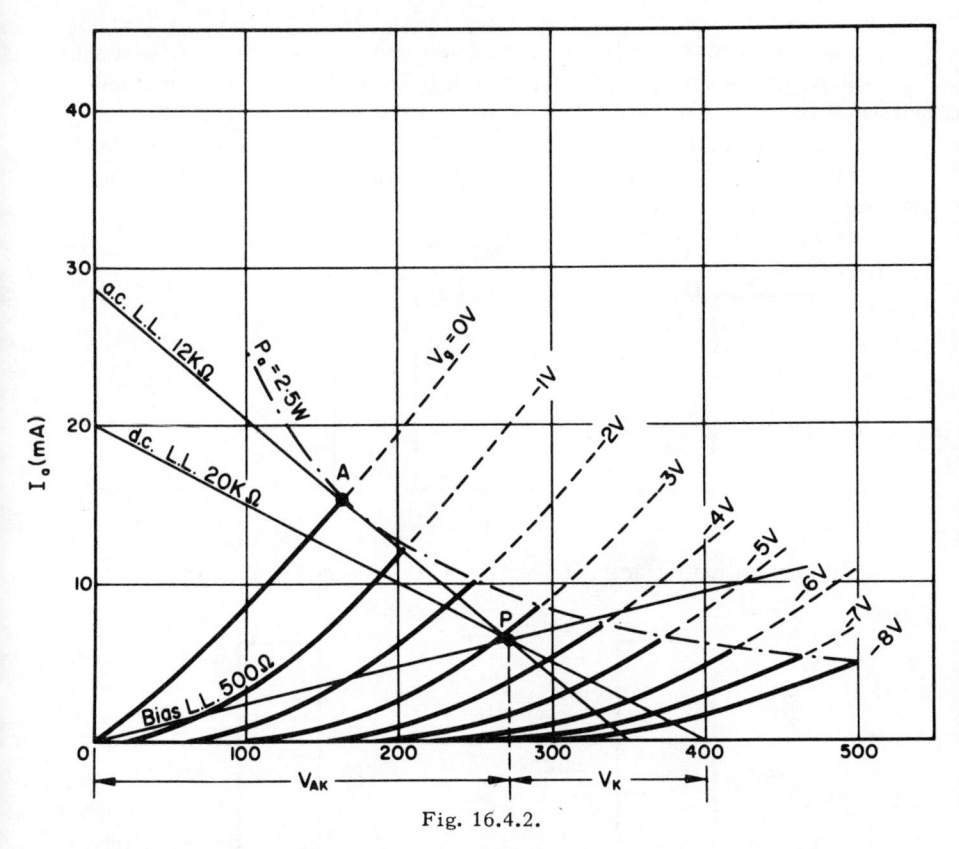

Fig. 16.4.2.

The a.c. load was chosen so as to cause the load line to just touch, but not cut, the p.a. curve. Cutting through the p.a. curve will result in the anode dissipating a wattage greater than that specified by the valve manufacturers. (In previous examples we have not touched the p.a. curve, just to be safe. This example is a little more practical). The load line should be completed by continuing the load line until it cuts the V_{ak} axis.

Pulse input

A positive going pulse with an extremely short rise time, may be considered as an instantaneous change in d.c. level. The time constants of the resistor–capacitor combinations are assumed to be such that the capacitors will not lose any appreciable charge during the steady state conditions of the input pulses. The no-signal levels have been established from the d.c. considerations. An a.c. load line has been drawn and when a signal is applied, the grid will 'move' along the a.c. load line in the 'positive' direction of the input signal. It now remains to 'apply' a positive going input signal of sufficient amplitude to cause the grid-cathode difference, V_{gk}, to become zero.

This is of course, the threshold of grid current. Any further increase in amplitude of input will cause grid current to flow. When the grid is at the point A ($V_{gk} = 0$) new values are given for V_g and V_k. It is seen that the new V_{ak} is 167 V. Therefore V_k must be 233 V. As V_{gk} is zero, V_g is at the same potential as V_k. Therefore V_g is 233 V. The grid and cathode waveforms are shown in figure 16.4.3.

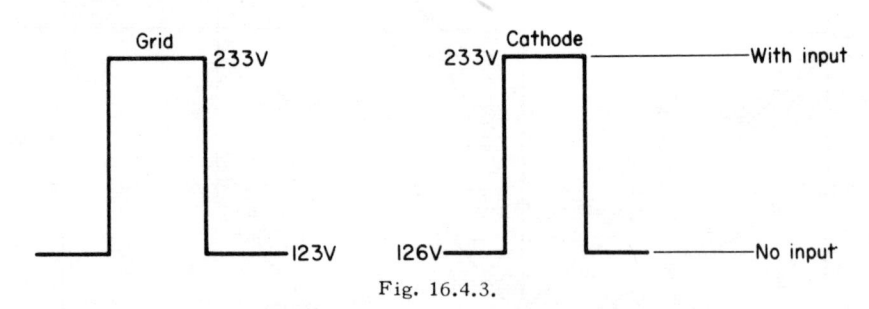

Fig. 16.4.3.

The following illustration in figure 16.4.4. completes the picture.

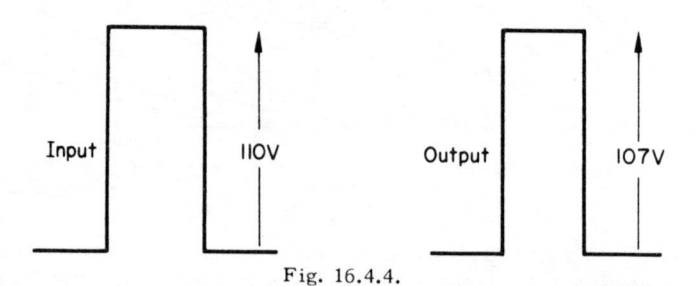

Fig. 16.4.4.

The gain of the cathode follower is seen to be 107.0/110.0 and is less the unity.

Although originally biassed at −3 V, the input signal required to take the 'amplifier' to the threshold of grid current is in the order of 110 V.

It may be seen that the output level is equal to the input less the original bias. The gain = 107/110. Under these conditions, the input required will be

$$v_{in} = \frac{\text{Bias (original)}}{\text{Gain} - 1}$$

FURTHER CONSIDERATIONS OF PULSE AND SWITCHING CIRCUITS 281

Noting that the bias is a negative quantity, the input in this case, will be

$$v_{in} = \frac{-3}{(107/110) - 1} = 110\,\text{V input}.$$

To cut the valve off, a V_{gk} of $-9\,\text{V}$ is required. The characteristics show that when $V_{gk} = -9\,\text{V}$, $V_{ak} = 350\,\text{V}$. Therefore $V_k = 50\,\text{V}$. $V_g = 50 - 9 = 41\,\text{V}$. An input signal of $-(123 - 41) = -82\,\text{V}$ is necessary to cut the valve off as shown in figure 16.4.5.

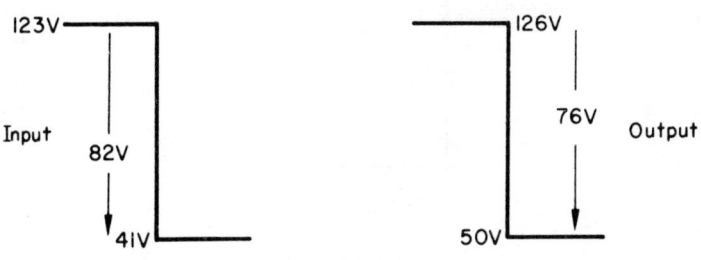

Fig. 16.4.5.

16.5. A phase splitter analysis

This is a double triode ECC81 connected as phase splitter circuit. The grid is held at a d.c. level of 50 V i.e.,

$$V_g = 400\,\frac{\text{load}}{\text{total}} = \frac{400 \times 50\,\text{V}}{400} = 50\,\text{V}.$$

A load line for $(10 + 10)\,\text{K}\Omega$ is constructed between points $(400, 0)$ and (0.20). (Remember that this load line really represents $-1/20\,\text{K}$ as the X Y axes are inverted.) Figure 16.5.1. shows the circuit to be analysed.

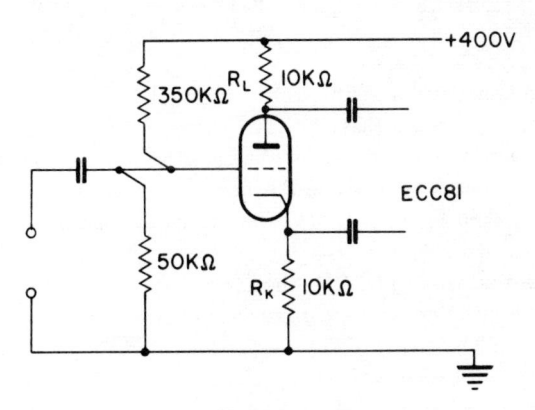

Fig. 16.5.1.

For the present example, consider only the d.c. conditions.

Graphical method

Figure 16.5.2. shows the static characteristic of the triode, and a load line is shown representing a total resistance of 20 KΩ ($R_L + R_K$).

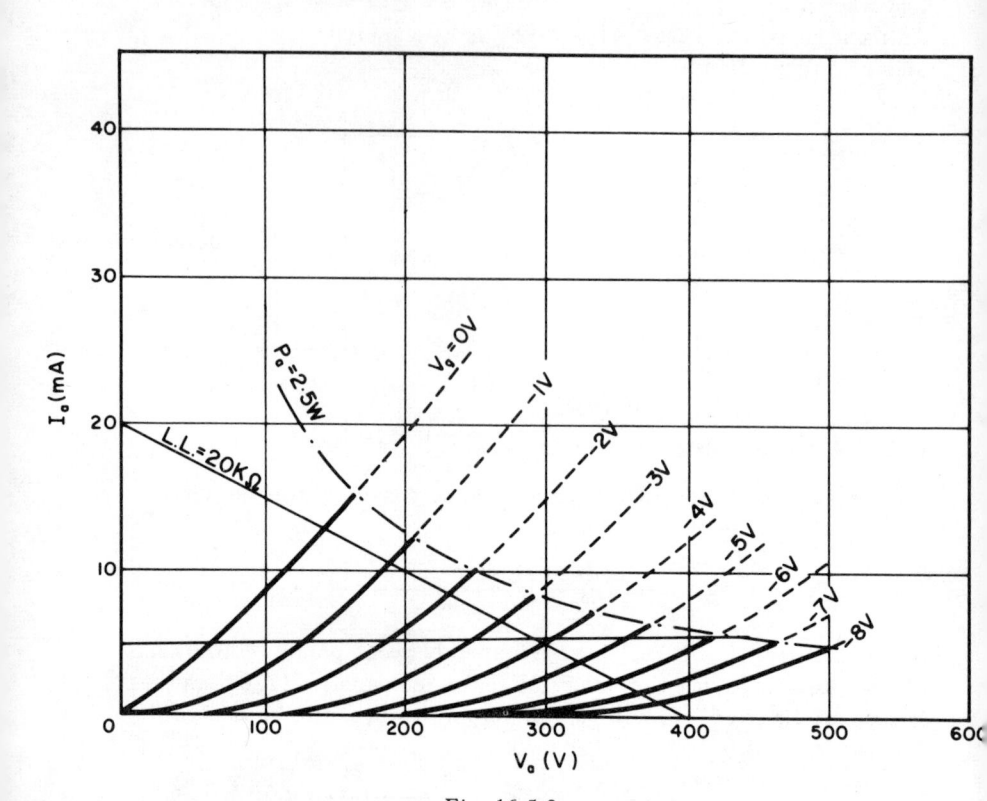

Fig. 16.5.2.

It is required to construct a simple table in order to find the quiescent operating point. It is known that $V_g = 50$ V. Assuming various values of bias, as before, and using simple Ohm's law, the anode (and hence cathode) current in R_K necessary to produce the assumed bias (V_{gk}) may be calculated. If V_{gk} were zero, then $V_k = V_g = 50$ V. The I_a must then be = 50 V/10 KΩ = 5 mA.

If we assume the bias to be 2 V, then the cathode will be 2 V positive to the grid, i.e. 52 V. The anode (or cathode) current necessary to produce 52 V across R_K = 52 V/10 KΩ = 5.2 mA.

The following table shows further values of calculated I_a according to the assumed values of bias.

V_g	Bias (V_{gk})	V_k (V)	I_a (mA)
50	0	50	5.0
50	−2	52	5.2
50	−3	53	5.3
50	−4	54	5.4
50	−5	55	5.5
50	−6	56	5.6

These points are plotted in the normal way, for the subsequent bias load line. The intersection of both lines give the operating point as shown in figure 16.5.3.

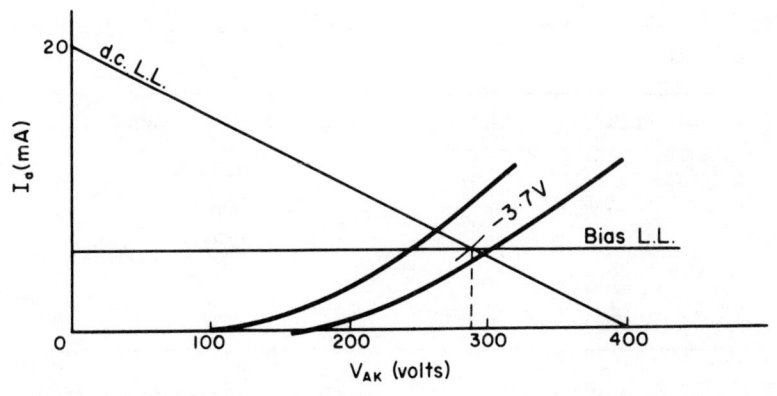

Fig. 16.5.3.

Bias = 3.7 V. I_a = 5.37 mA. V_{ak} = 292.6 V.
The remaining d.c. voltage (400 − 296 − 6) V must be divided between R_L and R_K, depending upon the ratio of the resistor values. As they are the same, 10 KΩ, each will have a p.d. of 53.7 V. The final picture is shown in figure 16.5.4. which is the original circuit diagram plus the voltages and current calculated above.

Simpler approach — non-graphical

Suppose we now apply the quicker but slightly less accurate method of determining the above d.c. conditions. As before, assume V_{gk} = 0 V.
If V_{gk} = 0 V, then as V_g = 50 V then V_K = 50 V. I_a necessary to produce 50 V across R_K = 50 V/10 KΩ = 5 mA. 5 mA flowing through R_L causes a p.d. across R_L of 50 V.
V_{ak} then = (400 − 50 − 50) V = 300 V.
$V_a = V_{ak} + V_K$ = (300 + 50) V = 350 V.

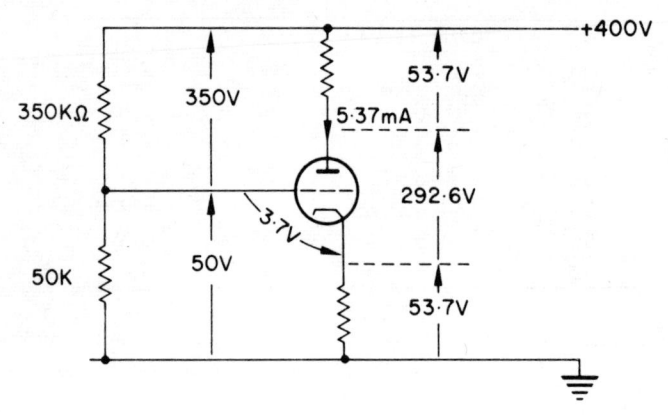

Fig. 16.5.4.

Now compare these results with the more accurate and lengthy method previously shown.

Graphical approach	Simple Ohms law approach
I_a 5.37 mA	5.0 mA
V_{ak} 292.6 V	300 V
V_{RL} 53.7 V	50 V
V_a 346.3 V	350 V
V_g 50 V	50 V

The errors are of the order of 6.5% which in practice might normally be swamped due to the tolerances not only of the valve but of the components also. This error could easily have been reduced by glancing at the characteristics where a -4 V bias could have been assumed.

It is suggested that the graphical method is ideal for detailed analysis or design and method two used for most normal practical work such as fault-finders, certainly in the first instance. For investigating a larger type valve with normal bias values of greater than 5 to 6 V, then instead of assuming $V_{gk} = 0$, assume $V_{gk} = 5$ V or so in order to obtain a much closer answer. (This will become evident from experience).

There are many variations to this circuit; figure 16.5.5. is offered as a further example.

Suppose we chose $R_L = 10 \, K\Omega$ as before, R_B is the bias resistor and R_K is the cathode resistor across which one output is developed.

Let $R_B + R_K = R_K'$.

Assume that the valve is to run at 5 mA. Assume also that

$$R_K' = (R_K + R_B) = 10 \, K\Omega$$

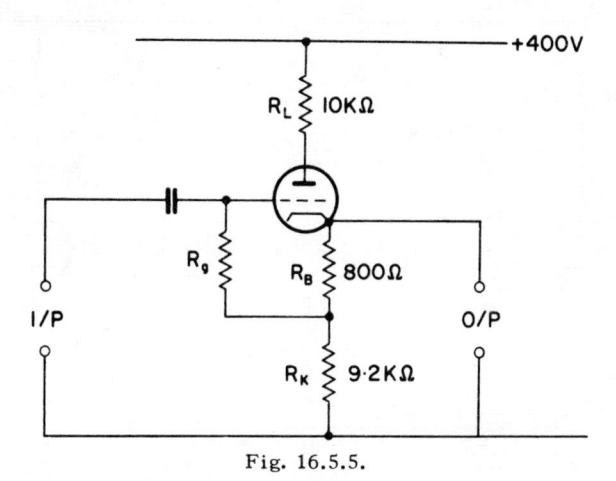

FURTHER CONSIDERATIONS OF PULSE AND SWITCHING CIRCUITS 285

Fig. 16.5.5.

in order to obtain approximately equal d.c. potentials across R_L and R_K'. Construct a load line for 20 KΩ as before i.e. $(R_L + R_B + R_K) = 20$ KΩ. Reference to the previous graph shows that at $I_a = -5$ mA a bias of 4 V is required (this is the operating point for this circuit). This means that the cathode end of R_g is to be 4 V positive to the grid end of R_g i.e. $V_{gk} = -4$ V. For 4 V to be developed across R_B, at 5 mA, $R_B = 4\text{V}/5\text{mA} = 800\,\Omega$.

$$R_K = 10\,\text{K}\Omega - 0.8\,\text{K}\Omega = 9.2\,\text{K}\Omega.$$

In practice R_K will become 9.1 KΩ and R_B 820 Ω, whilst R_L could remain at 10 KΩ. If equal outputs are required R_L must be 9.1 K also. This would mean a new load line for $(9.1 + 9.1 + 0.8)$ KΩ and a corresponding increase of 'short circuit' I_a to 21 mA (to position the load line). Further similar investigations will show that the quiescent I_a is 5 mA, and a bias of 4 V as before.

16.6. A linear analysis of a cathode coupled multivibrator

As V_2 grid is returned to the 300 V line, it is not unreasonable to assume that V_2 is ON and in grid current. Using a similar technique to that employed with the Clipper in chapter 13, an expression for the anode current of V_2, may be written

$$i_2 = \frac{300\,\text{V}}{r_a + R_K + R_3} = \frac{300\,\text{V}}{30\,\text{K}\Omega} = 10\,\text{mA}.$$

The cathode voltage is, then, $10 \times 5 = 50$ V. Now consider when V_1 is on but in a limited state due to unbypassed R_k. An expression for the anode current of V_1 is,

286   ELECTRONICS FOR TECHNICIAN ENGINEERS

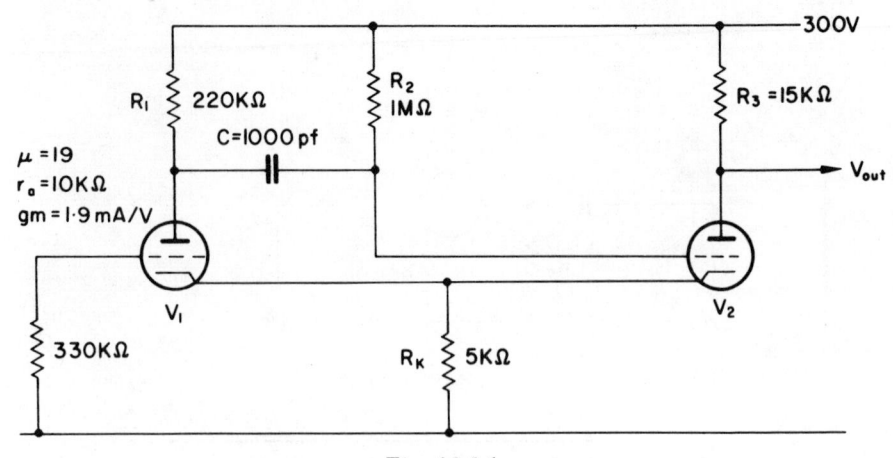

Fig. 16.6.1

$$i_1 = \frac{300\,V}{r_a + R_1 + R_k(1+\mu)} = \frac{300}{330} = 0.91\,\text{mA}.$$

The cathode voltage is $0.91 \times 5\,\text{K}\Omega = 4.55\,\text{V}$. The maximum signal that can develop across R_1 is $i_1 \times R_1 = 0.91 \times 220\,\text{K} = 200\,\text{V}$. This negative going 200 V is applied to the grid of V_2 and drags V_2 down below cut off. Just prior to the instant that the $-200\,\text{V}$ signal appears at V_2 grid, the grid was in grid current. As the cathode was at 50 V, the grid was also at 50 V.

The net effect of this 50 V (v_k) and the $-200\,\text{V}$ fall from V_1 anode is to cause the grid to be sitting at $+50 - 200 = -150\,\text{V}$. Once the grid of V_2 is sitting at $-150\,\text{V}$, the cathode is at $4.55\,\text{V}$ due to V_1 anode current only flowing through the cathode resistor, as V_2 is now cut off completely.

An expression for cut off is as follows. $-v_{gk}$ for cut off $= v_{ak}/\mu$. Therefore the grid cathode voltage for cut off is

$$v_{gk} = \frac{300 - 4.55}{19} = -15.6\,\text{V}.$$

As the cathode is sitting at 4.55 V (and v_{gk} must be $-15.6\,\text{V}$) the actual grid voltage for cut off must be $4.55 - 15.6 = -11.05\,\text{V}$. In the drawing shown, the $-11.05\,\text{V}$ cut off is in fact the 'cut-on' point where the pulse is completed.

V_2 grid waveform

The actual grid excursion is $-150 + 11.05 = 139\,\text{V}$. The 'aiming voltage' is 450 V. The rate of rise is

$$\frac{450}{C.R.} = \frac{450}{1\,\text{mS}}.$$

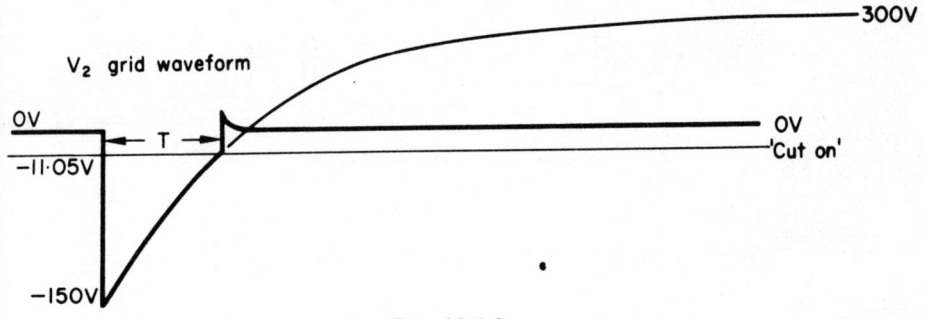

FURTHER CONSIDERATIONS OF PULSE AND SWITCHING CIRCUITS 287

Fig. 16.6.2.

The actual time T, for the pulse to form is

$$\frac{139\,\text{V}}{450\,\text{V}} \times \frac{1\,\text{mS}}{1} = 308 \text{ microseconds}.$$

(This assumes that the rate of rise is linear also.)

16.7. The diode pump

When a series of pulses is applied to a Diode Pump, the output will consist of a waveform known as a 'staircase'. This is shown in block schematic form in figure 16.7.1.

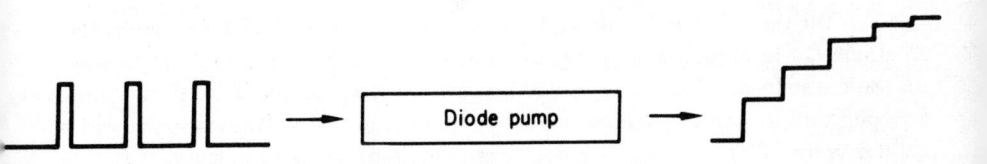

Fig. 16.7.1.

The amplitude of the steps will gradually decrease in an exponential manner. The device may be used for counting, discriminating and a whole host of other applications.

The circuit will be considered in stages, a capacitive divider will be examined first.

For the input of E volts, an output of

$$\frac{E \times C_1}{C_1 + C_2}\,\text{V}$$

will result.

U

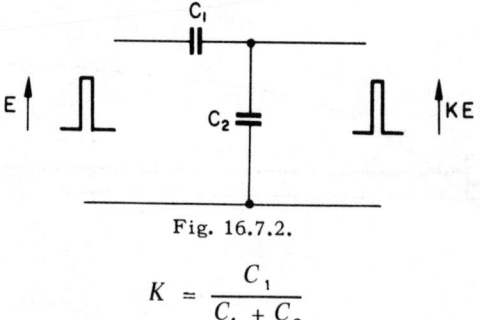

288   ELECTRONICS FOR TECHNICIAN ENGINEERS

Fig. 16.7.2.

Let
$$K = \frac{C_1}{C_1 + C_2}$$

The output for an input of E volts will be K volts (input) = KE volts. When the input falls to zero volts (during the lagging edge) the capacitor C_2 will discharge back through the input source. A diode, D_1 is inserted in order to overcome the 'leakage'. For one input, the output will now be as shown in figure 16.7.3.

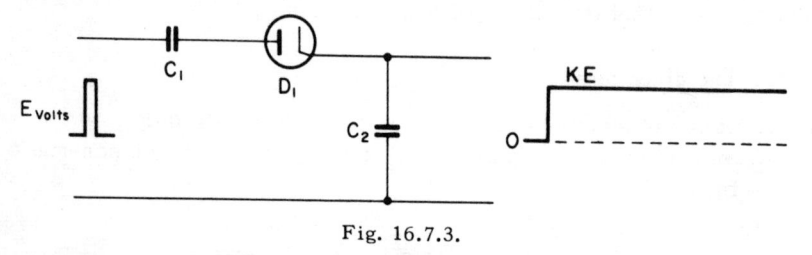

Fig. 16.7.3.

After the short pulse is applied and the lagging edge falls to zero, the diode D_1 is reversed biassed. Its anode will be at near zero volts whilst the cathode will be at the positive potential $K.E.$ volts. D_1 will be effectively open circuit and C_1 will be unable to discharge and will remain charged at KE volts. The capacitor C_1 will when charged, act as a battery. A further component is required to allow this capacitor to discharge after each pulse prior to the following input.

A second diode, D_1 is inserted as shown in figure 16.7.4.

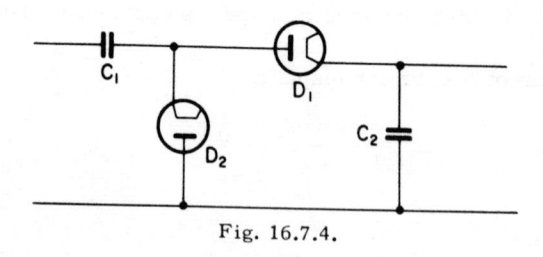

Fig. 16.7.4.

FURTHER CONSIDERATIONS OF PULSE AND SWITCHING CIRCUITS 289

During the leading edge of the first input pulse, D_1 is 'on' (switch closed) whilst the diode D_2 is 'off' (switch open). For this time interval only, the circuit will function as a simple capacitive divider as in figure 16.7.2.

When the input falls to zero, (this will be the lagging edge), the diode D_1 will be open circuit whilst the diode D_2 will be short circuit. C_2 will be charged to a potential KE volts and will remain at this potential until another input is applied. This is illustrated in figure 16.7.3.

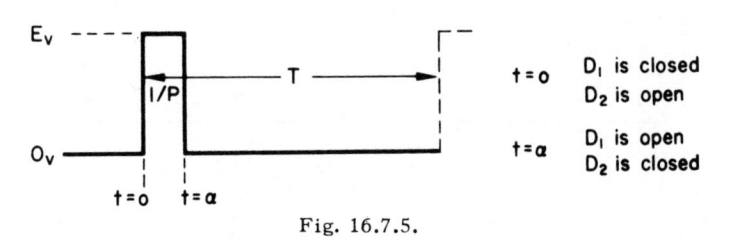

Fig. 16.7.5.

T is the period between input pulses.

Each subsequent pulse will cause C_2 to become charged a little more. After the first pulse is completed, D_1 cathode is sitting at $K.E.$ volts and is cut off. When the second pulse is applied, D_1 anode is dragged up from zero until its anode is slightly above $K.E.$ volts and D_1 conducts.

The input is therefore effectively $E - K.E.$ volts. This potential is reduced by the factor K before it is applied to C_2. This process continues and the potential across C_2 builds up in an exponential manner as seen in 16.7.6.

A load will need to be connected across the capacitor C_2 and if of a low resistance, will discharge C_2 resulting in a lowering of potential or, which may be worse, a fall in output level to zero between input pulses. Either will be undesirable and the need for a very high resistive load is obvious. A cathode follower connected as a 'bootstrap' will provide the buffer stage between the capacitor and the load as shown in figure 16.7.7. This is the complete circuit.

The input resistance to V_1 shown in figure 16.7.7. will be in the order of several megohms and will not under normal circumstances, affect the potential across C_2.

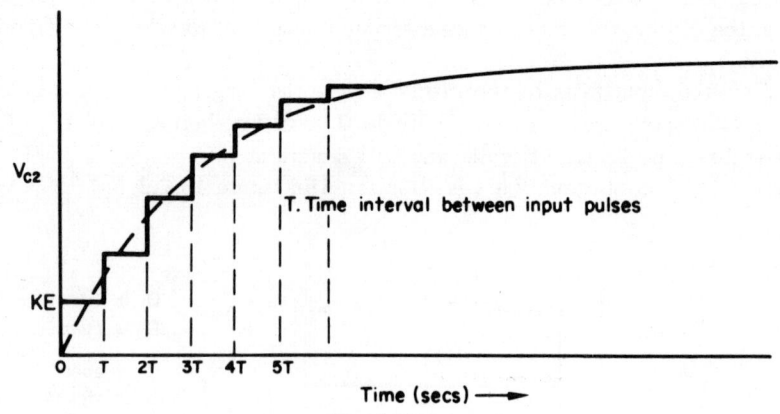

Fig. 16.7.6.

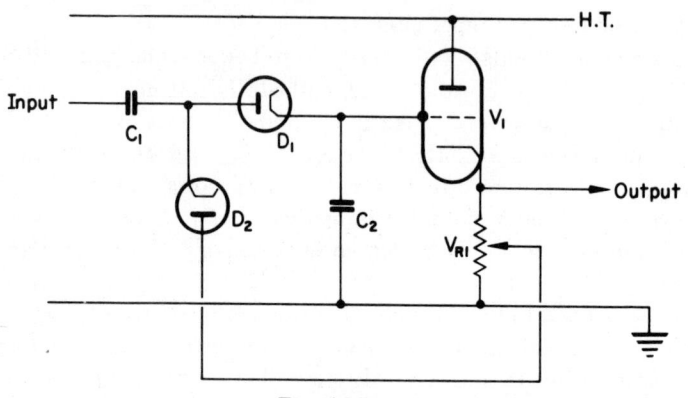

Fig. 16.7.7.

The bias for V_1 may be set by adjusting V_{R_1}.

The output from the diode pump may be used to, say, trigger a stage after 10 input pulses have been applied. Any number of pulses up to about 10 may be employed. After 10 pulses have been applied the rise in C_2 potential becomes less for each subsequent input and this results in a less discriminating output as the output flattens off.

CHAPTER 17

A delay line pulse generator

We will be looking at a somewhat different type of pulse generator in this chapter. An appreciation of some of the more basic properties of delay lines is essential of the pulse generator is to be understood.

This generator will be discussed in general terms only for the subject of delay lines alone is a vast complex subject.

We will see that when a step function input is applied to one end of a delay line, an impulse travels down the line and back again to the input.

The time taken for the impulse to travel in both directions determines the pulse duration of the generated pulse.

It follows therefore, that if the length of the dalay line is reduced, the pulse duration will be smaller. In this manner we are able to determine pulse durations quite accurately by means of trimming the delay lines.

The step function input can be the output from a valve anode, a thyratron output, a transistor collector output signal or a thyristor output. In all cases, the ouptuts — which provide an input to the line — result from rapidly switching the valve, etc., either on or off.

17.1. A simple pulse generator

Assume that a step function (0V to V volts) is applied to an oscillator circuit consisting of L and C in parallel (figure 17.1.1).

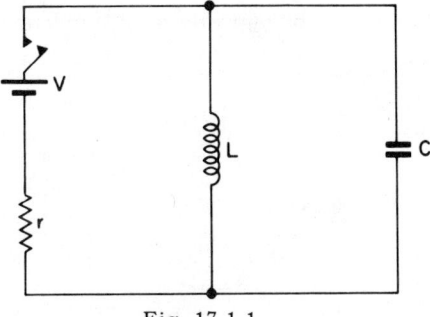

Fig. 17.1.1.

We will assume that there is no resistance in the coil. Both L and C have no voltage across them before closing the switch.

When the switch is closed then after a time, current will be flowing through the coil. (figure 17.1.2).

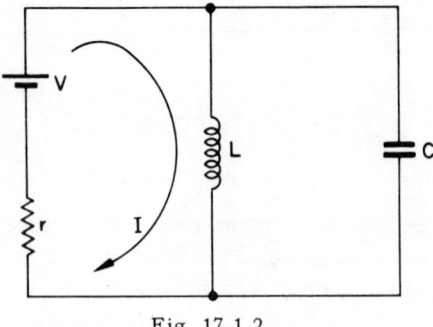

Fig. 17.1.2.

If the switch is now opened.

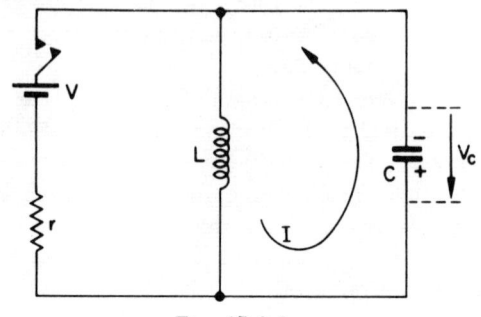

Fig. 17.1.3.

The coil will 'insist' on maintaining the current through itself but as it can no longer flow through r, it must flow into the capacitor C.

The capacitor charges up until the magnetic field has completely collapsed. The charged capacitor behaves as a battery and forces current through the coil, but now in the opposite direction (figure 17.1.4).

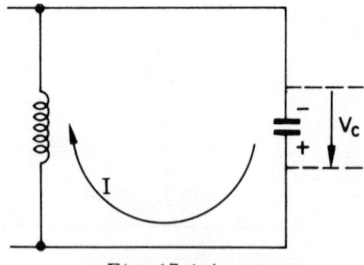

Fig. 17.1.4.

The current flows through the coil and the magnetic field builds up until finally the capacitor is discharged and the current tends to stop flowing: but the coil now 'insists' that the current continues to flow in the same direction, and its collapsing magnetic field ensures it does precisely that. (figure 17.1.5).

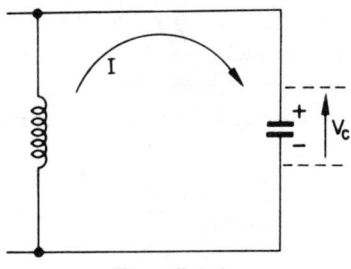

Fig. 17.1.5.

The current now enters the other plate of the capacitor and charges it up as before (figure 17.1.5) but with opposite polarity. This cyclic current continues indefinitely, and is sinusoidal in character. This is a basic oscillator circuit.

A practical coil, however, contains resistance; and as the current flows through the resistor, energy is dissipated and the amplitude of current gradually diminishes. This is known as a damped train, as shown in figure 17.1.6.

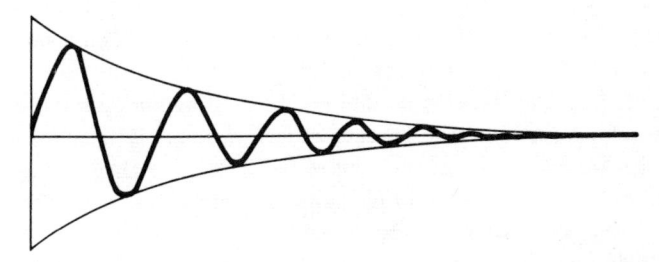

Fig. 17.1.6.

The frequency of a parallel circuit may be obtained by formula. (The coil resistance is ignored in this example).

At resonance $X_L = X_C$

$$\therefore 2\pi fL = \frac{1}{2\pi fC} \qquad \therefore f^2 = \frac{1}{4\pi^2 LC}$$

$$\therefore f = \frac{1}{2\pi\sqrt{LC}}$$

and this would be the frequency of oscillation in the above circuit. The frequency of oscillation could also have been obtained as in the following section dealing with delay lines, by equating energies.

The period of time between complete cycles = $1/f = 2\pi\sqrt{LC}$ sec
the period of time for half cycles = $\pi\sqrt{LC}$, say, T.

The voltage (V) may be replaced by a rectangular pulse. The LC circuit is connected as shown in figure 17.1.7 giving a very useful pulse generator. The rectangular pulse rapidly switches the valve amplifier.

The output consists of a single pulse the width of which is determined by the choice of L and C. Increasing C fourfold will double T.

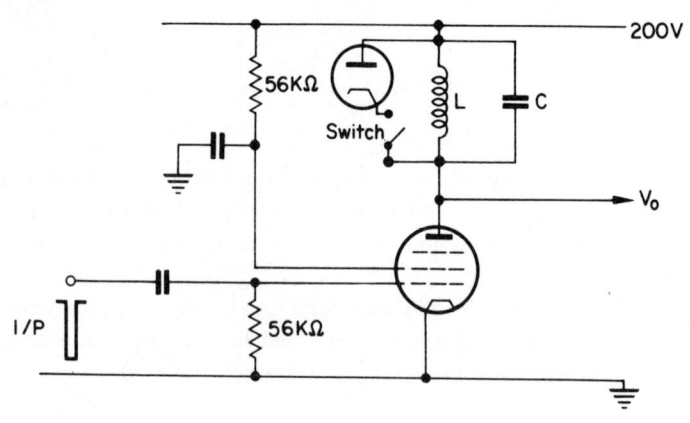

Fig. 17.1.7.

The valve would normally be passing heavy current. When a negative going input is applied, the anode current (which was steady in the coil) is suddenly switched off. Initially there is no appreciable voltage across the coil apart from the p.d. developed by the product of $i_a \times rL$. This will be very small.

The faster we switch it off (depending upon the rise time of the input pulse) the greater the output voltage initially across the coil. The anode circuit will ring, producing a damped train.

If the diode switch is closed, the foregoing applies, but immediately the anode falls negatively (the second half of the first cycle), the diode conducts and dissipates all of the energy in the tuned circuit. The output will appear as shown in figure 17.1.8.

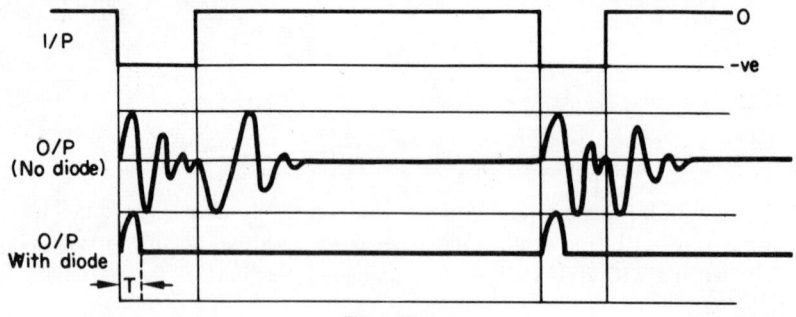

Fig. 17.1.8.

A short pulse or pip may be obtained from a very wide input pulse input.

17.2. Delay line equations.

A delay line is shown in a very simplified form in figure 17.2.1.

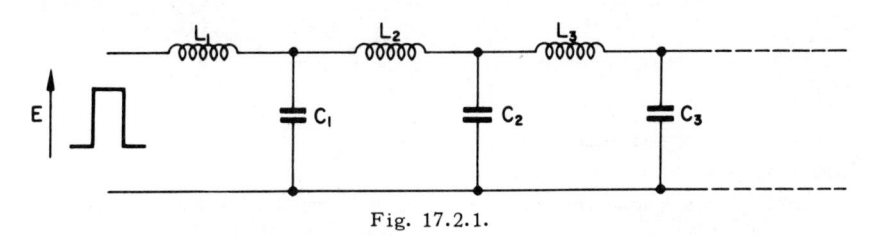

Fig. 17.2.1.

If a voltage is applied to the input of the line, a voltage e_1, and current i_1, impulse travels down the line.

The inductor L_1, resists the build up of current through itself but, as we discussed earlier, it eventually reaches maximum amplitude. As the current through L_1 builds up, C_1 begins to charge up and as the voltage builds up, so current attempts, and eventually succeeds, in flowing through L_2. Each one of these activities take some time and an impulse progresses along the line, but takes time to do so.

The line presents an impedance to the input generator and is known as the characteristic impedance of the line Z_0. We will later confine our discussions to resistance only and call the characteristic impedance R_0.

$R_0 = e/i$ and is true along the whole length of the line as shown in figure 17.2.2.

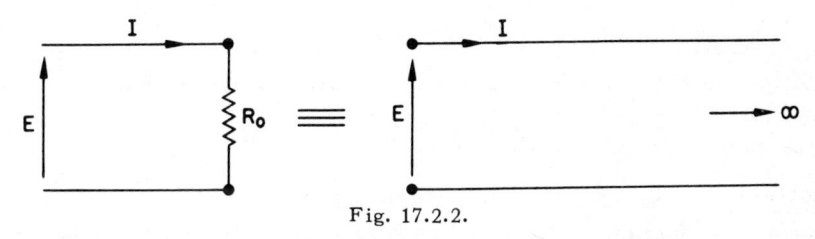

Fig. 17.2.2.

If the far end of the line is terminated with a load resistor R_L having a resistance equal to R_0, i.e. $R_L = R_0$, then some time after E is applied, all of the energy that travelled down the line is dissipated across R_L.

Delay line equations

Figure 17.2.3 shows a line with a forward travelling voltage and current impulse, e_1 and i_1.

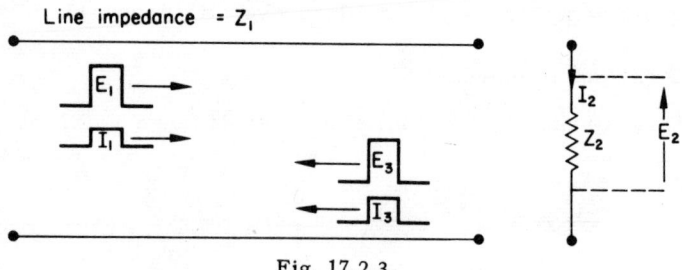

Fig. 17.2.3.

It also shows a reflected impulse e_3 and i_3, travelling in the reverse direction back towards the input.

It also shows a load resistor Z_2, a 'load' current i_2 and a 'load' voltage e_2.

Now $Z_1 = \dfrac{e_1}{i_1}$ $Z_2 = \dfrac{e_2}{i_2}$ and $Z_3 = \dfrac{e_3}{i_3}$ but Z_3 is equal to Z_1 and as e_3 and i_3 are travelling in the reverse direction, we can state that $\dfrac{e_3}{i_3} = -Z_1$.

Now $e_1 + e_3 = e_2$ (at a point about the load termination) (1)
and $i_1 + i_3 = i_2$ (2)

substituting for e_1, e_2 and e_3 in eq. (1)

$$i_1 Z_1 - i_3 Z_1 = i_2 Z_2 \tag{3}$$

multiplying (2) by Z_2, $i_1 Z_2 + i_3 Z_2 = i_2 Z_2$ (4)

Hence $i_1 Z_1 - i_3 Z_1 = i_1 Z_2 + i_3 Z_2$

and tidying up a little,

$$i_1 (Z_1 - Z_2) = i_3 (Z_1 + Z_2) \tag{5}$$

$$\dfrac{i_3}{i_1} = \dfrac{Z_1 - Z_2}{Z_1 + Z_2} \tag{6}$$

if $Z_2 = 0$ (a short circuit 'load' across the line output) then from (6) $i_3 = i_1$ and from (2),

$$i_2 = 2 i_1 \tag{7}$$

Also as $Z_2 = 0$, $e_2 = 0$ and from (1) $e_3 = -e_1$.

Hence for a short circuit at the far end of the line, twice the original current impulse flows through the short circuit and no voltage appears at the output.

When $Z = \infty$, i.e. when the line output is left open circuit with no load,

$$e_3 = e_1 \quad \text{hence} \quad e_2 = 2 e_1 \text{ and } i_2 = i_1 = 0.$$

Therefore for an 'open circuit' line, the voltage at about the termination is twice that of the voltage impulse travelling down the line and no current flows across the line output terminals.

A DELAY-LINE PULSE GENERATOR 297

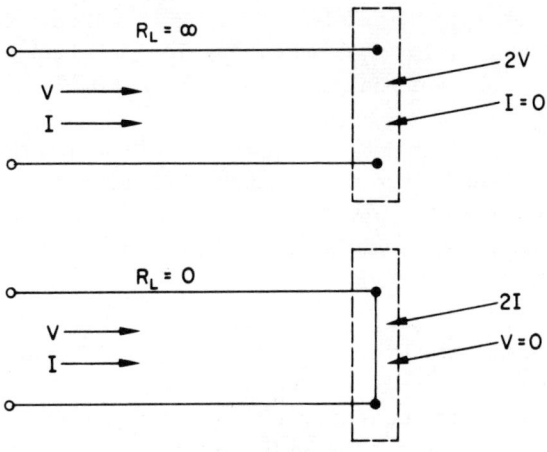

Fig. 17.2.4.

17.3. A delay line.

An equivalent circuit of part of a delay line is shown in figure 17.3.1.

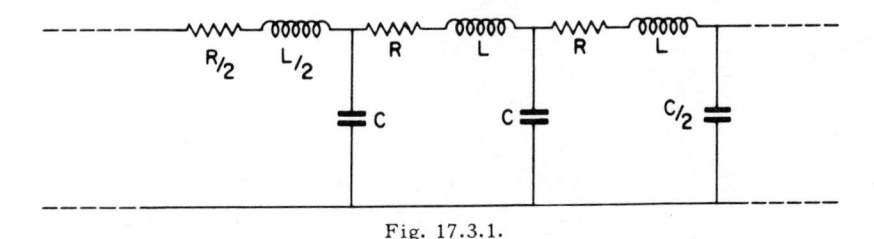

Fig. 17.3.1.

Delay lines are extremely complicated. Very complex mathematics would have to be employed to analyse them fully. In order to assist comprehension of a most difficult subject, a great deal of detail must at this stage, be ommitted. Sufficient details are supplied which will enable the reader to analyse this generator.

Delay lines may be produced in several forms; one of which is similar to a coaxial cable where the outer conductor is wound spiral fashion around the inner conductor, seperated by a dielectric. This delay cable is expensive, however, and as a general indication, to obtain a length sufficient for a delay of 1 second might cost well over £3,000,000. A second is of course a long time, whilst delays of 1 to 5μ s are common.

An artificial delay line is often produced, consisting of a tapped inductor with capacitors connected from each tap to chassis. R in figure 17.3.1 might be the d.c. resistance of the winding and will be ignored for this exercise. This artificial delay line would have very similar properties to the delay cable. There are many kinds of delay line; each has its own characteristics impedance R_0 (considering resistance only). Television coaxial cable is often said to be 75Ω cable; in this instance $R_0 = 75Ω$. This is assumed constant along the line independant of its length.

If the output of the line had been loaded with $R_L = R_0$, then the delay line would have been said to have been matched. Under these conditions a signal applied at the input would arrive at R_L T seconds later. T would of course, normally be expressed in μS. T is dependent upon the characteristics of the cable for a unit length.

Case 1. $R_L = R_0$

To terminate the output correctly with $R_L = R_0$, and apply a sudden voltage to the input, the voltage distribution will appear as follows. The voltage will cause a voltage and current impulse to 'move' down the line at a constant speed and the capacitors in the line will store electric field energy ($\frac{1}{2} Cv^2$) while the inductors will store magnetic field energy ($\frac{1}{2} L i^2$).

R_0 is the apparent resistance of the line as 'seen' by the incoming signal and its source. This could be represented by the simplified diagram; where R_s = source resistance and R_0 is the characteristics impedance of the line. v_s is the sudden input voltage (figure 17.3.3) or step function input.

By load over total $v = \dfrac{V_s \; R_0}{R_s + R_0}$ whilst $i = \dfrac{V_s}{R_s + R_0}$ as in figure 17.3.2.

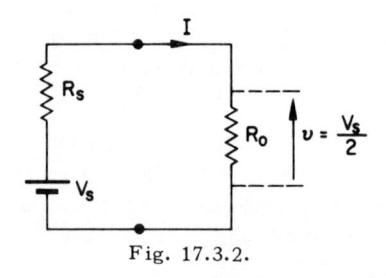

Fig. 17.3.2.

where v is the disturbance travelling down the line and in this case is $V_s/2$ as $R_0 = R_s$.

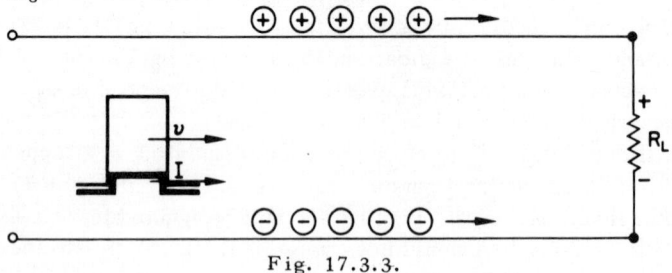

Fig. 17.3.3.

A DELAY-LINE PULSE GENERATOR

The actual length of this does not effect the initial distribution of the voltage and current; neither does the termination. At $t = 0$, the voltage across $R_L = 0$.

Summary

After a time T the 'disturbance' will reach the end of the line, and all the energy is dissipated across R_L (figure 17.3.4).

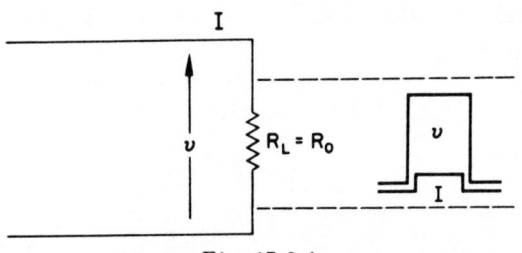

Fig. 17.3.4.

Case 2. $R_L = 0$

If we were to remove R_L and short circuit the end of the line, the electric field would collapse to zero when it reached the short circuit output.

Figure 17.3.5 shows positive charges leaving point A, travelling forwards towards the short circuited termination. Negative charges are seen leaving point B and also flowing forwards towards the short circuited termination.

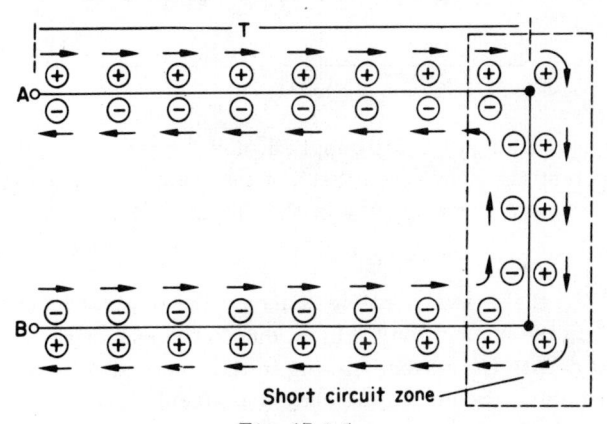

Fig. 17.3.5.

When the impulse, consisting of both positive and negative charges, reach the short circuit zone at time T, the positive charges flow clockwise round the short circuit and an identical number of negative charges flow anticlockwise through the short circuit.

Positive charges flowing in one direction is electrically identical to negative charges flowing in the opposite direction.

Hence the total current flowing through the short circuit is double that of either positive or negative charges flowing in the forward direction considered separately.

The charges continue flowing as shown, they move forward down the line, through the short circuit (this is where there is twice the current flowing) and on into the reverse direction back to the input.

After a time $2T$, negative charges arrive at point A and positive charges arrive at point B.

Figure 17.3.6 shows these charges that have arrived back at the input. It is seen that the potential at the input $= v + (-v) = 0$.

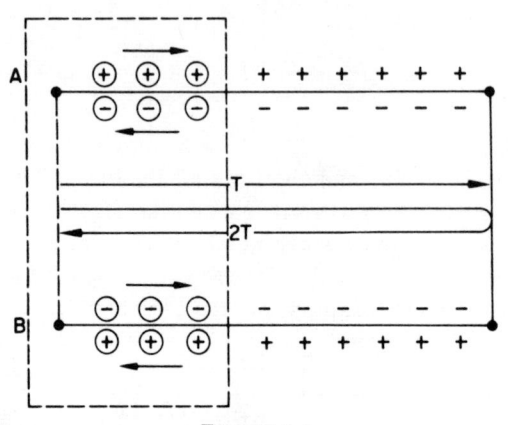

Fig. 17.3.6.

The original input, v, is still applied, and the reflected impulse has caused a potential $-v$ to appear back at the input after a time $2T$. Hence there is no voltage at the input — or the output — at $t = 2T$.

Summary

If $R_S = R_o$ the reflected voltage cancels the original input voltage, and the current in the short circuit region, due to the sum of positive and negative charges, is double the current flowing in the line just prior to the impulse 'hitting' the short circuit. As the current is double and as $R_S = R_o$,

$$i = 2\left(\frac{V_s}{R_o + R_s}\right) = \frac{2V_s}{2R_o} \quad \therefore \quad i = \frac{V_s}{R_o} \quad \text{which}$$

satisfies Ohm's law.
and as $R_L = 0$, $v_{RL} = 0$.
Equilibrium occurs after $t = 2T$.

A DELAY-LINE PULSE GENERATOR

The line is now storing magnetic field energy of magnitude $\frac{1}{2}L\,(2i)^2$ due to the inductance of the line.

Case 3 $R_L = \infty$

Now we remove the short circuit, and leave the line output terminals open circuit (figure 17.3.7).

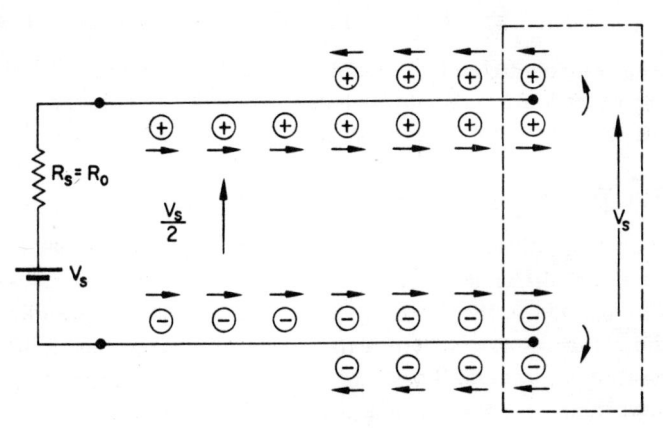

Fig. 17.3.7.

At $t = 0$, V_s is suddenly applied, $V_s/2$ appears at the input terminals and the charges move down the line. At $t = T$ the charges reach the open circuit; and as the current can no longer be supported, the magnetic field energy is changed to electric field energy, which results in the 'load' voltage doubling in value due to the doubling of like charges.

The charges move back as shown and after $t = 2T$, it may be seen that the input terminal voltage is twice the original input of $V_s/2$ to V_s. This satisfies Ohm's law.

Equilibrium is now reached, and the line is storing electric field energy to the capacitance of the magnitude $\frac{1}{2}C\,(2v)^2$.

Now when R_L is short circuit, energy stored after time $2T = \frac{1}{2}L\,(2i)^2$ and when $R_L =$ open circuit, energy stored after time $2T = \frac{1}{2}C(2v)^2$. $2T$ is the same for both, therefore by equating energies;

$$\tfrac{1}{2} L\,(2i)^2 = \tfrac{1}{2} C\,(2v)^2$$
$$\tfrac{1}{2} L\, 4 i^2 = \tfrac{1}{2} C\, 4 v^2$$
$$L i^2 = C v^2$$

and

$$\frac{v^2}{i^2} = \frac{L}{C} \quad \text{and} \quad \frac{v}{i} = \sqrt{\frac{L}{C}} \quad \text{and as} \quad \frac{v}{i} = R_0,$$

the characteristic resistance $= R_0 = \sqrt{\dfrac{L}{C}}$

302   ELECTRONICS FOR TECHNICIAN ENGINEERS

The energy supplied to the line in the time $2T$ must be supplied from the source; and as we need an expression containing L, C and T, we may equate energies.

$$(v.i.\ 2T) = [\tfrac{1}{2}C\ (2v)^2]\ [\tfrac{1}{2}L\ (2i)^2]$$
$$v^2 i^2 4T^2 = [2Cv^2][2Li^2]$$
$$v^2 i^2 4T^2 = 4CLv^2 i^2 \text{ and } T^2 = CL \therefore T = \sqrt{LC}.$$

Doubling the delectric of C would result in an increase of T to $\sqrt{2}T$.

One important feature should be noted: The output (delayed signal) is taken from the line *input* after time $2T$.

17.4. The Thyratron

A Thyratron is a valve which behaves as though it were an electronic switch but with a slight difference. When closed (or on), the device is essentially a short circuit, and is similar to a switch whose contacts are closed. But across the closed contacts there exists a small voltage (about 10V) whose source resistance is zero. Therefore, it is evident that the device acts as the two terminals of a low voltage battery whose resistance is zero ohms.

The switch takes a finite time to close completely; this time is known as the *ionisation* time, usually about $1\mu S$.

When the switch is open, the device presents no load at all as it is effectively open circuit. It takes a finite time to open completely, and this de-ionisation time is in the order of $50\mu S$.

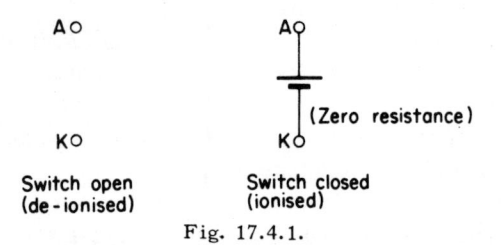

A○   A○
                    │
                    ─┬─ (Zero resistance)
                    ─┴─
K○   K○

Switch open   Switch closed
(de-ionised)   (ionised)

Fig. 17.4.1.

17.5. A delay line pulse generator

Consider the following circuit diagram in figure 17.5.1. A delay line pulse generator. Consider the circuit with the thyratron under 'open' and 'closed' conditions.

When the switch is open, the line charges towards 250V via the series resistors of 100KΩ. In a time $5.C.R.$ it would become charged to 250V. Before it becomes fully charged, an input pulse closes the switch and the 10V battery is effectively connected across the charged line. This is shown in figure 17.5.2.

A DELAY-LINE PULSE GENERATOR

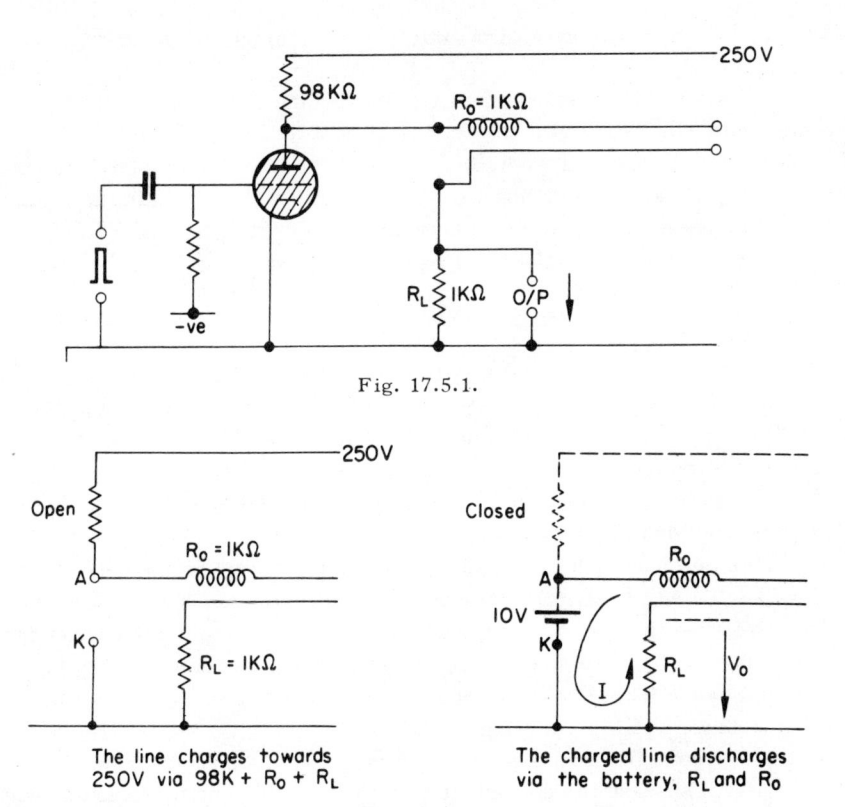

Fig. 17.5.1.

Fig. 17.5.2

The line subsequently discharges through the 'battery' (thyratron) R_L and R_0. The resultant current in R_L provides an output voltage which is negative with respect to chassis.

Figure 17.5.3 shows the line charging towards 250V.

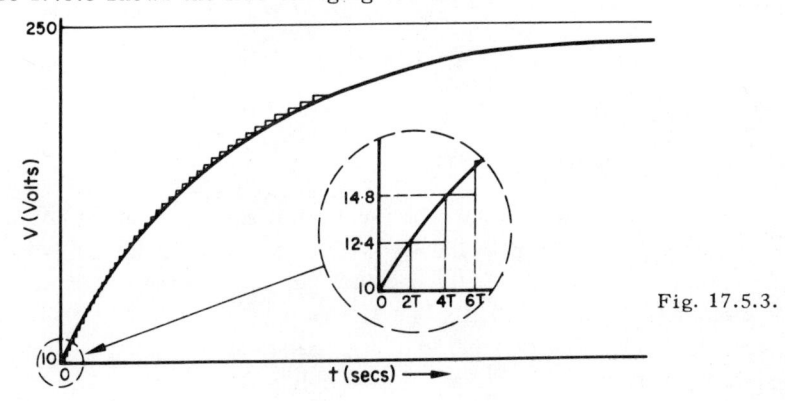

Fig. 17.5.3.

Figure 17.5.3 showing potential at line input increasing in approximately 2.4V steps every $2T$ μS.

At the instant the thyratron opens ($t = 0$) the line commences to charge towards 250V via 100KΩ at a time constant of C.R. where $R = 100K\Omega$ and C is the capacity of the line. As both ends of the line are effectively open circuit, energy oscillates to and fro. For each two way excursion the input potential increases by an amount determined by the 98K, R_0 and R_L. By load over total, it is seen that the line charges in steps of

$$\frac{240V \text{ load}}{\text{Total}} = \frac{240V \times 1K\Omega}{100K\Omega} \doteq 2.4 V$$

The input pulses occur at regular time intervals, and the frequency at which they occur is known as the p.r.f., (Pulse repetition frequency). The ionisation time is that for the thyratron to fully close. The de-ionisation time is that required for the device to fully open.

The maintaining voltage that exists across the 'closed switch' is shown by a zero resistance battery.

It should be clear that if we apply input pulses at pre-determined intervals, each one 'closing the switch', then we must calculate and be sure that the switch can close and open during the interval between input pulses. Secondly the line has to charge to some value whilst awaiting the next input pulse which will cause it to discharge via R_L, thus providing an output voltage of a predetermined width or duration.

Let us suppose the input pulses occur every $200 \mu S$. Assume the ionisation time $= 1 \mu S$ and the de-ionisation time $= 50 \mu S$. The forward (one way) delay of cable, $T = 2 \mu S$. The maintaining voltage $= 10V$.

We have mentioned the capacity of the cable, or line.

Let us consider C and evaluate it using the expressions we derived earlier on.

$$T = \sqrt{LC} \text{ and } R = \sqrt{\frac{L}{C}}$$

as $\dfrac{T_0}{R_0} = \dfrac{\sqrt{LC}}{\sqrt{(L/C)}}$ then $\dfrac{T^2}{R_0^2} = \dfrac{LC}{L/C}$ $\therefore$ $\dfrac{T^2}{R_0^2} = C^2$

Then $T/R_0 = C$
$\therefore$ $C = \dfrac{2 \mu S}{1 K\Omega} = 2000 \text{pF}$.

The line will charge (when the thyratron is open) towards 250V at a time constant $C.R.$, $2000 \text{p.F} \times 100 \text{ K}\Omega = 200 \mu S$. The line initially charged to 10V would be charged to 250V in 5 $C.R.$ or 1 mS, but the period of time between input pulses $200 \mu S$. The maximum permissible time (t) between input pulses for the line charge up $=$ [Period between pulses $-$ Ionisation $-$ de-ionisation $- 2T$] $= [200 - 1 - 50 - 4] = 145 \mu S$.

A DELAY-LINE PULSE GENERATOR

The line would charge eventually, to 250V, but the climb of 240V is interupted by an input pulse. The period of time during which it does charge is 145 μS.

From $V_c = V[1 - e^{-t/CR}]$ we require V_c.

Hence $V_c = 240[1 - e^{-145\mu/200\mu s}] = 240[1 - e^{-0.725}]$

Hence $V_c \doteqdot 125V$

The switch is now closed.

The circuit at this instant may be represented as shown in figure 17.5.4.

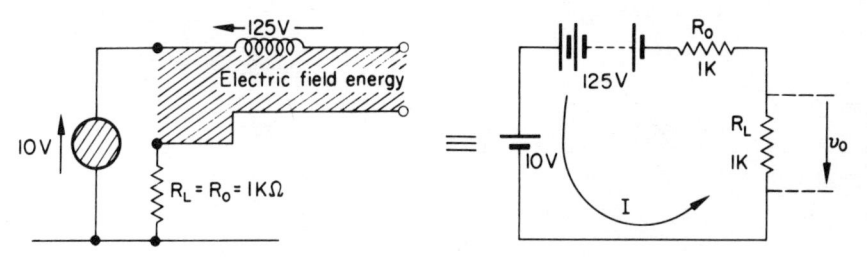

Fig. 17.5.4

$V_o = i \times R_L$ where $i = \dfrac{(125-10)V}{2K\Omega} = 57.5\text{mA}$

$\therefore V_o = 57.5V$ and is negative with respect to earth.

The voltage V_o is negative, due to the fact that at the instant the thyratron ionises, the p.d. across the load resistor R_L and the line drops suddenly from 125V to 10V (a change of 115V) and this change in voltage, 115V, is evenly distributed across R_o and R_L as they are both 1KΩ.

The impulse so caused at the input travels along the line and returns in a time $2T$; at the instant it returns to the input, it presents a $-125V$ to the anode of the thyratron and drags the positive ions from the valve thus causing the thyratron to de-ionise very rapidly by dragging the anode down below 10V. When this happens, the thyratron discharge can no longer be maintained. As the thyratron opens, the voltage across R_L rises rapidly towards zero, and the output pulse is complete. As T is directly proportional to the cable or line length, any pulse width may be achieved simply by choosing the appropriate length or number of sections of the line or cable.

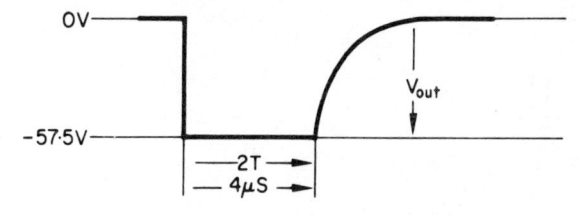

Fig. 17.5.5.

CHAPTER 18

Negative feedback and its applications

The output of an amplifier depends, to a great extent, upon the valve or transistor around which the amplifier stage is built. Due to manufacturing tolerances, a change of valve may well result in a marked difference in the circuit characteristics. As the valve ages, it is probable that the output from the stage will change accordingly. One method of overcoming these changes to a large extent is to provide negative feedback.

A common method is to tap off a portion of the output signal, and apply this back to the amplifier input. As the output signal is 180° out of phase with the input for a single stage circuit, the fraction of output fed back to the input subtracts from the input and reduces the overall gain. This feedback, if out of phase with the input signal, is known as negative feedback. Feedback is often complex, so are amplifier gains, but this chapter will deal with non complex examples only. On many occasions in this book it has been stressed that valves and transistors play a rather subservient role in circuits; the determining components are chosen by the designer. The valve or transistor may often be 'ignored', and by using simple Ohm's Law the circuit will often behave almost exactly as predicted.

When using negative feedback, the amplifier circuit gain is almost completely dependant upon the feedback components, and relies less and less upon valve changes and supply variations. The output signal will also be a much more faithful reproduction of the input signal. The major disadvantage, however is the reduction in gain. The new gain may be calculated and the methods of doing so will be discussed.

If the gain of an amplifier is shown as A, and with negative feedback, the new gain = $A/(1 + \beta A)$, then if the term $\beta A \gg 1$, the gain of the amplifier with feedback may be seen to be $1/\beta$. This result shows clearly that the gain depends upon $1/\beta$ and is independant of the original amplifier gain.

For example, if an amplifier had a gain of 10^4 and negative feedback introduced, where β was only 1%, i.e. $\beta = 1/100$, the gain with feedback would be 99. If the amplifier gain fell to 5000, i.e. a 50% reduction in original gain, the new gain with 1% feedback would become 98. Hence a 50% change in gain results in an overall drop in gain, with feedback, of about 1%.

18.1. Feedback and its effect upon the input-resistance of a single-stage amplifier

The effective input to an amplifier is that signal appearing between grid and cathode, v_{gk}. In a simple amplifier with no feedback and with the cathode adequately bypassed, $v_{in} = v_{gk}$ as $v_k = 0$. Should v_{gk} differ from v_{in}, the effective input to the amplifier is modified and may often be the result of feedback.

Let us consider the diagram shown in figure 18.1.1.

Example

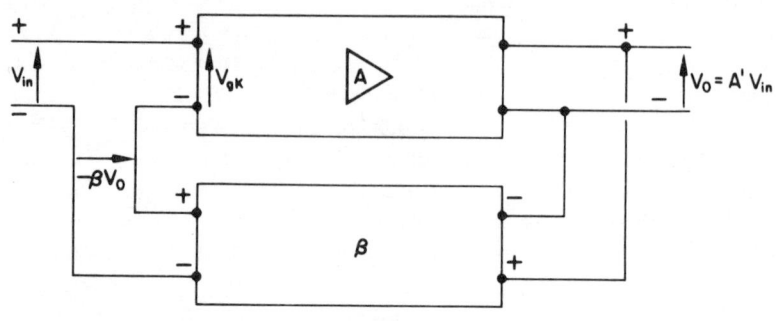

Fig. 18.1.1.

Where β is a fraction of the output voltage.

With series voltage feedback,

$$v_{gk} = v_{in} - \beta v_0 = v_{in} - \beta A v_{in} = v_{in}(1 - \beta A)$$

The gain with feedback $= \dfrac{v_0}{v_{in}} = \dfrac{A v_{in}}{(1 - \beta A)v_{in}} = \dfrac{A}{1 - \beta A}$.

If the feedback is negative, as in the figure, then the gain with feedback,

$$A' = \frac{A}{1 + \beta A}$$

Hence

$$v_{gk} = \left[v_{in} \; 1 - \frac{\beta A}{1 + \beta A} \right] = v_{in} \left[\frac{1 + \beta A - \beta A}{1 + \beta A} \right] = \frac{v_{in}}{1 + \beta A}$$

The input resistance to an amplifier without feedback,

$$R_{in} = \frac{v_{in}}{i_{in}}$$

Therefore $i_{in} = \dfrac{v_{in}}{r_{in}}$ but with no feedback, $v_{in} = v_{gk}$.

NEGATIVE FEEDBACK AND ITS APPLICATIONS 309

Therefore
$$i_{in} = \frac{v_{gk}}{R_{in}}$$

With feedback
$$V_{gk} = \frac{v_{in}}{1 + \beta A}$$

hence R' (R_{in} with feedback) =

$$\frac{v_{in}}{i_{in}} = \frac{v_{in}}{v_{gk}/R_{in}} = \frac{v_{in}}{v_{gk}} \cdot R_{in} = \frac{v_{in} \cdot R_{in} (1 + \beta A)}{V_{in}}$$

$$\therefore R' = R_{in}(1 + \beta A).$$

The input resistance of an amplifier with negative feedback is increased, irrespective of how the feedback voltage is derived.

18.2. Feedback in multistage amplifiers

The system gain of a multistage amplifier with voltage feedback.

The amplifier system is shown in figure 18.2.1.

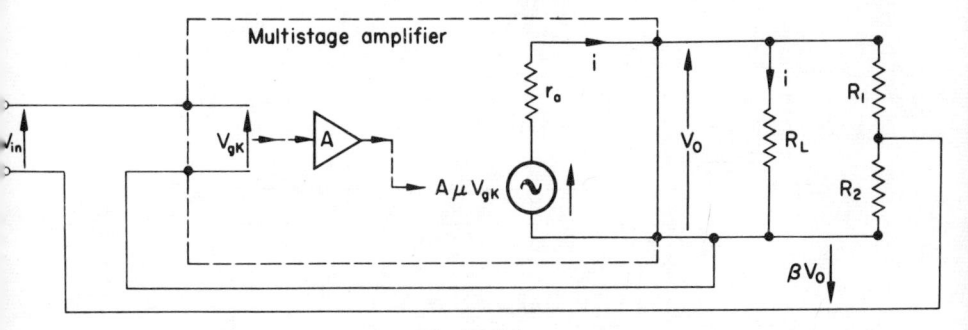

Fig. 18.2.1

A is the gain of all stages except the last. μ and r_a are the amplification factor and anode slope resistance for the last stage.

The input to the amplifier is shown as v_{gk}. v_{in} is the input to the complete system including feedback components.

$$\beta = \frac{R_2}{R_1 + R_2}$$

The system gain overall, without feedback, =

$$\frac{v_o}{v_{in}} = \frac{A\mu R_L}{r_a + R_L}$$

Assuming that $(R_1 + R_2) \gg R_L$, we will determine the gain with feedback. Applying Kirchhoff's law to the output circuit,

$$A\mu \, v_{gK} = i \, (r_a + R_L) \tag{1}$$

For negative feedback,

$$v_{gK} = v_{in} - \beta v_0 \quad \text{and} \quad i = \frac{v_0}{R_L}.$$

Thus (1) becomes,

$$A\mu \, (v_{in} - \beta v_0) = \frac{v_0 \, (r_a + R_L)}{R_L}$$

$$\therefore \quad A\mu \, v_{in} \, R_L = v \, (r_a + R_L + \beta A\mu \, R_L)$$

$$\therefore \quad \frac{v_0}{v_{in}} = \frac{A\mu \, R_L}{r_a + R_L \, (1 + \beta A\mu)}$$

which simplifies to

$$\frac{v_0}{v_{in}} = \frac{A\mu \, R_L}{1 + \beta A\mu} \bigg/ \left(\frac{r_a}{1 + BA\mu} + R_L \right)$$

A simplified equivalent circuit is shown in figure 18.2.2

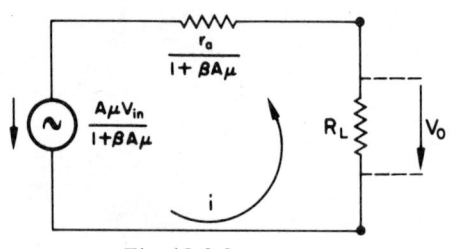

Fig. 18.2.2

The system gain of a multistage amplifier with current feedback

The amplifier system is shown in figure 18.2.3.
All amplifier stages except the last are shown by the symbol A. The last stage is represented by the generator $A\mu \, v_{gK}$ and the anode slope resistance r_a

$$\beta = \frac{i \, r}{i \, R_L} = \frac{r}{R_L}.$$

NEGATIVE FEEDBACK AND ITS APPLICATIONS

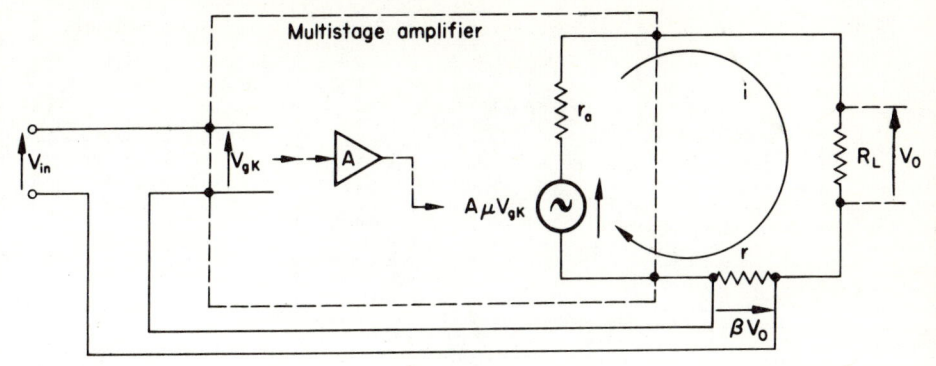

Fig. 18.2.3

With no feedback
$$\frac{v_o}{v_{in}} = \frac{v_o}{v_{gK}} = \frac{A\mu R_L}{r_a + r + R_L}$$

and with feedback.
$$v_{gK} = v_{in} - \beta v_o$$

Hence
$$A\mu v_{gK} = i(r\mu + r + R) \quad \text{but } i = \frac{V_o}{R_L}$$

$\therefore$
$$A\mu v_{gK} = \frac{v_o}{R_L}(r_a + r + R_L)$$

hence
$$A\mu (v_{in} - \beta v_o) = \frac{v_o}{R_L}(r_a + r + R_L) \quad \text{but } \beta = \frac{r}{R_L}$$

therefore
$$A\mu v_{in} - A\mu \frac{r}{R_L} v_o = \frac{v_o}{R_L}(r_a + r + R_L)$$

$\therefore$
$$A\mu v_{in} = \frac{v_o}{R_L}(r_a + r + R_L + A\mu r)$$

$\therefore$
$$A\mu v_{in} = \frac{v_o}{R_L}[r_a + R_L + r(1 + A\mu)]$$

hence the overall gain with feedback is given as

$$\frac{v_o}{v_{in}} = \frac{A\mu R_L}{r_a + r(1 + A\mu) + R_L}$$

The simplified equivalent circuit is given in 18.2.4.

312   ELECTRONICS FOR TECHNICIAN ENGINEERS

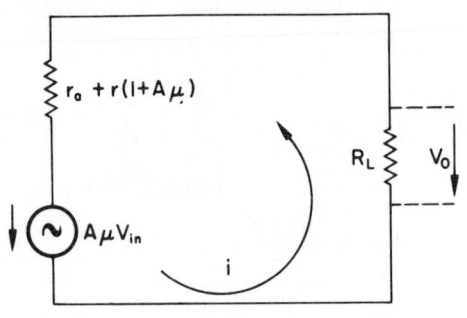

Fig. 18.2.4

18.3. Composite feedback in a single stage amplifier.

Series Voltage negative feedback in a valve voltage amplifier will reduce the output resistance by a factor $1/(1 + \mu B)$.

Series Current negative feedback will increase the output resistance by an amount $(1 + \mu) R_K$. We will discuss the proof of these statements and show how both may be combined in one circuit. It is possible, by choice of feedback to employ both voltage and current feedback and to determine the output resistance and to maintain other requirements at the same time.

The input resistance will still be $(1 + \beta A)$ times that with no feedback as shown in 18.1. Consider figure 18.3.1.

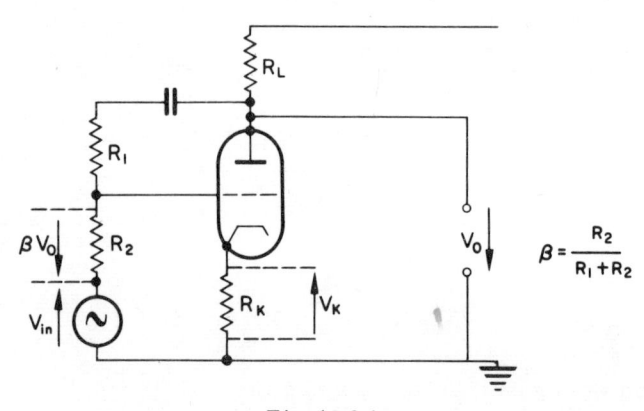

Fig. 18.3.1

R_k provides current feedback whilst $R2/(R_1 + R_2)$ provides voltage feedback.

The generator, V_{in}, is assumed to have zero internal resistance (in practice it should be $< (R_1 + R_2)/20$. $R_1 + R_2$ should be sufficiently high to allow us to ignore any signal current flowing through R_2.

NEGATIVE FEEDBACK AND ITS APPLICATIONS

An equivalent circuit may be drawn for this composite feedback circuit shown in figure 18.3.1. This is shown in figure 18.3.2.

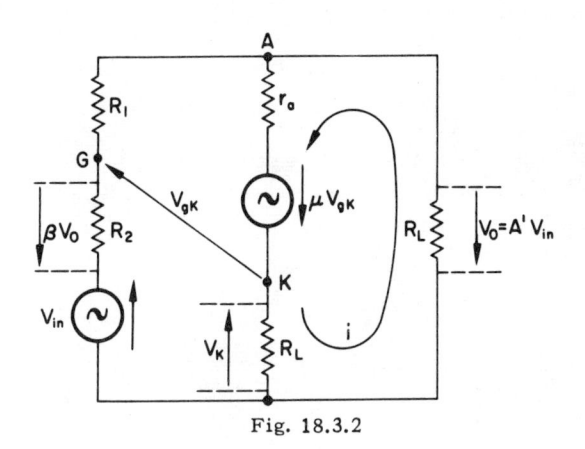

Fig. 18.3.2

(When drawing equivalent circuits, care should be taken to avoid using double headed arrows; the arrows shown in equivalent circuits in this book are single ended and allow for inherent 180°C phase shift within the valve).

The grid voltage v_g in this example is the algebraic sum of v_{in} and the feedback voltage, hence

$$v_g = v_{in} + \beta v_0 = v_{in} - \beta i R_L \quad \text{(as } v_0 \text{ is negative going).}$$

Hence

$$v_{gk} = v_{in} - \beta i R_L - v_k = v_{in} - \beta i R_L - i R_K = v_{in} - i[\beta R_L + R_K]$$

and by Kirchhoff's law.

$$\therefore \quad \mu(v_{in} - i[\beta R_L \; R_K]) = i(R_K + r_a + R_L)$$

$$\therefore \quad \mu v_{in} = i[r_a + R_K + R_L + \mu(\beta R_L + R_K)]$$

$$\therefore \quad \mu v_{in} = i[r_a + R_K(1+\mu) + R_L(1+\beta\mu)]$$

$$v_0 = i R_L$$

$$\therefore \quad \mu R_L v_{in} = v_0 [r_a + R_K(1+\mu) + R_L(1+\beta\mu)]$$

$$\frac{v_0}{v_{in}} = A' = \frac{-\mu R_L}{r_a + R_K(1+\mu) + R_L(1+\beta\mu)}$$

and is negative going compared to the input, v_{in}.

18.4. Effects of feedback on parameters μ and r_a due to composite feedback

From the expression

$$A' = \frac{\mu R_L}{r_a + R_L(1 + \mu\beta) + R_K(1 + \mu)}$$

we can draw a simple equivalent circuit using modified parameters.

The parameters are modified after comparing the expression for A' with the expression

$$A = \frac{\mu R_L}{r_a + R_L}$$

which is the expression for gain without feedback. We need to derive an expression similar to

$$\frac{\mu R_L}{r_a + R_L}$$

where R has no 'coefficient' (other than unity). Hence if we divide top and bottom of the expression

$$\frac{\mu R_L}{r_a + R_L(1 + \mu\beta) + R_K(1 + \mu)}$$

by the 'coefficient' of R_L in the denominator, i.e. $(1 + \mu\beta)$

Hence $\quad A' = \left(\dfrac{\mu}{1 + \mu\beta} \cdot R_L\right) \Big/ \left(\dfrac{r_a + R_K(1 + \mu)}{1 + \mu\beta} + R_L\right)$

The modified internal generator and anode slope resistance become

$$\mu' = \frac{\mu}{1 + \beta\mu} \quad \text{and} \quad r_a' = \frac{r_a + R_K(1 + \mu)}{1 + \mu\beta}$$

A simplified equivalent circuit is shown if figure 18.4.1.

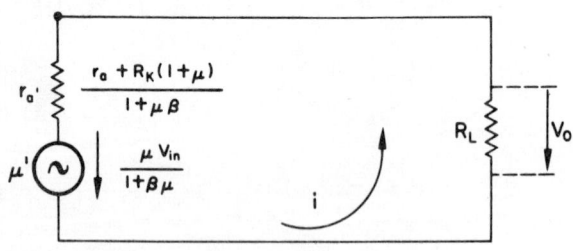

Fig. 18.4.1

The output resistance has become
$$\frac{r_a + R_K(1+\mu)}{1+\mu\beta}$$

The generator will generate an e.m.f. $= \dfrac{\mu V_{in}}{1+\beta\mu}$

18.5. The effects of feedback on output resistance

Output resistance is given by e/i where e is an external generator and i the current taken from the generator. Compare the effects upon output resistance of (a) no feedback,
 (b) current feedback,
 (c) voltage feedback,
 (d) composite feedback.

(a) *No feedback* (figure 18.5.1).

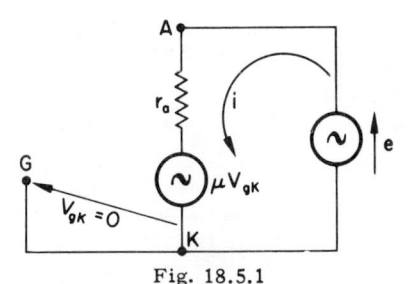

Fig. 18.5.1

$V_{gk} = 0, \quad \mu V_{gk} = 0$
The generator μV_{gk} is therefore effectively short circuit.

Rout $= \dfrac{e}{i} = r_a$

(b) *Current feedback* (Cathode resistor unbypassed) figure 18.5.2.

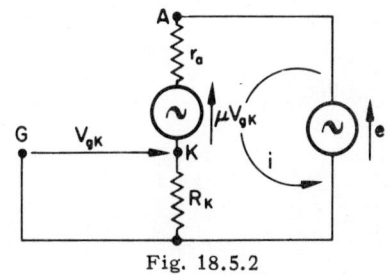

Fig. 18.5.2

$$v_{gk} = iR_K \quad \text{hence} \quad \mu v_{gk} = \mu i R_K$$

$$e - \mu i R_K = i(r_a + R_K)$$

$$e = i(r_a + R_K + \mu R_K)$$

$$\frac{e}{i} = \text{Rout} = r_a + R_K(1+\mu)$$

(note that this is one way of obtaining current feedback)

(c) *Voltage feedback* (R_1 and R_2 provide fraction of output to grid) figure 18.5.3.

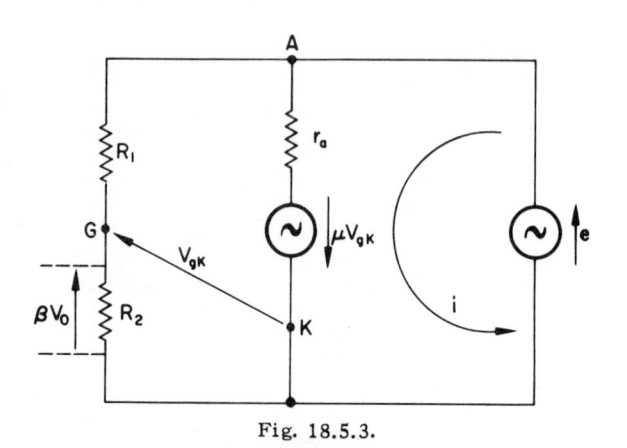

Fig. 18.5.3.

Assume $R_1 + R_2$ are very high resistance.

$$v_{gk} = \beta v_o = \beta e$$

$$\therefore \quad \mu v_{gk} = \mu \beta e$$

hence
$$e + \mu \beta e = i(r_a)$$

$\therefore$
$$e(1+\mu\beta) = i(r_a)$$

thus
$$\frac{e}{i} = \text{Rout} = \frac{r_a}{1+\mu\beta}$$

(d) *Composite feedback* (Fraction of v_o feedback and cathode unbyassed) figure 18.5.4.

$$v_{gk} = (\beta v_o - iR_K) = (\beta E - iR_K)$$

$\therefore$
$$e + \mu(\beta e - iR_K) = i(r_a + R_K)$$

NEGATIVE FEEDBACK AND ITS APPLICATIONS 317

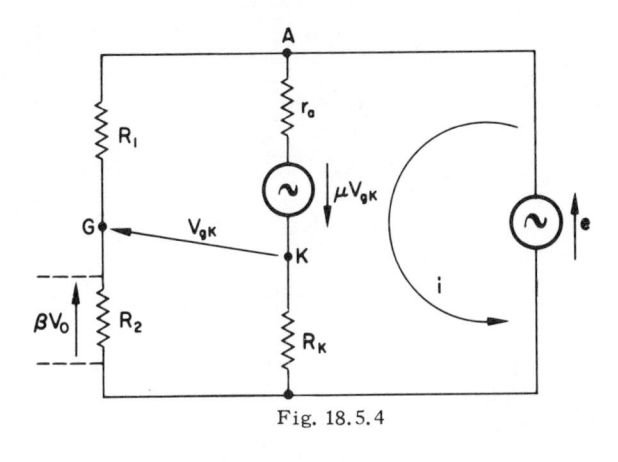

Fig. 18.5.4

$$e + \mu\beta e = i(r_a + R_K) + \mu i R_K$$

$$e(1 + \mu\beta) = i(r_a + R_K(1 + \mu))$$

$$\frac{e}{i} = \text{Rout} = \frac{r_a + R_K(1 + \mu)}{1 + \mu\beta}$$

which agrees with the result derived earlier in a different manner.

Note that with current feedback, the output resistance is increased (the numerator becomes larger) whilst with voltage feedback, the output resistance becomes smaller (due to the denominator becoming larger).

With composite feedback, both effects can clearly be seen as in the last example.

It is suggested that the reader practices drawing equivalent circuits for various amplifier configurations and to derive the appropriate formulae. (Hint: When v_{gk} indicates that the *grid* is instantaneously *rising*, the generator representing μv_{gk} must be shown instantaneously *falling* in order to allow for the phase shift of the amplifier stage).

18.6. Voltage and current feedback in a phase splitter

If we do not bypass the cathode resistor R_K, it will provide feedback proportional to the output current in one instance, and feedback proportional to the output voltage in another.

These differences are apparent when we consider the output resistance of a phase splitter looking into (a) the anode and (b) the cathode.

Figure 18.6.1 shows the circuit under investigation.

Fig. 18.6.1.

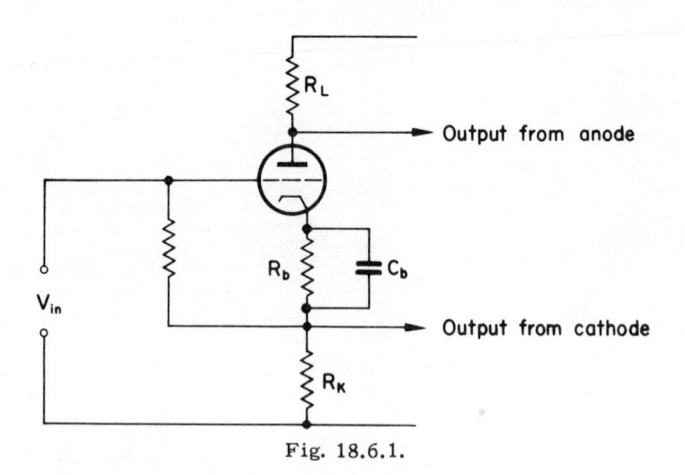

The equivalent circuits for deriving the output resistances is given in figure 18.6.2. (a) for the anode and (b) for the cathode.

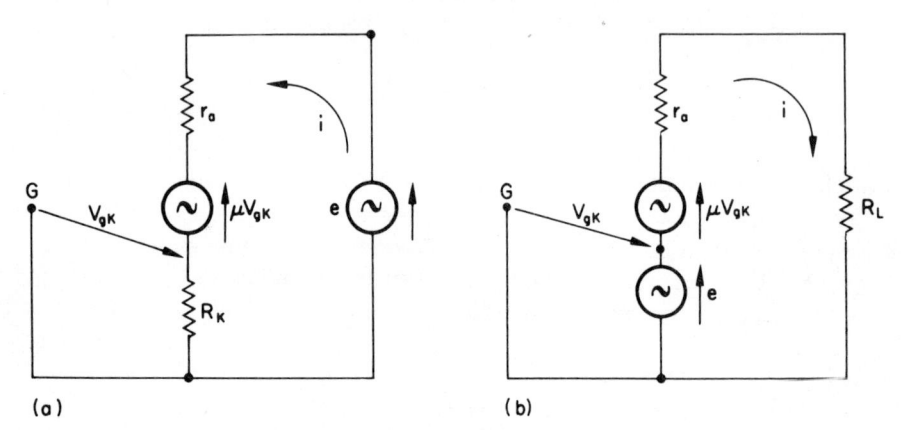

Fig. 18.6.2.

(a) $$v_{g_k} = i R_K \quad \therefore \quad e - \mu i R_K = i(r_a + R_K)$$

$$\therefore \quad e = i[r_a + R_K(1+\mu)]$$

$$\therefore \text{ Rout from the anode } = \frac{e}{i} = r_a + R_K(1+\mu)$$

(b) $$v_{g_k} = e \quad \therefore \quad e + \mu e = i(r_a + R_L)$$

$$\therefore \quad e(1+\mu) = i(r_a + R_L)$$

$$\therefore \quad \frac{e}{i} = \text{Rout from the cathode} = \frac{r_a + R_L}{1+\mu}$$

In (a) the cathode resistor provided current feedback whilst in (b) it provided a voltage feedback. It is seen that current feedback raised Rout whilst voltage feedback lowered Rout. The result in (a) would be in shunt with R_L and the result in (b) would be in shunt with R_K.

Positive feedback

Positive feedback may be employed to increase gain and to raise the input resistance. This method should be carried out with great care as the circuit is very liable to become unstable. Although unstable circuits are dealt with in detail elsewhere in this book, a brief look at one form of positive feedback may be desirable to illustrate the danger of instability.

Suppose it were necessary to cause the input resistance of a two valve amplifier to become effectively infinite at a given frequency. The input grid leak may be connected between input and output of the amplifier as shown in figure 18.6.3.

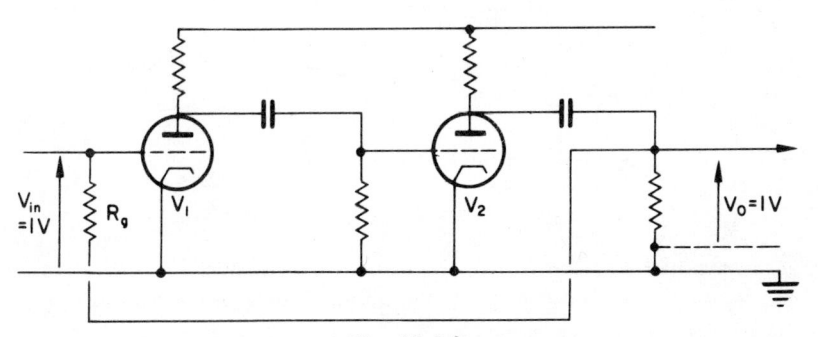

Fig. 18.6.3.

If the gain of the amplifier was adjusted to unity and having no phase shift, then for $a + 1v$ signal to the grid of V_1, $a + 1v$ signal would appear at the lower end of R_g.

The potential across $R_g = 1 - 1 = 0V$.

The current flowing through $R_g = \dfrac{v_{r_g}}{R_g} = \dfrac{0V}{R_g} = 0$.

The current flowing in R_g would be the input current and would flow from signal source v_{in} Hence as i in $= 0$.

$$R_{in} = \dfrac{v_{in}}{i_{in}} = \dfrac{v_{in}}{0} = \infty$$

The circuit is not reliable because if the gain increased above unity even by a little, the input current would flow back into the generator V_{in},

Y

as the lower end of R_g would be most positive due to v_o being greater than v_{in}. Should this occur, the circuit would almost certainly continue to oscillate even with v_{in} removed. The circuit is in fact, a multivibrator and is one basic circuit of a whole family of non stable circuits covered in detail earlier on.

Example

Figure 18.6.4 shows a single voltage amplifier with composite feedback. We need to adjust the feedback so as to cause the effective output resistance to become 10KΩ.

Fig. 18.6.4.

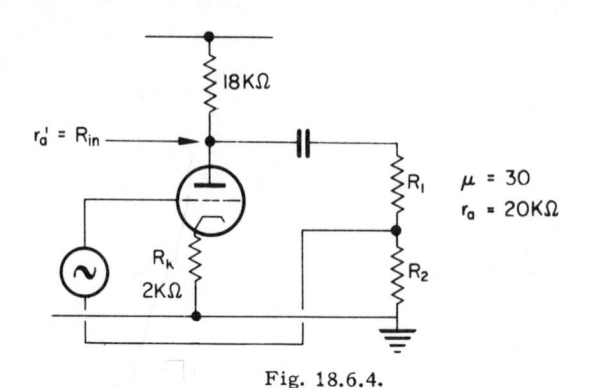

R_K is determined so as to ensure the correct d.c. conditions prevail. R_1 and R_2 form a potential divider, the output from which is the voltage feedback. This will have an amplitude $\beta = R_2 / (R_1 + R_2)$

Current feedback due to an unbypassed cathode resistor will raise the ra to an effective value of $r_a + (1 + \mu) R_K = 20 + (31)2 = 82$KΩ.

We will now ignore R_k in relation to 82K and assume $R_k = 0$. R_{out} needs to be 10 KΩ. Therefore $r'_a \mathbin{/\mkern-5mu/} 18K = 10K$. Hence $r'_a = 22.5$ KΩ

and as $\quad r'_a = \dfrac{r_a}{1 - \mu\beta} \qquad 22.5K = \dfrac{82K}{1 - 30\beta}$

∴ $\qquad\qquad\qquad 1 - 30\beta = \dfrac{82}{22.5}$

∴ $\qquad\qquad\qquad 1 - \dfrac{82}{22.5} = 30\beta$

$\qquad\qquad\qquad\qquad\quad \beta = \dfrac{22.5 - 82}{(22.5)(30)}$

∴ $\qquad\qquad\qquad \beta = \dfrac{-59.5}{675} = -0.882$

and is negative hence this is negative feedback. We must make this divider very high to ensure that it does not affect the output resistance.

$$\text{let } R = 820\text{K}\Omega \text{ say, then } \frac{820\text{K}}{R + 820\text{K}} = 0.882.$$

$$\therefore \quad 820 = 0.882 R_2 + 722 \qquad \therefore \quad R_2 \div \frac{98\text{K}}{0.882} = 111\text{K} \simeq 120\text{K}\Omega$$

to the nearest standard value.

18.7. Voltage series negative feedback. Large signal analysis

It has been shown that the gain of an amplifier will be reduced when negative voltage feedback is applied. The amplitude distortion that might otherwise exist due to the non-linearity of the grid curves are also reduced. Figure 18.7.1. shows a load line drawn on the characteristics representing 66.6KΩ which is the anode load of an EF86 Pentode.

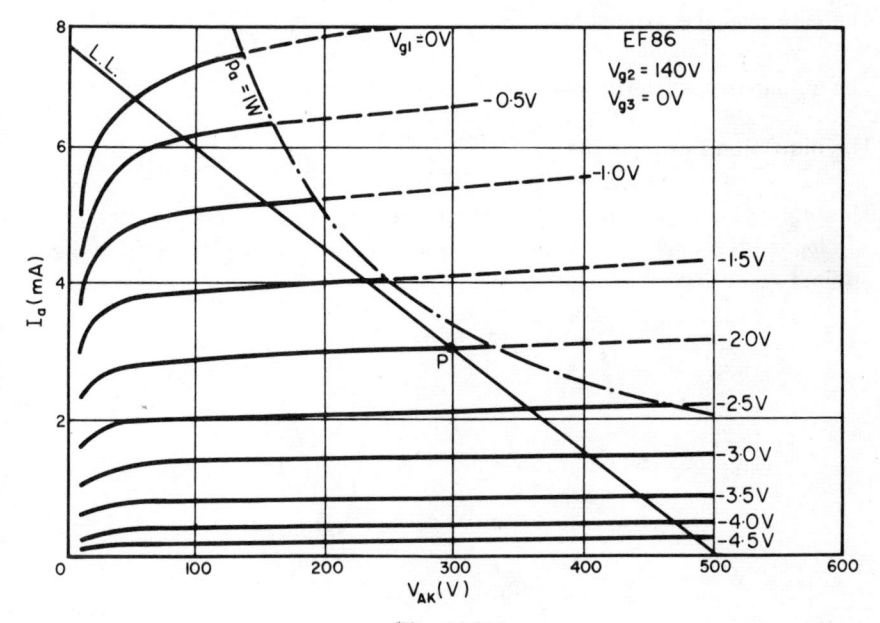

Fig. 18.7.1.

Assuming an operating point represented by $V_{g_k} = 2\text{V}$, an input signal of 4V peak to peak is applied. It is seen that the change in anode voltage for a 2V input is $300 - 50 = 250\text{V}$. For a -2V input, the change in anode voltage is seen to be $475 - 300 = 175\text{V}$.

ELECTRONICS FOR TECHNICIAN ENGINEERS

Figure 18.7.2. shows the output waveform relative to the input.

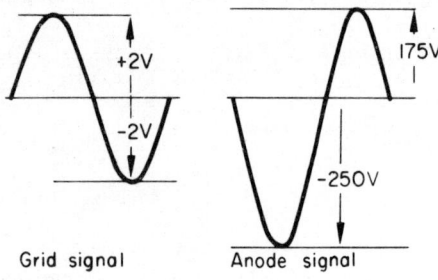

Grid signal   Anode signal

Fig. 18.7.2.

The gain $= \dfrac{175 - (-250)}{2 - (-2)} = \dfrac{425}{4} = 106.$

The gain over the first half cycle $= \dfrac{-250 + 175}{2} = \dfrac{-75}{2} = -37.5$

The gain over the second half cycle $= \dfrac{250 + 175}{-2} = \dfrac{425}{-2} = -212.5$

The output is assymetric and is therefore distorted.

The distortion in the output $= \dfrac{37.5 \times 100}{212.5} \% = 17.6\%$

This distortion can be reduced with feedback by a factor $1 + A\beta$.

Suppose $\beta = 1/10$. Then the input required, with feedback, to give the original output, is shown in figure 18.7.3.

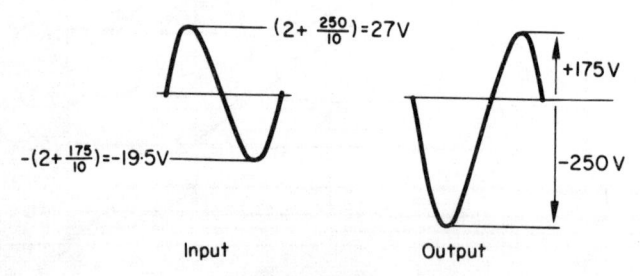

Input   Output

Fig. 18.7.3.

The positive going input during the first half cycle is seen to be 27V.
The gain therefore $= (-250)/27 = -9.26.$

The input during the second half cycle is seen to be -19.5V.
Hence the gain $= 175/(-19.5) = -8.98.$
The distortion therefore $= 0.14/9.12 \times 100\% = 1.54\%.$
The original distortion is reduced theoretically by a factor $1 + \beta A$.

NEGATIVE FEEDBACK AND ITS APPLICATIONS

Putting in known values, we get a reduction in original distortion of
$1 + 212/2 \cdot 1/10 = 11.6$.

Hence the distortion with feedback theoretically becomes $17.6\%/11.6 = 1.52\%$ which agrees very closely with the values obtained from the illustration in figure 18.7.3.

If a sinusoidal input were theoretically applied as shown in figure 18.7.4, a mean gain of 9.12 would result.

The original gain of 106 would be reduced by $1 + A\beta = 106/11.6 = 9.14$.

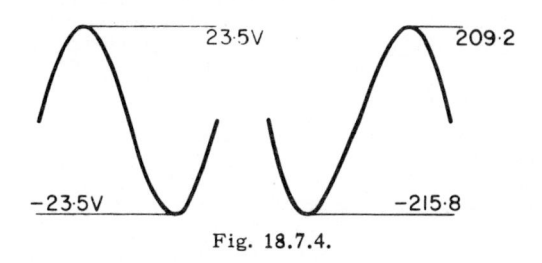

Fig. 18.7.4.

We will take this discussion a little deeper and investigate the effects of feedback upon our pentode amplifier.

We will see how a much larger input is required to maintain the output without feedback.

We will also see how the pentode constant-current characteristics are changed by means of feedback, to those of a triode.

The circuit diagram shown in figure 18.7.5. is that of the amplifier with negative feedback applied.

The voltate generator e' is seen to be connected in series with the feedback voltage βV_a.

The output is seen to be asymmetrical about the operating point even through the input voltage was quite symmetrical. Negative feedback will reduce this distortion to a nimimum although this will entail a very much larger input signal for the same output. The circuit diagram shown in figure 18.7.5 is that of the amplifier with negative feedback applied. The voltage generator e', is seen to be in series with the feedback voltage.

It is possible to analyse the circuit under d.c. conditions only and the circuit in figure 18.7.6 shows the d.c. version of figures 18.7.5.

Note that there is no d.c. blocking capacitor and that the input 'signal' E' is connected in a sense that it opposes the positive d.c. potential that must exist due to βV_a. A simplified version of the 'signals' in the circuit is shown in figure 18.7.7.

A glance at the circuit in figure 18.7.7 will show that $\beta V_a = E' - V_{g_k}$.

Therefore
$$V_a = \frac{E' - V_{g_k}}{\beta}$$

Fig. 18.7.5.

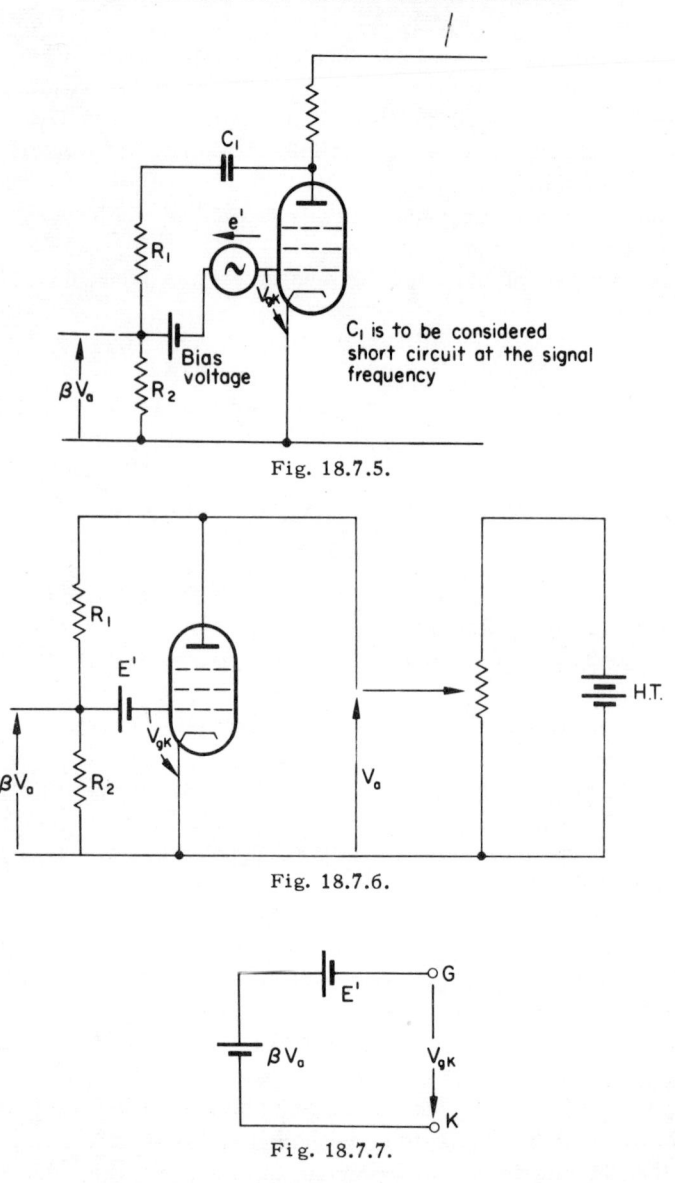

Cᵢ is to be considered short circuit at the signal frequency

Fig. 18.7.6.

Fig. 18.7.7.

This expression is required in subsequent analysis. The object of the analysis is to derive a new set of characteristics for a pentode with voltage negative feedback. The feedback will lower the effective valve parameters whilst the resultant characteristics will be similar to those of a triode.

Before proceeding with the analysis, it may be of interest to show that the actual effective input, V_{g_k}, is very much lower than the input to the

NEGATIVE FEEDBACK AND ITS APPLICATIONS

grid from the external generator due to the feedback voltage which appears in opposition to the input by an amount βV_a. From the formula derived, $V_{g_k} = E' - \beta V_a$ which shows how the grid voltage V_{g_k} is reduced. If the output voltage, V_a, is to be maintained, an additional input equal in amplitude but of opposite sense to the feedback, must be applied. The output $V_a = A \cdot V_{g_k}$.* Substituting this for V_a in the previous expression, $E' = [V_{g_k}(1 + A\beta)]$.

The object of the following is to derive a set of tables from which new characteristics can be plotted from the valve with feedback. Once completed, a normal load line may be drawn for the anode load and the gain established in the usual manner.

These characteristics will show that, for a given output, the input must be very much larger than the input without feedback. A comparism will be made between the two amplifiers, with and without feedback.

For the purpose of this example, let β be 1/10. A table to be drawn up will show the relationship between the anode voltage and the grid to cathode voltate. V_{g_k} for a given E'. The formula

$$V_a = \left(\frac{E' - V_{g_k}}{\beta}\right)$$

derived from figure 18.7.7 will provide sufficient information for this purpose. A value of 5V will be assumed for E', grid to cathode voltages which are on the original characteristics will be assumed and V_a will be calculated. A table will result.

Let $E' = 5V$.

Assumed V_{g_k} (V)	$V_a = \frac{1}{\beta}(E' - V_{g_k})$ (V)	Coordinates (V_{g_k}, V_a)
0	10 (5 − 0) = 50	0,50
−0.5	10 (5 − 0.5) = 45	−0.5,45
−1.0	10 (5 − 1.0) = 40	−1.0,40
−1.5	10 (5 − 1.5) = 35	−1.5,35
−2	10 (5 − 2.0) = 30	−2.0,30
−2.5	10 (5 − 2.5) = 25	−2.5,25
−3	10 (5 − 3.0) = 20	−3.0,20
−3.5	10 (5 − 3.5) = 15	−3.5,15
−4	10 (5 − 4.0) = 10	−4.0,10
−4.5	10 (5 − 4.5) = 5	−4.5,5

* where A is the amplifier gain, i.e. gain $= \dfrac{0/P}{1/P} = A$

Further tables need to be constructed for values of E' in increments of 5V up to a maximum of 50V. These tables are shown below.

V_{g_k}	Let $E' = 10V$. V_a (V)	Let $E' = 15V$. V_a (V)	Let $E' = 20V$. V_a (V)	Let $E' = 25V$. V_a (V)	Let $E' = 30V$. V_a (V)
0	100	150	200	250	300
−0.5	95	145	195	245	295
−1.0	90	140	190	240	290
−1.5	85	135	185	235	285
−2.0	80	130	180	230	280
−2.5	75	125	175	225	275
−3.0	70	120	170	220	270
−3.5	65	115	165	215	265
−4.0	60	110	160	210	260
−4.5	55	100	155	205	255

V_{g_k}	Let $E' = 35V$. V_a (V)	Let $E' = 40V$. V_a (V)	Let $E' = 45V$. V_a (V)	Let $E' = 50V$. V_a (V)
0	350	400	450	500
−0.5	345	395	445	495
−1.0	340	390	440	490
−1.5	335	385	435	485
−2.0	330	380	430	480
−2.5	325	375	425	475
−3.0	320	370	420	470
−3.5	315	365	415	465
−4.0	310	360	410	460
−4.5	305	355	405	455

It is seen that the higher the value of E', the higher the value of V_a will be necessary to obtain the appropriate grid to cathode voltage V_{g_k}. Had the fraction V_a fed back to the grid β, been smaller, the anode voltage in the tables would have been that much larger. For example, if β had have been 1/20th, then all of the values for V_a in the tables would have been twice the values shown.

Upon completion of the tables, the coordinates (V_{g_k}, V_a) may be plotted. Consider first, the table based upon E' of 5V as a constant. A point corresponding to V_{g_k} = 0V and V_a = 50V, is plotted on the original EF86 characteristics. Working down the table, a series of points are plotted for V_a against V_{g_k}. Figure 18.7.8 shows the *first* grid characteristic for E' = 5V superimposed upon the original graph. This feedback will cause

NEGATIVE FEEDBACK AND ITS APPLICATIONS 327

the Pentode to behave as a Triode, therefore the new characteristics will reflect this behaviour as seen in figure 18.7.10.

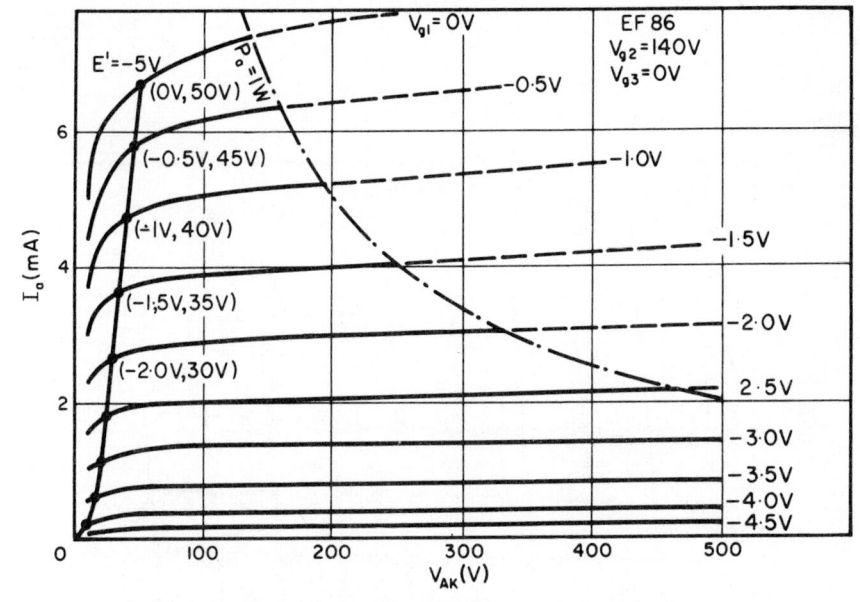

Fig. 18.7.8.

The new characteristics represent the family of grid curves for the original Pentode valve with voltage feedback. The value of β is 1/10th of the output.

The d.c. operating point is seen to be –2V. The grid will 'sit' at this point in the absence of a signal. When a signal is applied, the grid will traverse the load line around an operating point of $E' = -32V$. This corresponds to the point $V_{g_k} = 2V$. For an input of 18 volts, the change in anode potential is seen to be $300 - 132 = 168V$. For an input of –18V, the output is seen to be $462 - 300 = 162V$. The linearity has been improved considerably. The waveforms are shown in figure 18.7.9.

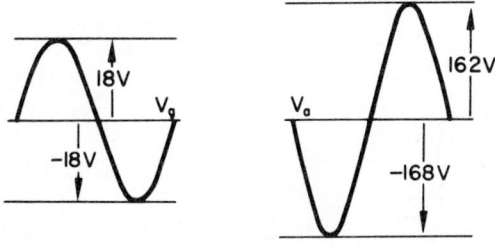

Fig. 18.7.9.

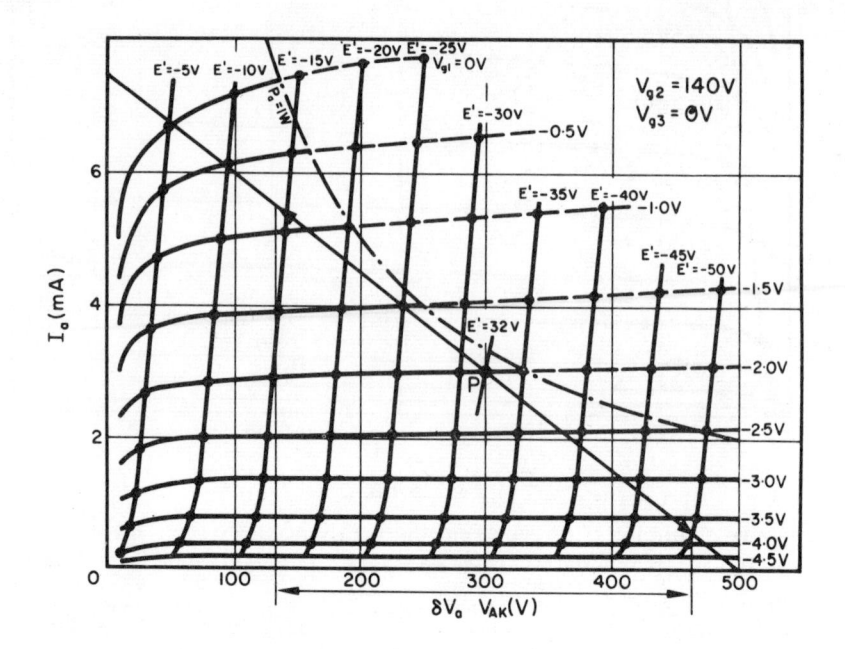

Fig. 18.7.10.

From the characteristics,

the gain of the circuit without feedback is $\dfrac{\text{Pk to Pk output}}{\text{Pk to Pk input}} = \dfrac{425\text{V}}{4\text{V}} = 106$

The gain of the circuit with feedback is $\dfrac{\text{Pk to Pk output}}{\text{Pk to Pk input}} = \dfrac{330\text{V}}{36\text{V}} = 9.4$

From the formula for the gain of an amplifier with feedback, $A' = \dfrac{A}{1 + \beta A}$

and substituting the values obtained, A' becomes $\dfrac{106}{1 + 10.6} = \dfrac{106}{11.6} = 9.14$

Increasing the feedback would result in an even more accurate result. It would also result in the gain figure being almost completely dependant upon the feedback components and very little upon the valve.

18.8. Stabilised power supplies

It will be shown that, when included in a feedback loop of gain A, the output resistance of a cathode follower is reduced by a factor of $1/(1 + A)$. If the original output resistance is given as r_a/μ, then with maximum feedback i.e. $\beta = 1$, the output resistance will fall to the value

$$\frac{r_a}{\mu (1 + A)}$$

If an amplifier of gain A, is connected between output and input of a cathode following having an output resistance r_o, negative feedback will cause the cathode follower output resistance to fall to $r_o/(1 + A)$.

If the cathode follower, r_o, is 500Ω, and the amplifier has a gain of 49, then the modified output resistance is reduced to $500\Omega/(1 + 49) = 10\Omega$.

Such is the case with series regulated power supply units. An example of this nature is given.

Cathode follower output resistance. The effects of further amplification upon the output resistance.

It has been shown that the output resistance of a cathode follower is given as

$$\frac{r_a}{1 + \mu} \text{ and when } \mu \gg 1, \text{ is } \simeq \frac{r_a}{\mu} \simeq \frac{1}{gm}$$

Now consider the circuit in figure 18.8.1. which shows a cathode follower plus a further amplifier connected between the cathode and the grid. The amplifier will provide a signal from the cathode back to the grid, this will be in antiphase with the cathode signal and this gives a degree of negative feedback.

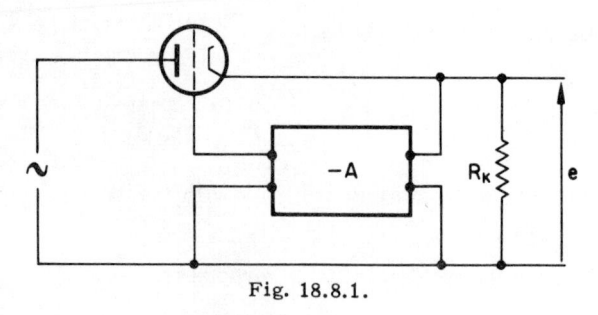

Fig. 18.8.1.

Figure 18.8.2. shows the equivalent circuit for figure 18.8.1.

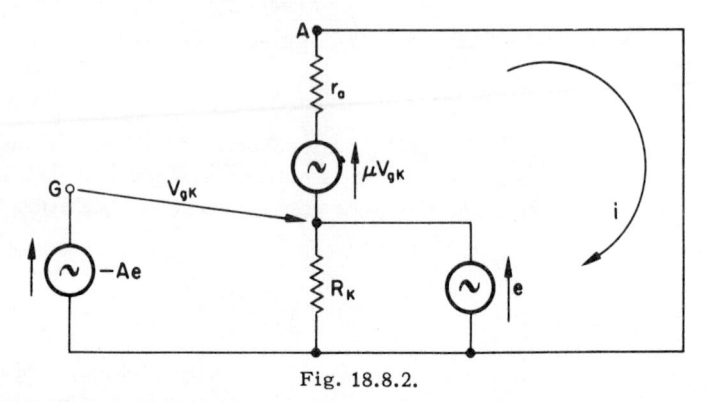

Fig. 18.8.2.

The signal at the cathode will be amplified, the phase changed, and fed back to the grid of an amplitude -A times that of the cathode signal.

The output resistance is required. Application of the e.m.f. e, will cause a current to flow. The ratio of the e.m.f., e, and the current i, will determine the output resistance. It has been stated that for a given cathode voltage, there will be a quantity -A times the cathode voltage applied to the grid.

The effective V_{g_k} will be the assumed input, e, plus the signal -AV_k or -AB volts. From figure 18.8.2.,

$$v_{g_k} = e + Ae \quad \therefore \quad \mu v_{g_k} = \mu(e + Ae)$$

$$\therefore \quad \mu(e + eA) = i(r_a) \text{ ignoring } R_K \text{ for the moment,}$$

$$\mu(e[i + A]) = i(r_a) \text{ and Rout} = \frac{e}{i}$$

$$\therefore \quad \frac{e}{i} = \frac{r_a}{\mu(1 + A)} \simeq \frac{1}{g_m(1 + A)}$$

This justifies the statement that the original output resistance

$$\frac{r_a}{\mu} = \frac{1}{g_m}$$

is decreased by a factor $1/(1+ A)$.

18.9. Series regulator

Figure 18.9.1. shows the full circuit diagram.

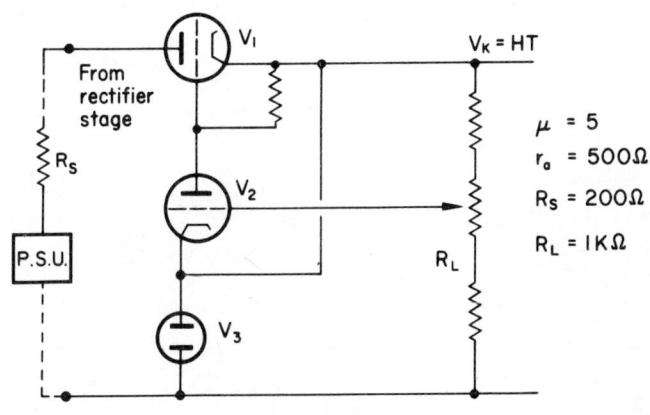

Fig. 18.9.1.

The circuit above may be represented in a simpler fashion as shown in figure 18.9.2. R_S is the source resistance of the power supply unit.

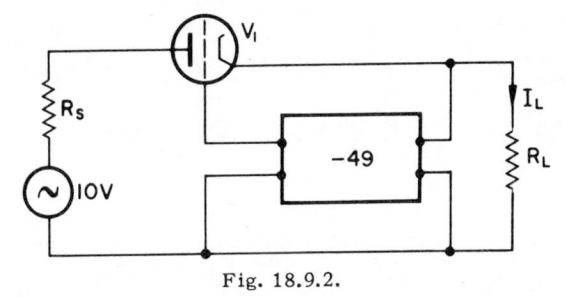

Fig. 18.9.2.

The 'black box' represents V_2 as an amplifier having a gain of 49. The output from the amplifier is fed back to the grid of V_1 in antiphase to the amplifier input. This represents the change in level of the stabilised output. The resultant negative feedback reduces any such variation across the load.

The amplifier also reduces the output resistance by a factor of $\frac{1}{1 + A}$.

The equivalent circuit to determine the change in output voltage (δv) for a 10v change in mains voltage is given in figure 18.9.3.

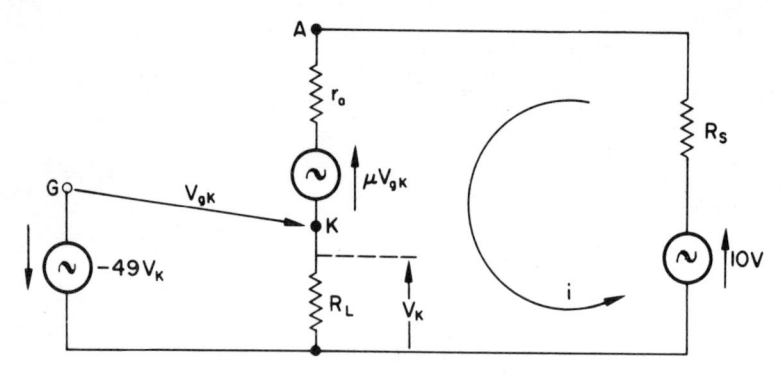

Fig. 18.9.3.

The reader should recall that it is the *change in grid to cathode* voltage that becomes the effective input to a valve.

Problem

By how much would the output vary if the mains input varied by 10Ω? Consider the equivalent circuit. We must first express v_{g_k} in terms of i.

$$v_{g_k} = (49v_K + v_K) \text{ and } v_K = 1000\,i \quad \therefore \quad v_{g_k} = 50{,}000\,i$$

Applying Kirchoff's first law to the anode circuit.

$$10 - \mu\,v_{g_k} = i\,(r_a + R_K + R_S)$$
$$\therefore \quad 10 - 5\,(50000\,i) = i\,(1700)$$
$$\therefore \quad 10 = i\,(1700 + 250{,}000)$$
$$\text{hence} \quad i = 10/129000 \text{ but } O/P = 1000\,i \text{ (volts)}$$
$$\therefore \quad \delta v = \frac{10\,000}{251700}\text{V} = 39.6\text{mV change.}$$

Fig. 18.9.4.

Unmodified circuit

Modified circuit

NEGATIVE FEEDBACK AND ITS APPLICATIONS

Sometimes it is necessary to shunt V_1 with a high wattage resistor in order to keep the valve within its power rating. This shunt resistor modifies μ and r_a. This is shown in figure 18.9.4.

Suppose the shunt resistor to be, say 500Ω, then from the valve equivalent circuit the new values may be derived.

μ' has a new value of 2.5 whilst r'_a becomes 250Ω. Putting these values in the previous equation.

$$10 - 2.5 \,(50{,}000 \, i) = i \,(1000 + 200 + 250)$$
$$\text{and } 10v = 2.5 \,(50000)\, i + i \,(1450)$$
$$\therefore \quad 10v = i \,(125 + 1.45)$$

and multiply by R_K to obtain volts,

hence $\qquad \delta v = \dfrac{10\,000}{126.45} V = 78.5\,\text{mV change}$

Without the shunt resistor, the stabilised output presents an output resistance R_o to the 1KΩ load.

As usual, if we need to find R_{out}, a voltage is applied to the output and the current that would flow into the output terminals is calculated.

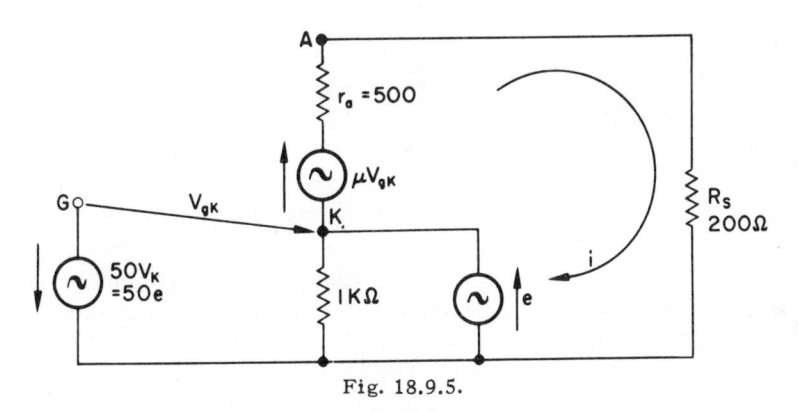

Fig. 18.9.5.

(Ignore the 1KΩ load, as this is external).

$$v_{g_k} = 51e; \text{ and using Kirchhoff's laws.}$$
$$e + 5\,(51e) = 700i$$
$$e\,(1 + 255) = 700i$$
$$\text{Rout} = \frac{e}{i} = \frac{700}{255} = \underline{2.74\,\Omega}$$

But the effect of the shunt resistor is to raise the output resistance:

If we again insert the modified values of μ and r_a given in figure 18.9.4., the modified output resistance can be determined.

Equating e.m.f.'s to p.d.'s, $e + (2.5 \times 51)e = i(200 + 250)$

and $\quad e(1 + 2.5 \times 51) = i(450)$

$\therefore \quad e(127.5) = 450i$

and $\quad \text{Rout} = \dfrac{e}{i} = \dfrac{450}{127.5} = \underline{3.54\Omega}$

Hence by shunting the series regulating valve with a resistor raises the output resistance and reduces the effectiveness of the circuit to minimise load voltage changes due to input fluctuations from the unstabilised supply.

18.10 Shunt type stabiliser circuit

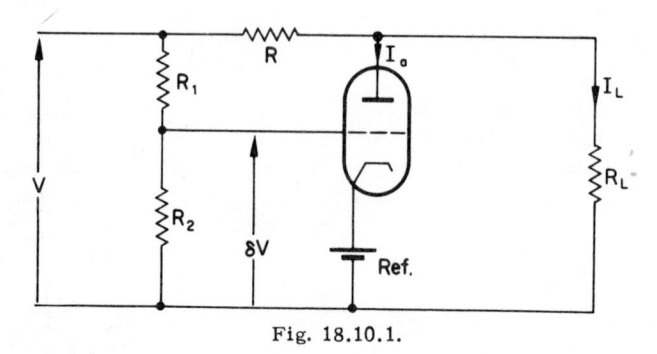

Fig. 18.10.1.

Assume the reference voltage to be constant at all times. R_L is a fixed load, and the load current and voltage needs to be constant. A mains variation causes the output from the rectifier stage to vary, and this is shown as v. The anode current Ia, should vary in order to accomodate any current changes due to any input voltage changes thus maintaining I_L at a constant value. We need to determine the ratio R_1/R_2 to provide a constant load current.

$$\delta v = \dfrac{R_2}{R_1 + R_2} \times v$$

the extra anode current (I_a) will be given as

$$gm \times v_{g_k} = gm \cdot \delta v = gm \cdot \dfrac{R_2 \, \mu}{R_1 + R_2}$$

but $\delta I_a = \dfrac{v}{R}$ and $\dfrac{v}{R} = gm \dfrac{R_2 v}{R_1 + R_2}$

$\therefore \quad \dfrac{1}{R} = \dfrac{gm \, R_2}{R_1 + R_2}$

and
$$R = \frac{R_1 + R_2}{gm\, R_2} \quad \therefore \quad gm\, R = \frac{R_1 + R_2}{R_2}$$

hence
$$gm\, R = \frac{R_1}{R_2} + 1$$

and
$$\frac{R_1}{R_2} = gm\, R - 1$$

R is determined by the load current I_L, the required load voltage, the quiescent anode current and the supply voltage.

This particular circuit is **restricted** to a valve having a suitable gm, this restricting its use to rather special circumstances.

18.11. Negative output resistance

A negative resistance will aid current flow rather than impede it. A typical example of negative resistance action and how to overcome the undesirable effects in this circuit, is shown.

Suppose a stabilised p.s.u. had a certain built in circuit of which one function would be to reduce the output resistance to zero.

If this compensation were 'overdone', it could cause the output resistance to become a negative quantity. Any reactive component used as a load would cause the p.s.u. to oscillate.

Example

A power supply unit shown in figure 18.11.1. has the following characteristics.

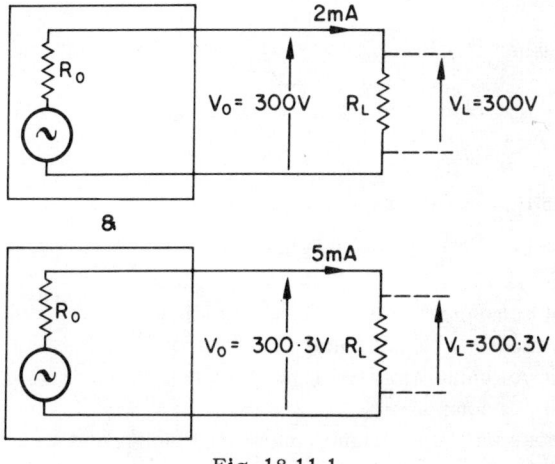

Fig. 18.11.1.

Note that for an *increase* in I_L, the terminal p.d. *increases* instead of the decrease one would expect due to the extra current through the internal resistance of the generator, R_0.

R_0 must be negative and has the following value

$$R_0 = \frac{\delta v_0}{\delta i_0} = \frac{300 - 300.3}{5 - 2}$$

$$R_0 = \frac{-0.3V}{3mA} = -100\Omega$$

If we need a voltage across the load that is to remain constant at the two load currents shown, we simply insert a resistor in series with the negative R_0, having the same ohmic value. The two conditions are examined to see whether V_L is constant.

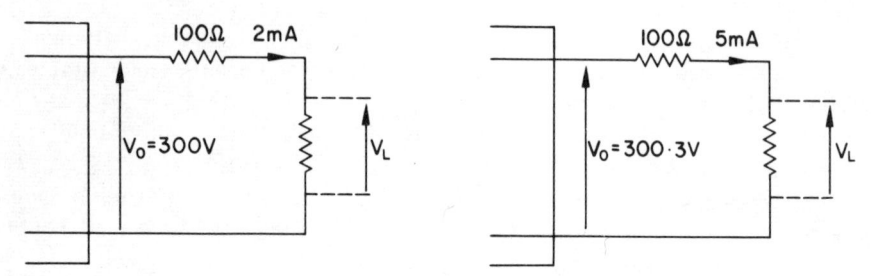

Fig. 18.11.2.

$$V_L = 300 - \frac{2 \times 100}{1000} = \underline{299.8V}$$

$$V_L = 300.3 - \frac{5 \times 100}{1000} = 299.8V$$

If $R_s = R_0$ (where R_0 is negative), v_L remains constant under varying i_L conditions.

The reader might check that at $I_L = 6$ mA, $V_0 = 300.8$ and V_L remains at 299.8V.

18.12. A stabilised power supply unit

The circuit of the stabilised power supply unit we intend discussing is shown in figure 18.12.1.

A number of components are marked with known values. We will assume that these are readily available and will be used in conjuction with those components as yet unknown. During the discussion we will determine the values of all other components.

We will choose an EL84 (triode connected) for V_1, and EF86 (triode connected) for V_2, and an 85A2 for V_3.

The rectifier will be an EZ81.

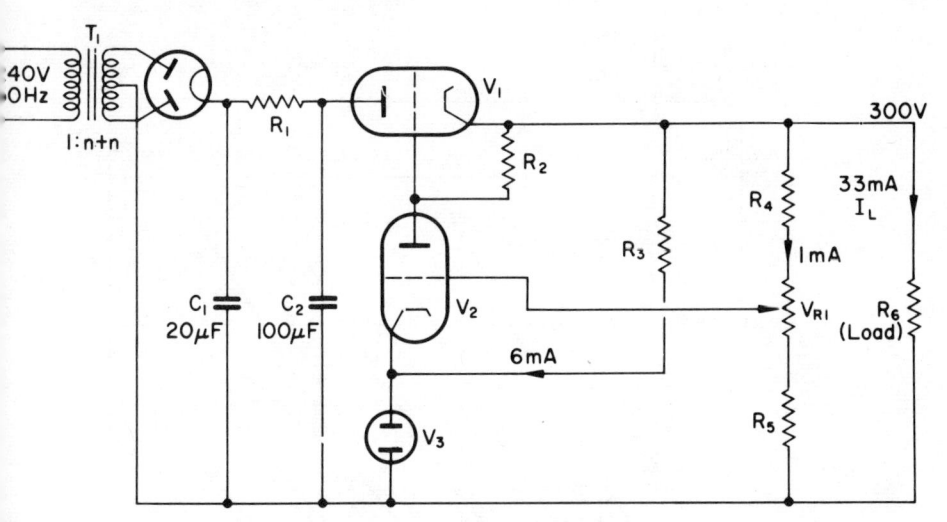

Fig. 18.12.1.

We will discuss each section of the circuit separately and determine values and operating points for the valves, as we go along. An assumption is initially made for the 'bleed' current of 1 mA and 6 mA current for V_3.

These currents are shown on the circuit diagram.

The p.s.u. is to provide 300V average d.c. at a load current of 33 mA. The output ripple is to be less than 100 mV. Some adjustment must be included to allow for initial setting up, thus 'overcoming' circuit component tolerances.

The series valve, V_1

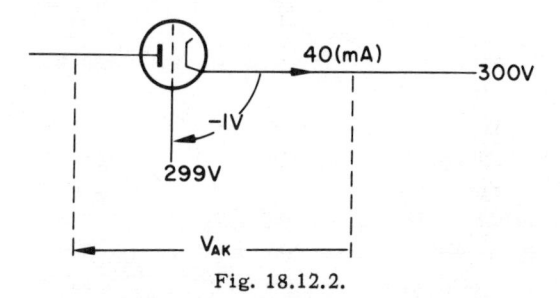

Fig. 18.12.2.

The manufacturer's data shows that $R_{g\,max,} = 300K\Omega$. We will therefore ignore any current flowing through $R_g, = R_2 = 300K\Omega$.

From the characteristics in figure 18.12.3, and for $I_a = 40$ mA, $V_{gk} = -1$, it is seen that $V_{ak} = 114V$. This is the minimum value including the negative going peak ripple across C_2 and assuming the mains supply to be at 94% of its nominal voltage.

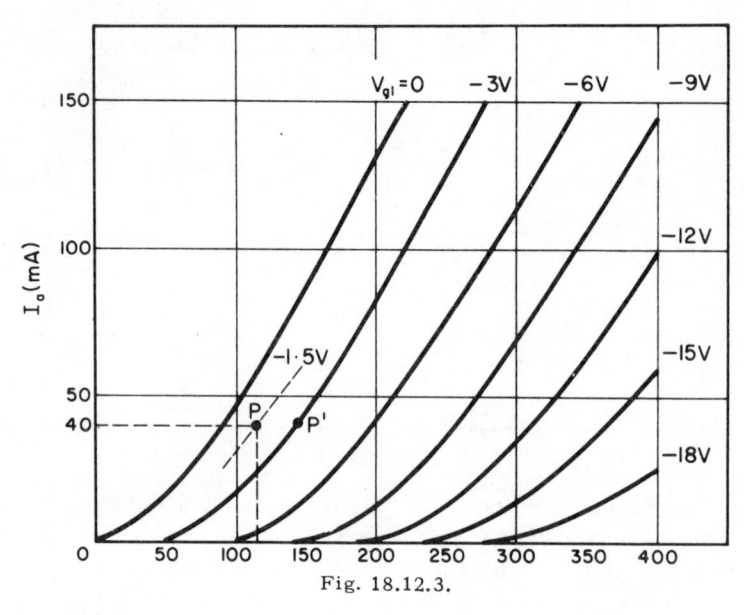

Fig. 18.12.3.

Amplifier valve, V_2

The voltage across V, i.e. $V_{ak} = (300 - 1) - (85) = 214V$. The grid of V_1, at –1V bias, is at 299V. The cathode of V_2 is at 85V due to V_3, hence the V_{ak} of $V_2 = 214V$.

The anode current is very very small and can be ignored.

V_2 has an anode load of $R_2 = 300K\Omega$. The gain of V_2 will be approximately 20, call this A_2.

The grid of V_2 will be a few volts negative with respect to its cathode, which in turn is sitting at 85V

Let us arbitrarily assume the grid will be at 80V.

The fraction of output ripple to appear at V_2 grid via the bleeder network will be 80/300 of the output ripple.

This is shown simply in figure 18.12.4.

The output ripple will be reduced due to V_1 acting alone. This reduction will be ignored so as to err on the safe side.

The ripple reduction due to V_2 as an amplifier with a gain of 20, and with an input of 80/300 of the output ripple is given as follows;

$$\frac{20 \times 80}{300} = 5.$$

NEGATIVE FEEDBACK AND ITS APPLICATIONS

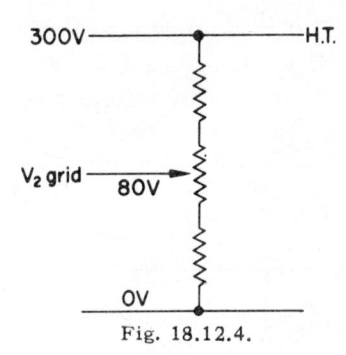

Fig. 18.12.4.

With a ripple reduction of five times, and as the output ripple must not exceed 100 mV, the tolerable input ripple across $C_2 = 5 \times 100$ mV. This can be much larger if the reduction due to V_1 alone is considered also.

Derivation of the value of R_3

The 85A2 in its preferred working condition, requires 6 mA burning current to maintain 85V reference voltage.

This current will flow through R_3.

The value of $R_3 = \dfrac{300 - 84}{6.10^{-3}} = 36$ KΩ.

Derivation of the value of R_1

The ripple voltage across C_1 is determined quite simply as we know its discharge (or load) current. This is the sum of all circuit currents = 33 + 6 + I bleed, where *I* bleed is the current through the bleeder chain. We have already assumed this to be 1 mA.

Hence the total average current drawn from C_1 = 33 + 6 + 1 = 40 mA.

Using the approximations covered previously, the ripple across C_1 becomes
$$\dfrac{1.T.}{C} = \dfrac{40\text{mA} \times 10 \text{ mS}}{20\mu\text{ F}} = 20\text{V } P - P.$$

There is therefore a peak voltage of approximately 10V peak.

We require 500 mV peak across C_2 hence the ripple reduction factor due to R_1 and $C_2 = 10/0.5 = 20:1$

R_1 must therefore be approximately $20 \times XC_2$, say 24 to be sure.

Hence
$$R_1 = \dfrac{24 \times 1}{2\pi f C} = \dfrac{24 \times 0.159}{100.\ 100.\ 10^{-6}}$$

R_1 = 380, say 390Ω as a standard value.

The voltage drop across $R_1 = 40 \times 0.39 = 16V$.

The mean volts required across $C_1 = 300 + 114 + 16 = 430V$.

We now need to establish the secondary voltage from the transformer that will provide an output from the EZ81 of 430V d.c.

From the characteristics of the EZ81, in figure 18.12.5. the transformer voltage required for a 40 mA load is approximately 350 – 0 – 350V r.m.s. We require $350 \times 100/94 = 375 - 0 - 375V$ r.m.s. to allow for nominal mains input.

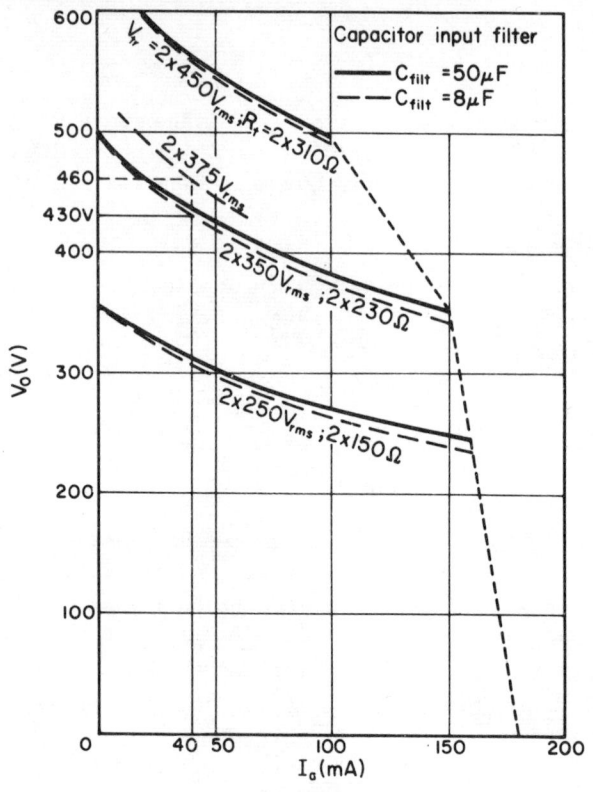

Fig. 18.12.5.

At 375 – 0 – 375V, the d.c. output is seen to be 460 V.

The final circuit becomes as shown in figure 18.12.6.

T.R.1. = 375 – 0 – 375V r.m.s. R_a to be determined.

$$C_1 = 20\mu F. \qquad C_2 = 100\ \mu F.$$

V_1 = EL84. V_2 = EF86. V_3 = 85A3. V_4 = EZ81.

R_1 = 390Ω R_2 = 300KΩ R_3 = 36KΩ

The d.c. output voltage with nominal transformer voltage of 375V at 40mA,

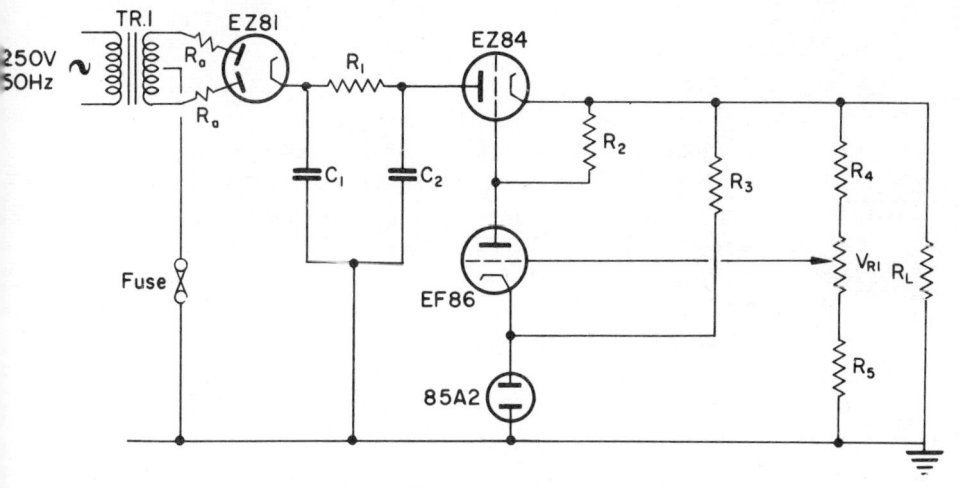

Fig. 18.12.6.

given from the manufacturer's data in figure 18.12.5. is approximately 460V d.c., and is shown as a dotted line.

The anode voltage of V_1 = 460 − 16 − 444V.

The V_{ak} of V_1 = 444 − 300 = 144V.

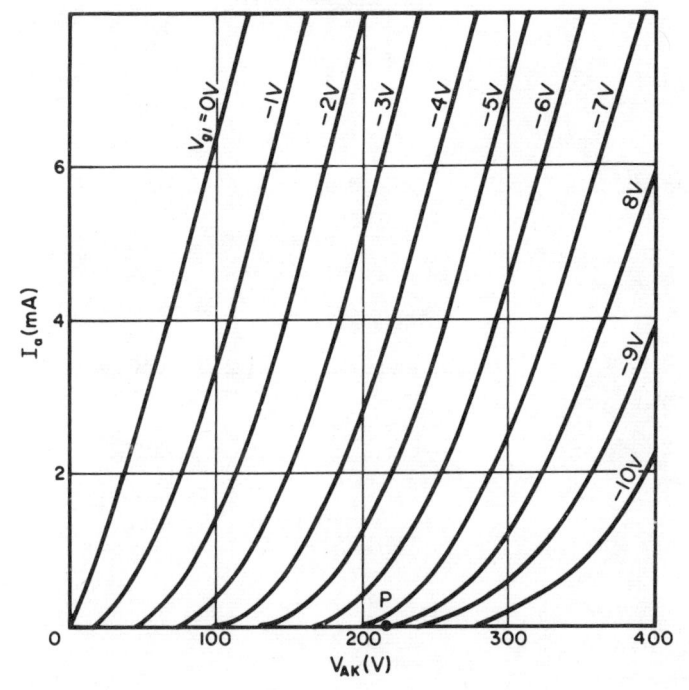

Fig. 18.12.7.

At an anode current of 40 mA, a bias of −3V is required. This is shown as p' in figure 18.12.3.

Hence the anode current of $V_2 = 3V/300K\Omega = 10\mu A$.

With a $V_{ak} = 214V$ and an anode current of $10\mu A$, a bias is required of approximately −8V as shown in figure 18.12.7.

The anode current of V_2 is so small that any variation in anode current will not affect the neon potential. Any slight change in $10\mu A$ through V_3 will be adequately swamped by the 6 mA flowing through it via R_3.

The grid voltage of $V_2 = 85 - 8 = 77V$.

We have assumed 1 mA through the bleeder chain and are in a position to determine the value of R_4, V_{R_1} and R_5.

We will determine these values so that V_{R_1} has a voltage swing of ± 10% to allow for 'setting up'.

Figure 18.12.8. shows a 10% increase in h.t. due to the slider of V_{R_1} set to its most negative position.

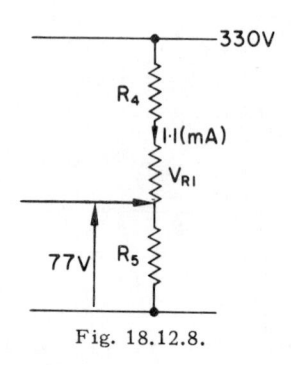

Fig. 18.12.8.

The bleed current is also increased by 10% of its nominal 1 mA.

$$R_4 + V_{R_1} = \frac{(330 - 77)V}{1.1mA} = \frac{253}{1.1} = 230K\Omega$$

$$R_5 = \frac{77}{1.1} = 70K\Omega$$

Figure 18.12.9. shows the chain and an h.t. of 270V and the slider at its most positive position.

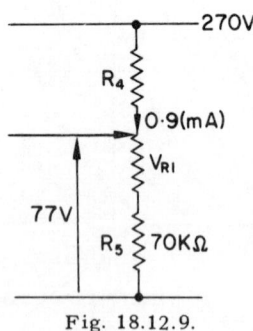

Fig. 18.12.9.

$$R_4 = \frac{(270-77)}{0.9\text{mA}} \text{V} = \frac{193}{0.9} = 215 \text{ K}\Omega.$$

Hence $\quad V_{R_1} = 230 - 215 = 15\text{K}\Omega.$

R_L, shown as R_6, this resistor may be required as a load when testing the p.s.u. $= 300\text{V}/33\text{mA} = 9.1\text{K}\Omega$.

The data in figure 18.12.10. advises a 250 Ω limiting resistor to be connected in series with each anode when an input of $2 \times 375\text{V}$ is applied.

This 250Ω is made up of the secondary winding resistance R_s, $n^2 R_p$, the reflected primary resistance, and an additional resistor (if required) R_a.

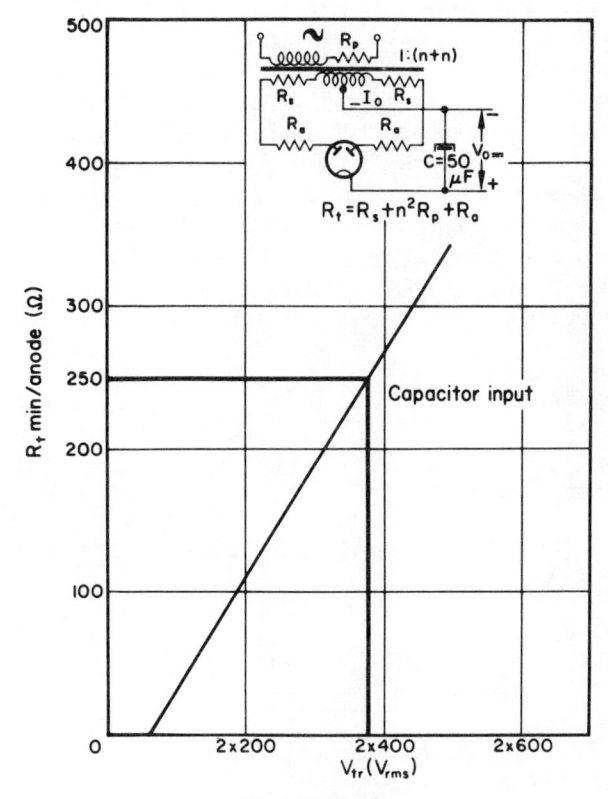

Fig. 18.12.10.

This value is for a reservoir capacitor of $50\mu\text{F}$. We are using a $20\mu\text{F}$ and the current pulses from the rectifier will be slightly less therefore the 250Ω resistor value errs on the safe side.

18.13. Attenuator compensation

Almost every piece of electronic equipment has an attenuator as part of its circuit. The attenuator may be labelled as such or it may be hidden within the equipment as part of one of the stages. Wherever the attenuator may be situated, it will be vitally necessary to correctly compensate it if the equipment is to function correctly over a wide frequency range and in particular, if the circuit is to function well with pulses.

Stray capacity will always exist in a circuit no matter what elaborate precautions are taken. A valve or transistor will have some input capacity.

The most important part of a pulse is often the leading edge. The function of the circuit will often depend upon the rate of rise of a pulse. If the edge is 'slanted' or in other words, the rise time too great, then the circuit may not function correctly, if it functions at all.

There are several direct reading capacity meters available. These meters are simple to use and the actual input capacity of a network, valve or transistor may be easily established. Once this known, or calculated, attenuators in the circuit should be correctly compensated. For simple attenuators, the method of compensation is quite easy. This chapter is included so as to show just why and how attenuators should be compensated.

A correctly compensated attenuator will give a faithful response to high frequencies in the same manner as it does to d.c.

A simple yet typical compensated attenuator is shown in figure 18.13.1.

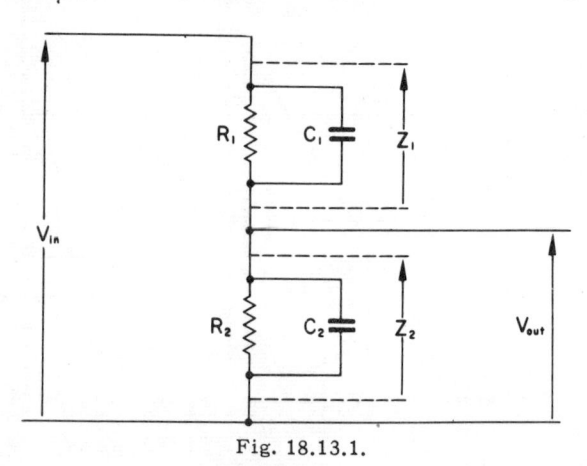

Fig. 18.13.1.

We need an attenuator that retains its d.c. value of attenuation at all frequencies. The output at d.c. $= R_2 \cdot V_{in}/(R_1 + R_2)$ (load over total) as each capacitor is 'open circuit'. Now if a pulse input is to be applied, we have to consider the frequency dependent components. Some capacity shown as C_2 may exist across say, R_2; it is then necessary to compensate for this by adding C_1 across R_1. Briefly, if the time constant $C_1.R_1$. is made equal to the time constant $C_2 R_2$, the load over total formula will apply for all frequencies. In other words, we must effectively remove all reactive component effects from the attenuator by a suitable choice of C_1

A simple proof of this requirement is as follows :

$$Z_1 = \frac{1}{Y_1} = \frac{1}{\frac{1}{R_1} + jwc_1} \quad \text{and} \quad Z_2 = \frac{1}{Y_2} = \frac{1}{\frac{1}{R_2} + jwc_2}$$

$$\frac{v_o}{v_{in}} = \frac{\text{Load } Z}{\text{Total } Z} = \frac{Z_2}{Z_1 + Z_2} = 1 \bigg/ \frac{\frac{1}{R_2} + jwc_2}{\frac{1}{R_1} + jwc_1} + \frac{1}{\frac{1}{R_2} + jwc_2}$$

Now tidying each term a little by multiplying top and bottom of each term, by R_1 in the terms containing R_1, and R_2 in the terms containing R_2

$$\frac{v_o}{v_{in}} = \frac{\frac{R_2}{1 + jwT}}{\frac{R_1}{1 + jwT} + \frac{R_2}{1 + jwT}}$$

Now if we let $C_1 R_1 = C_2 R_2$ and call the common time constant T, we may multiply top and bottom by $(1 + jwT)$

This gives

$$\frac{v_o}{v_{in}} = \frac{\frac{R_2}{1 + jwc_2 R_2}}{\frac{R_1}{1 + jwc_1 R_1} + \frac{R_2}{1 + jwc_2 R_2}}$$

$$\frac{v_o}{v_{in}} = \frac{R_2}{R_1 + R_2}$$

which is frequency independent, and is the 'load over total' expression for resistor networks.

Example

If $R_1 = 6\text{K}\Omega$, $R_2 = 4\text{K}\Omega$ $c = 60\text{pF}$, find C_1 for the attenuator to be properly compensated for optimum pulse response.

The time constants must be equal, $C_1 R_1 = C_2 R_2$

$$C_1 = \frac{C_2 R_2}{R_1} = 60 \text{ pF} \times \frac{4\text{K}\Omega}{6\text{K}\Omega} = 40 \text{ pF}.$$

The attenuator will present a capacity to the input (the input must therefore be taken from a low impedance source), but we are not concerned with this at the moment. It is unfortunately easy to overcompensate or undercompensate an attenuator: Figure 18.13.2. shows the effect of so doing.

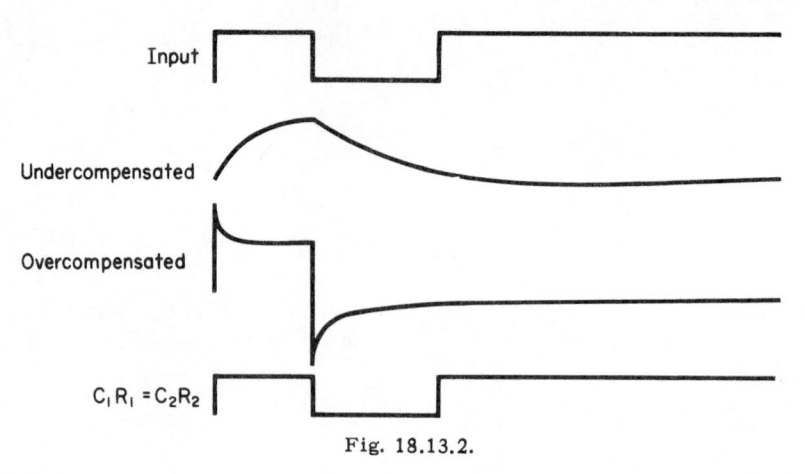

Fig. 18.13.2.

A simple application is that of a probe attenuator as shown in figure 18.13.3.

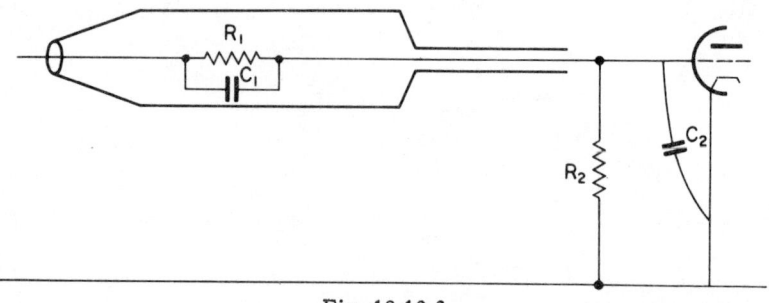

Fig. 18.13.3.

R_2 and C_2 are the grid leak and stray capacitance of a valve amplifier. R_1 is calculated to give the necessary attentuation. C_1 is chosen such that $R_1 C_1 = R_2 C_2$ for optimum pulse response.

18.14 Deriving values of components in an impedance convertor

This section will provide a very useful revision exercise.

The contents give a guide as to the application of several techniques including simple Ohm's law. Even at this rather advanced stage, it will become apparent that Ohm's law plus a little thought, will enable the technician to analyse, or derive component values for, many complicated circuits.

The technician should be able to examine a circuit and decide upon the d.c. conditions without resorting to advanced mathematics.

We will not attempt to decide upon optimum design values. We will keep this revision as simple as we can.

As an exercise, the following circuit contains items of interest such as feedback, input resistance, output resistance, attenuator compensation, neon valves, triode valves, thus providing a useful discussion on many of the topics covered so far.

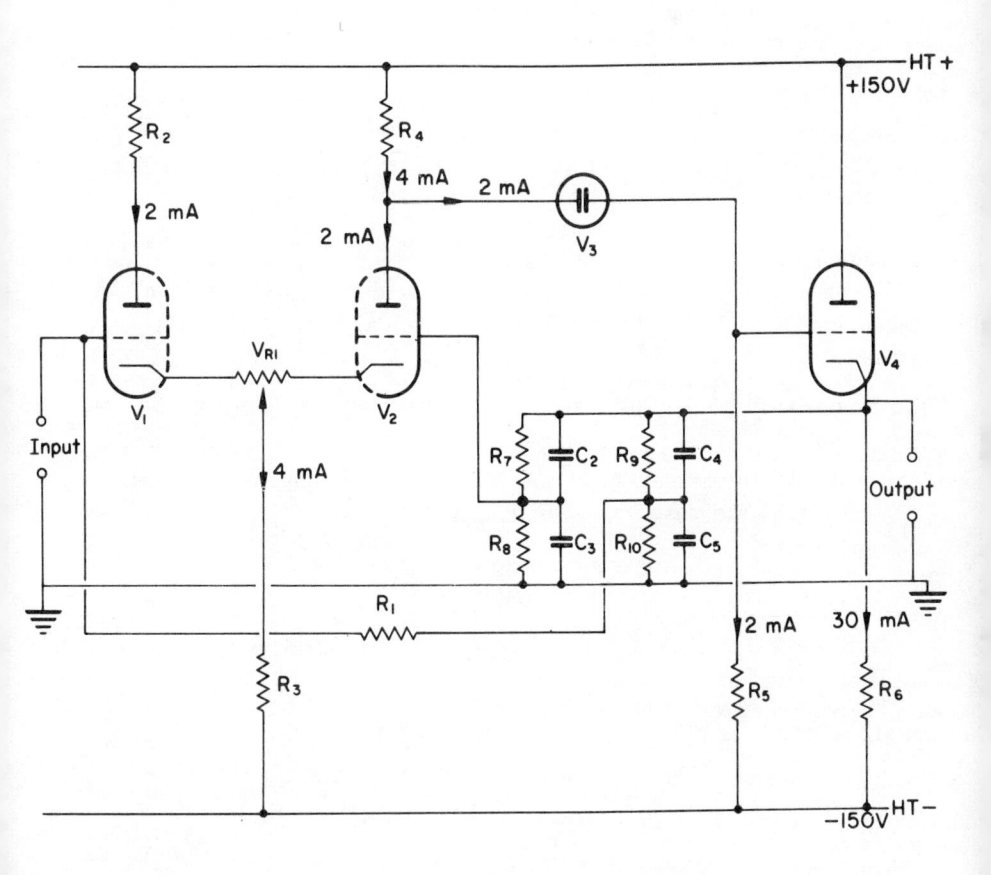

Fig. 18.14.1. Impedance Convertor.

The circuit diagram shown in figure 18.14.1. is that of an impedance convertor, and is intended to have the following characteristics; gain of unity, no overall phase shift, a low output resistance, and a high input resistance. With no input, the output must be at zero d.c. volts.

We will decide upon the d.c. circuit conditions, assume that the circuit will behave exactly as predicted and then we will calculate the component values.

If we require 0 volts output, with no input, we must have a negative h.t. rail, as due to the current in V_4 there must be a positive voltage drop across R_6. If we 'lower' the bottom end of R_6 to a negative rail, instead of earth, and provided that this negative rail potential exactly equals the voltage across R_6, then the output will be zero.

Suppose we choose the valves as follows:-

V_1, Mullard type EF86 (Triode connected)
V_2, Mullard type EF86 (Triode connected)
V_3, Mullard type 85A2
V_4, Mullard type EL84 (Triode connected).

A glance at the valve characteristics might show that a reasonable quiescent anode current for each valve, allowing a decent 'swing' with a signal in, $V. = 2$ mA, $V_2 = 2$ mA, $V_4 = 30$ mA, while V_3 needs 2 mA to provide approximately 85V across itself.

The characteristics for these valves are given at the end of this chapter and the reader should refer to these as we take the discussion further and ensure that he understands just how our numerical values have been obtained.

d.c. conditions

V_4. EL84.

The anode current is to be 30mA. We need zero voltage at the output terminal, hence we must drop 150V across R_6.

$$R_6 = \frac{150V}{30mA} = 5K\Omega.$$

V_4 has a V_{ak} of 150V and at an anode current of 30mA, a bias of about −4V as seen in figure 18.14.3.

The grid of V_4 will therefore be at a potential of −4V also.

We have chosen to pass 2mA through V_3, hence 2mA will flow through R_5. There will be $150 - 4 = 146$V across R_5 therefore $R_5 = 146V/2mA = 73$, say 75KΩ.

V_2. EF86.

The anode current in V_2 is to be 2mA. There will be 2mA flowing through V_3. Hence the total current flowing through $R_4 = 4$mA.

NEGATIVE FEEDBACK AND ITS APPLICATIONS 349

The neon valve has 85V across itself, hence the anode potential of $V_2 = 85 - 4 = 81V$. The voltage across $R_4 = 150 - 81 = 6.9V$. R_4 therefore 69V/4mA = 17.25 say, 18KΩ.

V_1. FF86.

Allowing for symmetry of the long tailed pair, we will assume that V_1 anode potential is at the same potential as V_3, i.e. 81 volts. We will assume that the anode current of V_1 is also 2mA. Hence $R_2 = 69V/2mA = 34.5$ say 33KΩ.

V_1 grid is to be at zero volts. Assuming zero bias, the common cathodes must also be at zero volts. (The graph in 18.14.2 shows this bias −1V).

The potential across $R_3 + \frac{1}{2} V_{R_1} = 150V$.

The current flowing through this combination is 4mA.

Hence $R_3 + \frac{1}{2} V_{R_1} = 150V/4mA = 37.5$ say 38KΩ.

Let $V_{R_1} = 5K$, therefore $R_3 = 38 - 2.5 = 35.5$ say 36KΩ

V_{R_1} is a 'set zero' control. It allows for tolerances in the circuit in that it can restore the output to zero potential, with the input earthed, by varying the bias on V_1 and V_2 until the required condition is obtained.

Let $R_1 = 1MΩ$. R_7 and R_8 form a potential divider which, if we consider d.c. only, must provide a fraction of the output back to the grid of V_2 in the form of negative feedback thus reducing the overall gain to unity.

The gain from V_1 grid to V_2 anode, with V_2 grid earthed, is given as

$$\frac{1}{2} \frac{\mu R_L}{r_a + R_L}$$

Assume that this gain is 10.

This can be accurately determined graphically or worked out by small signal formulae. This gain of 10 is the 'open loop' gain, i.e., without feedback, call this A.

The gain with feedback $A' = \dfrac{A}{1 - \beta A}$

As negative feedback is required, β is negative.

The expression therefore becomes $A' = \dfrac{A}{1 + \beta A}$

Only β is unknown and has to be found.

$$A'(1 + \beta A) = A \quad \text{and} \quad A' + A'\beta A = A.$$

$$A'\beta A = A - A' \quad \text{hence} \quad \beta = \frac{A - A'}{A' A} = \frac{10 - 1}{10} = \frac{9}{10}$$

By load over total
$$\frac{R_8}{R_7 + R_8} = \frac{9}{10}$$

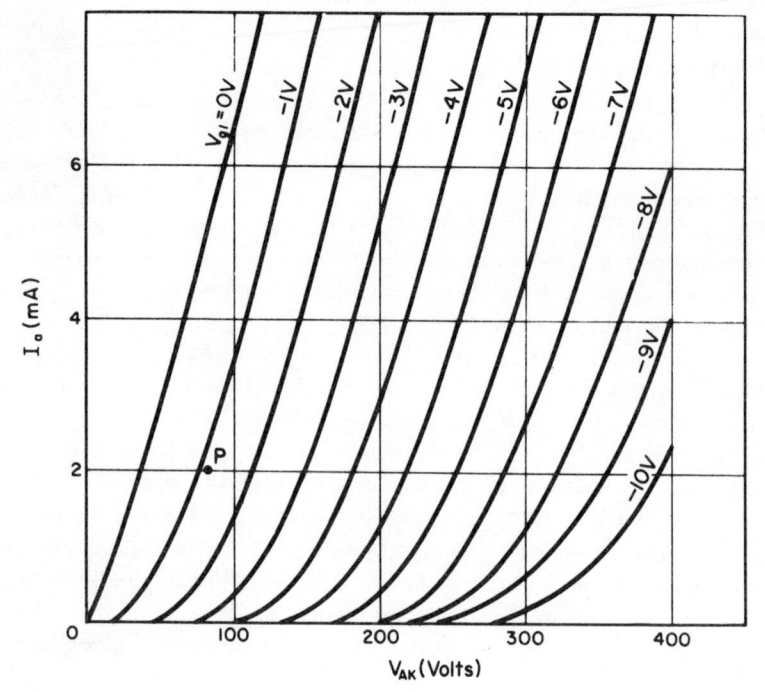

Fig. 18.14.2.

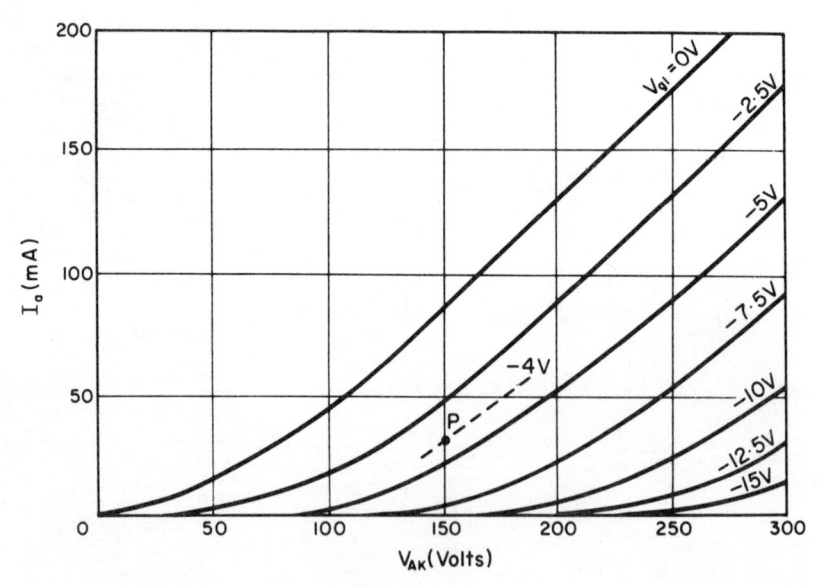

Fig. 18.14.3.

If we let $R_7 = 20K$, then $R_8 = 180$ K.
A check on the gain, which must be unity, might be carried out before we proceed further.

$$A' = \frac{A}{1 + \beta A} = \frac{10}{1 + 10 \times \frac{9}{10}} = 1. \quad \text{Q.E.D.}$$

If we assume that 5pF exist across R_8, represented by C_3, then for optimum pulse response when pulses are applied to the circuit input, $R_7 C_2 = R_8 C_3$.

$$C_2 = \frac{R_8}{R_7} \times C_3. = \frac{180}{20} \times 5pF = 45pF$$

This could be a 33pF fixed capacitor with a 3 – 30pF trimmer capacitor connected in shunt.

Suppose we decide upon an input resistance of 100mΩ. This will be so if when we connect 1V to the input and ensuring that the p.d. across R_1 is 1V/100 this causing the input resistance to become effectively $100 \times 1MΩ$.

The overall gain however, is unity.

The 1V input will appear at the output and across the divider $R_9 + R_{10}$. The output from this divider, to which the lower end of R_1 is connected, must provide $99/100 \times 1$ volt or simply, an attentuated output of 99/100.

Therefore
$$\frac{R_{10}}{R_9 + R_{10}} = \frac{99}{100}$$

If we let $R_{10} = 100KΩ$ say, then $R_9 = 1.01$ KΩ.

In practice, R_9 could be partially a preset, one that is adjusted and preset to give the required input resistance.

If we assume, measure or calculate, the value of C_5 to be say, 2pF, then optimum conditions as before, $C_4 = 100/1.01 \times 2$ pF and could be an 82 pF plus a trimmer of 3 – 30 pF as before.

Application of negative feedback results in a lowering of the gain, an increase in bandwidth, and a more faithful reproduction of the input signal.

CHAPTER 19

Locus diagrams

Of the many varied tasks performed by electronic Technician Engineers, one is to take a whole series of measurements. These are often compared with theoretical design values.

When considering the comparison between actual results and theory for any two components at right angles, and where one is constant and the other variable, many calculations may be necessary.

Locus diagrams provide an excellent tool enabling us to obtain rapid answers; the diagrams discussed in this chapter include a series circuit with one variable component and a series circuit having constant value components and a variable frequency input signal.

This by no means exhausts the subject of locus or circle diagrams, but is included here so as to give the technician engineer an insight into this most useful 'tool'.

Some further examples of circuitry, using operator j, are given so as to give even greater breadth of approach towards circuit analysis. The elementary discussion of the operator j is once again included for the home-study student who might not yet have covered this subject. Readers that are familiar with this subject might ignore this section.

19.1. Introduction to Locus (or circle) diagrams for series circuits

In order to present these diagrams in a manner easily understood they have been drawn to show only the information generally needed by technicians and technician engineers.

All voltages and currents will, as usual, be in r.m.s. *values.*

Figure 19.1.1. shows a circuit consisting of a variable resistor in series with a capacitor; across this circuit is a supply voltage

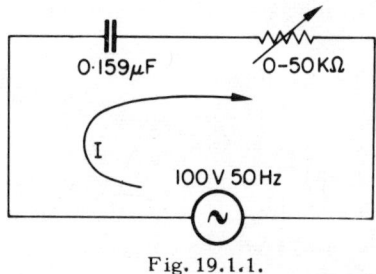

Fig. 19.1.1.

Assume V to be 100V at a frequency of 50Hz.
Assume C to be a 0.159 ufd capacitor.
Assume R to be a variable resistor (or rehostat) $(0-15)\,K\Omega$.

Before constructing the diagram, let us solve a problem or two using conventional, a.c. theory so as to provide a datum for subsequent comparison.

Suppose we need to know the current flowing when $R = 0$. The first step is to calculate X_c, the reactance of the capacitor for a 50Hz supply.

$$X_c = \frac{1}{2\pi fc} = \frac{1}{2.\pi.50.\ 0.159.10^{-6}}$$

as $\frac{1}{2\pi} = 0.159$, substitute and $X_c = \frac{0.159}{50.\ 0.159.\ 10^{-6}}$

$$= \frac{0.159.\ 10^6}{50\ \ \ 0.159} = \frac{10^6}{50} = \frac{10^2.\ 10^4}{50} = 2 \times 10^4 = 20\,K\Omega$$

This capacitor has a reactance of $20\,K\Omega$ at 50 Hz only.

It may be noted that 2π is a constant. Once C has been chosen it will be a constant, and therefore if $X = 20K\Omega$, $R = 0$ hence $I = V/|Z|$ where $|Z|$ is the total effective resistance to a.c.

when $R = 0$, $Z = \sqrt{R^2 + X_c^2} = \sqrt{(0^2 + X_c^2)} = 20\,K\Omega$ of course.

when $R = 20\,K\Omega$, $Z = \sqrt{(R^2 + X_c^2)} = \sqrt{(20^2 + 20^2)} = \sqrt{800} = 28.3\,K\Omega$.

when $R = 50\,K\Omega$, $Z = \sqrt{50^2 + 20^2} = 53.8\,K\Omega$

Suppose we need to know VR for the given values of R? Then from the load over total,

$$R = \frac{V \times R}{|Z|}$$

Hence when $R = 0$, $VR = 0$.

when $R = 20K\Omega$ $\quad VR = \frac{100 \times 20\,K\Omega}{28.3K\Omega} = 70.7V$

and when $R = 50\,K\Omega$, $VR = 93.0V$.

We could have established VR by first deriving an expression for the circuit current, I.

For instance, when $R = 50\,K\Omega$, $I = \frac{100V}{|Z|} = \frac{100V}{53.8\,K\Omega} = 1.86\,mA$

Hence $VR = IR = 1.86mA \times 50K\Omega = 93.0V$.

It will be seen that when adding R to X, we are dealing with vectorial addition, not algebraic; and for different values of R, the impedance, Z has to be calculated each time before the current can be calculated. (This will

LOCUS DIAGRAMS

not be new to the reader, but is included for the home study student why may not have dealt with simple series networks).

What of V_c? Once I is known, all that needs to be done in order to find V_c is to multiply X_c by I.

At $R = 0$: $\quad V_c = I \times X_c = 5\text{mA} \times 20\text{K}\Omega = 100\text{V}$ (of course).
At $R = 20\text{K}\Omega$ $V_c = I \times X_c = 3.535\text{ mA} \times 20\text{K}\Omega = 70.7\text{V}$
At $R = 50\text{K}\Omega$ $V_c = I \times X_c = 1.86\text{ mA} \times 20\text{K}\Omega = 37.2\text{V}$

These may be represented by vectors as shown in figure 19.1.2.

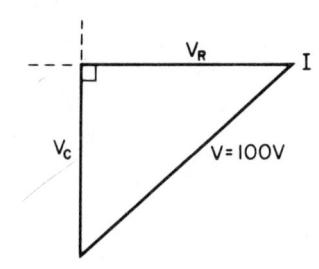

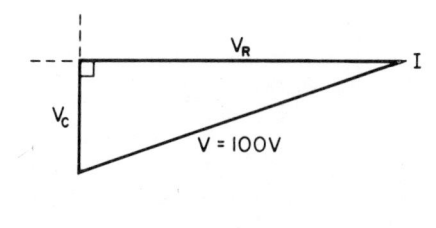

Fig. 19.1.2.

As a check: $\quad V = \sqrt{V_R^2 + V_c^2}$

$(R = 20\text{K}\Omega)\ 100 = \sqrt{70.7^2 + 70.7^2} = \sqrt{10,000} = 100\text{V}$
$(R = 50\text{K}\Omega)\ 100 = \sqrt{93^2 + 37.2^2} = \sqrt{8650 + 1350} = 100\text{V}$

The power dissipated in the circuit is given as

$$P = \frac{VR^2}{R} \quad \therefore \quad P, (R = 20\text{K}\Omega) = \frac{VR^2}{R} = \frac{70.7^2}{20\text{ K}\Omega} = \frac{500\text{V}}{20\text{K}\Omega}\text{mW} = 250\text{ mW}$$

and $\quad P\ (R = 50\text{K}\Omega) = \dfrac{VR^2}{R} = \dfrac{93^2}{50\text{K}\Omega} = \dfrac{8650}{50}\text{mW} = \underline{173\text{ mW}}$

Lastly, power factor (pf). This may be given as the cosine of the angle between current and applied voltage, or simply $R/|Z|$.

$$\text{Pf at } R = 20\text{K}\Omega = \frac{R}{|Z|} = \frac{20\text{K}\Omega}{28.3\text{K}\Omega} = \underline{0.707}\ \text{(leading)}$$

$$\text{and Pf at } R = 50\text{K}\Omega = \frac{R}{|Z|} = \frac{50\text{K}\Omega}{53.8\text{K}\Omega} = \underline{0.93}\ \text{(leading)}$$

We have now briefly revised Z, I, Power, pf, VR, VC,.
Let us now repeat the foregoing example, using a circle diagram.

19.2. Plotting the diagram

The first step is to draw two lines at right angles as shown in figure 19.2.1.

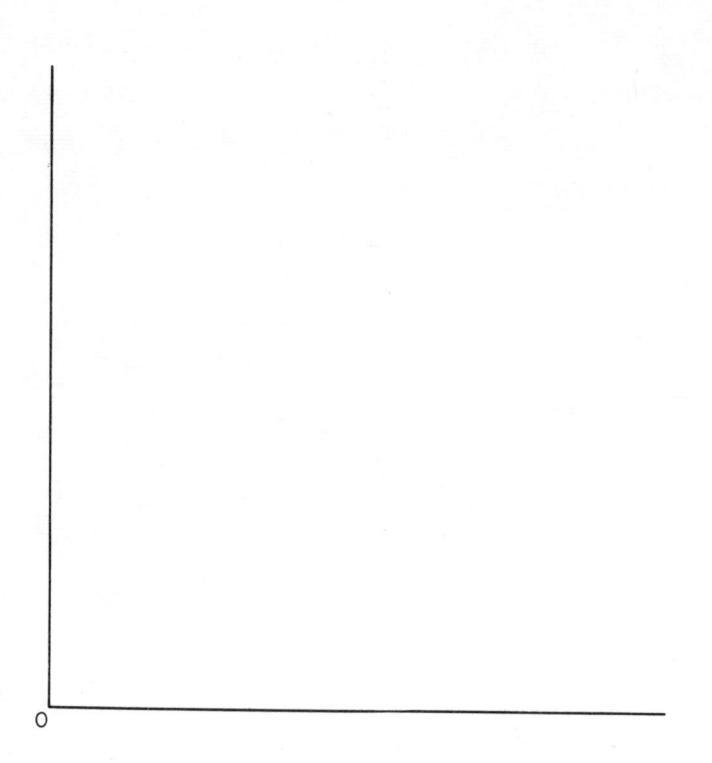

Fig. 19.2.1.

Mark the point (0,0) as 'O' as shown in figure 19.2.1. This point is a common reference from which most of our subsequent measurements will be made.

19.3. Resistance

The second step is to decide upon a scale for ohms. Suppose that, as we have a maximum resistance of 50,000Ω, we might decide that on this size of paper, we will have 10KΩ to the inch. It follows therefore that a line drawn for the 50KΩ resistor, will be 5 inches in length. The line representing the capacitive reactance of 20KΩ will accordingly be 2 inches in length.

The actual scale is unimportant, it can be in inches, centimetres, feet (if the paper is large enough) or any other acceptable unit of length. Once the scale has been chosen however, it must be strictly kept to when drawing any further lines, and subsequent measurement, when dealing with resistance, reactance and impedance.

Plot a point 2 inches up from 'O' on the y axis and mark this $X_c = 20\text{K}\Omega$. The resistance is accounted for by the line drawn parallel to the x axis at a point level with X_c.

This is shown in figure 19.3.1.

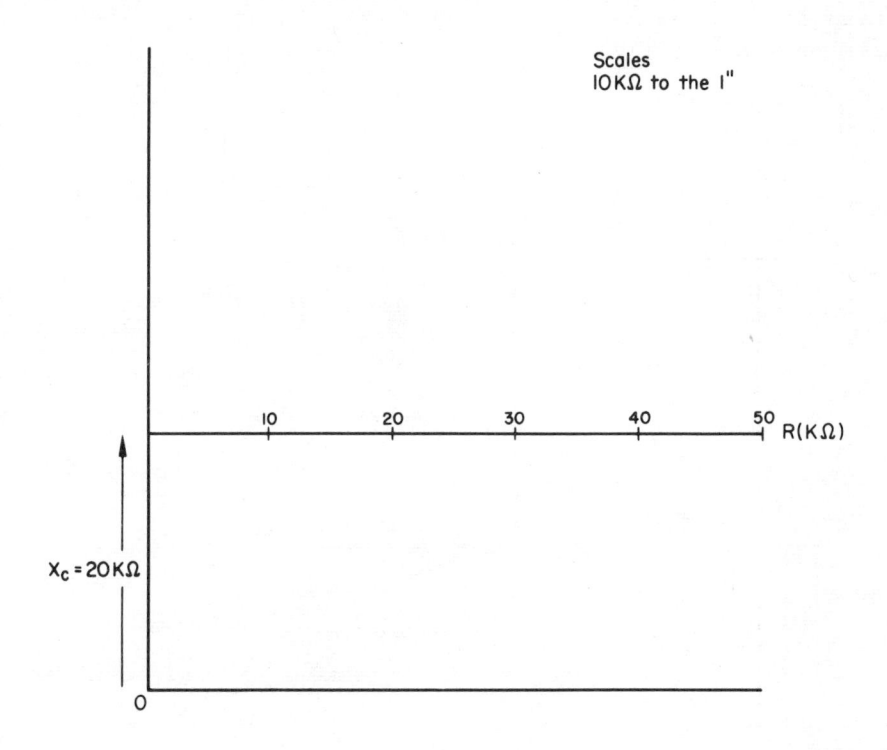

Fig. 19.3.1.

It is important to note the chosen scale units in the table on the drawing as shown. We will have a number of different scales in this table by the time we have completed our diagram. If we complete the table as we go along, there is less chance that different scales will be chosen for the same identities, i.e. we need one scale for ohms, one for volts, etc.

Each scale is determined individually for ohms, volts, current, etc. The separate scales may all be in inches or centimetres, or they can be different. It does not matter if volts are shown in inches and current in centimetres, it is purely a matter of convenience or preference. The size of the paper upon which the diagram is to be drawn, is often the governing factor.

19.4. Voltages.

The next step is to decide upon a scale for volts. Suppose we choose to let the applied 100V to be represented by a length, of say, 10 cm.

We have therefore decided for this example that we will have a scale of 10V to the cm. (Remember, we could have said 10V to the inch, etc, the choice is ours entirely.)

Draw a semicircle having a diameter of 10 cm in length as shown in figure 19.4.1. and identify it as the "volts locus". Record in the table, that 1 cm corresponds to 10V.

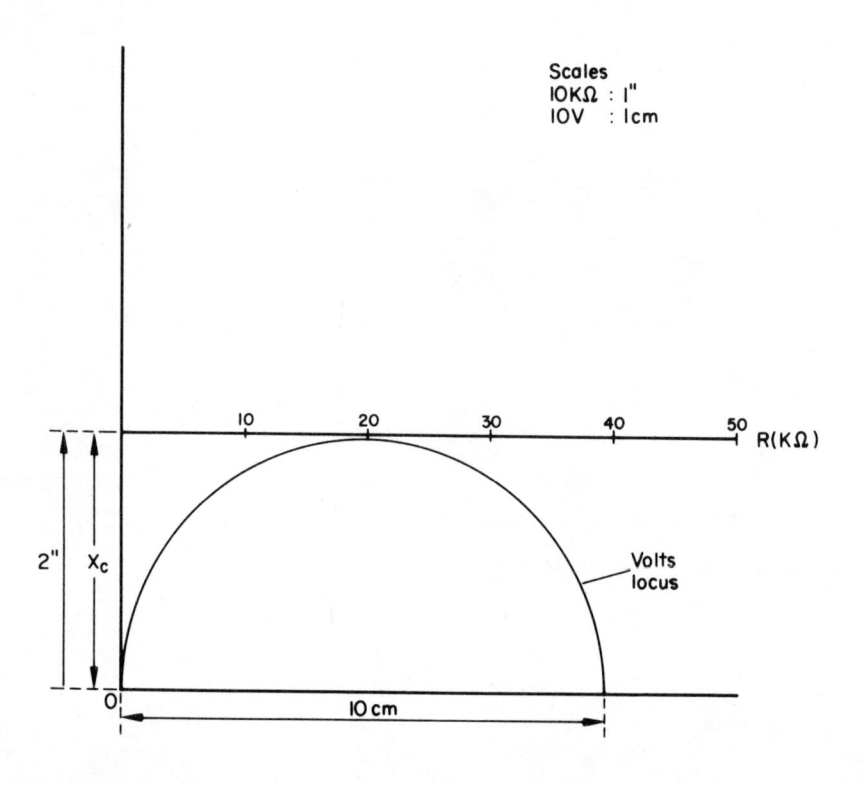

Fig. 19.4.1.

19.5. Current.

We need now to plot a semicircle for current. Again we need to decide upon a scale for current. We can choose any scale we like provided we can accomodate our next semicircle on our existing drawing.

We need to determine the maximum possible current that would flow with $R = 0$. The current will be at its maximum and will determine both our scale units of length and the size of our semicircle.

The maximum possible current when $R = 0$, is given by $I_{max} = V/|Z|$ and as

$$R = 0, \quad I_{max} = \frac{V}{X_C} = \frac{100V}{20K\Omega} = 5 \text{ mA}.$$

Suppose for this example, we let 5 mA be shown by a length of 5 inches. Hence 1 mA to the inch will result. This should be recorded in our table for future reference and a semicircle drawn as shown in figure 19.5.1. having a diameter of 5 inches.

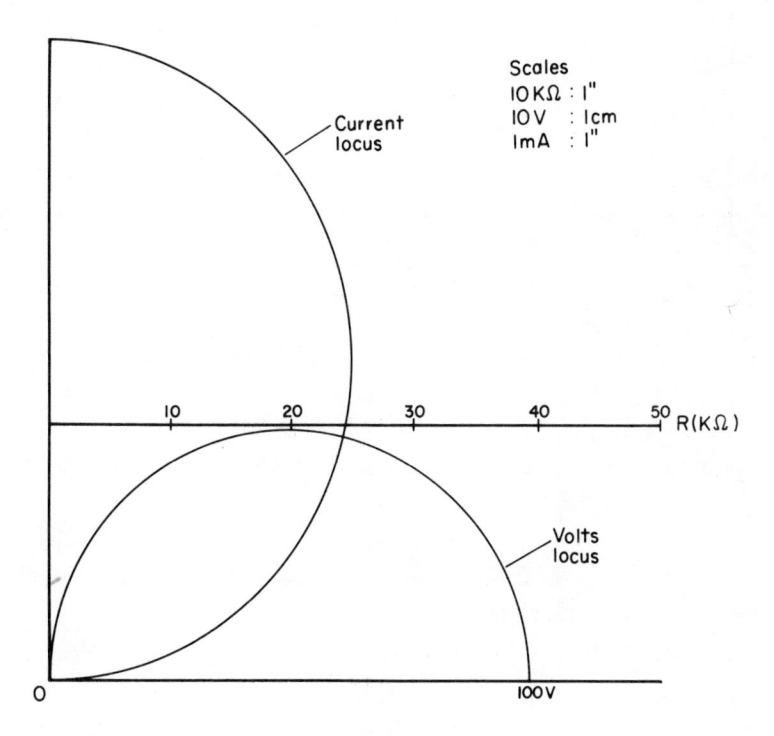

Fig. 19.5.1.

The locus diagram is not yet complete, but it might be desirable to consolidate our position by taking a few measurements and comparing them with mathematically derived values.

Let us 'measure' with our rule, the impedance of the circuit, the circuit current, and voltage distributions when $R = 0$, $20K\Omega$, and $50K\Omega$.

19.6. Measurements.

The lines O – A, O – B, O – C, or figure 19.6.1. represent our measurements when $R = 0$, 20 and 50KΩ.

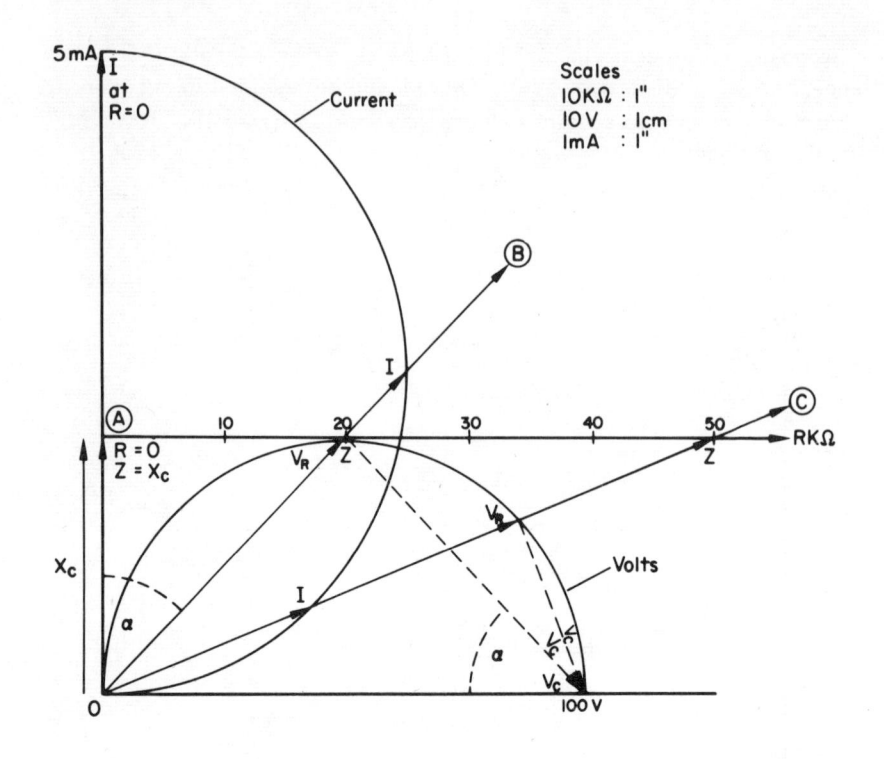

Fig. 19.6.1

Mathematically derived values

When $R = 0$, $Z = \sqrt{R^2 + X_C^2} = \sqrt{0 + 20^2} = 20\text{K}\Omega$.

$$I = \frac{V}{|Z|} = \frac{100\text{V}}{20\text{K}\Omega} = 5\,\text{mA}.$$

$$V_R = IR = \frac{5}{1000} \times 0 = 0\text{V}.$$

From

$$V = \sqrt{V_R^2 + V_C^2} \quad \therefore \quad V_C = V$$

$$V_R = \sqrt{V^2 - V_C^2} = 0.$$

The result is hardly surprising as $R = 0$.

LOCUS DIAGRAMS

If R is zero, then obviously all of the applied 100V will exist across C. Hence $V_C = 100$V.

When $R = 20\text{K}\Omega$. $|Z| = 28.3\text{K}\Omega$. $I = 3.535$ mA.
$V_R = 70.7$V $V_C = 70.7$V.

When $R = 50\text{K}\Omega$. $|Z| = 53.8\text{K}\Omega$. $I = 1.86$ mA.
$V_R = 93$V. $V_C = 37.2$V.

Measured values

Determining impedance

$R = 0$ (O–A)

Draw a line from O through $R = 0$, as shown in figure 19.6.1. This line is vertical. The line should 'cut' both semicircles as shown. The length of the line from 'O' to the point where it 'cuts' the 'R' line gives a measurement of 2 inches. Referring to the table for Ohms, we see that 2 inches corresponds to $20\text{K}\Omega$. (This is simply the capacitive reactance of course.)

$R = 20\text{K}\Omega$ (O–B)

Draw a line from 'O' through $R = 20\text{K}\Omega$ as shown. The impedance, $|Z|$ is seen to be represented by 2.83 inches. The impedance is therefore $28.3\text{K}\Omega$.

$R = 50\text{K}\Omega$ (O–C)

The line passing through $R = 50\text{K}\Omega$, from 'O' is seen to be 5.38 inches. Hence the impedance is $53.8\text{K}\Omega$.

Voltage

$R = 0$ (O–A)

The first line passing through $R = 0$, cuts the voltage locus at a zero length from 'O'. Hence $V_R = 0$.

$R = 20\text{K}\Omega$ (O–B)

The second line passing through $R = 20\text{K}\Omega$ m cuts the voltage locus at a length of 7.07 cm from 'O'. The table shows that 1 cm corresponds to 10V, hence $V_R = 70.7$V.

$R = 50\text{K}\Omega$ (O–C)

The third line is 9.3 cm in length from 'O' to the voltage locus. Thus $V_R = 93$V.

V_C in every case, is determined by measuring from the point at which the line cuts the voltage locus (marked (D)) to the point marked (E) in figure 19.6.2.

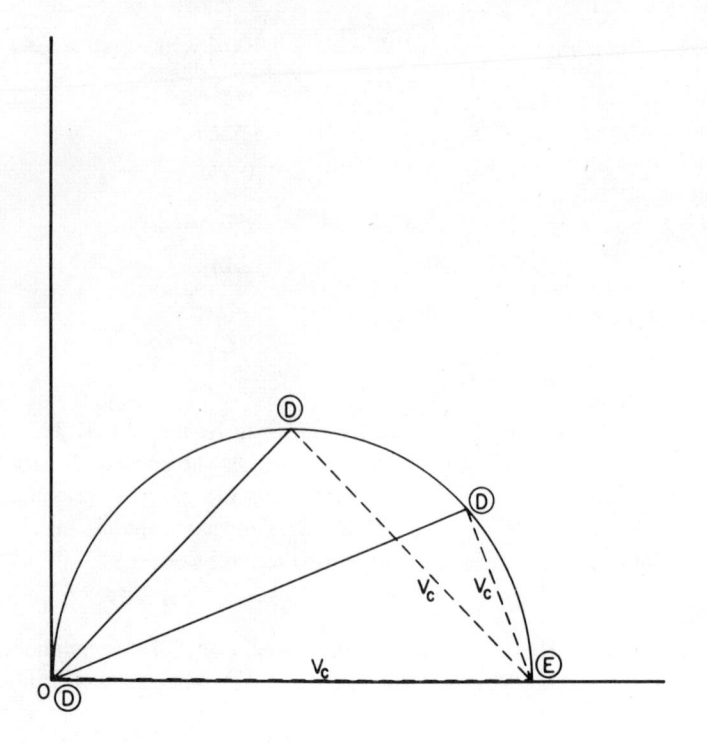

Fig. 19.6.2.

It would be useful to isolate from figure 19.6.2 a voltage 'triangle' as shown in figure 19.6.3.

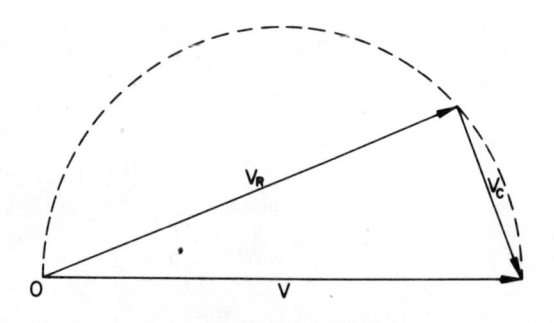

Fig. 19.6.3.

This is seen to be the normal triangle with the exception that it is in the incorrect quadrant. (We will discuss this further in a while, but ignore it for now so as to avoid undue complications).

9.7 Power Factor

Power factor may be expressed as $\text{pf} = R/|Z|$ and as $|Z|$ varies with R, the power factor varies also.

We need not calculate anything in order to obtain the power factor for different values of R. We can draw a power factor quadrant on our diagram as shown in figure 19.7.1.

This is drawn by simply choosing a 'one unit radius' for the quadrant, i.e. 1 inch, 10 inches, etc., in fact any value that is easily divisible by 10. We have chosen 1 inch.

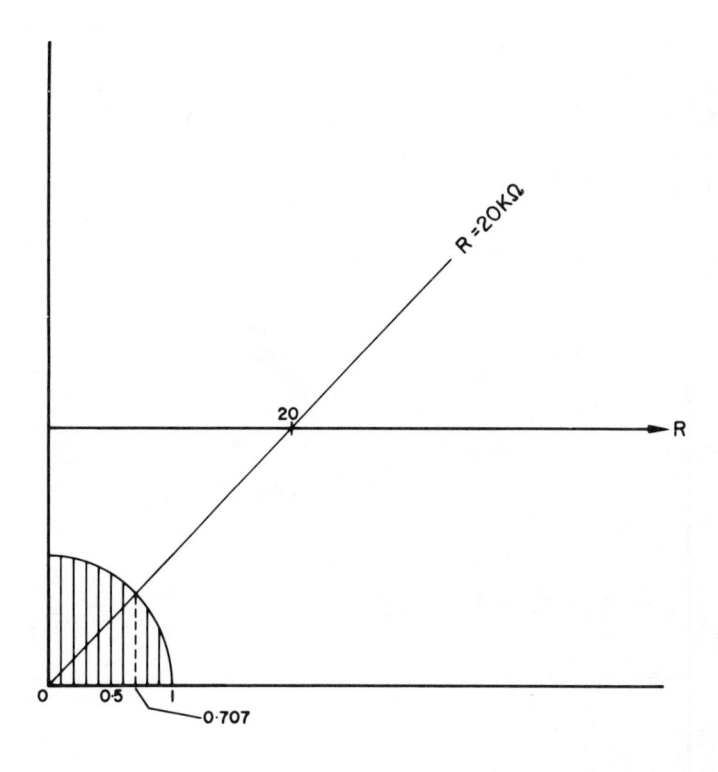

Fig. 19.7.1.

We must divide the unit (1 inch in our case) along the x axis, into 10 equal spaces, then mark as shown from 0 to 1.0. Next we must draw vertical lines above each point between 0 and 1.0 and terminate them when they touch the quadrant.

Measurement of power factor.

For say, $R = 20K\Omega$, the power factor of the circuit $R/|Z| = 20K\Omega/28.3K\Omega$

$$\frac{R}{|Z|} = \frac{20K\Omega}{28.3K\Omega} = 0.707.$$

Draw a line from 'O' to $R = 20K\Omega$, and note the point on the quadrant through which the line passes; cast your eye straight down and this will be seen to give a pf of 0.707 as shown heavy in the figure 19.7.1.

19.8. Power

The power in the circuit may be expressed either as $p = \dfrac{V_R^2}{R}$ or $P = V.1.\cos\theta$.

Where θ is the angle between the line O–D and the x axis as shown in figure 19.8.1.

We cannot decide upon a scale for Power; we must take an existing length and fit a scale to it.

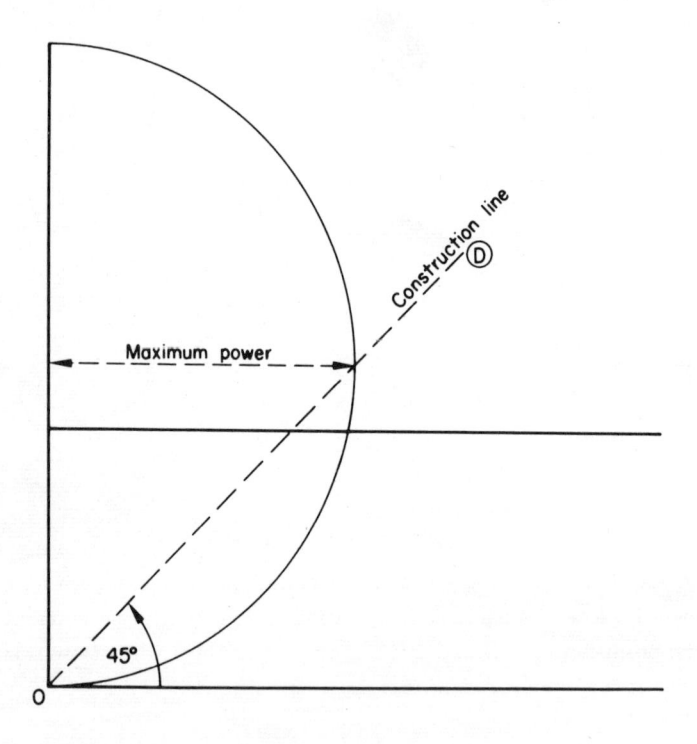

Fig. 19.8.1.

LOCUS DIAGRAMS

We discussed earlier, the maximum power theorem. We saw that for this simple circuit, the maximum power occurs when $X_C = R$. The angle of the line cutting R (when $R = X$) will be 45°. Figure 19.8.1. shows this *construction line* drawn from 'O' to 'D', drawn at 45° from the x axis. This construction line will not be required when the diagram is complete, hence it should be drawn very feintly.

Where its line cuts the current locus, a line should be drawn horizontally to the y axis. This length is the maximum power line.

If we now calculate the actual maximum power for this particular circuit, we can 'fit' the scale length to our answer and calibrate this length.

$$P_{max} = \frac{V_R^2}{R} \quad \text{when } R = X_C. \text{ hence } P = \frac{70.7^2}{20K\Omega} = \frac{5000 \, mW}{20} = 250 \, mW.$$

The line length representing $P_{max} = 2.5''$.

Hence the power scale is 100 mW to the inch.

In each case where we have (or will have) drawn lines through R, we can measure the power in the circuit by noting the length of the lines as shown in figure 19.8.2.

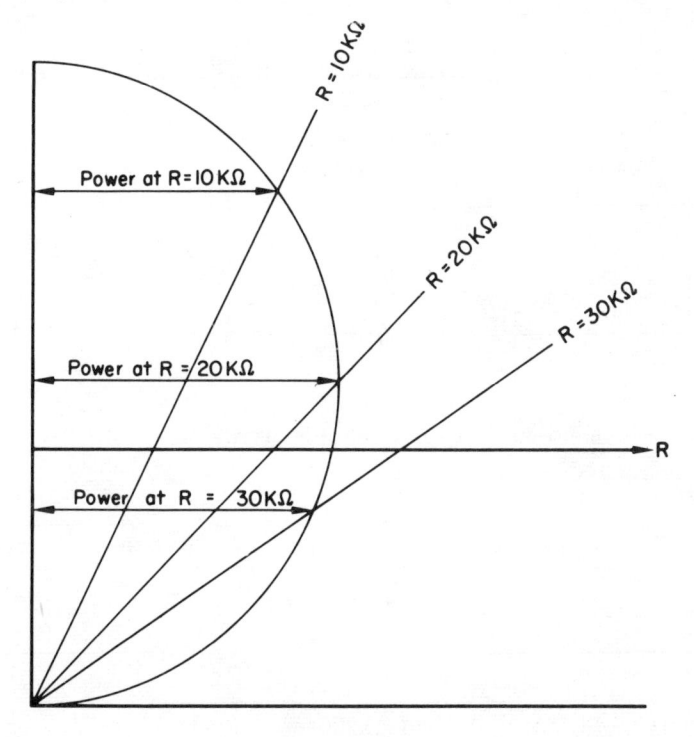

Fig. 19.8.2.

A sample set of measurements for $R = 40\text{K}\Omega$ have been taken in figure 19.8.3.

The current loccs may be calibrated in admittance, Y, as $Y = I/V$ and as V is a constant, $Y \propto i$.

This is our completed locus diagram.

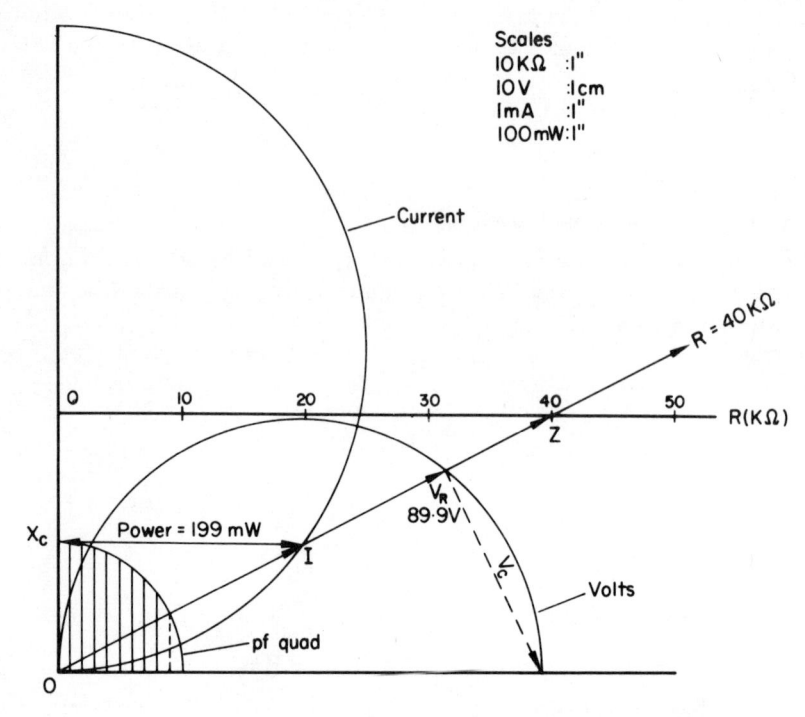

Scales
10KΩ :1"
10V :1cm
1mA :1"
100mW:1"

Fig. 19.8.3.

Note:- *All* measurements (except power and pf) are taken from 'O'

R	$|Z|$	I	P	pf	V_R	V_C
40KΩ	44.6KΩ	2.23mA	199mW	0.896	89.9V	45V

Calculated values

R	$|Z|$	I	P	pf	V_R	V_C
40KΩ	44.7KΩ	2.24mA	200mW	0.895	89.5V	45V

LOCUS DIAGRAMS

With a little care it is possible to easily obtain answers which are perfectly adequate, and certainly have less errors than the possible tolerance of actual component values. Before continuing, it is necessary to briefly discuss operator, j in order to establish in which quadrant we should really be working.

Series a.c. circuits.

19.9. Use of the operator j

This brief introduction is for the reader that has not yet dealt with operator j.

j is a vector operator. By placing a j with a term, the term is 'operated upon' and rotated anticlockwise through $90°$.

Example: A 10Ω resistor could be shown as $10\angle o$ but $j10 = \angle 90°$.

A resistor value is shown simply as a quantity of ohms. It has no phase angle, and is not affected by varying frequencies.

Let us consider a resistor of 1Ω. An a.c. voltage of 1V would cause 1 amp to flow, $I = V/R$. The current would be in phase with the voltage. To a.c. however, there can be other components which would effect the current flow, and these must be taken into consideration before calculations are made. One of these is the inductor, or coil. Suppose the reactance of a coil (X_L) to be 1Ω at a given frequency; then with 1V applied, 1 amp will flow. $I = V/X_L$. The current will lag the voltage by $90°$, as outlined earlier in this book.

The other component is the capacitor. Assume this to have a reactance (X_C) of 1Ω at a given frequency. Then with 1V applied, 1A will flow, $I = V/X_C$.

The current will lead the voltage by $90°$. Therefore the voltage V_L will lead the voltage V_R by $90°$, and V_C will lag V_R by $90°$. V_C lags V_L by $180°$ (see diagram in figure 19.9.1).

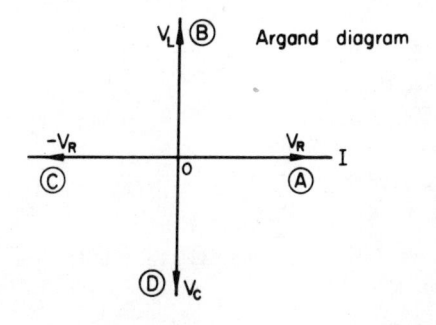

Fig. 19.9.1

Point A represents V_R (in phase with circuit current which is common to all components).
Point B represents V_L (90° ahead of the circuit current).
Point C represents V_C (90° lagging the circuit current).
Point D again represents resistance, but is seen to be negative.

If point A were to represent a 1Ω resistor, the point B would represent a coil of reactance 1Ω, whilst point C would represent a capacitor of reactance 1Ω. Point D would represent a resistance of -1Ω (negative resistance). The diagram is reproduced once more, but this time with different identities.

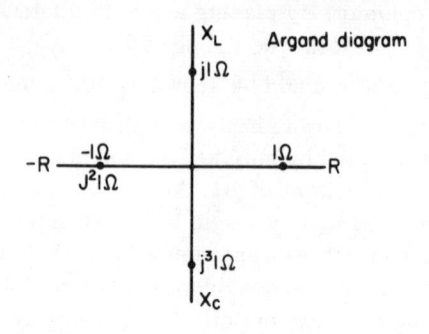

```
                          X_L       Argand diagram
                        • j1Ω

             -1Ω                         1Ω
    -R ─────────────────┼───────────────•─── R
            j²1Ω

                        • j³1Ω
                          X_C
```

Fig. 19.9.2

R (1Ω) is seen to be simply 1Ω.
X_L (1Ω) is seen to be $j\,1\Omega$.
$-R$ (-1Ω) is seen to be $j^2 1\Omega$.
X_C (1Ω) is seen to be $j^3 1\Omega$.

To 'convert' a resistance of 1Ω to a coil of 1Ω reactance, we simply 'operate' upon the resistance value by j, or in this case $j\,1\Omega$; this has the effect of rotating the point on the vector anticlockwise, by 90°. Operating upon $j\,1\Omega$ by a further j gives $j^2 1\Omega$, and the second j rotates the term by a further 90°, giving $j^3 1\Omega$. This represents a capacitor whose reactance is 1Ω.

Let us sum up.
For R = 1Ω write 1Ω.
For X_L = 1Ω write $j\,1\Omega$.
For R = -1Ω write $j^2 1\Omega$.
For X_C = 1Ω write $j^3 1\Omega$.
and for R = 1Ω write $j^4 1\Omega$.

but $j^4 1\Omega$ is exactly the same as 1Ω, they both represent $R = 1\Omega$. also
$j^2 1\Omega = -1\Omega$ (for $R = -1\Omega$)

$\therefore$ j^2 must represent -1 and j represents $\sqrt{-1}$.

LOCUS DIAGRAMS

The two terms (j and non j) in each of the expressions obtained in this way are in fact, the coordinates for a point on our diagram indicating the impedance Z and the phase angle.

Consider the $16 + j\,11$ on the diagram in figure 19.9.3.

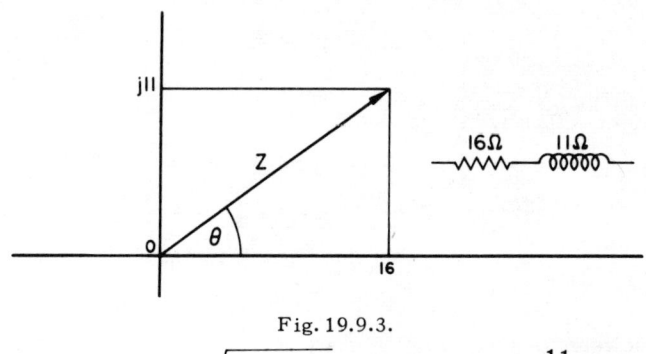

Fig. 19.9.3.

$$Z = \sqrt{16^2 + 11^2} \text{ and } \theta = \tan^{-1} \frac{11}{16}$$

where tan $11/16$ means 'the tangent whose angle corresponds to' $11/16$. It is a simple matter to evaluate $11/16$; look in the table of natural tangents and find the angle corresponding to the value $11/16$.

The impedance of a series circuit is generally written as $a \pm jb$, where a is the value of ohms for a resistor and b is the value of ohms corresponding to the reactance of a capacitor or inductance (given by the sign). Consider the following circuit in figure 19.9.4.

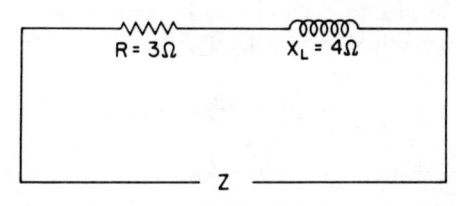

Fig. 19.9.4.

We need the impedance Z.

$$Z = 3 + j\,4$$

this may also be expressed as $|Z| = \sqrt{3^2 + 4^2} = 5\Omega$.

What of $j^3\,1$? This is really $(j^2 1)\,(j)$; and as $j^2 = -1$, j^3 may be written as $(-1)\,(j) = -j1$.

In the series $C-L-R$ circuit, R may be represented as $(x\,\Omega)$, with an inductance of $j\,(x\,\Omega)$ and a capacitor of $-j\,(x\,\Omega)$. The total impedance may be represented as a single value of resistance to a.c. at one particular frequency.

Example 1.

R = 3Ω   XL = 4Ω

Fig. 19.9.5.

Using operator j, Z may be expressed as $(3 + j4)\,\Omega$.

Example 2.

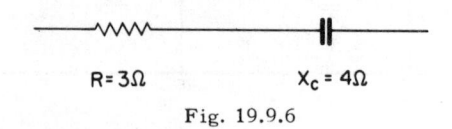

R = 3Ω X_c = 4Ω

Fig. 19.9.6

This may be represented, by using operator j, as $Z = (3 - j4)\,\Omega$.

Example 3.

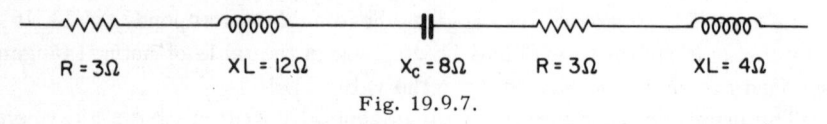

R = 3Ω   XL = 12Ω X_c = 8Ω   R = 3Ω   XL = 4Ω

Fig. 19.9.7.

or using operator j method, $Z = (3 + j12 - j8 + 3 + j4) = (6 + j8)\,\Omega$. Using operator j, it is easy to see whether L or C predominates; all that needs be done is to look at sign preceeding the j quantity. Remember in a series circuit j is inductive and $-j$ is capacitive.

One final point of interest: $-j$ may be written as $1/j$.

The proof is simple. Let us take $1/j$. Now multiply top and bottom by j.

$$\frac{1}{j} \times \frac{j}{j} = \frac{j}{j^2} = \frac{j}{-1} = -j$$

When arriving at an expression containing resistance (real) and reactive (imaginary) components, simply add all the real terms seperately and add the real terms seperately and add the imaginary terms in order to simplify the expression. An example may be helpful, all values being given in ohms.

Simplify
$$3 + j4 + 7 - j8 + j20$$
$$= 3 + 7 + j(4 - 8 + 20)$$
$$= 10 + j(16) = 10 + j16$$

representing

R = 10Ω   XL = 16Ω

Fig. 19.9.8.

LOCUS DIAGRAMS

Quadrants

There are, of course four quadrants. The expressions $\pm A \quad \pm jB$ give co-ordinates enabling us to position a point in one quadrant.

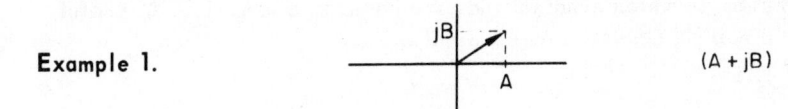

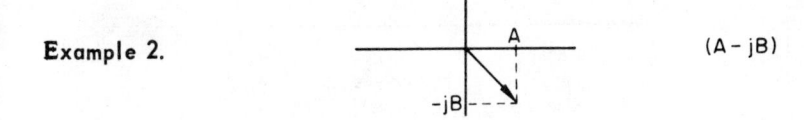

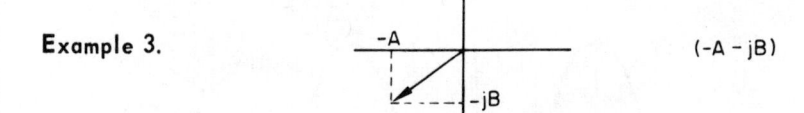

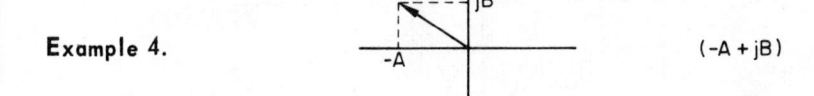

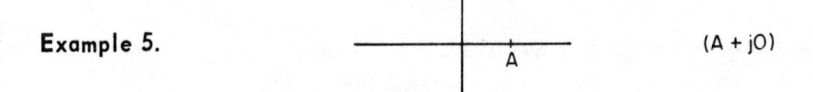

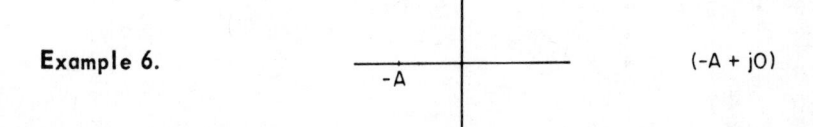

Example 1. $(A + jB)$

Example 2. $(A - jB)$

Example 3. $(-A - jB)$

Example 4. $(-A + jB)$

Example 5. $(A + j0)$

Example 6. $(-A + j0)$

Examples of
Locus Diagrams.

Fig. 19.9.9.

We have drawn our locus diagram in the 1st quadrant only but in many ways this is not unhelpful as it eliminates the need to reproduce drawings in other quadrants.

When looking up sine, cosine and tangents in the appropriate tables, it is useful to know in which quadrant they are positive or negative. A useful rule is *Old San Ta Clause*, i.e. *All Sin Tan Cos*.

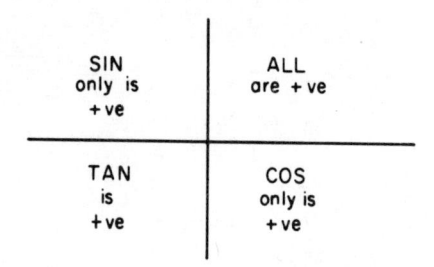

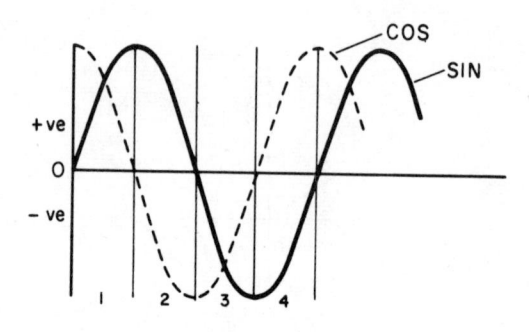

SIN only is +ve	ALL are +ve
TAN is +ve	COS only is +ve

Fig. 19.9.10

Sin is seen to be positive in quadrants 1 and 2 only.
Cosine is seen to be positive in quadrants 1 and 4 only.

Example.

Sin 270° = −1
Sin 90° = 1
Cos 0° = 1
Cos 180° = −1

If it is desired to find the quadrant in which the admittance should lie, sketch an argand diagram, and with simple use of operator j, determine the correct quadrant. The angles shown in the diagrams are correct always, but the current may be lagging, say, instead of leading.

19.10. A series L−R circuit

Consider a circuit as shown in figure 19.10.1.

LOCUS DIAGRAMS

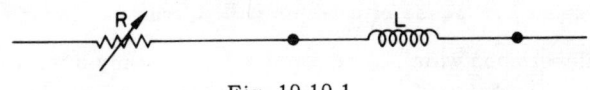

Fig. 19.10.1

An inductor has some d.c. resistance and this must be shown on the circuit separately, as in figure 19.10.2.

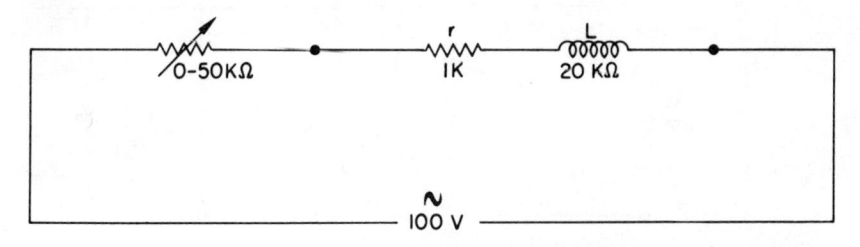

Fig. 19.10.2.

The approach to constructing a locus diagram is exactly the same as before, except that L should be written instead of C. X_c reads X_L, etc. Suppose that L has reactance of 20KΩ (as did C) and rL is $1K\Omega$. R is (0–50) KΩ as before.

The final locus diagram will be identical except for the minimum value of R. R in the circuit can never be less than 1KΩ as we cannot short circuit the d.c. winding resistance of the inductor, figure 19.10.3 illustrates a complete circle diagram, plus a dotted line showing the minimum resistance of 1KΩ.

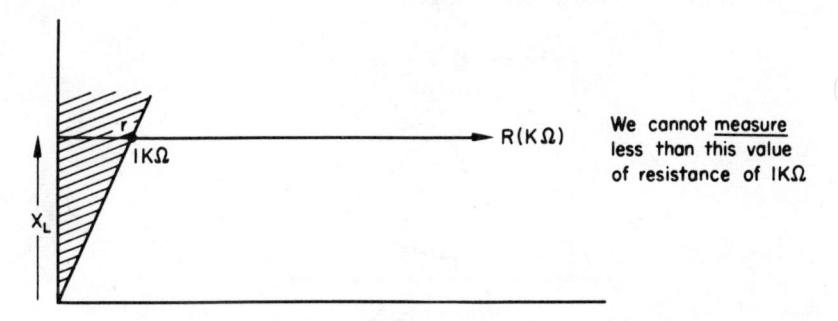

We cannot <u>measure</u> less than this value of resistance of 1KΩ

Fig. 19.10.3.

Note: When calculating maximum current in order to determine the size of semicircles, it is necessary to assume, as before, that $I_{max} = V/XL$ when $R_{Total} = 0$, which in this case will give 5 mA, where R_T is the total resistance in the circuit including r_L. In practice, however, the minimum value is of course 1KΩ; the entire variation in resistance in practice is (1 – 51) KΩ.

19.11. Frequency response of a series C.R. circuit.

This is a subject upon which much may be written, only a brief outline will be given here, together with an example of a basic circuit which is commonly met in electronics. We will investigate the frequency response of the circuit (Figure 19.11.1).

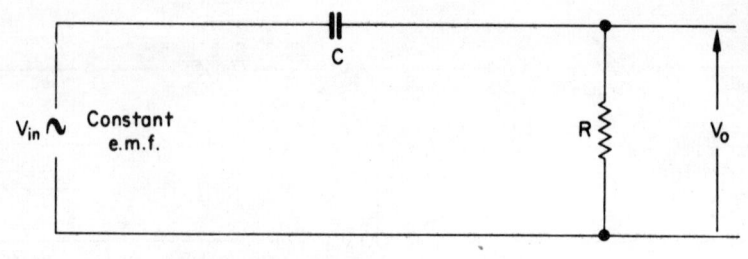

Fig. 19.11.1.

It is desired to measure the output voltage, assuming a constant input *voltage* at a varying frequency. Normally, X_c would be calculated for a particular frequency, Z would then be calculated. The admittance would then be determined. The current I is given as $I = V_{in} \times Y$. Then, lastly, $V_o = I \times R$. This is a lengthy process for one frequency, but to repeat the above calculations for many different frequencies could be extremely labourious.

A very simple locus diagram may be drawn to represent circuits of this kind. The resistor R might be the input resistance to an amplifier stage, etc. The step by step construction is given.

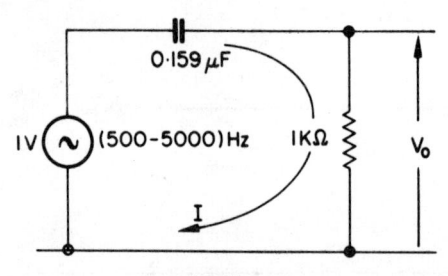

Fig. 19.11.2.

The input is kept at a constant amplitude of 1 volt whilst the frequency is varied from 500 Hz to 5 Hz.

We need to know V_o at 500, 1000, 1500 Hz.

Step 1.

Draw two lines at right angles (Fig. 19.11.3).

LOCUS DIAGRAMS

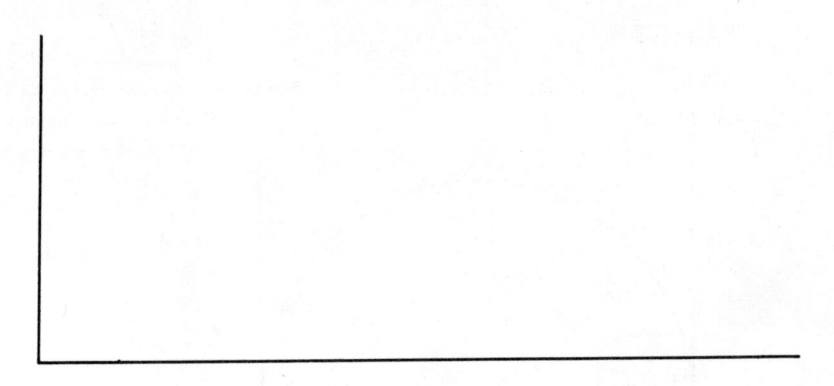

Fig. 19.11.3.

Step 2.

Choose a scale for resistance, say 1000Ω to 10 cm. (Choose any scale but do not subsequently alter it.) Draw a line representing a fixed resistance of 1 KΩ (R_L) as shown in figure 19.11.4.

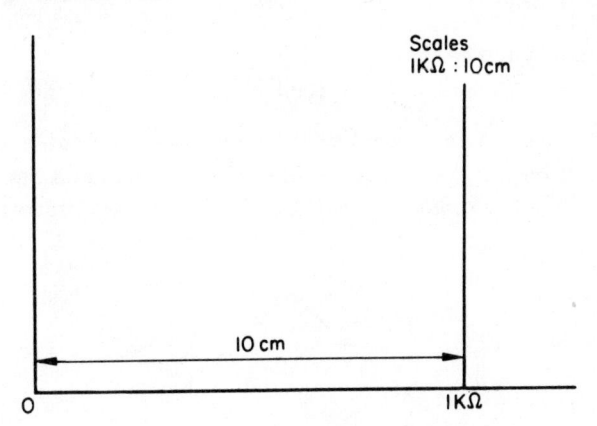

Scales
1KΩ : 10cm

10 cm

O   1KΩ

Fig. 19.11.4.

Step 3.

We now have to draw a semicircle representing the input voltage and once again we choose a scale, say 1V to 10 cm. (figure 19.11.5).

The size of the semicircle with respect to other lines is unimportant. The fact that the voltage locus coincides with the line representing $R = 1000Ω$, is coincidental. *It can be of any size* (it is more accurate, of course, if made as large as possible). One calculation only need to be made in order to establish the frequency scale.

Calculate the frequency at which $X_c = R_L$

In the present case, as $R = 1 KΩ$.

```
                    Scales
                    1KΩ :1·0 cm
                    1V  :10 cm
```

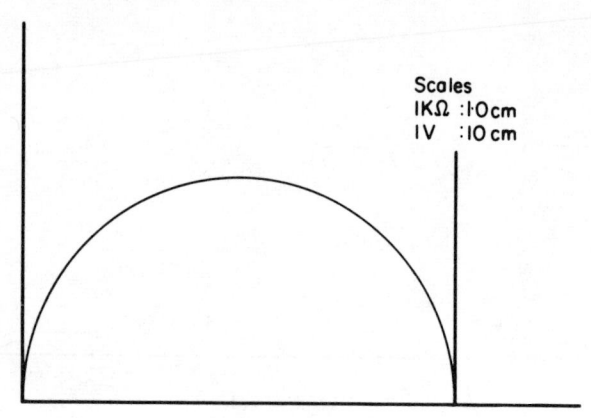

Fig. 19.11.5.

From $X_c = R_L$

$X_c = 1000\,\Omega$

$\dfrac{1}{2\pi fc} = 1000\,\Omega$

$1000 = \dfrac{0.159}{fc}$ $\therefore$ $f = \dfrac{0.159 \cdot 10^6}{1000 \times 0.159} = 1000\,\text{Hz}.$

If the reactance of the capacitor has the same ohmic value as R_L at 1000 Hz, then, as the circuit current is common to both components, the voltage across both components will be of the same amplitude as shown in figure 19.11.6.

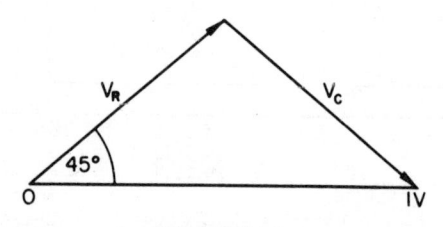

Fig. 19.11.6.

(ignoring the fact that again we have the wrong quadrant).

It is evident that at 1000 Hz there is a 45° phase angle when $V_R = V_C$. Therefore we draw a *feint* line at 45° as shown in figure 19.11.7.

A frequency scale must now be selected. It is known that at $X_c = R_L$ the frequency is 1000 Hz. Choose a scale for frequency say 1 inch to 1000 Hz. In the present case this corresponds to 1.0″. Then position the rule horizontally such that 0 on the rule is in coincidence with the Y axis until the 1000 Hz point on the rule (1.0″) intersects the 45° constructional line.

LOCUS DIAGRAMS

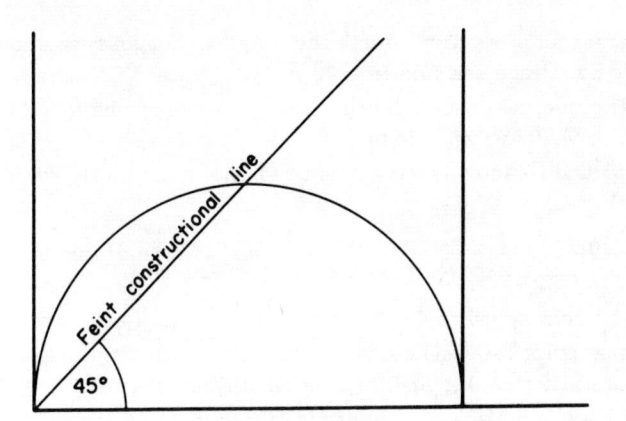

Fig. 19.11.7

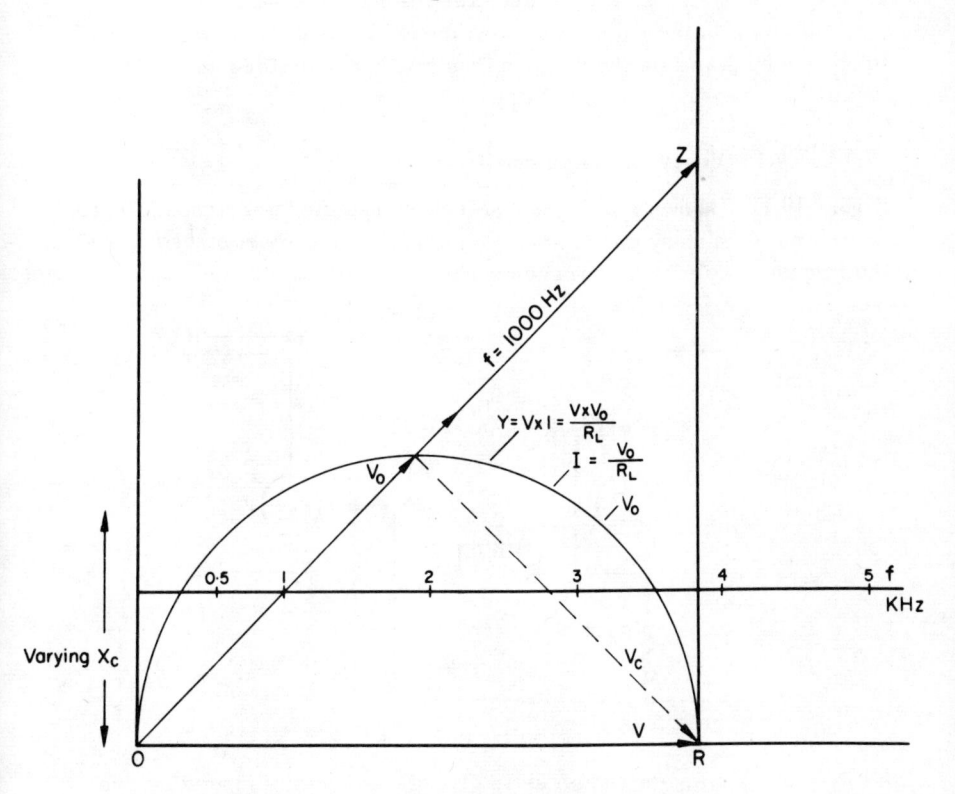

Fig. 19.11.8.

The line representing frequency is than drawn.

Then mark the voltage locus V_0, as in figure 19.11.8.

Now we are in a position to use the diagram. Suppose we choose to find V_0 at 1000 Hz. Place the rule from '0' to 1000 Hz and measure V_0 from '0' to where the rule cuts the V_0 semicircle; it is seen to be 0.707 V. V_c is also measured, as 0.707 V. $Z = 1414\,\Omega$

If we needed to know the admittance (Y) then the circle could be calibrated as Y.

At $X_c = 0$, $Y = \dfrac{1}{z} = \dfrac{1}{R} = \dfrac{1}{1\,\text{K}\Omega} = 1\,\text{m}\mho$ and the diameter of the circle, 10 cm would correspond to $1\,\text{m}\mho$. To find I: $I = V_{in} \times Y = 1\text{V} \times 1\,\text{m}\mho$

The same point for $1\,\text{m}\mho$ could become 1 mA corresponding to 10 cm and it follows naturally that V_0 (under these conditions when $X_c = 0$) is in fact $I \times R_L = 1\,\text{mA} \times 1\,\text{K}\Omega = 1\text{V}$. In effect, the circle has been calibrated as V_0 (1V = 10 cm diameter). Choose other frequencies and measure the above; then calculate in the normal way, and check the results. Several 'frequency' lines may be drawn on the diagram thus giving a wider frequency range if required.

19.12. A frequency selective amplifier

Figure 19.12.1 shows a frequency selective amplifier. A cathode follower is connected as in many cases, the loading factors of further stages may shunt the frequency selective components thus changing the desired characteristics.

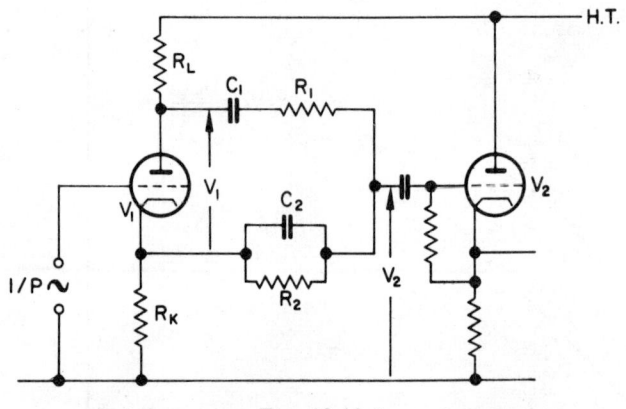

Fig. 19.12.1.

R_L and R_K are selected so as to give the correct d.c. operating conditions in the normal manner as shown in figure 19.12.2. (R_L and R_K have the same ohmic value).

The circuit may be simplified as shown in figure 19.12.3. This is seen to be a Wien Bridge.
$$Z_1 = R_1 + \frac{1}{j\omega C_1} \quad Z_2 = \frac{1}{\dfrac{1}{R_2} + j\omega C_2}$$

LOCUS DIAGRAMS

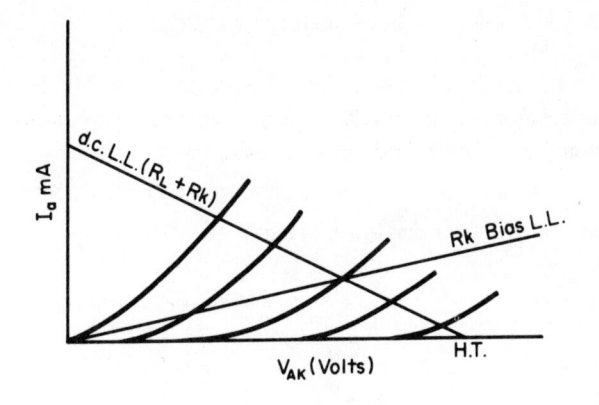

Fig. 19.12.2

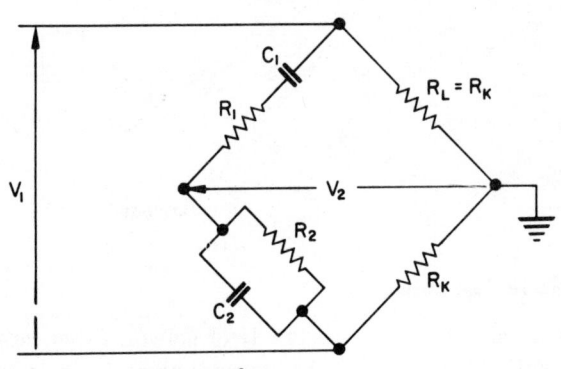

Fig. 19.12.3.

At balance, $Z_1 Z_4 = Z_2 Z_3$. but as $Z_3 = R_L$ $Z_4 = R_K$ and as $R_L = R_K$

$$\frac{Z_1}{Z_2} = 1$$

hence

$$\left(R_1 + \frac{1}{jwc_1}\right)\left(\frac{1}{R_2} + jwc_2\right) = 1$$

$$\therefore \frac{R_1}{R_2} + \frac{C_2}{C_1} + j\left[WC_2 R_1 - \frac{1}{W C_1 R_2}\right] = 1 \tag{1}$$

Considering active parts; $\dfrac{R_1}{R_2} + \dfrac{C_2}{C_1} = 1$

It will simplify matters to let $\dfrac{R_1}{R_2} = \dfrac{C_2}{C_1} = \dfrac{1}{2}\left(\text{i.e., } \dfrac{1}{2} + \dfrac{1}{2} = 1\right)$

Therefore, $R_2 = 2R_1$ and $C_1 = 2C_2$.

Hence $R_1 = \tfrac{1}{2}R_2$ and $C_2 = \tfrac{1}{2}C_1$. Substituting in equation (1) and equating the reactive part to zero, an expression for the desired frequency to be selected, is given as;

$$f_0 = \frac{1}{2\pi C_2 R_2}$$

Figure 19.12.4. shows the completed bridge.

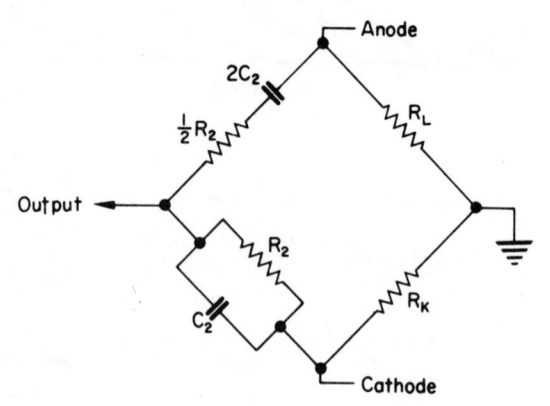

Fig. 19.12.4.

19.13. The Twin Tee network

The Twin Tee is yet another in the family of networks that, at one special frequency f_0, will exhibit a special electrical characteristic.

The circuit of the Twin Tee is shown in figure 19.13.1.

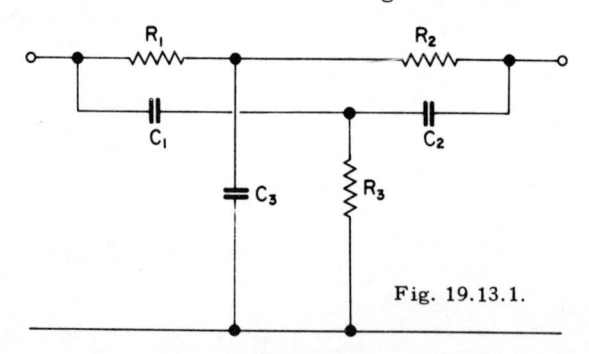

Fig. 19.13.1.

The output, V_o from the network, with an applied input signal, V_{in}, of constant amplitude and varying frequency, will be shown in figure 19.13.2. It should be noted that the curve is non-symetrical about f_0.

The network, as its name implies, consists of two Tee networks

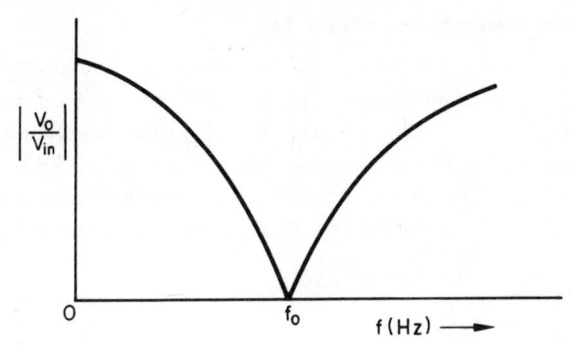

Fig. 19.13.2.

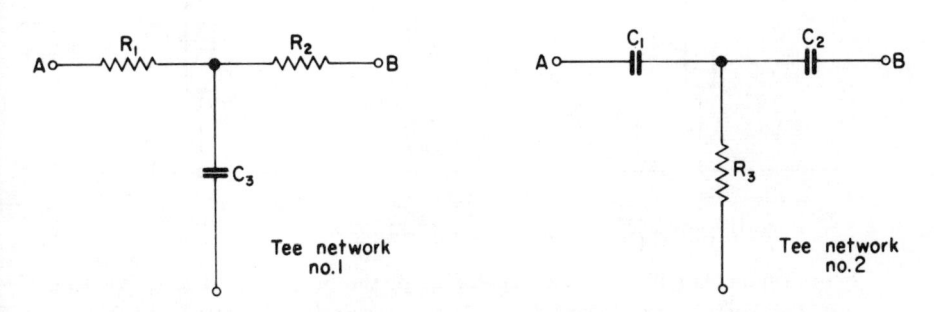

Fig. 19.13.3.

connected in parallel. These are shown separately in figure 19.13.3.

We intend to examine the networks and to derive a formula for the frequency at which the 'null' occurs. The frequency, f_0, may be expressed in terms of the resistors and capacitors and by suitably choosing these components, any desired 'null' frequency may be obtained.

Referring back to figure 19.13.2, it may be clearly seen that at the frequency f_0 the output voltage will be zero. The shape of the curve other than that of the 'null' point, is determined by all components in the complete network. From these, the Q or the selectivity may be determined. For the purpose of this exercise however, the null point only, will be investigated.

Assuming a single Tee network, for the moment, it may be seen that the Tee may be transformed to a Pi network. This is shown in figure 19.3.4

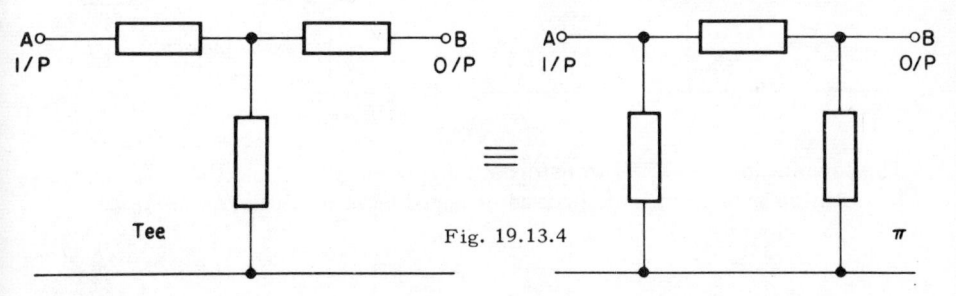

Fig. 19.13.4

A single Pi may result from a single Tee

If the output is to be made zero at the frequency f_0 this may be accomplished in a number of ways. There is perhaps no better way of causing the output to become zero than removing the branch $A-B$ in the π network. If this branch is removed, there can obviously be no reasonable doubt that the output will be zero. This, then, will be the approach to derive the formula for the null point in terms of C and R. This is illustrated in figure 19.13.5.

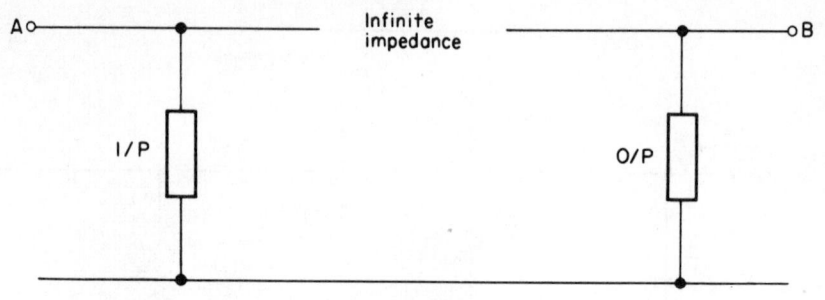

Fig. 19.13.5.
With the branch $A-B$ removed, the output will be zero.

It is argued that if at f_0 the impedance of the branch $A-B$ must be open circuit, the impedance of the branch $A-B$ must be infinite. If the branch $A-B$ is considered in terms of admittance, then for the same conditions the admittance of the branch $A-B$ must be short circuit or have zero admittance (If $Z = \infty$, then as $Y = 1/Z$, $Y = 1/\infty = 0$).

Before proceeding with the transformation of the Tee networks 1 and 2, it must be very desirable to demonstrate a Tee to Pi transformation.

Example of Tee to Pi transformation

Figure 19.13.6 shows a Tee and Pi network.

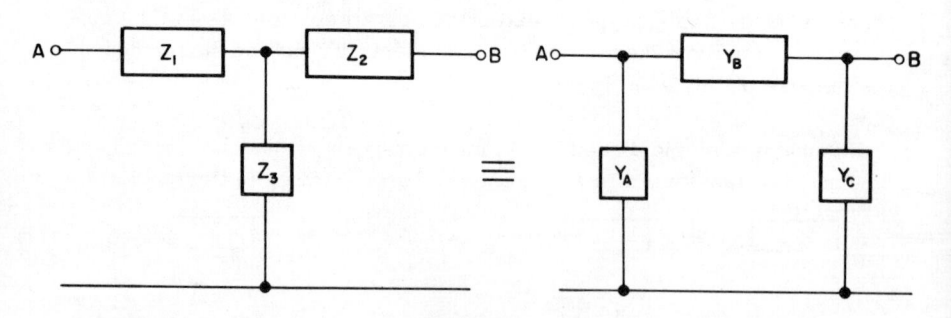

Fig. 19.13.6

The formula for Tee to Pi transformation is identical to the Pi to Tee formula, with the exception that, instead of impedance, admittances are used.

$$Y_A = \frac{1}{Z_A} = \frac{\frac{1}{Z_1}\frac{1}{Z_3}}{\frac{1}{Z_1}+\frac{1}{Z_2}+\frac{1}{Z_3}} \qquad Y_B = \frac{1}{Z_B} = \frac{\frac{1}{Z_1}\frac{1}{Z_2}}{\frac{1}{Z_1}+\frac{1}{Z_2}+\frac{1}{Z_3}}$$

$$Y_C = \frac{1}{Z_C} = \frac{\frac{1}{Z_2}\frac{1}{Z_3}}{\frac{1}{Z_1}+\frac{1}{Z_2}+\frac{1}{Z_3}}$$

Considering the original problem, the Tee network (1) will be transformed. The Tee network (1) and the equivalent Pi is shown in figure 19.13.7.

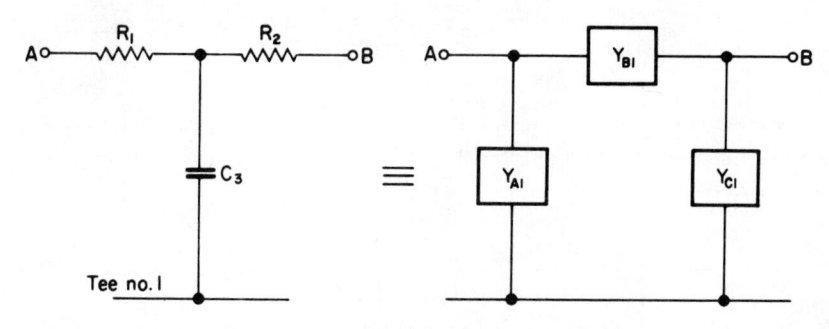

Fig. 19.13.7.

$$YA_1 = \frac{\frac{1}{R_1}\,jWC_3}{\frac{1}{R_1}+\frac{1}{R_2}+jWC_3} \qquad YB_1 = \frac{\frac{1}{R_1}\frac{1}{R_2}}{\frac{1}{R_1}+\frac{1}{R_2}+jWC_3}$$

$$YC_1 = \frac{\frac{1}{R_2}\,jWC_3}{\frac{1}{R_1}+\frac{1}{R_2}+jWC_3}$$

The Tee network (2) will be similarly transformed as in figure 19.13.8.

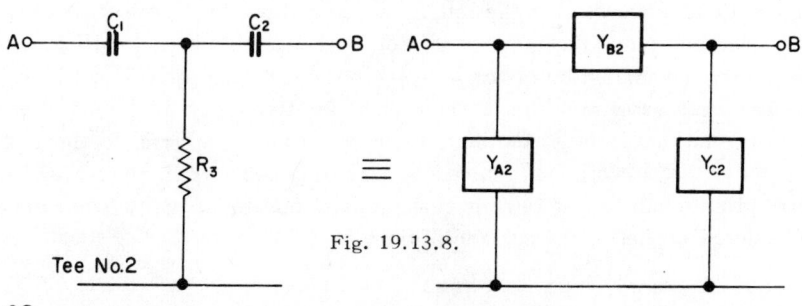

Fig. 19.13.8.

Considering next, the Tee network (2), the equivalent Pi becomes;

$$Ya_2 = \frac{\frac{1}{R_3} \; jwC_1}{\frac{1}{R_3} + jWC_1 + jWC_2} \qquad YB_2 = \frac{jWC_1 \; jWC_2}{\frac{1}{R_3} + jWC_1 + jWC_2}$$

$$YC_2 = \frac{\frac{1}{R_3} \; jwC_2}{\frac{1}{R_3} + jWC_1 + jWC_2}$$

hence,
$$YB_2 = \frac{-W^2 \; C_1 \; C_2}{\frac{1}{R_3} + jWC_1 + jWC_2} \qquad (\text{as } j^2 = -1)$$

Both networks are re-connected in parallel. The complete original Tee network has now been transformed to a Pi network. Figure 19.13.9 shows the combined network.

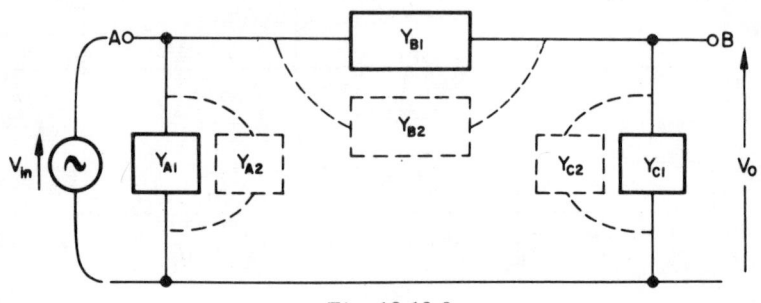

Fig. 19.13.9.

It should now be obvious why the original Tee network was separated into two parts. The transformation of the seperate networks was an easy task, but if the original Tee has been left as a single unit, the mathematics could have been much more involved. It is, of course, an easy matter to replace the seperate networks in shunt once each has been transformed. Figure 19.13.9 shows that the left hand vertical portion of the Pi network is connected across the input generator. Although this branch will play a part in the determination of the 'skirts' of the response, it will not be considered when examining the attenuation at the frequency f_0. It has already been argued that in order to cause the output to become zero, the horizontal portion of the network, (Yb_1 and Yb_2,) is to be 'removed'. Figure 19.13.10 shows this condition. It is seen that the only portion of the network to be considered is that of the horizontal branch.

LOCUS DIAGRAMS

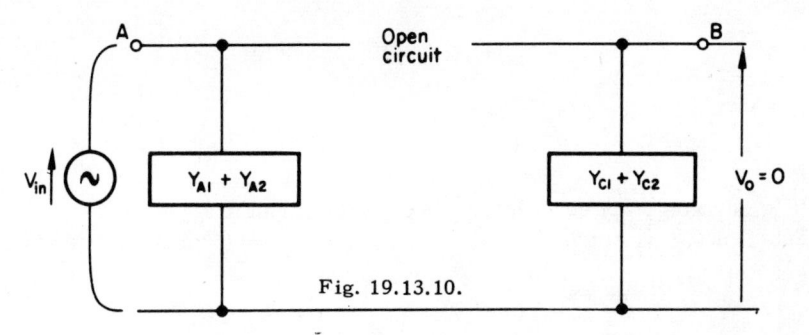

Fig. 19.13.10.

If the horizontal branch were to be physically removed, there would be no output. As this is neither possible or desirable, the admittance of the branch at f_0 must, when Y_{b_1} and Y_{b_2} are added, become zero. This will mean that the impedance will be infinite, The next step is therefore, to add the admittances and equate them to zero. This is shown in figure 19.13.11.

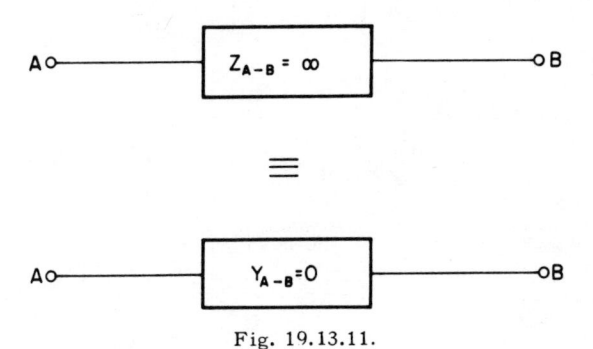

Fig. 19.13.11.

Adding Y_{b_1} and Y_{b_2} and equating this sum to zero.

$$\frac{\frac{1}{R_1} \cdot \frac{1}{R_2}}{\frac{1}{R_1} + \frac{1}{R_2} + jWC_3} + \frac{-W^2 C_1 C_2}{\frac{1}{R_3} + jWC_1 + jWC_2} = 0$$

$$\therefore \frac{\frac{1}{R_1} \cdot \frac{1}{R_2}}{\frac{1}{R_1} + \frac{1}{R_2} + jWC_3} = \frac{W^2 C_1 C_2}{\frac{1}{R_3} + jWC_1 + jWC_2}$$

Thus, $\left(\frac{1}{R_1} \cdot \frac{1}{R_2}\right)\left(\frac{1}{R_3} + jWC_1 + jWC_2\right) = \left(\frac{1}{R_1} + \frac{1}{R_2} + jWC_3\right)(W^2 C_1 C_2)$

Separating the active and reactive terms, and considering the reactive parts only,

$$W^3 C_1 C_2 C_3 = \frac{1}{R_1} \cdot \frac{1}{R_2} (W)(C_1 + C_2)$$

$$\therefore W^2 C_1 C_2 C_3 = \frac{1}{R_1} \cdot \frac{1}{R_2} (C_1 + C_2) \quad (1)$$

It is common practice to let $C_1 = C_2 = \tfrac{1}{2} C_3$ and to let $R_1 = R_2 = 2R_3$. Substituting R_1 for R_2 and R_3 and C_1 for C_2 and C_3 and equation (1) we get from (1)

$$W^2 C_1 C_2 C_3 = \frac{1}{R_1} \cdot \frac{1}{R_2} (C_1 + C_2)$$

$$\therefore W^2 C_1 C_2 C_1 = \frac{1}{R_1} \cdot \frac{1}{R_1} (C_1 + C_2)$$

thus,
$$2W^2 C_1^3 = \frac{2C_1}{R_1^2}$$

$$\therefore W^2 C_1^2 = \frac{1}{R_1^2}$$

thus,
$$W^2 = \frac{1}{C_1^2 R_1^2}$$

and
$$W = \frac{1}{C_1 R_1}$$

and as $W_0 = 2\pi f_0$, where f_0 is the null frequency,

$$\therefore f_0 = \frac{1}{2\pi C_1 R_1}$$

At the frequency at which the output falls to zero, the combined $A–B$ branch admittances become zero. The combined $A–B$ branch impedance is infinite (figure 19.13.12).

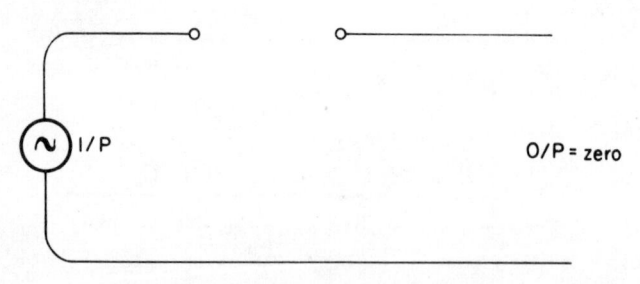

Fig. 19.13.12.

Figure 19.13.13 shows a circuit complete with values of capacitors and resistors which when placed into the formula, will give a null at a frequency of 100 Hz.

The frequency $f_o = \dfrac{1}{2\pi C_1 R_1} = \dfrac{0.159}{C_1 R_1}$ $\left(\text{where } 0.159 = \dfrac{1}{2\pi}\right)$

In practice, the components C_3 and R_3 will need to be trimmed or finely adjusted, in order to obtain the maximum attenuation at f_o.

This network may be used in a feedback path of a selective amplifier, become the F.D.N. in an oscillator circuit, or with several connected in series, may be used say to examine a sinewave from a squarewave input, In the latter case each Twin Tee should have an f_o of an odd harmonic of the fundamental as shown in figure 19.13.14.

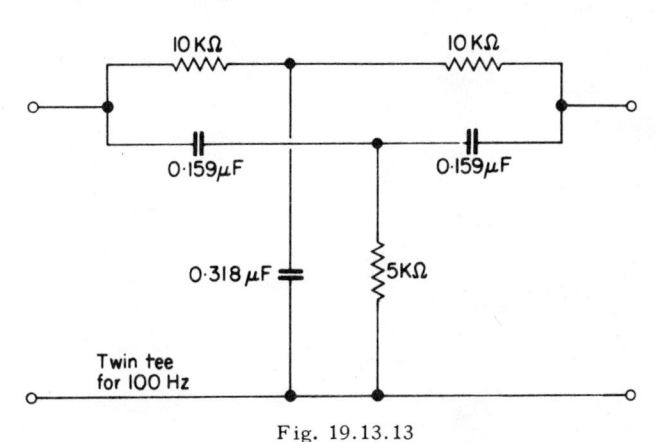

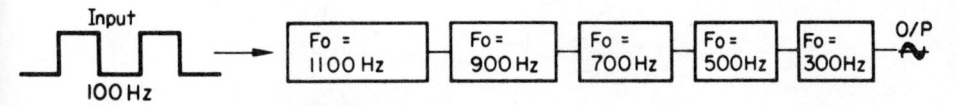

Fig. 19.13.13

Fig. 19.13.14.

It is an interesting exercise to examine with an oscilloscope, the change in waveshape, at the various junctions between Twin Tees. The effect of removing various odd harmonics is quite evident. The amplitude of the sinewave output will be very much smaller than the original input. For precise results, each stage must be isolated from each other to prevent loading effects.

The values of the capacitors and resistors were chosen to have certain values relative to C_1 and R_1, this was done so, as the ratios chosen are providing that the following formula is satisfied.

$$\left(\dfrac{C_1 + C_2}{C_3}\right)(R_3)\left(\dfrac{1}{R_1} + \dfrac{1}{R_2}\right) = 1$$

Placing the values chosen in the example, we get

$$\left(\frac{C_1 + C_2}{2C_1}\right)\left(\frac{R_1}{2}\right)\left(\frac{2}{R_1}\right) = 1$$

$$(1) \times (1) = 1$$

The Twin Tee is a common network, it may be found in many electronic equipments.

CHAPTER 20

Simple mains transformers

In common with the other subject matter in this book, the approach will be as simple as possible. A number of factors will be ignored and only those considered absolutely essential for the trainee technician engineer will be discussed. We will be discussing the practical losses of a transformer and how to optimise the efficiency of the end product.

The quiescent state of a transformer may, as with valves and transistors, be either in the centre of their linear characteristics, or in a state analogous to a binary system; either in one or two stable states with nothing stable in between.

The latter category would normally employ square loop material whilst the former as in this case, is analogous to a class A amplifier.

There are load line techniques that can be employed in transformer design and most manufacturers publish information, from which sets of tables can be drawn. Several tables have been drawn from values which were derived from certain parameters. These parameters will be kept constant, e.g., the flux density, will remain at $1.1 \, \text{Wb/m}^2$. The current density of the copper wire used for the windings will remain at 1000A/inch^2 Two graphs have been plotted; Figure 20.4.4. shows % regulation against secondary volt amps; whilst Figure 20.4.3. shows % efficiency against secondary volt amps.

Both graphs will be used during the design stages of the transformer. When the % efficiency of the completed transformer is measured, figure 20.4.3. will indicate whether or not, a reasonably efficient design has been accomplished.

20.1. A simple design. (1)

Let us then, consider figure 20.1.1.; this is the transformer to be designed; a simple mains transformer having a tapped input of 0–200–220–240 V at 50 Hz. The secondary will give 100 V at 5 A on full load.

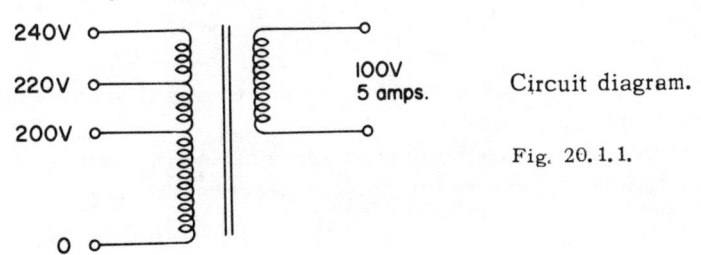

Circuit diagram.

Fig. 20.1.1.

Our earlier elementary theory assumed a perfect transformer, but in these examples allowances will be made for factors which, in practice, cause the transformer to be far from perfect. The reader will recall that, true power is active (non-reactive), but when a reactive component is present, the power P, is expressed as $V.I. \cos \phi$, where V is the voltage applied to the circuit, I is the current flowing in the circuit, ϕ is the angle, in degrees, or phase difference between the current and voltage. P is the power in watts.

The product of Volts and Amps is expressed as Voltamperes and from $P = V.I. \cos \phi$, we see that when $\phi = 0°$, $\cos \phi = 1$. Hence the power, $P = V.I.$ In other words, when no reactive components is present, the Voltamps equal the power.

This transformer has a secondary Voltamps of 100V·5A = 500VA. An approximate formula for finding the cross sectional area (c.s.a.), of the core when the secondary VA is known, is $\sqrt{\text{Secondary VA}}/1.1$ for transformers between 600 and 1000 VA; for transformers with a secondary VA between 50 and 250 the denominator becomes 0.9. Between 250 and 600 it becomes 1.0. (These figures result from practical experience). Substituting in the formula, the known secondary VA of 500, we get
c.s.a. $= \sqrt{500}/1 = 22.4 \, \text{cm}^2$.

From table 1 it can be seen that a c.s.a. of 22.85 cm² is obtained from a 2¼ inch pile of 437 A laminations, or a c.s.a. of 21.7 cm² is obtained from a 2½ inch pile of 435 A laminations. Comparing the two laminations, from the dimensions given in table 3, we see that the 437 A is ½ inch longer and 1½ inches taller than the 435 A. Thus if the overall size is important, then the smaller of the two laminations could be used.

It is assumed that the overall size is important, therefore we will use the type 435 A laminations and in table 1, column 4, it is seen that for a 2½" pile, the Volts per turn (V/T) are 0.53 V/T. $\beta = 1.1 \, \text{Wb/m}^2$. The frequency of the British mains supply is 50 Hz. From the % efficiency curve we see that for a secondary VA of 500, an efficiency of 94% can be expected, thus the primary input VA will be

$$\frac{500 \times 100}{94} = 532 \, \text{VA}.$$

For an applied voltage of 200 V, the primary current will be

$$\frac{VA}{V} = \frac{532}{200} = 2.66 \, \text{A}.$$

(The lowest input voltage has been chosen as this will result in the highest primary current.)

To avoid any possibility of an appreciable temperature rise, when fully loaded, the copper wire will be run at a current rating, not exceeding

1000 A per square inch. From table 4, column 2, 17 s.w.g. will take 2.463 A at this rating. Therefore 17 s.w.g. has a suitable rating, and will be chosen for the primary winding. Column 4, table 2, gives the outside diameter of 17 s.w.g. as 0.059 inch.

Knowing the V/T for the particular pile of laminations chosen, and knowing also, the value of the voltages to be applied to the primary winding, (in this case 200, 220, 240) the number of turns required will be given by

$$\frac{200\,V}{0.53/T} + \frac{20\,V}{0.53/T} + \frac{20\,V}{0.53/T} = 377 + 38 + 38 = 453 \text{ Turns.}$$

The type of lamination to use and the number of turns of wire of a particular gauge that we must wind on to the bobbin or former must be known in order to complete the primary winding. Therefore we must examine the 'window' dimensions of the laminations and from these determine the total winding height available, also the winding width of the bobbin or former to be used.

Referring to figure 20.4.5. and table 3, we see that the width of the former (dimension 'E') is 3¾ inch. This is the maximum width of the former but allow, say, ¼ inch at each end of the former for insulating purposes. Thus the effective winding length will be 3.25 inches (or 3¼ inches). The number of turns that can be wound on a length of 3.25 inches, using 17 s.w.g., is

$$\frac{3.25 \text{ inches}}{0.059 \text{ inches}} = 55.2.$$

We should at this stage allow two turns for winding tolerances, (although once a winding skill is developed, this allowance can be reduced or ignored) and consequently this leaves 53 turns per layer (T.P.L.). The number of turns to allow on each layer is 53. The total turns is 453 T. The number of layers therefore must be

$$\frac{453\,T}{53\,T/L} = 8.55\,L$$

which of course must be 9 layers.

A layer of paper 2 mils thick is inserted between adjacent layers of wire in order to assist in insulating each winding from its neighbour; upon completion of the whole primary winding, three layers of Empire cloth are wound tightly over the final layer, and secured very tightly before the secondary winding is started. Knowing the number of layers and the thickness of the insulating material, the overall primary winding height including the Empire cloth can now be calculated. At this point we must add all of the layers and see just how much room is required for the primary complete.

	9 layers of wire at 0.059"	0.531"
	8 layers of paper at 0.002"	0.016"
	3 layers of Emp. cloth at 0.010"	**0.030"**
	Winding height required for the primary	**0.577"**

Table 3 shows the winding height as 1.625" and if 0.1" is allowed for the bobbin, or former, this leaves 1.625" − 0.1" = 1.525".

We now take 80% of this value which will leave 1.22" and assume the primary and secondary windings will occupy approximately the same amount of space, which becomes 1.22"/2 = 0.61". This is the winding height we can allow for the primary and secondary windings plus the inter-winding insulation. (The 80% allows for inexperienced winding, this will increase as the winder gains skill.)

0.577" is required for the primary, and there is 0.61" available, it can be seen that the primary will fit in quite well.

It might be very useful to construct a table, at this time, showing the results of the various calculations, as they are made.

PRIMARY	s.w.g.	Primary Turns	T.P.L.	No. of Layers	Total Winding height
WINDING	17	377 38 38	53	9	0.577"

The table is incomplete, but this can be added to as further results are obtained.

The secondary winding

The secondary full load current is 5 A; table 4 shows either 14 s.w.g. or 14½ s.w.g. as a suitable gauge at the rating of 1000 A/inch².

As this secondary is an outside winding, 14½ s.w.g. will be chosen. This has been selected as a suitable gauge, as the secondary is not covered by a further winding and will not run hot. Figure 20.4.4. shows the approximate % voltage regulation for values of output VA and, for an output VA of 500, the voltage regulation is given as approximately 4%. This means that the fully loaded secondary voltage will be 100 V, whilst the off-load secondary voltage will be approximately 104 V.

The volts per turn are the same, of course, i.e. 0.53 V/T. As the secondary volts are 104 V, the number of turns required are

SIMPLE MAINS TRANSFORMERS

104 V/53 V/T = 197 T. The winding width is 3.25" as before, and the outer diameter (o.d.) of 14½ s.w.g. is given as 0.080".

The number of turns per layer
$$\frac{3.25}{0.08} = 40.$$

As before, and for the same reason, we will subtract 2 turns, giving, 40 − 2 = 38 T.P.L.

The number of layers required is
$$\frac{197}{38} = 5.2,$$

but of course this must be 6 layers. (With winding skill, this will become 5 layers).

The winding height of the secondary is as follows:

6 layers of 14½ s.w.g. at 0.08"	0.48"
5 layers of paper at 0.002"	0.010"
2 layers of cloth at 0.010"	0.020"
The total height for the secondary winding is	0.51"

As 0.61" was allowed for each winding, the secondary will fit in nicely.

The following table has been drawn up for the secondary and will be referred to when making subsequent calculations.

SECONDARY WINDING	s.w.g.	Turns.	T.P.L.	Layers	Total winding height.
	14½	197	38	6	0.51"

Most of this has been straightforward so far, and it is now necessary to determine the d.c. resistance of both windings. The weight of the copper will also be required.

A method of obtaining the MEAN LENGTH of the primary winding

If the weight, or resistance, of the winding under construction is required, the length of an average turn of wire in that winding must be found.

Once the length of an average turn is found, then knowing the number of turns, the total length of wire, and from this, the approximate resistance and its weight can be calculated.

Consider figure 20.4.2. There is a former of side x'' upon which a coil is wound to a total height of h''. Then $h/2$ will give the position of a turn of wire halfway between the former and the total winding height, h.

If the sides of the former of length x'' are moved outwards, until they rest on the line showing the position of the average turn, it is seen that they do not increase or decrease their length but in figure 20.4.2. it is seen that four quadrants of a circle are left. (These are shaded).

The radius of the 'circle' is $h/2$ inches. Therefore the circumference can be calculated. This circumference plus the perimeter of the bobbin will give the average length of a turn in that winding.

In practice, $2\pi r/0.8$ is used to find the circumference, as the circle is not quite a circle and some allowance must be made for its shape.

When a second winding is placed on top of the first winding, a slightly different method is used. The length of the perimeter of the former is still required but this time the sides are moved outwards until they rest in the position of the average turn of the secondary half way across that winding. It is now required to find the total height of the primary winding plus half of the secondary winding.

This will give the radius of the second circle and the circumference may once again be calculated. This value plus the perimeter of the former will give the length of the average turn of wire on the secondary.

Using the above methods and substituting the values obtained, we get

Height of primary winding h inches $= 0.577''$

Then $h/2 = 0.288'' =$ the radius of the circle.

The circumference $\dfrac{2\pi r}{0.8} = \dfrac{2 \times 3.14 \times 0.288}{0.8} = 2.27''$

The perimeter of the former, from table 1, column 5, is $9.06''$

The length of an average turn $= 9.06'' + 2.27'' = 11.13''$

The resistance of the winding is obtained from,

$$R = \frac{L \times n \times r}{36 \times 1000}$$

where R is the resistance of the winding and

where r is the resistance per 1000 yards of wire to be used

and n is the number of turns on the winding

and L is the length of the average turn of wire, in inches.

Substituting, we get

$$R = \frac{11.13 \times 453 \times 9.747}{36 \times 1000} = 1.37\,\Omega.$$

The approximate resistance of the primary winding is $1.37\,\Omega$.

SIMPLE MAINS TRANSFORMERS

The weight of the copper wire required for the primary winding is obtained from:

$$\frac{\text{Resistance of winding}}{\text{Resistance per lb (table 1)}} = \frac{1.37\,\Omega}{0.3422\,\Omega/\text{lb}} = 4.16\,\text{lb}$$

As this amount is only just enough, it might be advisable to have a little in hand; therefore, we might obtain 4¼ lb of wire to make sure.

The average length of a turn of wire for the secondary winding is given by:

height of primary plus half height of secondary winding = radius of circle for secondary winding.

$$= 0.577'' + \frac{0.510''}{2} = 0.832''.$$

The circumference becomes

$$\frac{2\pi r}{0.8} = \frac{2 \times 3.14 \times 0.832''}{0.8} = 6.55''$$

and the average length of a turn of wire on the secondary is:

$$6.55'' + 9.06'' = 15.61'' \text{ and from } R = \frac{L \times n \times r}{36 \times 1000}$$

we get

$$\frac{15.61'' \times 197 \times 5.292}{36 \times 1000} = 0.454\,\Omega.$$

The weight of wire required for the secondary winding is,

$$\frac{0.454}{0.1} = 4.54\,\text{lb}$$

but as before it is better to have a little more, just in case. Therefore we might allow 4¾ lb of wire for the secondary winding.

It is now necessary to complete the table of winding data for the complete transformer, as follows.

	s.w.g.	PRI. Turns.	T.P.L.	Layers	Wdg. height.	Resistance and weight
PRI.		377				
		38 = 453	53	9	0.577"	1.37 Ω 4¼ lb
	17	38				
SEC.	14½	197	38	6	0.51"	0.454 Ω 4¾ lb

We must allow for the 3 layers of 10 mils Empire cloth over the primary. There are also 2 layers of 10 mils Empire cloth over the secondary.

The calculated total winding height allowed is 1.220″

The total height including insulation is 1.087″.

It is clear then, that our windings and insulation can be accomodated nicely, without unnecessary waste.

20.2. Transformer losses

The losses in the total primary winding is, $I_p^2 R_p = (2.22)(1.37) = 6.8$ W.
The losses in the secondary winding is, $I_s^2 R_s = (5)^2 (0.454) = 11.35$ W.
The total copper losses = 18.15 W and if we assume maximum efficiency, i.e. the copper losses equal the iron losses, we have a total loss of 36.30 W.
From efficiency, $\eta = \dfrac{\text{Power output} \times 100}{\text{Power output plus losses}} = 93.5\%$.

Thus our calculated efficiency is very close to the estimated value of 94%.

Estimated voltage regulation

The total primary voltage drop is $I_p R_p = 2.22 \times 1.37 = 3.02$ V.
The total secondary voltage drop is $I_s R_s = 5.0 \times 0.454 = 2.27$ V.
The primary voltage drop referred to the secondary is:

$$\frac{V_p n_s}{n_p} = \frac{3.02 \times 197}{453} = 1.32 \text{ V}.$$

The total effective secondary voltage drop is $1.32 + 2.27 = 3.59$ V.
Assume there is 103.59 V on open circuit secondary, and 100 V on fully loaded secondary, the voltage regulation is

$$\frac{(103.59 - 100)(100)}{103.59} = 3.47\%.$$

Thus the calculated voltage regulation is very close to the estimated value of 4%. In this case it is an improvement which is all to the good. The foregoing figures have assumed a power factor of unity. It is assumed that no phase angle existed between the applied voltage and current. In practice, however, there would almost certainly be a reactive component present.

As an academic example, we will assume a phase angle of 30°. From the approximate voltage regulation formula, we have:

$$VR\% = I_p \frac{(R_e \cos \phi + X_e \sin \phi)(100)}{V_p}$$

Where R_e the effective resistance of the primary
and X_e the effective reactance of the primary,
and ϕ the angle of lag of current on the voltage.

$$R_e = R_p + R_s \frac{(n^p)^2}{(n_s)^2} = 1.37 + 0.454 \frac{(453)^2}{(197)^2} = 1.37 + 2.40 = 3.77 \ \Omega.$$

Assuming the voltage regulation is 4%, then from the formula for $VR\%$, we get:

$$4 = \frac{2.22\,(377 \cos 30° + X_e \sin 30°)\,(100)}{240}.$$

Then $\quad \dfrac{4 \times 240}{222} - 3.27 = \dfrac{X_e}{2}$. Then $X_e = 2.12\,\Omega.$

Thus $\qquad Z_e = 3.77 + j2.12$
and $|Z| = \sqrt{(3.77)^2 + (2.12)^2} = \sqrt{14.22 + 4.46} = \sqrt{18.68} = 4.32\,\Omega.$

If we short circuit the secondary by connecting a current meter across the output terminals, then applying a low voltage to the primary and slowly increasing this voltage until full load secondary current results, it will be seen that the applied voltage is given by $Z \times I_p$.
I_p is the full load primary current.

Thus this voltage $= 4.32 \times 2.22 = 9.6\,V$. This voltage is known as the impedance voltage of the transformer.

This parameter is very useful to the designer in a number of ways. The open circuit test applied to a transformer will give the iron losses, whilst the short circuit test gives the copper losses. These two tests will provide sufficient information to assess the performance of the transformer quite accurately and will indicate, whether or not a reasonably successful design has been accomplished.

There are, of course, many ways one might commence to design a transformer. The ways outlined here are not claimed to be the best, or the most economical; this has been a further example in the practical application of theory.

20.3. A design of a simple transformer (Second example)

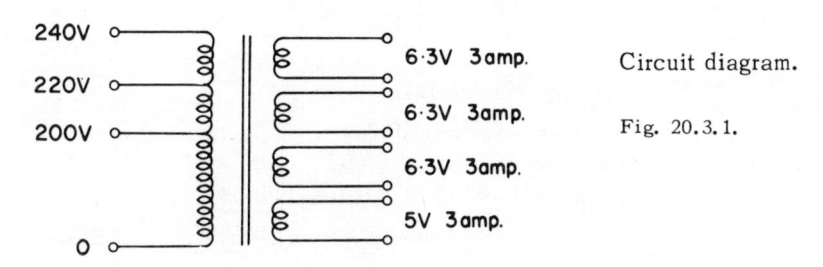

Circuit diagram.

Fig. 20.3.1.

The secondary VA is $3(6.3 \times 3) + (5 \times 3) = 71.7$, say 72 VA.
The c.s.a. of the core is:
$$\frac{(72)^{\frac{1}{2}}}{0.9} = \underline{9.45\,\text{cm}^2}.$$

This transformer has a number of separate windings and, as each windings must be insulated from its neighbour, we must lose a considerable amount of the available winding space occupied by the insulation. In order to compensate for this loss we will increase the c.s.a. of the core by about 20% and from table 1 we see that we require a 1¾" pile of 475 A laminations to give us a c.s.a. of $10.15\,\text{cm}^2$.

Only experience will enable the reader to decide upon the amount by which he must increase the c.s.a. as shown above; but this will come to him at a later stage.

The chosen core will give us a V/T of 0.248. From the tables as before, the estimated efficiency is 85% and VR is 6.8%.

Obviously if we use a slightly larger core, the efficiency and voltage regulation must get a little better. We may find that our final calculations are even better than our estimated values at this time.

$$\text{Primary turns} = \frac{200}{0.248} + \frac{20}{0.248} + \frac{20}{0.248} = 808 + 81 + 81 = 970\,\text{T}.$$

$$\text{The primary VA} = \frac{72 \times 100}{85} = 85\,\text{VA} \quad \text{and} \quad \frac{85\,\text{VA}}{220\,\text{V}} = 390\,\text{mA}.$$

(We chose the 220 V tapping point for the current calculation as this will give a variation either way when using the 200 and 240 V tap.)

From table 4 we choose 24 s.w.g. for the primary. From table 2 we see that this has an outer diameter of 0.024". The width of the former is $2\frac{3}{8}"$ and so the width of the winding is to be $2\frac{3}{8}" - \frac{1}{4}" = 1\frac{7}{8}"$ (or 1.875").
The height of the winding $= 1" - 0.1" = 0.9"$.
If we take 80% of this, as before, we are left with $\underline{0.72"}$.

The primary T.P.L. is
$$\frac{1.875}{1.034} = 78 - 2 = 76.\,\text{T.P.L.}$$

The number of layers $= \frac{970}{76} = 12.8$ this must be 13 layers.

The height of the primary winding is:

13 layers of wire at 0.024"	0.312"
12 layers of paper at 0.002"	0.024"
3 layers of cloth at 0.010"	0.030"
Total height	0.366".

This is approximately half of the total winding space available. Let us consider now, the 6.3 V secondaries.

Each 6.3 V winding will be considered separately. What we decide for one will apply to the others. We expect a VR of 6.8%. The off load voltage must be 6.3 plus 6.8% of 6.3 V. This is equivalent to 6.72 V.

$$\text{The turns} = \frac{6.72}{0.248} = 27 \text{T}.$$

Therefore each 6.3 V winding consists of 27 turns of 16½ s.w.g. of o.d. 0.063".

$$\text{The T.P.L.} = \frac{1.875}{0.063} = 30 - 2 = 28. \text{ T.P.L.}$$

As we require only 27 turns, we can accomodate these on one layer. The height for one layer is 0.063". The insulation is paper, as before. The paper has a thickness of 0.010".

The actual overall height, including paper is

1 layer of wire at 0.063"	0.063"
2 layers of paper at 0.010"	0.020"
Total height for one winding	0.083"

The height for three windings is, of course, three times that of one winding, this is then the total height of the three 6.3 V windings, 0249". We have now to deal with the 5 V winding.

Adding the expected *VR%* figure, we require 5 V plus 6.8% of 5 V. This sums to 5.34 V. The turns are:

$$\frac{5.34}{0.248} = 22 \text{ T}.$$

Reference to table 4 suggests that 16½ s.w.g. is an appropriate wire to use. The o.d. of 16½ s.w.g. is 0.063".

We require 22 turns and as we can fit this into one layer, we can calculate the winding height.

1 layer of wire at 0.063"	0.063"
2 layers at 0.010"	0.020"
This sums to a total of	0.083"
The total secondary winding height is	0.366"
	0.249"
	0.083"
giving a total of	0.698".

2D

This should fit quite nicely into our estimated winding height of 0.72". We could add one more turn of 0.010" insulation between the outer 6.3 V winding and the 5.0 V winding as an added precaution.

This would increase our total height to 0.708", which should still fit in very nicely in the maximum space we have allowed ourselves.

We have reached the point where we have to calculate the mean turn for the primary. The table 1 shows that the perimeter of the bobbin is 6.19". The height of the primary winding is 0.336". (The 3 layers of cloth is NOT included).

The mean turn, then, is $\frac{2\pi}{0.8} \times \frac{0.168}{1} + 6.19 = 7.51''$.

The total length of the winding in yards in $\frac{7.51 \times 970}{36} = 202$ yards.

The resistance of the primary is $\frac{202 \times 63.16}{1000} = 12.8\,\Omega$.

(The 63.16 was obtained from the table showing the resistance of copper wire).

The weight of the copper wire in the secondary is $\frac{12.8}{14.37} = 0.9$, say 1 lb.

The mean length of the secondary turn of the 6.3 V windings

The total height is 0.042". The length of the mean turn is found in the same manner as before.

The mean length is $\frac{0.491 \times 2\pi}{0.8} + 6.19 = 10''$.

The total length of wire is $\frac{10 \times 27}{36} = 7.5$ yards.

The resistance of the 6.3 V windings is $\frac{7.5 \times 8.49}{1000} = 0.066\,\Omega$.

This will give an average value for the three windings.

The weight of wire required is $\frac{0.066 \times 3}{0.26} = 0.76$ lb.

The mean length of the 5 V winding

The radius is the sum of the (a) height of the primary, (b) height of the three 6.3 V secondaries, and (c) half the height of the 5 V winding. This sums to $0.366 + 0.249 + 0.032 = 0.647''$.

The mean length is $\frac{0.647'' \times 2\pi + 6.19''}{0.8} = 11.27''$

The total length of the 5 V winding is $\frac{11.27 \times 22}{36} = \underline{6.9 \text{ yards}}$.

The resistance of the 5 V winding is $\dfrac{6.9 \times 8.49}{1000} = \underline{0.0585\,\Omega}$.

The weight of the wire is $\dfrac{0.0585}{0.26} = \underline{0.225\,\text{lb}}$.

The total weight therefore, is for the 16½ s.w.g. = 0.985 lb. We will call this 1 lb.

Voltage regulation

I_p at 240 V input, = 85/240 = 355 mA.

The primary volt drop is $I_p R_p = 0.35 \times 12.8 = 4.55\,\text{V}$.

The primary volt drop referred to the 6.3 V secondary is $\dfrac{4.55 \times 27}{970} = 0.127\,\text{V}$.

The total secondary volt drop is 0.127 3(0.066) = 0.325 V.

The VR is $\therefore \dfrac{0.325 \times 100}{6.3} = 5.2\%$, which is better than estimated.

The total copper losses are as follows:

Primary winding copper losses $I_p^2 R_p = 0.39^{2''} \times 12.8$ = 1.93 W

6.3 V winding copper losses $I_s^2 R_s = 3(3^2 \times 0.066)$ = 1.79 W

5.0 V winding copper losses $I_s^2 R_s = 3^2 \times 0.0585$ = 0.53 W

Total copper losses   = 4.25 W

Assuming maximum efficiency, copper losses = iron losses = 8.5 W

$$\%\text{ eff.} = \frac{\text{Pout}}{\text{Pout + loss}} = \frac{72 \times 100}{72 + 2(4.25)} = \frac{7200}{80.5} \simeq 90\%$$

and is better than estimated.

20.4. Simple practical test of a transformer. To establish V_Z.

Circuit diagram.

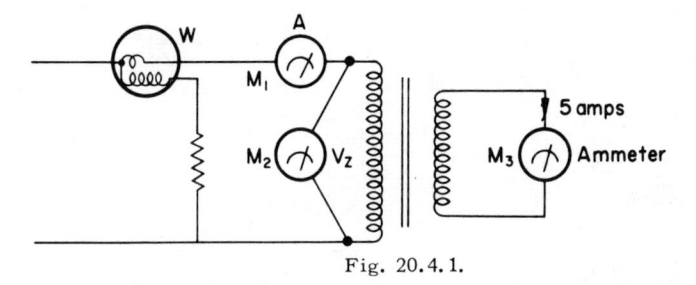

Fig. 20.4.1.

The d.c. resistance may be measured in the normal manner. The secondary may be referred to the primary. M_2 will read the impedance voltage which will cause the secondary current to become 5 amps. M_3 reads the 5 amps. The wattmeter (W) measures the input power. M_1 and M_2 read the primary

VA. From these measurements, $\dfrac{P}{VA} = \dfrac{V.1.\cos\theta}{V.1.}$ may be obtained. ϕ may be obtained from the tables.

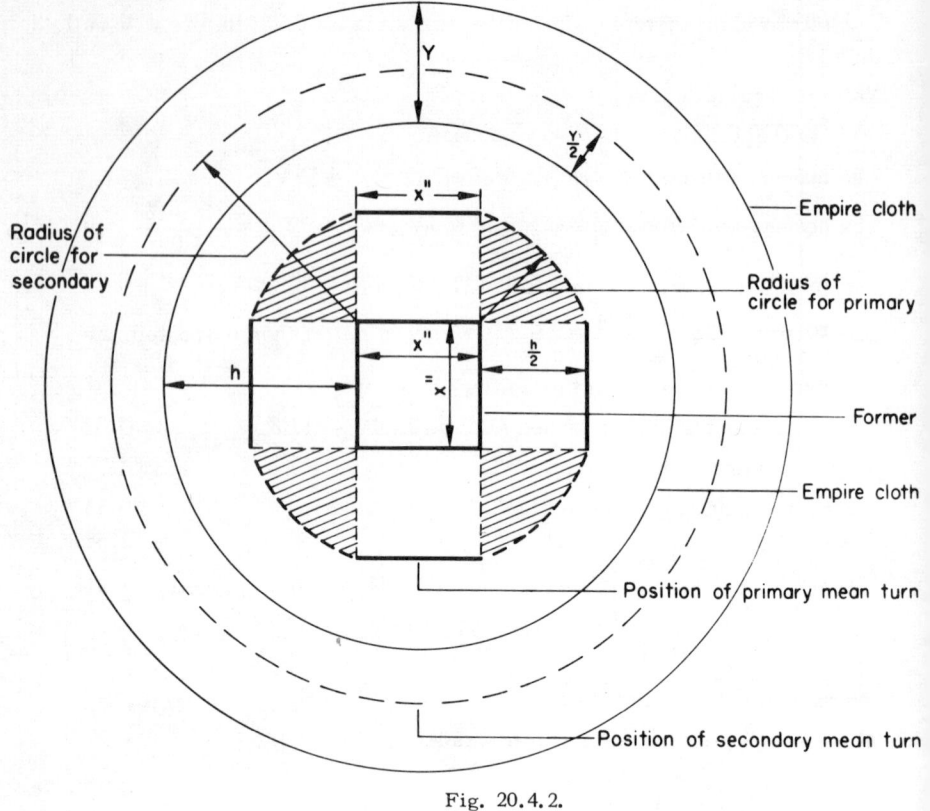

Fig. 20.4.2.

h is the height of the primary winding.

Y is the height of the secondary winding.

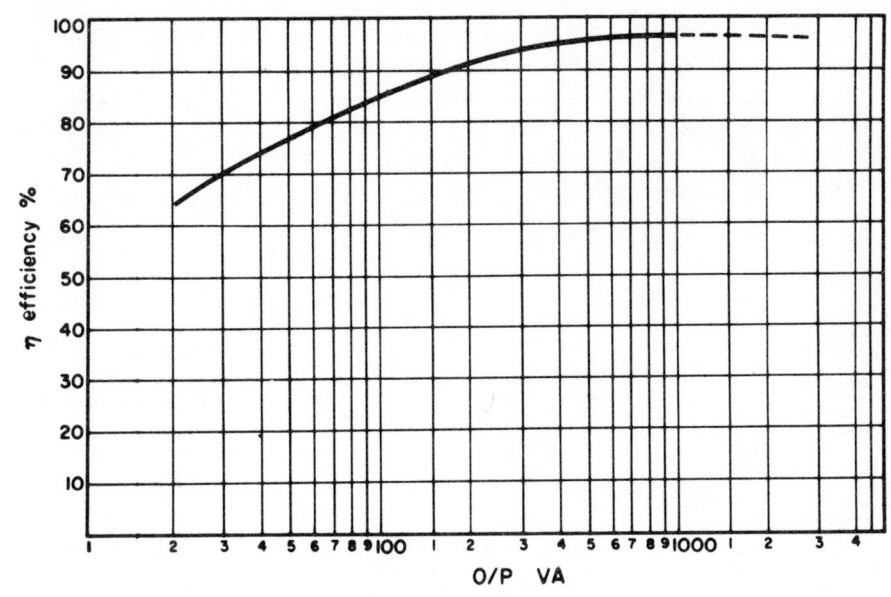

Fig. 20.4.3.

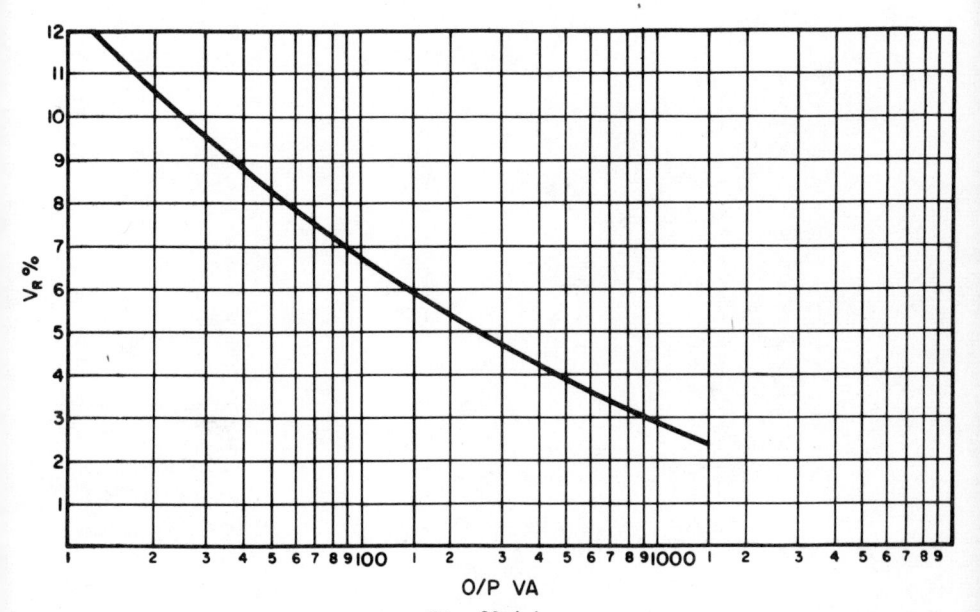

Fig. 20.4.4.

From $E_p = 4.44 N_p f \cdot \Phi_m$, we can substitute Φ_m for $\beta \times a$ where $\beta = 1.1\,\text{Wb/m}^2$ and a is the nett area of the core to be used. Hence $E_p = 4.44 N_p f \cdot \beta \cdot a$ and as $f = 50\,\text{Hz}$, $\beta = 1.1\,\text{Wb/m}^2$ and assuming in each case that $N_p = 1$, the Volts/Turn for the primary may be written as $\text{V/T} = 4.44 \times 50 \times 1.1 \times a$. This applies for secondaries also.

Example: $a = 12.69$, determine the V/T.

Hence $\text{V/T} = 4.44 \times 50 \times 1.1 \times 12.69 = 0.31\,\text{V/T}$. This example is included in the table and shown dotted in columns 3 and 4.

TABLE 1.

LAMS	Pile (in)	Nett area of core (sq. cm)	V/T $B = 1.1\,\text{Wb/m}^2$ $F = 50\,\text{Hz}$	Perimeter of former (in)
403A	.625	2.27	.0553	3.13
403A	.75	2.72	.0662	3.38
403A	1.00	3.62	.0885	3.88
403A	1.25	4.53	.1105	4.38
401A	.75	3.26	.0796	3.69
401A	1.00	4.35	.1062	4.19
401A	1.25	5.44	.1325	4.69
401A	1.5	6.53	.1592	5.19
82A	1.00	5.45	.1326	4.38
82A	1.25	6.81	.1658	4.88
82A	1.5	8.16	.199	5.38
440A	1.00	5.06	.1235	4.5
440A	1.25	6.34	.158	5
440A	1.5	7.61	.186	5.5
440A	1.75	8.87	.2165	6
404A	1.00	5.44	.133	4.55
404A	1.25	6.8	.166	5
404A	1.5	8.16	.199	5.5
404A	1.75	9.54	.233	6
475A	1.00	5.8	.1415	4.68
475A	1.25	7.25	.177	5.19
475A	1.5	8.7	.212	5.69
475A	1.75	10.15	.248	6.19
475A	2.00	11.6	.284	6.69
460A	1.25	9.06	.221	5.69
460A	1.5	10.86	.265	6.19
460A	1.75	12.69	.31	6.69
460A	2.00	14.45	.351	7.19
428A	1.25	8.86	.216	5.69
428A	1.5	10.6	.259	6.19
428A	1.75	12.38	.302	6.69
428A	2.00	14.15	.346	7.19
428A	2.25	15.9	.388	7.69
428A	2.5	17.7	.432	8.19

contd.

Table 1 contd.

LAMS	Pile (in)	Nett area of core (sq. cm)	V/T $B = 1.1 \text{ Wb/m}^2$ $F = 50 \text{ Hz}$	Perimeter of former (in)
435A	1.5	13.05	.318	7.06
435A	1.75	15.2	.371	7.56
435A	2.00	17.4	.425	8.06
435A	2.25	19.55	.477	8.56
435A	2.5	21.7	.53	9.06
435A	2.75	23.9	.584	9.86
435A	3.00	26.1	.636	10.06
437A	1.5	15.25	.372	7.75
437A	1.75	17.8	.434	8.25
437A	2.00	20.3	.496	8.75
437A	2.25	22.85	.558	9.25
437A	2.5	25.4	.62	9.75
437A	2.75	27.95	.682	10.25
437A	3.00	30.5	.745	10.75
437A	3.25	33	.806	11.25
437A	3.5	35.55	.86	11.75
41A	2.5	36.3	.886	11.63
41A	2.75	40	.978	12.13
41A	3.00	43.6	1.065	12.63
41A	3.25	47.2	1.15	13.13
41A	3.5	50.5	1.24	13.63
41A	3.75	54.4	1.325	14.13
41A	4.00	58.2	1.42	14.63
122A	2.75	47.8	1.168	13.13
122A	3.00	52.2	1.272	13.63
122A	3.25	56.6	1.385	14.13
122A	3.5	61	1.49	14.63
122A	3.75	65.4	1.595	15.13
122A	4.00	69.6	1.7	15.63
147A	.825	4.44	.1083	4.25
147A	1.00	5.06	.1235	4.5
147A	1.25	6.34	.1548	5.0
147A	1.5	7.61	.1858	5.5
147A	1.75	8.87	.217	6.0
147A	1.825	9.51	.307	6.25
29A	1.00	5.8	.1415	4.69
29A	1.25	7.25	.177	5.19
29A	1.5	8.7	.212	5.69
29A	1.75	10.15	.248	6.19
29A	2.00	11.6	.284	6.69
196A	1.125	7.35	.1793	5.19
196A	1.25	8.17	.199	5.44
196A	1.5	9.77	.239	5.69
196A	1.75	11.43	.279	6.19
196A	2.00	13.06	.3185	6.69
196A	2.25	14.7	.359	7.19

contd.

Table 1 contd.

LAMS	Pile (in)	Nett area of core (sq. cm)	V/T $B = 1.1\,\text{Wb/m}^2$ $F = 50\,\text{Hz}$	Perimeter of former (in)
78A	1.25	9.06	.221	5.69
78A	1.5	10.86	.265	6.19
78A	1.75	17.69	.31	6.69
78A	2.00	14.5	.351	7.19
78A	2.25	16.32	.398	7.69
78A	2.5	18.1	.442	8.19
120A	1.5	13.05	.318	
120A	1.75	15.2	.371	
120A	2.00	17.4	.425	
120A	2.25	19.55	.477	
120A	2.5	21.7	.53	
120A	2.75	23.9	.584	
120A	3.00	26.1	.636	
248A	1.75	17.8	.434	4.37
248A	2.00	20.3	.496	4.87
248A	2.25	22.85	.558	5.37
248A	2.5	25.4	.62	5.87
248A	2.75	27.9	.682	6.37
248A	3.00	30.5	.745	6.87
248A	3.25	33.00	.806	7.37
248A	3.5	35.5	.86	7.87

TABLE 2.

s.w.g.	Wire dia. (in)	Ohms per 1000 yd.	o.d. EN (in)
4	.232	.56790	
5	.212	.68010	
6	.192	.82920	
7	.176	.98680	
8	.160	1.1941	
9	.144	1.4741	
10	.128	1.8657	.134
11	.116	2.2720	.122
12	.104	2.8260	.110
12½	.098	3.195	.104
13	.092	3.6120	.098
13½	.086	4.1390	.092
14	.080	4.7760	.085
14½	.076	5.2920	.080
15	.072	5.8970	.076
15½	.068	6.6110	.072
16	.064	7.4630	.0675
16½	.060	8.4910	.063

contd.

Table 2 contd.

s.w.g.	Wire dia. (in)	Ohms per 1000 yd.	o.d. EN (in)
17	.56	9.7470	.59
17½	.052	11.305	.055
18	.048	13.267	.0508
18½	.044	15.789	.0465
19	.040	19.105	.0425
19½	.038	21.170	
20	.036	23.590	.0384
20½	.034	26.440	
21	.032	29.85	.0343
21½	.030	33.96	
22	.028	38.99	.0302
22½	.026	45.22	.0280
23	.024	53.07	.0261
23½	.023	57.78	
24	.22	63.16	.024
24½	.021	69.31	
25	.020	76.42	.022
25½	.019	84.68	
26	.018	94.35	.0198
27	.0164	113.65	.0181
28	.0148	139.55	.0164
29	.0136	165.27	.0151
30	.0124	198.80	.0138
31	.0116	227.2	.0129
32	.0108	262.1	.0121
33	.0100	305.7	.0112
34	.0092	361.2	.0103
35	.0085	433.2	.0095
36	.0076	529.2	.0086
37	.0068	661.1	.0078
38	.0060	849.1	.0069
39	.0052	1130.5	.0061
40	.0048	1326.7	.6340
41	.0044	1578.9	.0052
42	.0040	1910.5	.0048
43	.0036	2359	.0044
44	.0032	2985	.0039
45	.0028	3899	.0035
46	.0024	5307	.0030
47	.0020	7642	.0026
48	.0016	11941	
49	.0012	21230	
50	.0010	30570	

TABLE 3.

No.	Length A (in)	Height B (in)	Yoke C (in)	Window Height D (in)	Window Width E (in)	Tongue F (in)	Nett Weight (lb/in)	Mean Path (in)	(cm)
403A	$2\frac{3}{8}$	$1\frac{21}{32}$	$\frac{5}{16}$	$\frac{11}{16}$	$1\frac{1}{32}$	$\frac{5}{8}$	.728	4.69	11.9
401A	3	$1\frac{7}{8}$	$\frac{3}{8}$	$\frac{3}{4}$	$1\frac{1}{8}$	$\frac{3}{4}$	.974	5.25	13.35
82A	$2\frac{11}{16}$	$2\frac{1}{8}$	$\frac{13}{32}$	$\frac{15}{32}$	$1\frac{5}{16}$	$\frac{15}{16}$	1.2	5.118	13
440A	$3\frac{1}{4}$	$2\frac{1}{2}$	$\frac{7}{16}$	$\frac{3}{4}$	$1\frac{5}{8}$	$\frac{7}{8}$	1.41	6.5	16.5
442A	$3\frac{9}{16}$	$2\frac{1}{2}$	$\frac{7}{16}$	$\frac{7}{8}$	$1\frac{5}{8}$	$\frac{15}{16}$	1.5	6.78	17.2
404A	$3\frac{9}{16}$	$3\frac{3}{16}$	$\frac{7}{16}$	$\frac{7}{8}$	$2\frac{5}{16}$	$\frac{15}{16}$	1.8	8.16	20.75
475A	4	$3\frac{3}{8}$	$\frac{1}{2}$	1	$2\frac{3}{8}$	1	2.16	8.75	22.25
460A	$4\frac{1}{2}$	4	$\frac{5}{8}$	1	$2\frac{3}{4}$	$1\frac{1}{4}$	3.1	10	25.4
428A	5	$4\frac{1}{4}$	$\frac{5}{8}$	$1\frac{17}{64}$	3	$1\frac{7}{32}$	3.39	11	27.95
435A	$6\frac{1}{4}$	$5\frac{1}{4}$	$\frac{3}{4}$	$1\frac{5}{8}$	$3\frac{3}{4}$	$1\frac{1}{2}$	5.11	13.8	35
437A	$6\frac{3}{4}$	$6\frac{3}{4}$	$\frac{7}{8}$	$1\frac{5}{8}$	5	$1\frac{3}{4}$	7.2	16.8	42.6
441A	$8\frac{1}{2}$	$7\frac{1}{4}$	$1\frac{1}{4}$	$1\frac{3}{4}$	$4\frac{3}{4}$	$2\frac{1}{2}$	11.1	18	45.7
248A	$9\frac{1}{2}$	$9\frac{1}{2}$	$1\frac{1}{2}$	$1\frac{3}{4}$	$6\frac{1}{2}$	3	16.7	22.5	57.2
122A	$9\frac{1}{2}$	11	$1\frac{1}{2}$	$1\frac{3}{4}$	8	3	19	25.5	64.7
147A	$2\frac{5}{8}$	$2\frac{3}{16}$	$\frac{7}{16}$	$\frac{7}{16}$	$1\frac{5}{16}$	$\frac{7}{8}$	1.14	5.25	13.3
29A	3	$2\frac{1}{2}$	$\frac{1}{2}$	$\frac{1}{2}$	$1\frac{1}{2}$	1	1.48	6	15.25
196A	$3\frac{3}{8}$	$2\frac{13}{16}$	$\frac{9}{16}$	$\frac{9}{16}$	$1\frac{11}{16}$	$1\frac{1}{8}$	1.89	6.75	17.12
78A	$3\frac{3}{4}$	$3\frac{1}{8}$	$\frac{5}{8}$	$\frac{5}{8}$	$1\frac{7}{8}$	$1\frac{1}{4}$	2.3	7.5	19
120A	$4\frac{1}{2}$	$3\frac{3}{4}$	$\frac{3}{4}$	$\frac{3}{4}$	$2\frac{1}{4}$	$1\frac{1}{2}$	3.35	9	22.85
248A	$5\frac{1}{4}$	$4\frac{3}{8}$	$\frac{7}{8}$	$\frac{7}{8}$	$2\frac{5}{8}$	$1\frac{3}{4}$	4.55	10.5	26.7

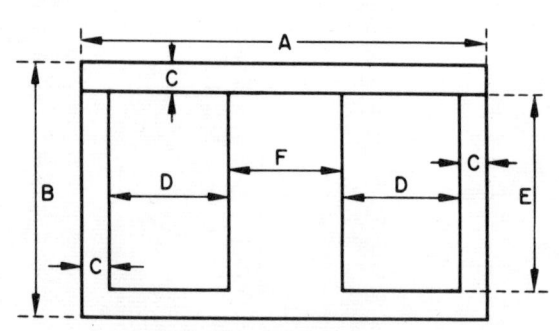

Fig. 20.4.5.

TABLE 4.
Based on 1000 A per square inch.

s.w.g.	Nett area (sq. in)	s.w.g.	Nett area (sq. in)
4	.04227	28	.00017203
5	.03530	29	.00014527
6	.02895	30	.00012076
7	.02433	31	.00010568
8	.02011	32	.00009161
9	.016286	33	.00007854
10	.012868	34	.00006648
11	.010568	35	.00005542
12	.008495	36	.00004536
13	.006648	37	.00003632
13½	.005811	38	.00003827
14	.005027	39	.00002124
14½	.004536	40	.00001809
15	.004072	41	.00001520
15½	.003632	42	.00001256
16	.003217	43	.00001017
16½	.002827	44	.00000804
17	.002463	45	.00000615
17½	.002124	46	.00000452
18	.0018096	47	.000003142
18½	.0015205	48	.00000201
19	.0012566	49	.000001131
19½	.0011341	50	.000000785
20	.0010179		
20½	.0009079		
21	.0006042		
21½	.0007069		
22	.0006158		
22½	.0005309		
23	.0004524		
23½	.0004155		
24	.0003801		
24½	.0003464		
25	.0003142		
25½	.0002835		
26	.0002545		
27	.0002112		

Example

For 1000 A per sq. in, the area of wire e.g. 0.012868 sq. in will take 12.86 A at the above rating.
(Simply multiply the values in the column 'nett area' by 1000 to get the maximum current to allow through the particular gauge of wire.)

CHAPTER 21

Semiconductors

In this chapter we will concern ourselves with a detailed examination of these devices. Although transistors are used extensively throughout industry, there is still a role to be played by valves. The manufacturing techniques of transistors are improving so rapidly, that it is almost impossible in any book, to discuss these devices and remain up to date.

Junction transistors, although improving in performance, will under certain circumstances, cause answers in practice to differ from those theoretically predicted. The M.O.S.T., however, is a device that has some of the desirable properties of both valves and transistors.

These devices are playing an ever increasingly important role, particularly in integrated circuitry.

The half-life of electronic technology — and its associated industrial techniques — is about 5 years or so.

It follows therefore, that the whole subject of transistorised circuitry needs to be re-examined at least once in every 5 years if one is to keep up to date with the subject.

There have been many different symbols used to denote the current gain of transistors. The 'alpha' convention is retained in this book for ease of presentation.

This will become evident in later chapters.

21.1. Junction transistors

A single atom of germanium has a nucleus which is positively charged to 32 electron units surrounded by 32 negatively charged electrons, causing the net charge of the atom to be zero.

Only four of these electrons play any part in the electrical properties of germanium as a conductor; the remaining 28 are tightly bound to the nucleus. These 4 electrons are called the valence electrons. Therefore germanium (and silicon) are said to be tetravalent.

A single atom of germanium

A single crystal of germanium has atoms arranged in a regular pattern; this is a lattice structure, the distance between any two neighbouring atoms being the same. Each atom is linked to its neighbour by sharing valence electrons; therefore each atom is associated with 8 valence electrons.

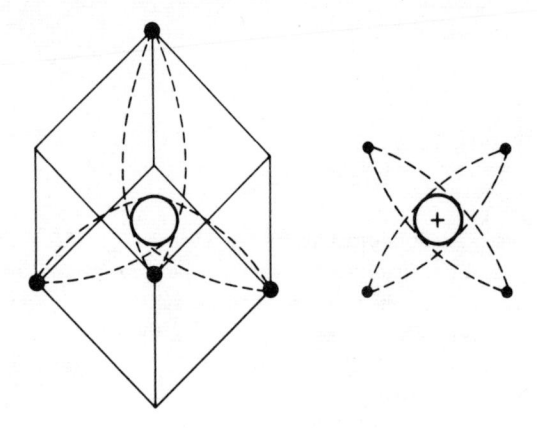

Fig. 21.1.1.

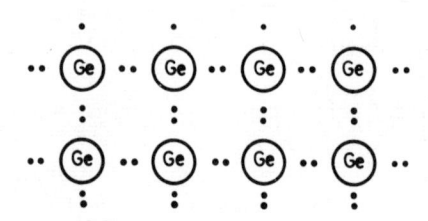

Fig. 21.1.2.

These form covalent bonds, and at low temperatures are fully occupied in binding the atoms. The crystal is an insulator, because there are no free holes or electrons to conduct.

If the temperature is increased, the material absorbs energy: the lattice structure vibrates, and subsequently disturbs the uniformity, so that at any instant a few atoms will have lost an electron whilst others will have gained one. (An atom that has lost an electron has 'vibrated', and literally 'shaken' an electron out of its orbit.)

The 'holes' left by the lost electrons, and also the electrons themselves move about in a random manner. A hole is caused by the loss of an electron, and as an electron has a negative charge a hole may be said to have a positive charge.

If an electric field is applied to the crystal, for example by connecting a battery across the material, the holes and excess electrons drift in opposite directions, and both contribute to conduction of electricity through the crystal. At any instant, the number of holes flowing equals the number of free electrons. These are called hole—electron pairs. At room temperature the lifetime of a hole is approximately $100 \mu s$ before it recombines with an electron.

When used for transistors, germanium contains a small percentage of impurities (1 part in 10^8).

21.2. *n*-type material

If antimony, which has 5 outer-shell or valency electrons and is termed pentavelent, is introduced to germanium, these atoms fit well into the general structure except for one electron (as the original lattice structure needs only 4). This excess electron is very loosely bound as only 4 are needed to keep the structure uniform. Even at room temperature, this excess electron is free to move around through the crystal, and is available for conduction purposes.

The conductivity of this germanium is now somewhat 20 times greater than that of pure germanium at room temperature. This germanium is said to be *n* type, since it has an excess of electrons. (*n* for negative).

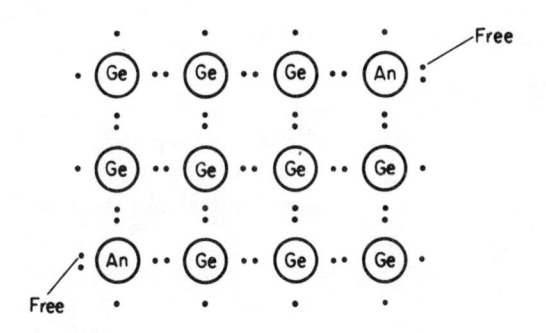

Fig. 21.2.1.

21.3. *p*-type material

If indium, which is trivalent, (group 3) is added to pure germanium, as the impurity, it will lock in quite well with the crystal lattice structure, but will be deficient of one electron for each atom of indium. This has an excess of holes. This is then a *p*-type material. At room temperature, these free holes are available for conduction (*p* for positive).

21.4. Energy level

Germanium has a narrow 'forbidden' band (0.76 e/v).

Normally, all the valence band levels are fully occupied, and all the conduction band levels are completely empty. Since the conduction band has no electrons, it contributes nothing to conduction, and the valence band cannot conduct. One way to make it conduct is to heat it up, when a number of electrons will acquire sufficient energy to fly up to the conduction band.

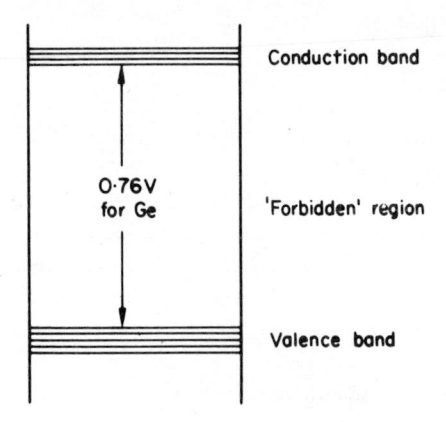

Fig. 21.4.1.

21.5. *n*-type germanium — donor atoms

Consider germanium (group 4), which has been contaminated with a pentavalent material such as arsenic (group 5). The introduction of these foreign atoms modifies the energy level diagram insomuch that an extra permissible level occurs just below the normal conduction band. At low temperatures, this level is occupied by one of the electrons of the foreign atoms; the remaining four electrons of any foreign atom link with the neighbouring germanium atoms as in the pure material. Even at normal room temperatures, the loose electron has sufficient energy to move away from its parent atom, i.e. it moves from its special level up to the conduction band. The electrons normally in the valence band need have less energy in order to detach themselves from the valence band. The germanium will become a much better conductor because of the present of the free electrons in the conduction band. The arsenic atom is called a donor atom.

21.6. *p*-type germanium — acceptor atoms

In this case the germanium has been contaminated with a trivalent material such as gallium (group 3). At increased temperature it is possible that electrons from the new level, will reach the conduction band as the result of the hole current in the valence band. The conduction is due to the moving holes in the valence band. Gallium is an acceptor atom, as it accepts an electron, being in itself deficient of one electron in its neutral state so that four are needed. It may be shown that the holes are less mobile than the electrons, the ratio of mobility being about 2:1 approximately. It is important to realise that *p*-type material possesses a few free electrons (minority carriers), while *n*-type material possesses a few free holes (minority carriers).

21.7. PN junction

If a piece of n-type germanium is brought into intimate contact with a piece of p-type germanium to form a junction in which the main lattice structure is continuous, then the free electrons in the n material will diffuse into the p material, whilst the free holes in the p-type material will spread across the junction into the n section. This migration results in the n section acquiring a positive charge relative to the p section and potential difference or potential gradient will be produced across the junction. This potential difference tends to oppose further migration, and eventually a state of equilibrium will be reached when the net flow of current across the junction is zero.

There will always be some electrons and holes crossing the junction in a random manner and the hole flow will equal the electron flow; therefore the net current flow will be zero.

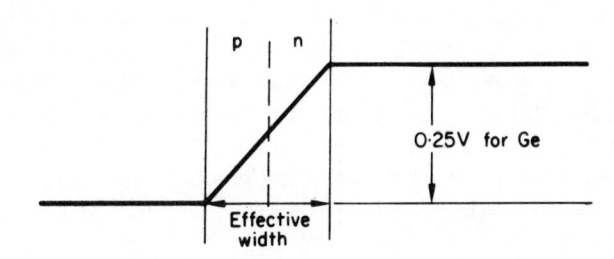

Fig. 21.7.1.

The effective width of the junction is called the depletion layer.

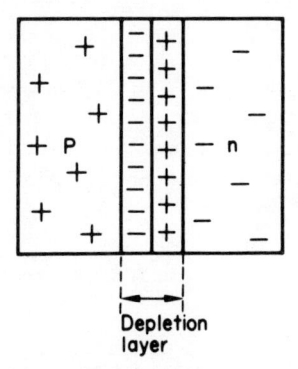

Fig. 21.7.2.

The depletion layer is such that only electrons in the n region having energies greater than a certain value will cross it.

2E

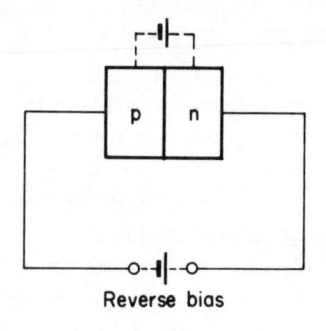

Fig. 21.7.3.

21.8. Reverse bias

The reverse bias will increase the size of the potential barrier. The larger the reverse bias, the steeper the slope of the voltage gradient, and minority carriers will more easily 'slide' down the 'potential hill', but the majority carriers find it hard to climb the steep hill. The external current flowing then is due to minority carriers. The explanation is that to the minority carriers present in the two materials, the potential hill (or gradient) is not a hill to be climbed, but a slope down which they may 'fall'. The minority carriers are therefore attracted across the junction, and an external current flows. This current will necessarily be small, and will be very temperature-dependent.

The energy gained by the minority carriers will be obtained at the expense of the external battery energy. With small reverse bias values, saturation occurs, and further voltage increases do not result in increased reversed current.

The potential difference produced by the reverse bias appears almost entirely across the junction. The field strength at the junction is comparatively large, and if this exceeds a certain value, some valence bonds may be broken and the reverse current rises rapidly. The particular reverse bias value at this point is known at the zener voltage.

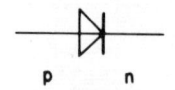

Fig. 21.8.1.

The analogy to a metal rectifier is $p\alpha$ anode and $p\alpha$ cathode.

SEMICONDUCTORS 417

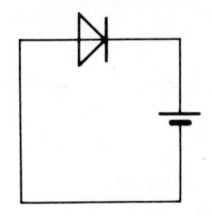

Fig. 21.8.2.

The reverse current is affected by temperature and could be represented by the following graph.

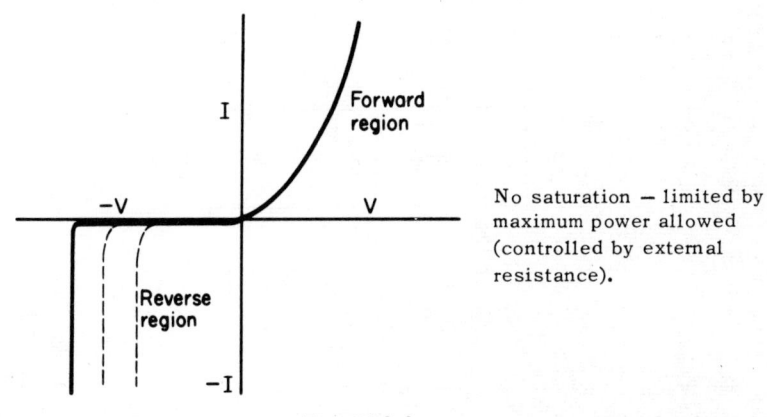

No saturation — limited by maximum power allowed (controlled by external resistance).

Fig. 21.8.3.

Temperature affects reverse current, as current is determined by the generation of hole—electron pairs due to thermal agitation.

21.9. Forward bias

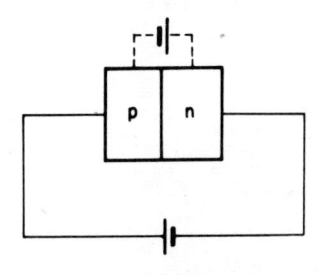

Fig. 21.9.1.

The potential barrier is lowered, and the chances of a hole in the p section crossing the barrier are now very much greater. The forward current is therefore much greater. The reverse hole current will be unaffected.

Quite a small bias, as shown, may increase the forward hole current many times.

In all cases, an equivalent electron flow must pass across the barrier. A typical value of bias necessary to 'wipe out' the voltage gradient and thus cause current to flow, would, for germanium, be slightly in excess of 0.2 V. Therefore the forward bias has lowered the potential gradient set up at the interface due to the migration of holes and electrons, thus allowing more electrons to leave the n material and more holes to leave the p type. This further migration of charges does not raise the barrier again, because the materials are continuously re-supplied by charges at the external connection from the battery.

21.10. The junction transistor

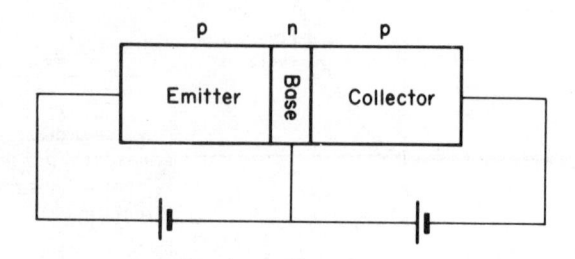

Fig. 21.10.1.

Sandwich a piece of n material between two pieces of p material to form a *pnp* junction transistor. The block diagram is shown in figure 21.10.1; the physical appearance is shown in figure 21.10.2.

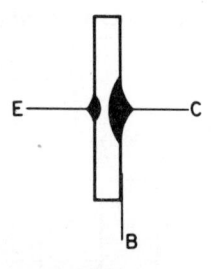

Fig. 21.10.2.

Due to the very heavy contamination of the emitter with respect to the base, holes leave the emitter quite readily and enter the base when the emitter is forward biassed; provided that they have sufficient momentum, they will overcome the potential barrier and enter the collector.

Due to re-combination in the base, a fraction of the holes neutralise electrons in the base; these electrons are subsequently replaced from the external battery. This replacement of electrons in the base constitutes the base current.

Consider figure 21.10.3. The current gain for all configurations is given as
$$\frac{I_{out}}{I_{in}}.$$
For a common base, the current gain is termed α. (Alpha.)

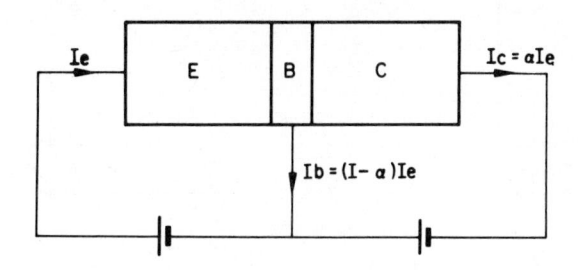

```
         Ie  ┌─────┬───┬─────┐  Ic = αIe
        ────▶│  E  │ B │  C  │────
             └─────┴─┬─┴─────┘
                     │ Ib = (1- α)Ie
                     ▼
        ────────────┤├─────────┤├────
```

Fig. 21.10.3.

From figure 21.10.3. it may be seen that if the base is common, then
$$\alpha = \frac{I_c}{I_e} \text{ say } 0.98$$
for a common emitter, the current gain is termed α'. (Alpha dash.) If, then, the emitter is common to both input and output,
$$\alpha' = \frac{I_c}{I_b} = \frac{I_e}{(1-\alpha)I_e} = \frac{\alpha}{1-\alpha} = 49 \text{ (for our given } \alpha.)$$

For a common collector, the gain is expressed by alpha double dashed,
$$\alpha'' = \frac{I_e}{I_b} = \frac{I_e}{(1-\alpha)I_e} = \frac{1}{1-\alpha} = 1 + \alpha' = 50.$$

It is important to appreciate that a tiny change in the value of α will result in a very large change in α'. For example, suppose $\alpha = 0.99$. This is a very slight change on the 0.98 used previously.
$$\alpha' = \frac{1}{1-\alpha} \quad \alpha' = \frac{1}{1-0.99}$$
hence $\alpha' = 100$. α' has doubled in value for a change in α from 0.98 to 0.99.

There is an analogy between valves and transistors as shown below in figure 21.10.4.

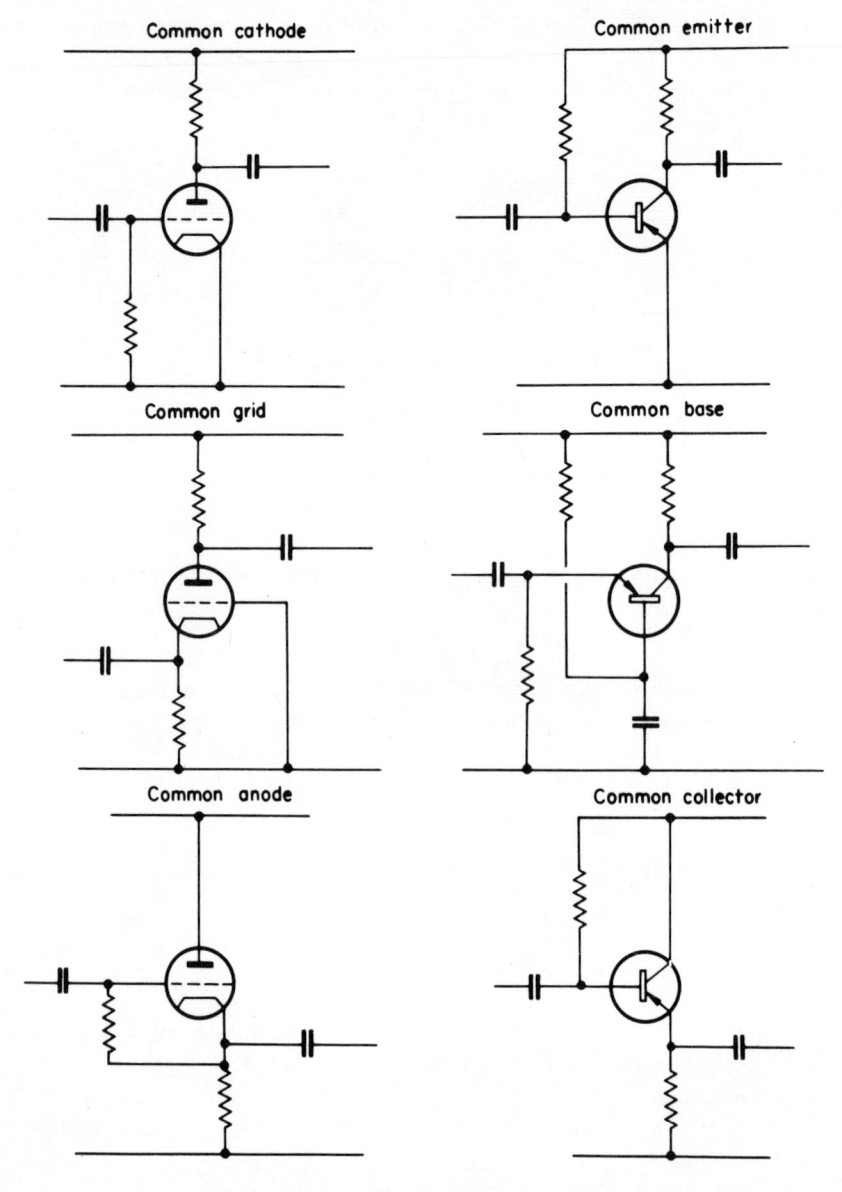

Fig. 21.10.4.

21.11. Input and output resistance. Common base configuration

Although there are many methods of obtaining an expression for the output resistance, this example is based upon the simple equivalent 'T' for the common base. The equivalent circuit is shown in figure 21.11.1.

SEMICONDUCTORS

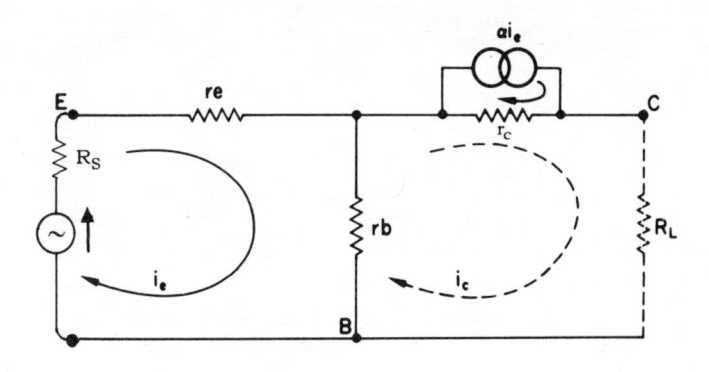

Fig. 21.11.1.

Applying the technique shown in 2.5.3., with $R_L = 0$, an expression for R_{in} becomes

$$R_{in} = \frac{r_e + r_b(r_c - r_m)}{r_b + r_c}$$

where $r_m = \alpha r_c$. The current generator will generate a current of an amplitude alpha times that of the emitter current. It follows therefore, that the current generator will generate a current alpha times that of the current flowing through the emitter resistance, r_e. (This is an important fact particularly at high frequencies as r_e is shunted by a capacitance and current flowing through the capacitor does not contribute to the effective input signal current).

Further, the generated current is seen to be flowing in antiphase through the collector resistance, r_c, with respect to the output current, i_c.

These rather obvious points of interest may be easily overlooked unless some special reference is made. These points should be fully appreciated before proceeding with the remainder of this example.

Before attempting to examine the output resistance, the input terminals should be short-circuited. The output resistance will be determined in the usual manner by applying an external current into the collector and calculating the resultant potential that will be developed between the collector and base. The output resistance will be given from Ohm's law as

$$r_o = \frac{v_o}{i_o}$$

where i_o is the externally applied current. The circuit showing these conditions is shown in figure 21.11.2.

The input is short-circuited. The 'input current' flowing through the emitter resistance, r_e, is shown as i_e. The value of i_e may be established in terms of i_o by the 'load over total' for currents.

$$i_e = i_o \times \frac{(r_b)}{r_b + r_e}$$

hence $v_1 = i_0 \times \dfrac{(r_e r_b)}{r_b + r_e}$

$$v_2 = (i_0 - \alpha i_e) r_c$$
$$= \left(i_0 - \dfrac{\alpha i_0 r_b}{r_b + r_e}\right) r_c$$
$$= i_0 \left(1 - \dfrac{\alpha r_b}{r_b + r_e}\right) r_c$$
$$= i_0 \left(r_c - \dfrac{\alpha r_c r_b}{r_b + r_e}\right) \text{ but let } \alpha r_c = r_m, \text{ therefore}$$
$$v_2 = i_0 \left(r_c - \dfrac{r_m r_b}{r_b + r_e}\right)$$

as $v_0 = v_1 + v_2$

Therefore, $v_0 = i_0 \left(\dfrac{r_e r_b}{r_e + r_b}\right) + i_0 \left(r_c - \dfrac{r_m r_b}{r_b + r_e}\right)$

$$= i_0 \left(\dfrac{r_e r_b}{r_e + r_b} + r_c - \dfrac{r_m r_b}{r_e + r_b}\right)$$

Thus $v_0 = i_0 \left(r_c + \dfrac{r_e r_b}{r_e + r_b} - \dfrac{r_m r_b}{r_e + r_b}\right)$

$$= i_0 \left(r_c + \dfrac{r_e r_b - r_m r_b}{r_e + r_b}\right)$$
$$= i_0 \left(r_c + \dfrac{r_b (r_e - r_m)}{r_e + r_b}\right)$$

and $R_{out} = \dfrac{v_0}{i_0}$

$$= i_0 \dfrac{\left[r_c + \dfrac{r_b (r_e - r_m)}{r_e + r_b}\right]}{i_0}$$

$\therefore \quad R_o = \dfrac{r_c + r_b (r_e - r_m)}{r_e + r_b}$

The expression for the input resistance, R_{in}, was shown to be

$$R_{in} = \dfrac{r_e + r_b (r_c - r_m)}{r_b + r_c} \text{ whilst the expression for}$$

$R_{out} = \dfrac{r_c + r_b (r_e - r_m)}{r_b + r_e}$, if we examine these side by side we can see a certain similarity.

$$r_e + \frac{r_b(r_c - r_m)}{r_b + r_c} \quad \text{and} \quad r_c + \frac{r_b(r_e - r_m)}{r_e + r_b}$$

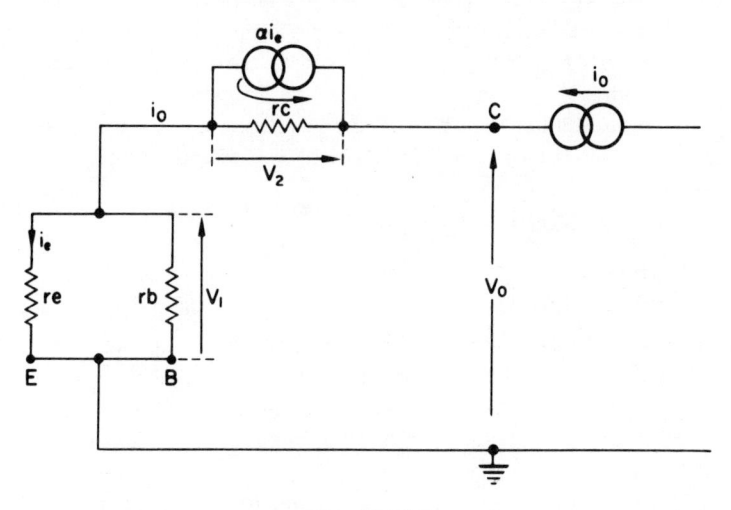

Fig. 21.11.2.

Examination of these formulae will reveal that r_e and r_c are interchanged in the formulae. All other terms remain unchanged. It follows therefore, that the expression for R_o could be written down by inspection of the expression for R_{in} for by duality, they have a common 'shape' depicted by the black boxes shown in figure 21.11.3.

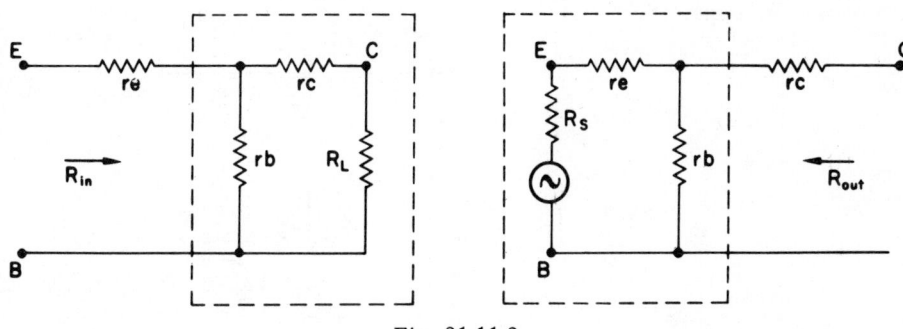

Fig. 21.11.3.

It may be seen that for R_{in}, the resistance r_e, is in series with the effective resistance within the black box, whilst for R_o, the collector resistance, r_c, is in series with the effective resistance, within its black box. Note that the 'fulcrum' represented by r_b is common to both circuits and is equally common in both formulae where it is centrally placed as a 'fulcrum'. Once the input (or output) resistance has been derived for any configuration, the other formula may be written down as by duality as it follows the same pattern as outlined above.

Should a load resistor be connected as shown in figure 21.11.3. the resistor R_L should be added to r_c wherever r_c appears in the expression, similarly, if an input generator is connected to the input, its source resistance R_S should be added to r_e wherever r_e appears in the expression for R_{out}. Nothing should ever be added to the term r_m as r_m is defined as αr_c, hence any resistance added to it would not make sense.

21.12. Bias stabilisation

Thermal runaway

If we consider the collector to base as a reverse biassed diode, we must consider the effects of the minority carriers, which cause the leakage current of the diode and are temperature dependent.

This leakage current can, under certain conditions cause damage to or even destruction of, a transistor, due to a regenerative effect known as 'Thermal Runaway'.

As the temperature is increased, more minority carriers are released, and I_c increases; this in turn raises the temperature of the collector junction, which releases more minority carriers, finally resulting in abnormally high I_c, damaging the transistor ($P = I^2 R$). Transistor circuits must be designed to prevent this thermal runaway by limiting the I_c to a safe value. As usual, the limiting factor will be determined by the values of external resistors in the circuits.

Common base amplifier

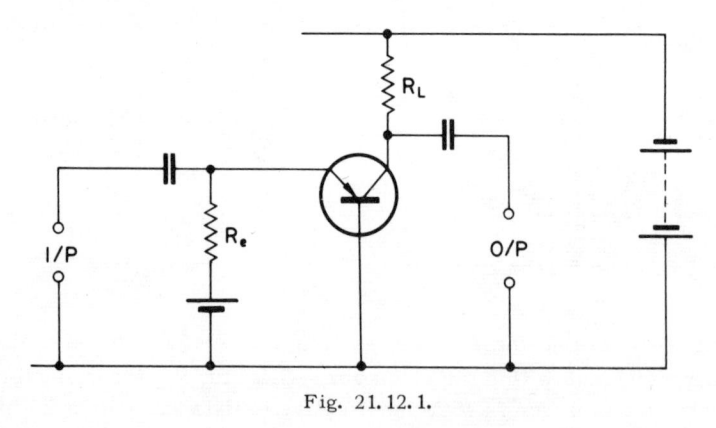

Fig. 21.12.1.

In this circuit, the current gain is α (just less than 1).

The total I_c can be given as

$$I_c = aI_e + I_{co}$$

Where I_{co} = leakage current collector to base. For small junction transistors I_{co} is very small, being about $5\mu A$, at $26°C$. Its rise is reasonably linear with temperature, and may reach $55\mu A$, at $55°C$, but even so this is still negligible compared with aI_e (approximately $1 mA$), and has little effect on the performance.

Provided that R_e and R_L are chosen to keep the collector dissipation below the safe value, the likelihood of 'Thermal Runaway' is reduced; a circuit of this nature is said to have a good d.c. stability.

Common emitter amplifier

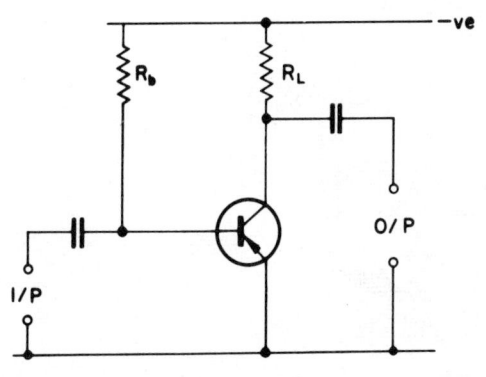

Fig. 21.12.2.

If the base bias resistor R_b is removed from the figure, there is a residual collector current as in the common base amplifier. The collector-base junction is reverse biassed, and hence a reverse current of I_{co} is present.

There is no external base circuit, hence no external base current; therefore it follows that I_e must equal I_{co} and this acts as an input signal which is amplified by the current gain of the common emitter.

$$a' = \frac{a}{1-a}$$

$$I_{co} + aI_{co} = I_{co}$$

Thus the total leakage current is $(l + a')$, usually represented by $I_{co'}$.

As a' can be as great as 50, $I_{co'}$ can be considerably greater than I_{co}. In fact, at $25°C$, $I_{co'}$ can be $250\mu A$ rising to $2.5 mA$ at $55°C$. Protective circuits must therefore be included.

In general, for a common emitter $I_c = a'I_b + I_{co'}$ where $a'I_b$ is the useful component of I_c and $I_{co'}$ the leakage current.

21.13. Stability factor (K)

Suppose there is a change of leakage current $I_{co'}$ due to a change in temperature in an unstabilised circuit. If I_b remains constant, the change in $I_{co'}$ will cause an equal change in I_c.

Now apply a stabilising circuit. Over the same temperature range, the change in collector current is reduced to a smaller value than in an unstabilised circuit.

The ratio of the two changes is known as the 'Stability Factor'.

The stability factor $K = \dfrac{\delta I_c \text{ in a stabilised circuit}}{\delta I_c \text{ in an unstabilised circuit.}}$

Therefore assuming $a' I_b$ is constant,

$$K = \frac{\delta I_c}{\delta I_{co'}}$$

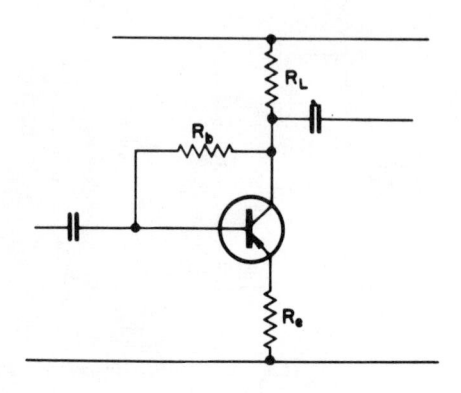

$a' = 50$ say
$R_L = 5\,\text{K}\Omega$
$R_b = 100\,\text{K}\Omega$

Fig. 21.13.1.

then
$$K = \frac{\delta I_c}{\delta I_{co'}} = \frac{1}{1 + \dfrac{a' R_L}{R_b + R_L}} = \frac{1}{3.5}$$

The improvement in temperature stability is $\simeq 3.1$ whereas with the circuit shown in figure 21.13.2. the stability factor is worsened.

$$K = \frac{\delta I_c}{\delta I_{co'}} = \frac{1}{1 + \dfrac{a' R_e}{R_e + R_b}}$$

and if $R_e = 1\,\text{K}\Omega$

$R_b = 100\,\text{K}\Omega$

$$\therefore K = \frac{1}{1 + \dfrac{50}{100}} = \frac{1}{1.5}$$

SEMICONDUCTORS

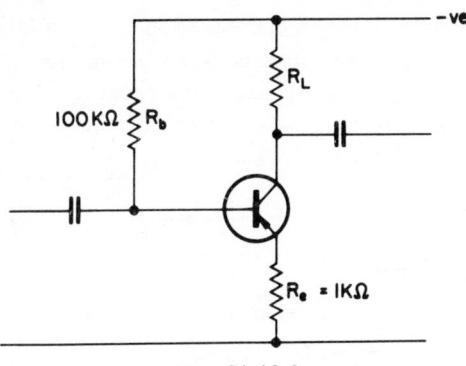

Fig. 21.13.2.

For an unstabilised circuit, $I_c = I_{co'}$ and $K = 1$. For a stabilised circuit δI_c is smaller than δI_{co} and K is less than 1. The smaller K becomes the better the stabilisation.

21.14. Protection circuits for a common emitter amplifier

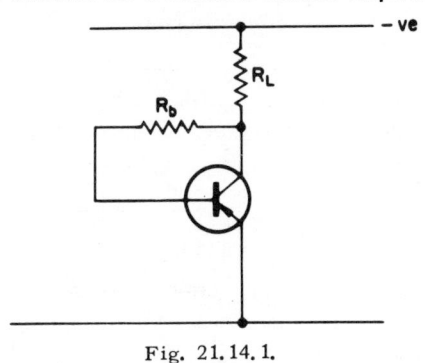

Fig. 21.14.1.

Figure 21.14.1. shows a simple method of improving the thermal stability of a common emitter amplifier.

Suppose the leakage current increased due to a temperature increase. This would cause an increase in the voltage drop across R_L and the collector would become less negative; the base current will fall, and so will the useful component of I_c. This in turn causes the collector to move to a more negative potential, and so some re-adjustment is made, resulting in a total I_c greater than the original value but not as great as if uncompensated.

From the examples previously given, it has been shown that the precise improvement in d.c. stability can be calculated and equal to

$$K = \frac{1}{1 + \dfrac{a' R_L}{R_b + R_L}}$$

As R_L, R_b, and a' are all positive integers, K must be less than unity.

The disadvantage of this system is that some of the output signal voltage is fed back into the input in antiphase, thus causing negative feedback, thus reducing the gain. A method of eliminating the negative signal feedback is shown below in figure 21.14.2.

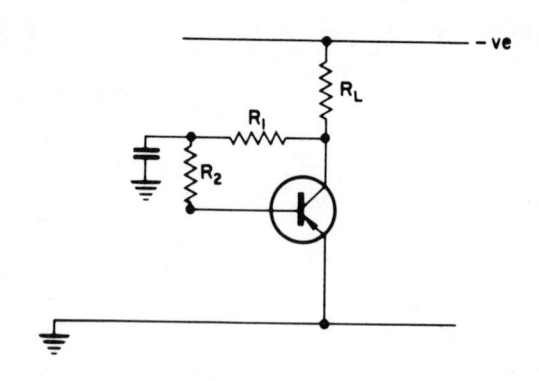

Fig. 21.14.2.

R_1 and R_2 are usually made equal in value, and C_1 should have a reactance which is small compared with the resistance value looking in at the junction of R_1 and R_2 at the lowest frequency to be amplified.

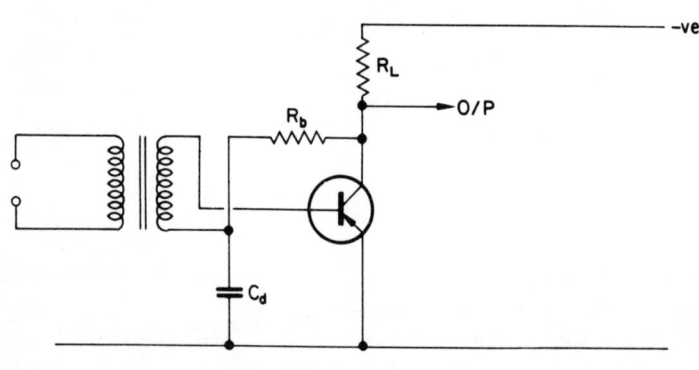

Fig. 21.14.3.

The diagram in figure 21.14.3. shows one method of stabilisation when transformer coupling is employed.

Use of a Potential Divider and Emitter resistor (figure 21.14.4.).

A better method of ensuring good d.c. stability is shown in figure 21.14.4. with a potential divider of R_1 and R_2 and emitter resistor R_e. Capacitor C_d is required for decoupling to prevent degeneration due to negative feedback.

Fig. 21.14.4.

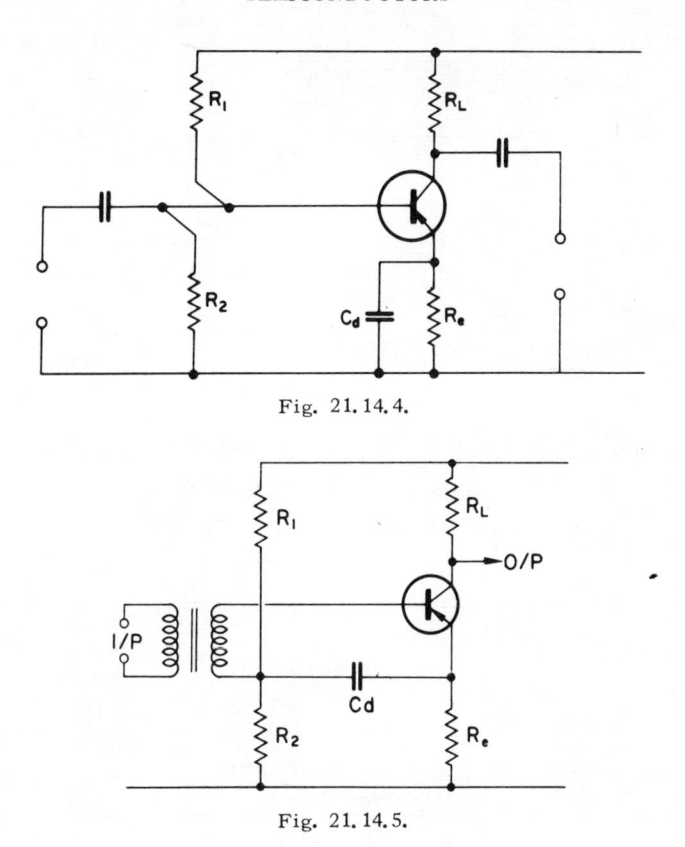

Fig. 21.14.5.

Figure 21.14.5. is similarly an improvement on the circuit diagram shown in figure 21.14.3.

21.15. R_{in} grounded emitter

If we now examine the output resistance looking back into the collector, we can *assume* that the transistor is connected in either the *common base* or *common emitter* configuration. The result will be *identical*. Let us examine this in more detail and prove that this statement is valid.

Figure 21.10.4. shows a common base circuit. We have previously quoted an expression for the input resistance

$$R_{in} \ (C. \ \text{Base}) = r_e + \frac{r_b \ (r_c - r_m)}{r_b + r_c} \tag{1}$$

Now let us derive an expression for the input resistance of a grounded emitter, the circuit for which is also given in figure 21.10.4.

Fig. 21.15.1.

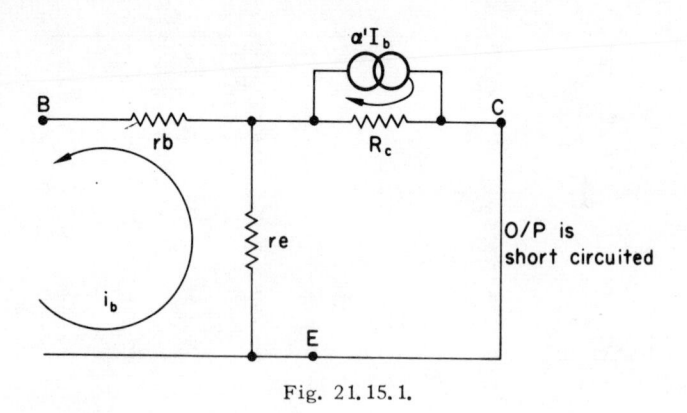

Using the technique discussed in section 2.5, r_b is seen to be constant and may be ignored, and added later. The circuit now becomes as shown in figure 21.15.2.

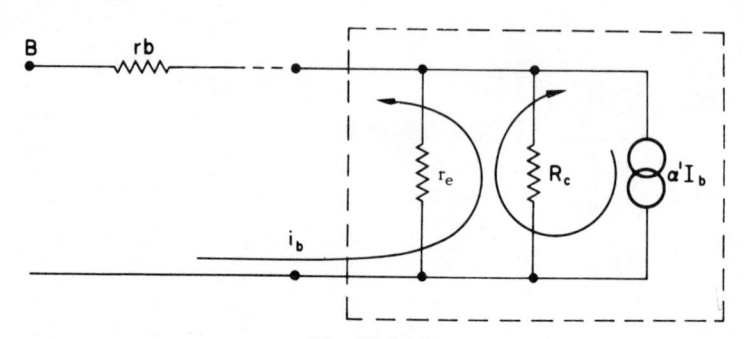

Fig. 21.15.2.

The p.d. across network = $i_b (1 + a') \left(\dfrac{r_e R_c}{r_e + R_c} \right)$

and input resistance = $\dfrac{i_b (1 + a') R_e R_c}{(i_b)(r_e + R_c)} = \dfrac{r_e (R_c + r_m)}{r_e + R_c}$ (where $a' R_c = r_m$)

True input resistance = $r_b + \dfrac{r_e (R_c + r_m)}{r_e + R_c}$ after replacing r_b. (2)

Relationship between a and a', r_c and R_c

Before proceeding further, we should discuss the current generators for both common emitter and common base circuits.

The collector circuit for both a common base and common emitter is given.

SEMICONDUCTORS

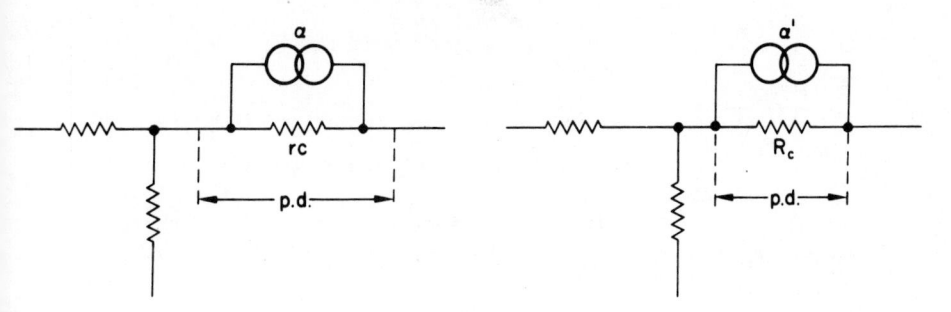

Fig. 21.15.3.

The common emitter current generator is labelled a' and the collector resistor R_c.

The expression $a' = a/(1 - a)$ has previously been discussed and allows us to relate a' in terms of a. The p.d. across each collector resistance must be the same for a given emitter current.

If we assume a current a flowing through r_c and assume a current of a' is flowing through R_c, and equate the resultant potentials, we will establish a relationship between r_c and R_c.

Hence
$$a r_c = a' R_c$$

$\therefore$
$$a r_c = \frac{a R_c}{1 - a}$$

hence $(1 - a) r_c = R_c$ therefore $R_c = (1 - a) r_c$.

Further, as $r_m = a r_c$, $r_m = \dfrac{a R_c}{1 - a} = a' R_c$.

Let us now return to our original problem.

If we now 'reverse' the equations (1) and (2), then by duality, we obtain expressions for output resistance,

$$R_{\text{out}} \text{ (C. Base)} = r_c + \frac{r_b (r_e - r_m)}{r_b + r_e} \tag{3}$$

and
$$R_{\text{out}} \text{ (C. Emitter)} = R_c + \frac{r_e (r_b + r_m)}{r_e + r_b} \tag{4}$$

we should equate (3) and (4) and show that they do equate. This will justify the statement that we can, when considering R_{out} of a collector, assume either a common base or common emitter configuration. Hence, by equating R_{out} for both a common base and a common emitter,

$$r_c + \frac{r_b (r_e - r_m)}{r_e + r_e} = R_c + \frac{r_e (r_b + r_m)}{r_e + r_b} \quad \text{but } R_c = (1 - a) r_c,$$

2 F

$$\therefore \quad r_c + \frac{r_b(r_e - r_m)}{r_b + r_e} = r_c - \alpha r_c + \frac{r_e(r_b + r_m)}{r_e + r_b} \text{ and } \alpha r_c = r_m,$$

$$\therefore \quad r_c r_b + r_c r_e + r_b r_e - r_b r_m = (r_c - r_m)(r_e + r_b) + r_e r_b + r_e r_m$$

$$\therefore \quad r_c r_b + r_c r_e + r_b r_e - r_b r_m = r_c r_e + r_c r_b - r_m r_e - r_m r_b + r_e r_b + r_e r_m$$

and as they equate exactly, we may use wither C.B. or C.E. configurations when dealing with the output resistance, looking back into the collector.

21.16. Grounded collector – input resistance with the output short circuit to a.c.

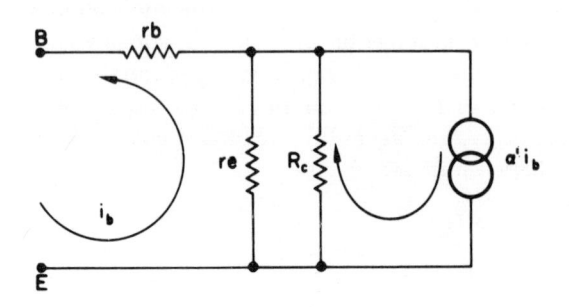

Fig. 21.16.1.

R_b is constant and will be ignored for now, and replaced later.

p.d. across $\quad \dfrac{R_c r_e}{r_c + r_e} = \dfrac{i_b(1 + \alpha') R_c r_e}{R_c + r_e}$

and dividing by input current i_b gives R_{in} at that point.

$$R_{in} = \frac{i_b(1 + \alpha') R_c r_e}{i_b(R_c + r_e)} = \frac{r_e(R_c + r_m)}{r_e + R_c} \quad \text{(where } \alpha' R_c = r_m\text{)}.$$

Now add r_b, which gives

$$R_{in} = \cdot r_b + \frac{r_e(R_c + r_m)}{r_e + R_c}$$

The term r_m in the common emitter and common collector expressions, is written as $+r_m$. This is due to the phase shift in a common emitter configuration and $\alpha' i_b$ flows in a direction opposite to that of the input current i_b, when related to a common base.

21.17. Variations in R_L. (Common base)

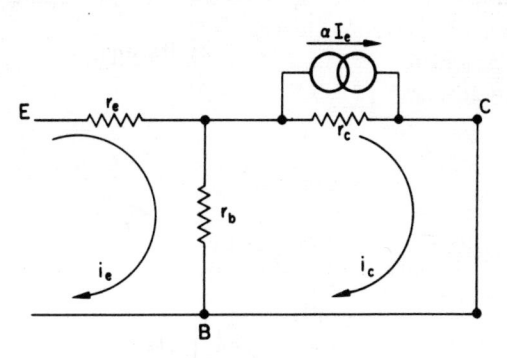

Fig. 21.17.1.

as alpha tends to unity and if $R_L = 0$, then i_c tends to i_e.

Hence $$R_{in} = r_e + \frac{r_b(i_c - i_e)}{i_e} \div r_e \text{ as } i_c = i_e$$

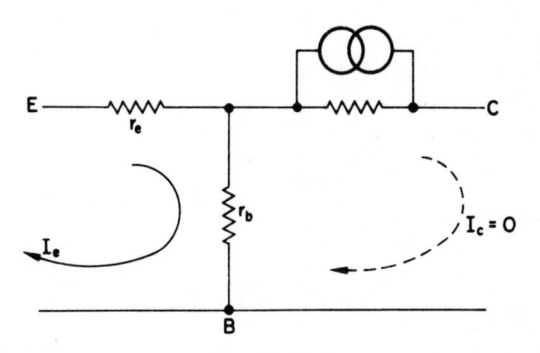

Fig. 21.17.2.

as $R_L \to \infty$, $R_{in} \to r_e + r_b$ (as $i_c \to 0$).

21.18. R_{out} — Grounded collector

Let us next examine R_{out} for a grounded collector.

A grounded base $$R_{in} = r_e + \frac{r_b(r_c - r_m)}{r_b + r_c}$$

Looking into the emitter (for R_{out}, grounded collector) is the same as looking into the emitter for R_{in}, grounded base (output short circuit), as the collector and base are in shunt, in both.

$$R_{out}, \text{ grounded collector} = r_e + \frac{r_b(r_c - r_m)}{r_b + r_c}$$

once more adding external resistors, if any.

Another method of tackling R_{in} is to use Kirchoff's law for closed loops. Consider R_{in} for a grounded collector.

The current generator $a'i_b$ is replaced by its equivalent voltage generator having an e.m.f. of $a'i_b R_C = r_m i_b$.

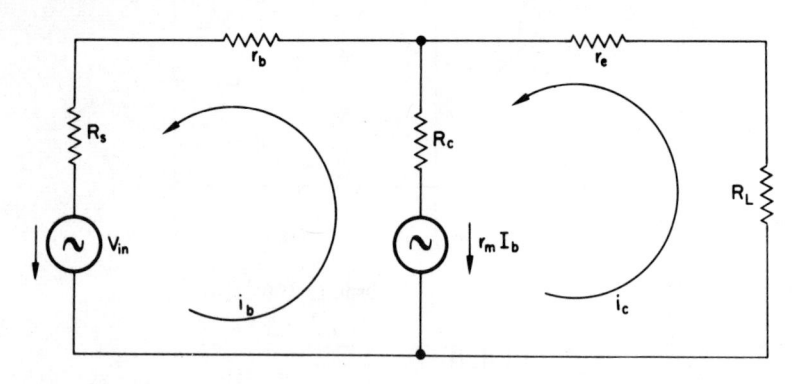

Fig. 21.18.1.

$$v_{in} - r_m i_b = i_b (R_s + r_b + R_c) - i_c (R_c) \qquad (1)$$

$$\therefore \quad r_m i_b = i_c (R_L + r_e + R_c) - i_b (R_c) \qquad (2)$$

from (2), $\quad i_c = \dfrac{(r_m + R_c) i_b}{R_c + r_e + R_L}$ and substituting in (1),

$$v_{in} - r_m i_b = \dfrac{i_b (R_s + r_b + R_c)}{1} - \dfrac{R_c (r_m + R_c) i_b}{R_c + r_e + R_L}$$

Collecting like terms,

$$v_{in} = \dfrac{i_b (R_s + r_b + R_c)}{1} - \dfrac{R_c (r_m + R_c) i_b}{R_c + r_e + R_L} + r_m i_b$$

Ignoring external resistors,

$$R_{in} = r_b + R_c - \dfrac{R_c (r_m + R_c)}{R_c + r_e} + r_m$$

This may be 'simplified' to,

$$R_{in} = r_b + \dfrac{r_e (R_c + r_m)}{R_c + r_e} \text{ or } r_b + \dfrac{(r_e r_c)}{r_c (1 - a) + r_e}$$

With a voltage input, we must remember to add R_s, which is in series with the input resistance to which it is connected. With a collector load, we must add R_L to r_c or R_C.

21.19. Expressions incorporating external resistors

The following circuits will be examined and using the formulae derived; expressions for the *whole circuit* will be written down.

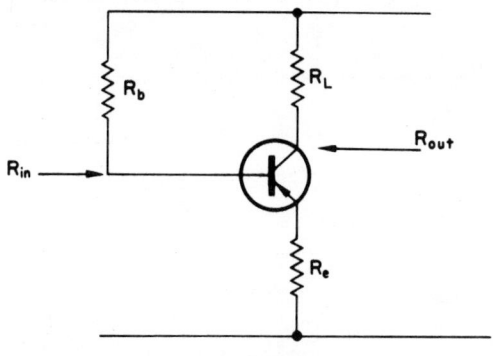

Fig. 21.19.1.

Grounded emitter

$$R_{in} = \left[(r_b) + \frac{(r_e + R_e) + (R_c + R_L + r_m)}{r_e + R_e + R_c + R_L}\right] /\!/ R_b$$

$$R_{out} = \left[(R_c) + \frac{(r_e + R_e) + (r_b + R_b + r_m)}{r_e + R_e + R_c}\right] /\!/ R_L$$

Grounded base

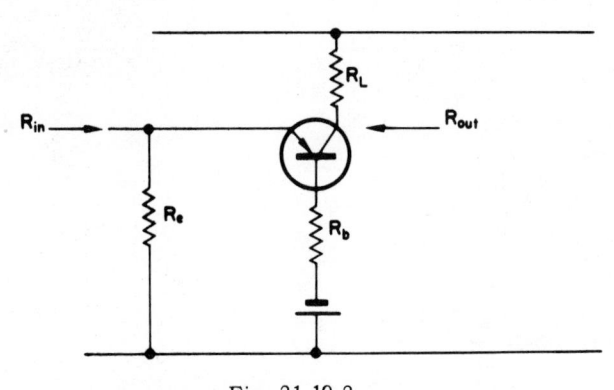

Fig. 21.19.2.

$$R_{in} = \left[r_e + \frac{(r_b + R_b) + (r_c + R_L - r_m)}{r_b + R_b + r_c + R_L}\right] /\!/ R_e$$

$$R_{out} = \left[r_c + \frac{(r_b + R_b) + (r_e + R_e - r_m)}{r_b + R_b + r_e + R_e}\right] /\!/ R_L$$

Grounded collector

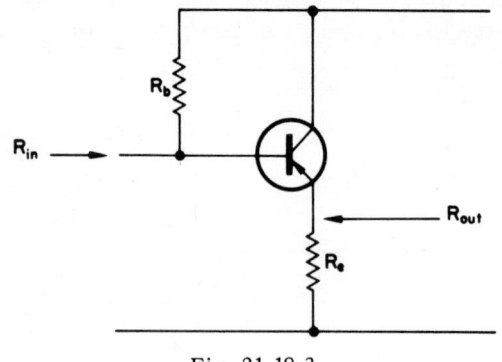

Fig. 21.19.3.

$$R_{in} = \left[r_b + \frac{(r_e + R_e) + (R_c + r_m)}{r_e + R_e + R_c} \right] \parallel R_b \quad (\text{No } R_L)$$

$$R_{out} = \left[r_e + \frac{(r_b + R_b) + (r_c - r_m)}{r_b + R_b + r_c} \right] \parallel R_e \quad (\text{No } R_L)$$

21.20. Voltage gain

Grounded base

The collector load is in shunt with the collector resistance. The effective resistance in collector circuit

$$= \frac{R_L \times r_c}{R_L + r_c}$$

and

$$v_{out} = \frac{i_c \times R_L \cdot r_c}{R_L + r_c}$$

Voltage gain is given by

$$\frac{v_0}{v_{in}} = \frac{i_c}{i_e} \frac{(R_L \cdot r_c)}{(R_L + r_c)} \frac{1}{R_{in}}$$

and as current gain $\frac{i_c}{i_e} = \alpha$

∴

$$\frac{v_0}{v_{in}} = \frac{\alpha R_L \cdot r_c}{R_{in}(R_L + r_c)}$$

SEMICONDUCTORS

Call this shunt combination of

$$\frac{R_L \cdot r_c}{R_L + r_c}, R_{out}, \text{ for instance.}$$

Then $\qquad v_{out} = i_c \times R_{out}$

and $\qquad v_{in} = i_e \times R_{in}$

hence the voltage gain $= \dfrac{i_c}{i_e} \times \dfrac{R_{out}}{R_{in}} = \alpha \dfrac{R_{out}}{R_{in}}$

(but as R_L is $\ll r_c$) say $5\,K\Omega$, as opposed to $1\,M\Omega$, ignore r_c.

$\therefore$ voltage gain $\doteqdot \alpha \dfrac{R_L}{R_{in}}$ and as $\alpha \to 1$ so the voltage gain is dependent upon R_L/R_{in}.

If $R_L = 5\,K\Omega$ and $R_{in} = 20\,\Omega$ and $\alpha = 1$, then the voltage gain

$$\doteqdot \frac{5000}{20} \doteqdot 250.$$

21.21. Power gain

Power gain $= \dfrac{P_{out}}{P_{in}}$

$$P_{in} = \frac{v_{in}^2}{R_{in}} \text{ and } P_{out} = \frac{v_{out}^2}{R_{out}}$$

Power gain $= \dfrac{P_{out}}{P_{in}} = \dfrac{(v_{out})^2}{R_{out}} \bigg/ \dfrac{(v_{in})^2}{R_{in}}$

Power gain $\therefore = \dfrac{v_{out}^2}{R_o} \times \dfrac{R_{in}}{v_{in}^2}$

$= \dfrac{v_{out}^2}{v_{in}^2} \times \dfrac{R_{in}}{R_o}$ and if we substitute $\dfrac{v_{out}^2}{v_{in}}$ from the

above formulae,

power gain $= \dfrac{\alpha^2 R_{out}^2}{R_{in}^2} \times \dfrac{R_{in}}{R_o}$ this may be reduced to

power gain $= \dfrac{\alpha^2 R_o}{R_{in}} \simeq \dfrac{R_{out}}{R_{in}}$ as $\alpha \to 1$.

Grounded emitter

Voltage gain $= \dfrac{v_0}{v_{in}}$

$v_0 = i_c (R_{out})$

and $v_{in} = i_b (R_{in})$

hence the gain $= \dfrac{i_c R_0}{i_b R_{in}} = a' \dfrac{R_0}{R_{in}}$ for most practical purposes.

Power gain $= \dfrac{P_{out}}{P_{in}} \simeq \dfrac{v_{out}^2}{v_{in}^2} \times \dfrac{R_{in}}{R_{out}} \simeq (a')^2 \dfrac{R_0}{R_{in}}$

Common collector

Voltage gain $= \dfrac{v_0}{v_{in}} = \dfrac{(1 + a') R_e'}{R_{in}}$ (where R_e' is the load resistor in shunt with the output resistance.)

This is approximately $\dfrac{a' R_e'}{R_{in}}$ and as $R_{in} \gg R_e'$ the voltage gain must be less than 1.

Power gain $= \dfrac{v_0^2}{R_e'} \times \dfrac{R_{in}}{v_{in}^2} \simeq \dfrac{R_{in}}{R_e'}$ which is approximately $a' \times R_e'$. as the voltage gain is almost unity.

21.22. Current gain

Consider a grounded base shown in figure 21.22.1.

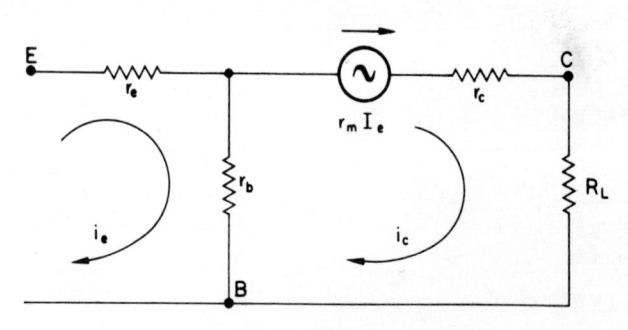

Fig. 21.22.1.

where $r_m i_e = a i_e \cdot r_c$. Then $r_m i_e = i_c (r_b + r_c + R_L) - i_e r_b$

hence $\dfrac{i_c}{i_e} = \dfrac{r_m + r_b}{r_b + r_c + R_L}$ (which is the current gain of the circuit.)

SEMICONDUCTORS

Now consider a grounded emitter:

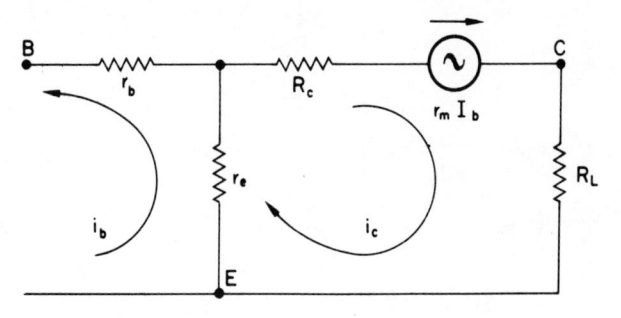

Fig. 21.22.2.

$$r_m i_b = i_c (r_e + R_c + R_L) + i_b r_e$$
$$\therefore \quad i_b (r_m - r_e) = i_c (r_e + R_C + R_L) \text{ and if } R_L = 0,$$

hence
$$\frac{i_c}{i_b} = \frac{r_m - r_e}{r_e + R_c}$$

and is the current gain of the circuit, and would be α' provided that $\alpha/(1 - \alpha) = \alpha'$, but α' is really $-\alpha/(1 - \alpha)$ due to phase reversal,

hence
$$-\frac{i_c}{i_b} = \frac{r_e - r_m}{r_e + R_c}$$

and is the current gain for a grounded emitter.

Current gain – grounded collector

The common collector gain may be calculated in several ways, either by the technique given above, or else by using the two expressions derived previously, containing i_e and i_b,

$$\left(\alpha'' = \frac{i_e}{i_b} \right),$$

21.23. A simple d.c. amplifier

A simple directly connected amplifier using a $p-n-p$ and an $n-p-n$ transistor is shown in figure 21.23.1.

A positive going input to V_{T_1} base causes V_{T_1} current to increase in the direction shown in the figure. This current is dragged from the base of V_{T_2} thus causing V_{T_2} collector current to increase. This increase causes a voltage drop across R_L thus the collector potential 'falls' in a positive going direction.

For a given input, R may be connected to divert some of V_{T_1} collector

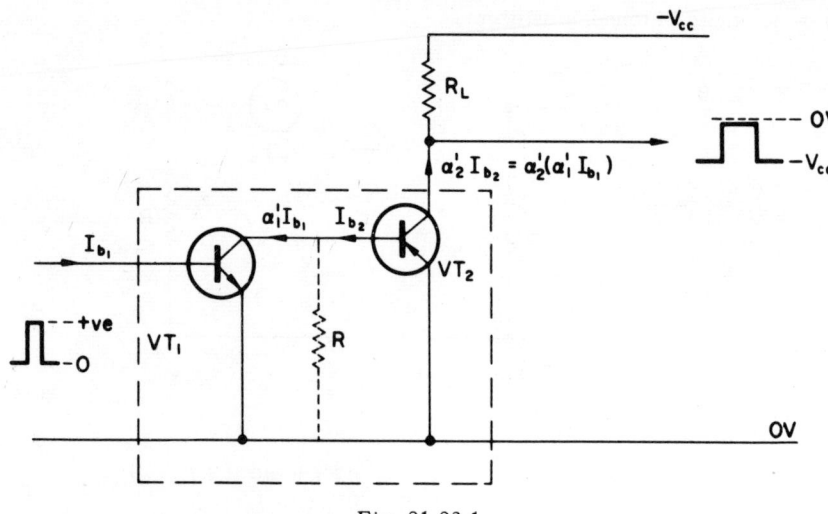

Fig. 21.23.1.

current thus ensuring that the maximum base current of V_{T_2} is restricted to a safe value.

This circuit is temperature conscious. Any d.c. drift in the V_{T_1} stage will be amplified by V_{T_2}.

21.24. Gain controls

A junction transistor requires a current input.

A valve requires a voltage input. Figure 21.24.1. shows two arrangements for gain control.

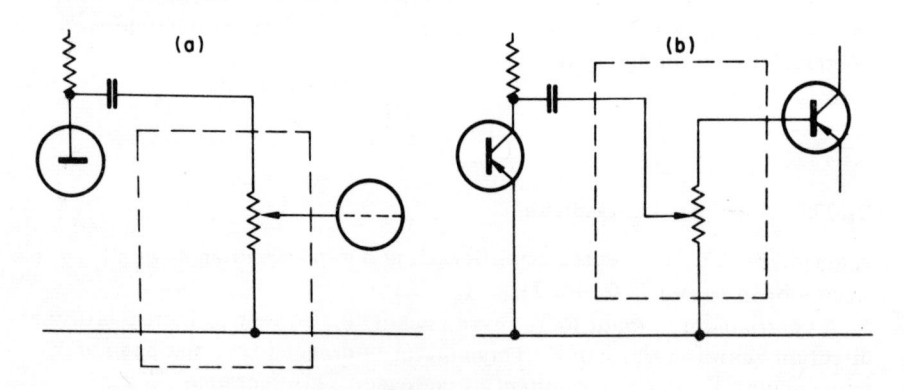

Fig. 21.24.1.

Network (a) is a potential divider and is commonly used in valve circuits. The source impedance must be low compared to the load impedance and thus the network provides a voltage output.

Network (b) is a current divider and is often used in transistor circuits. The source impedance must be high compared with the load impedance thus the network provides a current output.

21.25. Simple transistor amplifier considerations

Let us now consider a few practical methods of deriving resistor values in three basic amplifiers. Each amplifier will be slightly different and will provide us with some useful revision. Consider figure 21.25.1.

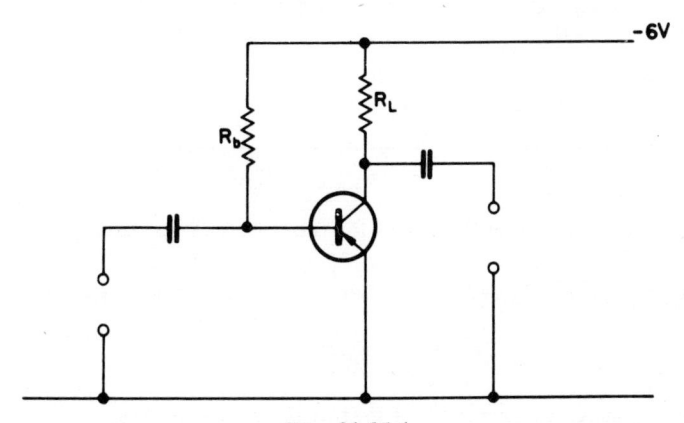

Fig. 21.25.1.

Suppose we were asked to run the device at 1 mA collector current and that we should have a collector–emitter potential of -3 V. The transistor is assumed to have an α of 50.

The base current $= \dfrac{I_c}{\alpha'} = \dfrac{1\,\text{mA}}{50} = 20\,\mu\text{A}.$

With -3 V across the device, and with a supply of -6 V, there must be -3 V across R_L.

The 1 mA collector current flows through R_L hence

$$R_L = \frac{3\,\text{V}}{1\,\text{mA}} = 3\,\text{K}\Omega.$$

Assuming a base-emitter voltage of zero (zero bias) there will be 6 V across R_b, and with $20\,\mu\text{A}$ flowing through it,

$$R_b = \frac{6\,\text{V}}{20\,\mu\text{A}} = 300\,\text{K}\Omega.$$

Now let us consider the amplifier in figure 21.25.2. This has R_b

connected between collector and base to provide compensation for temperature drift.

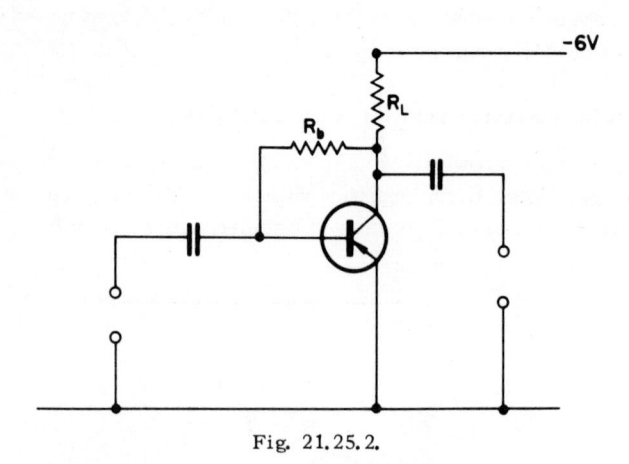

Fig. 21.25.2.

Assuming the same conditions as for the previous amplifier, $R_L = 3\,\text{K}\Omega$ as before.

Once more, assuming zero bias, there will be 3 V across R_b, hence

$$R_b = \frac{3\,\text{V}}{20\,\mu\text{A}} = 150\,\text{K}\Omega.$$

The fact that R_b is exactly half that of the previous example is coincidental. The value will depend of course, upon the collector-base potential.

The third amplifier is the most stable of all three. The circuit is given in figure 21.25.3.

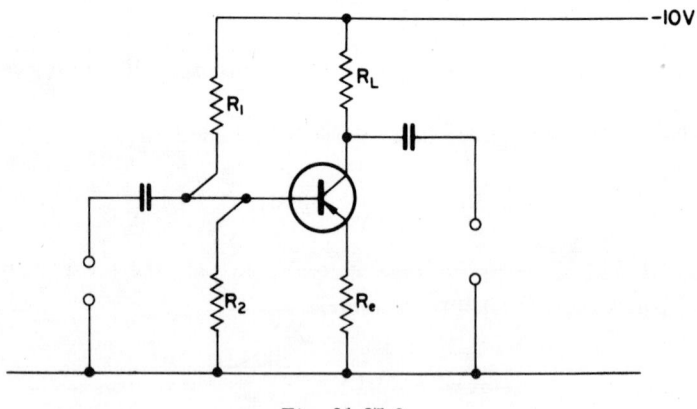

Fig. 21.25.3.

Let us assume similar conditions for this amplifier.

With $V_{ce} = 4\,\text{V}$, there will be 6 V to be shared amongst R_L and R_e. Suppose we decided to assume 1/5th of the supply across R_e, i.e. 2 V, then there would be 4 V across R_L.

Therefore $\qquad R_L = \dfrac{4\,\text{V}}{1\,\text{mA}} = 4\,\text{K}$, say $3.9\,\text{K}\Omega$.

Assuming $I_c = I_e$, $\quad R_e = \dfrac{2\,\text{V}}{1\,\text{mA}} = 2\,\text{K}\Omega$, say $1.8\,\text{K}\Omega$.

The bleed current through R_1 and R_2 should be $> 10 I_b$, hence assume I bleed $= 1\,\text{mA}$, as $I_b = 20\,\mu\text{A}$.

Assuming $V_{be} = 0$, i.e., zero bias, then there will be 2 V across R_2.

Hence $\qquad R_2 = \dfrac{2\,\text{V}}{1\,\text{mA}} = 2\,\text{K}\Omega$.

There will be $1\,\text{mA} + 20\,\mu\text{A}$ through R_1, say $1\,\text{mA}$. Hence R_1 must have 8 V across it therefore
$$R_1 = \dfrac{8\,\text{V}}{1\,\text{mA}} = 8\,\text{K}\Omega.$$

This has been a simple approach to deriving resistor values for a given d.c. condition. It also ignored the stability factor K.

21.26. Measurements of I_c/V_c characteristics

Figure 21.26.1. shows a typical arrangement for plotting the collector characteristics for a transistor operating in the common emitter configuration.

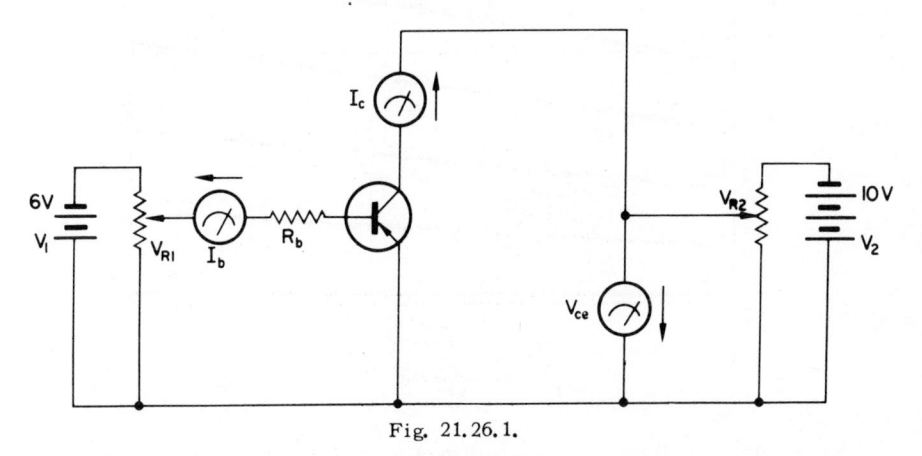

Fig. 21.26.1.

R_b should be calculated to ensure that I_b can never exceed the maximum rated base current.

$$R_b = \frac{V_1}{I_{b\,max}}$$

assuming zero input resistance of the transistor.

In the example shown, for an OC71, and with $V_1 = 6\,V$, R_b should be 50 KΩ minimum to restrict I_b to $120\,\mu A$.

V_{R_1} should be sufficiently low in value to ensure that any change in base current will have a minimal effect upon the potential between the slider and earth.

V_{C_e} is recorded on M_3, I_c on M_2 and I_b on M_1.

With I_b set to zero, V_{C_e} is increased in small steps from zero to $-9\,V$. I_c should be plotted for each value of V_{C_e}.

The above should be repeated for $I_b = 10\,\mu A$.

Repeating several times more for different values of I_b up to $120\,\mu A$ will provide a family of curves as shown in figure 21.26.2.

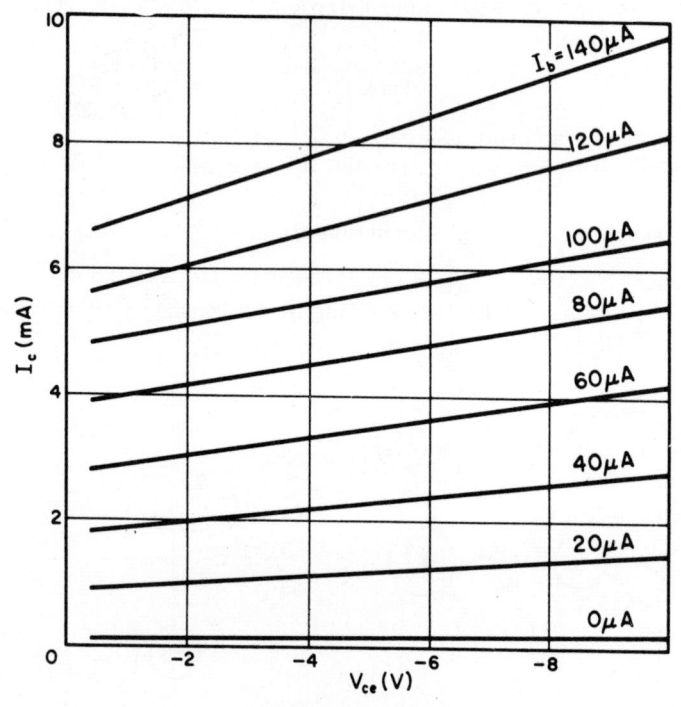

Fig. 21.26.2.

Below $V_{C_e} = -0.5\,V$, the knee of each characteristic is seen to be as shown in figure 21.26.3.

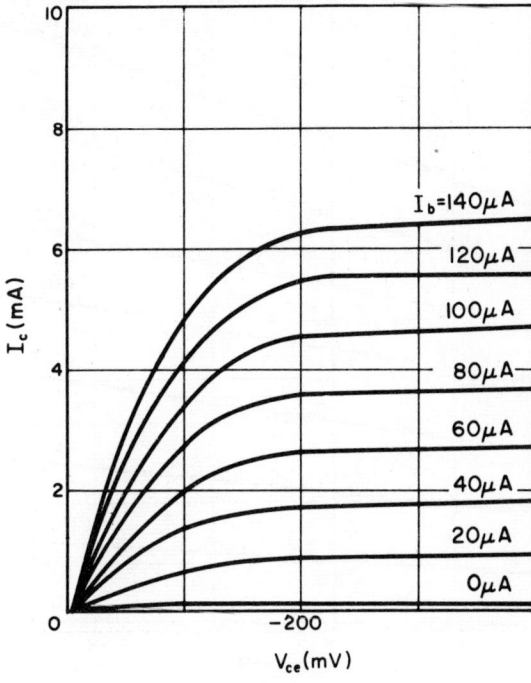

Fig. 21.26.3.

a' may be obtained from these characteristics, the method of doing is as follows:

Refer to figure 21.26.4.

Draw a feint vertical line at a given V_{C_e}, say -4 V.

Choose a suitable I_b, say $60\,\mu$A (point A) and draw a line as shown to $I_b = 20\,\mu$A (point B). This will be the change in I_b. Draw a horizontal line from point (A) to point (C) and from point (B) to point (D). The change from $C - D$ is the change in I_c as shown in figure 21.26.4.

$$a' = \left|\frac{\delta I_c}{\delta I_b}\right|_{V_{ce}} = \left|\frac{2\,\text{mA}}{40\,\mu\text{A}}\right|_{-4\,\text{V}} = \frac{2000\,\mu\text{A}}{40\,\mu} = 50.$$

We can use the characteristics to determine a suitable operating point for the amplifier shown in figure 21.26.5.

If operating conditions are not specified, and are to be chosen, it is important to draw a maximum power curve, as for valves, and ensure that subsequent load lines do not cut through the curve.

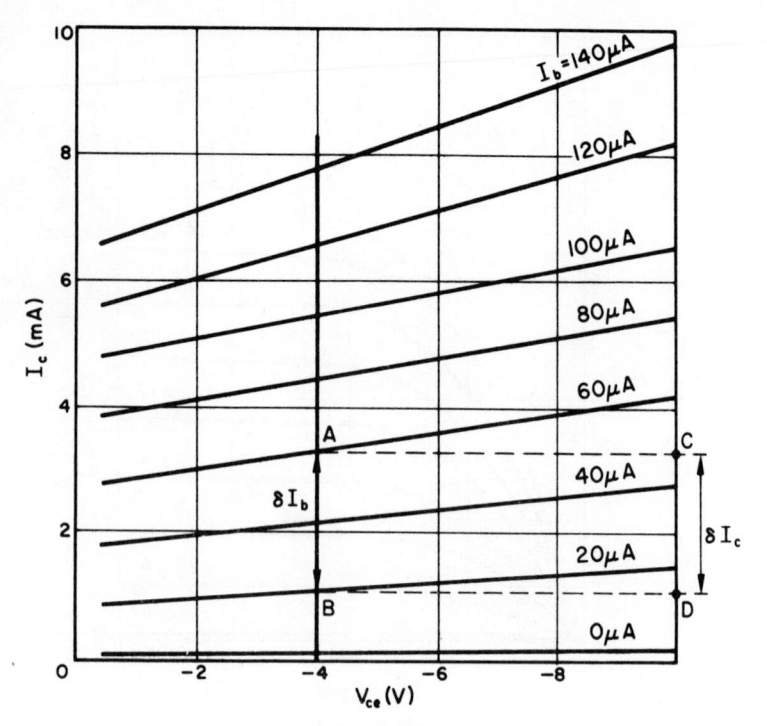

Fig. 21.26.4.

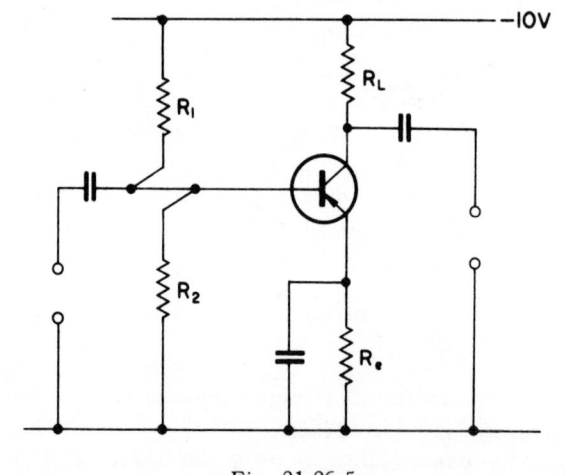

Fig. 21.26.5.

Let us assume the d.c. conditions of (3,1). Referring to figure 21.26.6., the operating point is marked at (−3, 1.0) and is shown as point P.

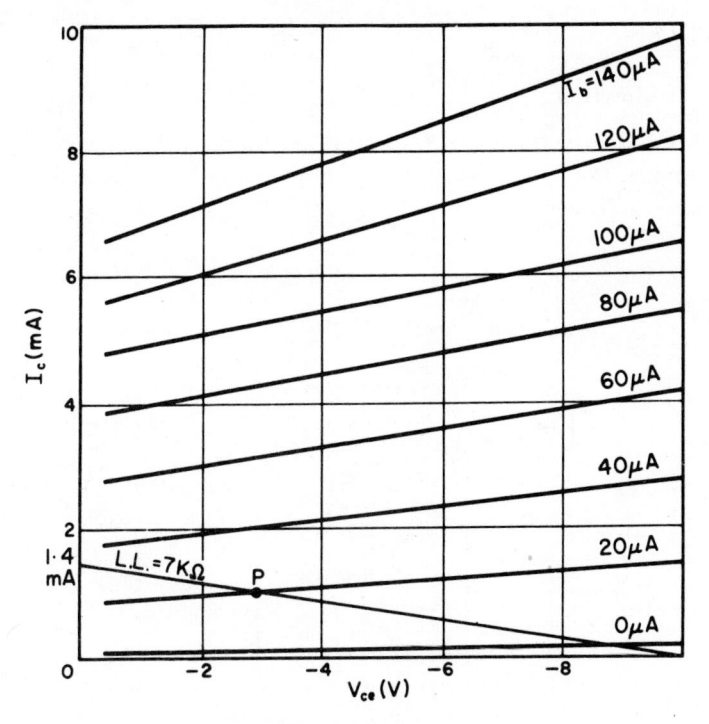

Fig. 21.26.6.

The supply is −10 V hence the load line should be drawn from −10 V, through point P and terminated at the I_c axis as shown.

The 'short circuit' current is seen to be 1.4 mA.

Thus $$R_L + R_e = \frac{10\,\text{V}}{1.4\,\text{mA}} \simeq 7\,\text{K}\Omega.$$

If we are to have 1/5th of the supply across R_e, then

$$R_e = \frac{2\,\text{V}}{1\,\text{mA}} = 2\,\text{K}\Omega. \text{ Hence } R_L = 5\,\text{K}\Omega.$$

I_b is seen to be 20 μA on the graph.

The input characteristics for this OC71 shows that for an I_b of 20 μA, a V_{be} of 0.1 V is required.

This is small and can often be ignored.

With a bleed current through $R_1 + R_2$ of 1 mA, ignoring the small I_b,

$$R_2 = \frac{2\,\text{V}}{1\,\text{mA}} = 2K \text{ and } R_1 = 8\,\text{K}\Omega.$$

Standard values would normally be used and the circuit conditions recalculated.

21.27. Clamping

When the output voltage of a transistor amplifier has to be restricted to within rather precise limits, a pair of diodes may be connected as shown in figure 21.27.1.

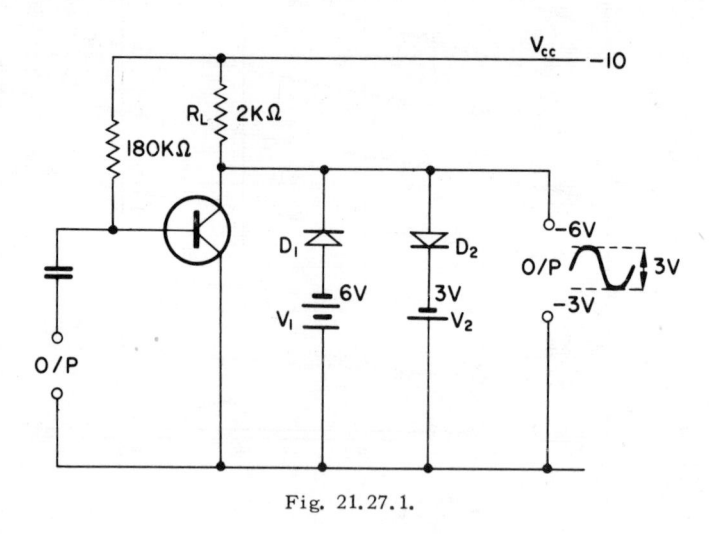

Fig. 21.27.1.

d.c. conditions

A 2 KΩ load line is drawn on the characteristics in figure 21.27.2. With approximately 10 V across R_b, $I_b \simeq 56\,\mu A$. The resultant operating point shows. $I_b = 56\,\mu A$. $I_c = 2.8\,mA$. $V_{R_L} = 5.6\,V$ and $V_{C_e} = 4.4\,V$.

Signal conditions

As the transistor collector rises towards V_{cc}, it reaches the potential V_1. The diode D_1 will conduct and clamp the collector to V_1 potential of $-6\,V$. D_1 will remain conducting until the collector potential falls below $-6\,V$ hence D_1 will be non-conducting.

When the collector falls to the level of V_2, the diode D_2 will conduct this effectively placing the battery V_2 between collector and earth, thus clamping the collector potential to the level V_2 of $-2\,V$. Any increase in collector current due to increase in base current cannot affect the clamped condition. The characteristics are shown in figure 21.27.2. complete with clamping conditions and output waveform.

A two stage phase invertor amplifier

Figure 21.27.3. shows the circuit of two transistors connected as an invertor stage. Both transistors are operating as common emitters.

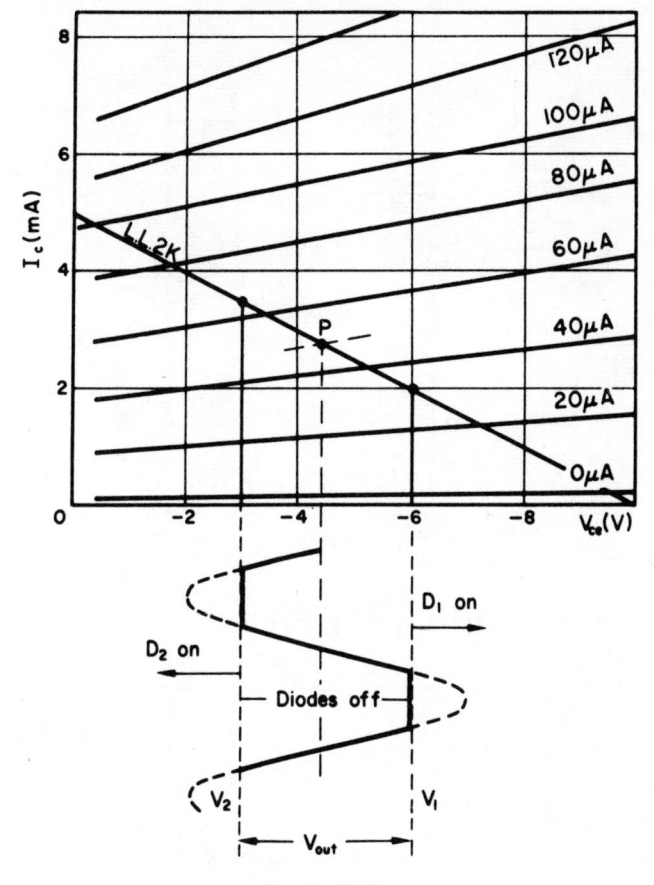

Fig. 21.27.2.

Assuming that the input signal takes V_{T_1} base negative. The collector of V_{T_1} moves in a positive direction and provides both (a) one output in antiphase to the input and (b) a positive going input to the base of V_{T_2}. This causes the collector of V_{T_2} to move negatively and provides the second output signal. This is in phase with the input.

The resistor R_1 provides d.c. bias for V_{T_1} as discussed earlier. R_5 provides similar bias for V_{T_2}. R_4 is chosen to provide the signal current drive for V_{T_2} base.

The emitter CR combination for both transistors provide d.c. bias and a.c. decoupling to prevent degeneration due to feedback.

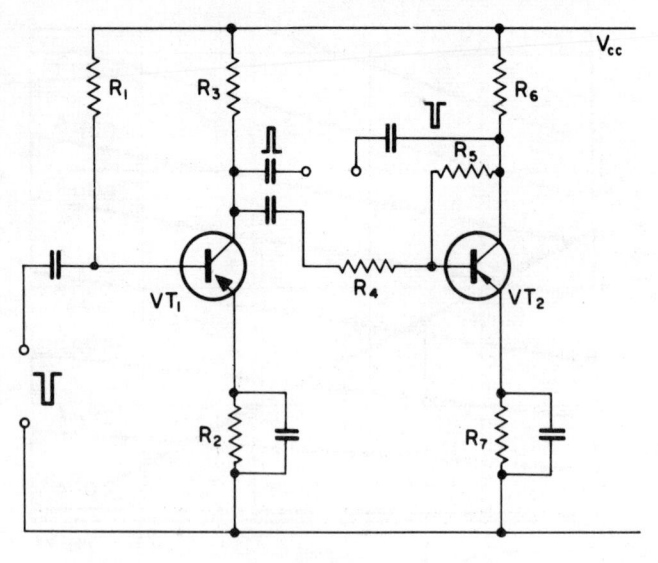

Fig. 21.27.3.

21.28. A small transformer-coupled amplifier

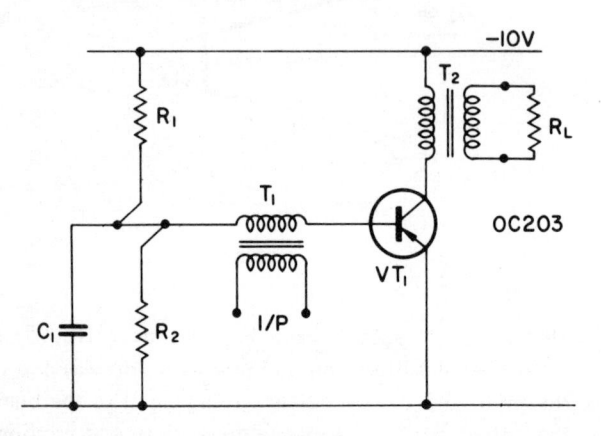

Fig. 21.28.1.

The theoretical efficiency of a resistive loaded amplifier in class A, is 25% max. With transformer coupling as shown in figure 21.28.1., the theoretical efficiency is 50% max.

The theoretical value for efficiency assumes ideal characteristics for valves and transistors. We will discuss the amplifier shown in the figure 21.28.1., and develop the circuit and attempt to reach near-maximum efficiency of 50%.

% Efficiency = $\dfrac{\text{output}}{\text{input}} \times 100$. Where the output is in r.m.s. values and the input is the steady d.c. power taken by V_{T_1} from the -10 V supply.

R_1 and R_2 form a potential divider, and as shown earlier, provides the necessary V_{b_e} for a given value of base current bias.

C_1 decouples R_2 and one side of the secondary of T_1, thus ensuring that a steady state is maintained at that point whilst signal currents are flowing in the circuit.

T_2 couples the transistor to the load resistor R_L and will be considered as having zero winding resistance.

The alternating component of the collector current flowing in the primary winding of T_2 will cause an alternating e.m.f. to be induced in the secondary winding across which the load resistor R_L is connected.

The resulting alternating current flowing through the load resistor, and this may be a loudspeaker in practice, produces the output voltage across it.

The transformer T_1 allows an input signal source to be connected without affecting the d.c. conditions and perhaps more important, is isolated from any d.c. thus ensuring no damage to the input signal generator.

The signal generator may be a previous amplifier stage where the T_1 primary may be the load of that stage.

The base-emitter potential of V_{T_1} is determined by the relative values of R_1 and R_2 and as T_1 is assumed to have negligible secondary winding resistance, is unaffected by the inclusion in the circuit of the input transformer.

Figure 21.28.2. shows a simplified and ideal situation allowing the operating point P to be positioned such that the peak change in collector voltage and collector current becomes twice that of the quiescent values.

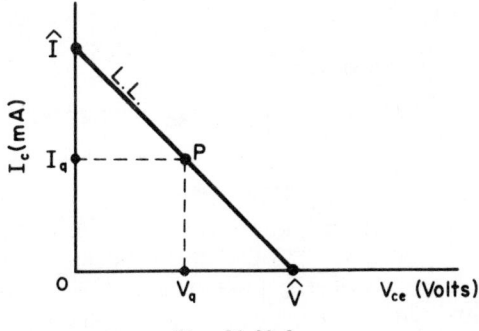

Fig. 21.28.2.

The power out for maximum input

$$= \frac{(V_{max} - V_{min})}{2\sqrt{2}} \times \frac{(I_{max} - I_{min})}{2\sqrt{2}}$$

$$= \frac{(V_{max} - V_{min}).(I_{max} - I_{min})}{8} \text{ W}$$

The power in $= V_q I_q$

and if $\quad V_{max} = 2v_q$ and $V_{min} = 0$
and if $\quad I_{max} = 2I_q$ and $I_{min} = 0$

then power out $\quad = \dfrac{(2V_q)(2I_q)}{8}.$

then % efficiency $= \dfrac{P_{out}}{P_{in}} \times 100$

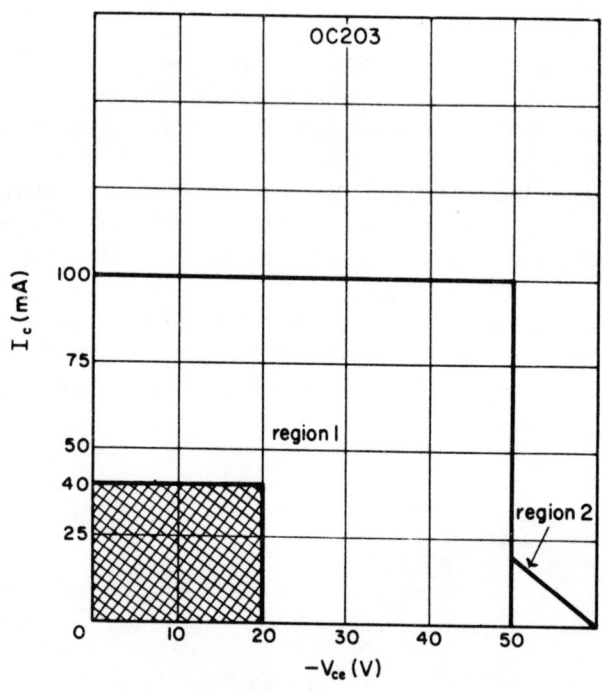

Fig. 21.28.3.

ABSOLUTE MAXIMUM COLLECTOR-EMITTER VOLTAGE PLOTTED
AGAINST COLLECTOR CURRENT

Region 1. Permissible area of operation under all conditions of base drive.
Region 2. For operation in this region the circuit must be capable of providing reverse current bias.

SEMICONDUCTORS

$$= \frac{(2V_q)(2I_q)}{8 \cdot (V_q I_q)} \times 100 = 50\%.$$

We intend now to examine a simple amplifier and compare its efficiency with the theoretical maximum of 50%.

Deriving component values

The amplifier is shown in figure 21.28.1.

Figure 21.28.3. shows the maximum V_{c_e} plotted against I_c. We must work within the area shown as Region 1.

As we have $-10\,\text{V}$ available for our supply, and as we need to 'swing' the collector from $0 - 20\,\text{V}$, in an attempt to obtain an efficiency near to 50%, we can be sure that we will run our transistor with the area shaded. This is a very approximate estimate of the proposed working area, but it does show that we are running well under the maximum values allowed. Figure 21.28.4. shows the I_c/V_{c_e} characteristics for the OC203.

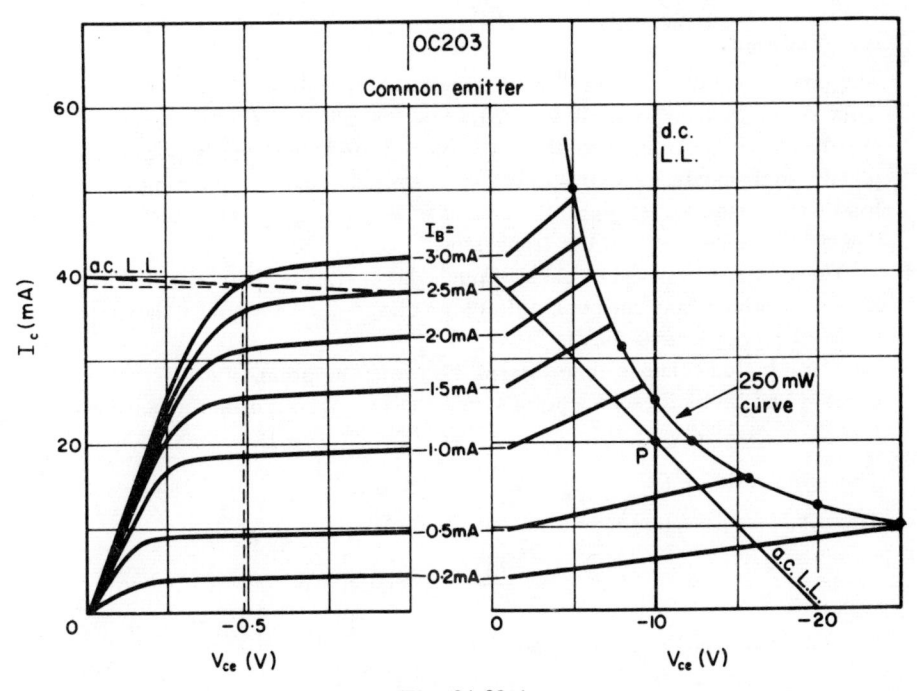

Fig. 21.28.4.

The illustration on the left is an enlarged view of that on the right, between $V_{c_e} = 0 - 1.0\,\text{V}$. The a.c. load line is shown on both characteristics. The dotted lines on the left illustration shows the peak value of current and the bottoming potential of the OC203 with the a.c. load chosen for this example.

Maximum power curve

The maximum total power allowed for the OC203 is 250 mW. This information is obtained from the manufacturer's data sheets.

A 'power max' curve is shown plotted on the characteristics in figure 21.28.4. This is plotted in precisely the same manner as for the valve versions discussed earlier on.

d.c. load line

The primary winding resistance of T_2 is assumed to be negligible and as there is no emitter resistor in the circuit, the d.c. load line will be vertical and positioned at the point corresponding to the supply, i.e. -10 V.

The operating point

The operating point must be chosen and of course must be on the d.c. load line below the 'power max' curve.

A point given by $(-10, 20)$ was arbitrarily chosen to allow for transistor production 'spreads'. This is shown as point P.

a.c. load line

The power consumed by V_{T_1} in the absence of a signal is useless power. This power is seen to be $10\,V \times 20\,mA = 200\,mW$ and is taken from the supply. If we apply signals that are not distorted, the supply power will remain unchanged. We will obtain power for subsequently passing on to the load when signals are applied, but this power will be subtracted from the 200 mW dissipated within the transistor.

The a.c. load line must pass through the operating point P. We have already decided that our peak collector voltage during signal conditions is to be -20 V, or $2 \times (-10\,V)$.

The a.c. load line is therefore plotted from the point $(-20, 0)$ through point P and terminates at a point twice the quiescent collector current, i.e., $2 \times 20 = 40\,mA$.

The turns ratio of T_2

The a.c. load line represents an a.c. load of

$$\frac{20\,V}{40\,mA} = 500\,\Omega.$$

Therefore from $n^2 R_L$ = a.c. load.

$$n^2 = \frac{\text{a.c. load}}{R_L}$$

and as the a.c. load is $500\,\Omega$ and $R_L = 3\,\Omega$,

$$n = \sqrt{\frac{500}{3}} = \sqrt{166.7}$$

thus $n \doteq 13$. Therefore T_2 will have a turns ratio of 13:1.

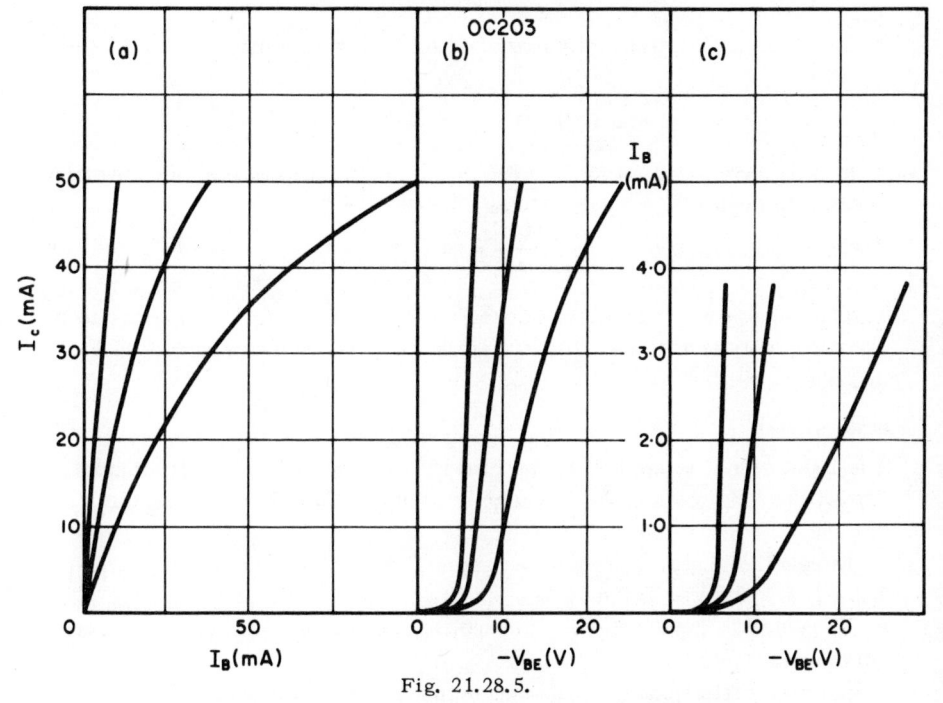

Fig. 21.28.5.

TRANSFER, MUTUAL AND INPUT CHARACTERISTICS.
COMMON EMITTER.

Each figure (a), (b) and (c) show three curves. These indicate the minimum, typically average and maximum values due to production tolerances. We will use the average, or centre curve in each instance.

Base current bias

Figure 21.28.5. (a) shows that for an I_c of 20 mA, we require an I_b of 0.8 mA approximately. Hence α' is approximately 25. Reference to the operating point shows this is of the right order of base current.

Figure 21.28.5 (c) shows that to cause a base current of 0.8 mA, a base-emitter voltage of $V_{be} = -0.8$ V is required.

Figure 21.28.5. (b) shows that for a V_{be} of -0.8 V, a collector current of -20 mA will flow. This graph however is valid for $V_{ce} = 4.5$ V only, but we can allow for higher values on a proportional basis. Hence we are able to determine the values of R_1 and R_2 to provide a base voltage of -0.8V.

The base current is to be 0.8 mA. Hence the current through the divider is assumed to be, say, 5 mA, although in practice it could be greater.

$$R_2 = \frac{0.8\,V}{5\,mA} = 0.16\,K\Omega = 160\,\Omega.$$

If we now use a standard value of $180\,\Omega$, the actual divider current must be

$$\frac{0.8\,V}{0.18\,K\Omega} = 4.45\,mA.$$

R_1 will have a p.d. of $10 - 0.8 = 9.2\,V$ and a current of $(4.45 + 0.8)\,mA$ passing through it.

Hence $$R_1 = \frac{9.2\,V}{5.25\,mA} = 1.75\,K\Omega \text{ and can be a } 1.8\,K\Omega\,//\,47\,K\Omega.$$

The capacitor C, should be chosen so that at the lowest frequency to be used, its reactance is $R_{in}/10\,\Omega$, where R_{in} is the input resistance of V_{T_1} in shunt with $R_1\,//\,R_2$.

Input signal

The input signal is applied to the primary winding of T_1. For full output across the load resistor R_L, we require a change of $\pm 20\,mA$ collector current.

From 21.28.5. (b) we require an approximate change in v_{be} of $1.2 - 0.4 = 0.8\,V = \pm 0.4\,V$.

Knowing the input source output voltage, the turns ratio of T_1 is easily determined.

Ignoring distortion, the maximum output power is given as:

$$\frac{(V_{max} - V_{min})(I_{max} - I_{min})}{8} = \frac{20 \times 40}{8} = 100\,mW.$$

The efficiency $= \dfrac{P_o \times 100}{P_{in}} = \dfrac{100}{200} \times 100\% = 50\%.$

However, the total voltage swing is seen from the graph to be 19.5 V and the current 39 mA. Hence the efficiency is approximately 47.5%.

There are many other factors to consider in the design of even a simple amplifier, but are too numerous to discuss here. This however, has been seen to be a further example of the basic theory discussed in earlier sections of the book.

CHAPTER 22

'h' parameters

22.1. Equivalent circuit

The use of h parameters enable one to determine the various conditions necessary when deriving input resistance, current gain, voltage feedback, ratio, (v.f.r.) and output conductance. These are valid for small signals only and at frequencies well below 'cut off frequencies'. The h parameters have been chosen in this chapter for the reason that they contain both Z and Y parameters and the subsequent investigations are intended to stimulate further thoughts on a wider range of simple network analyses.

Consider the h parameters for a grounded base transistor. The equivalent h circuit is as follows.

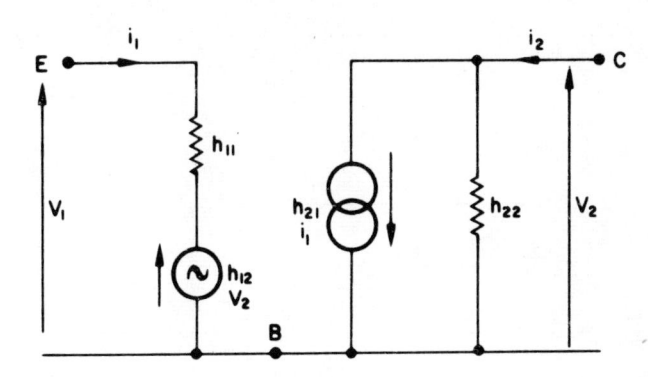

Fig. 22.1.1.

The suffix consists of two figures. The first figure denotes whether the particular h parameter is to be found in the input or output part of the circuit.

A suffix one (first figure) denotes input, whilst two denotes output. The second figure indicates whether the output circuit or input circuit has any influence upon the parameter under consideration.

Example

h_{11}. The first figure (1) shows that this parameter is in the input circuit. The second figure (1) shows that it is influenced by the input circuit only. h_{11} is the input resistance and is given as

$$\frac{v_1}{i_1}.$$

458   ELECTRONICS FOR TECHNICIAN ENGINEERS

h_{12}. The first figure (1) shows that this generator is situated in the input circuit whilst the second figure (2) indicates that the generator (h_{12}) is influenced by the output circuit.

h_{21}. Using the same argument, this current generator is found in the output and is influenced by the input circuit.

h_{22}. This is the output conductance and is influenced by the output circuit.

The suffixes appear in two equations for figure 22.1.1. The equations are as follows:

$$v_1 = h_{11} i_1 + h_{12} v_2$$

$$i_2 = h_{21} i_1 + h_{22} v_2$$

The suffixes also show the order in which they fit into these equations: The first number indicates the equation row number whilst the second number indicates the column in which it is written: i.e., h_{12} is written in the 1st row equation 1 and column 2.

If the input circuit is examined and Ohm's law applied, it may be seen that

$$i_1 = \frac{v_1 - h_{12} v_2}{h_{11}}$$

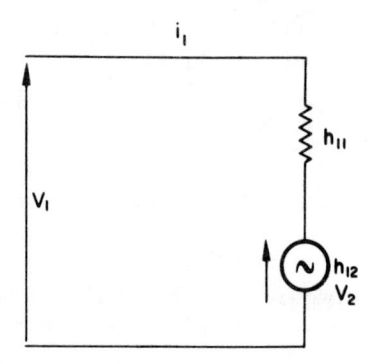

Fig. 22.1.2.

This may be rearranged thus,

from $\quad v_1 = h_{11} i_1 + h_{12} v_2, \; i_1 = \dfrac{v_1 - h_{12} v_2}{h_{11}}$

Then $\quad h_{11} = \dfrac{v_1 - h_{12} v_2}{i_1}$ \hfill (1)

Consequently from (1) $h_{12} v_2 = v_1 - h_{11} i_1$ or

$$h_{12} = \frac{v_1 - h_{11} i_1}{v_2} \quad (2)$$

If the output is examined, an expression for the corresponding parameters may be derived.

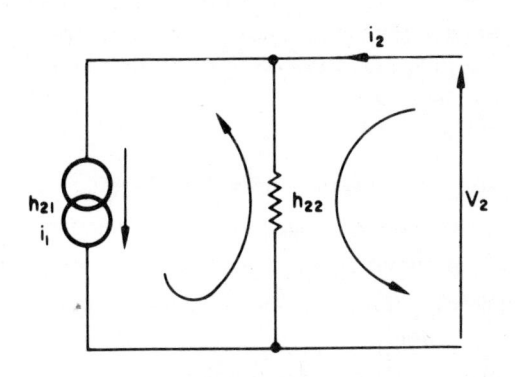

Fig. 22.1.3.

The conductance
$$h_{22} = \frac{i_2 - h_{21} i_1}{v_2} \quad (3)$$

∴
$$i_2 = h_{22} v_2 + h_{21} i_1$$

consequently
$$h_{21} i_1 = i_2 - h_{22} v_2$$

or
$$h_{21} = \frac{i_2 - h_{22} v_2}{i_1} \quad (4)$$

It is important to appreciate that when placing a short circuit across a current generator, all of the current generated will flow in the short circuit. The current generator will not collapse although its associated shunt resistance will play no further part in the current distribution.

If it is required to solve for h_{11}, then from (1), v_1/i_1 is required. The term $h_{12} v_2$ must therefore be zero. As $h_{12} v_2$ is proportional to v_2, v_2 must be zero, hence $h_{12} v_2 = 0$.

This is easily achieved as a voltage is zero to a.c. when kept at a constant value. This is arranged by placing a large capacitance across the output. This does not affect the necessary d.c. conditions but prevents any a.c. from appearing at the output terminals.

Consider h_{12}, v_1/v_2 is required, consequently in (2) it is required to cause $h_{11} i_1$ to become zero. One cannot short circuit h_{11} but it is possible to cause i_1 to become zero by 'open circuiting' the input so that i_1 cannot flow. Hence $h_{11} i_1$ becomes zero. The open circuit (to a.c.) in the input is arranged by positioning a large inductance in series with the input. This allows the transistor to be properly biassed with the relevant d.c. current, but prevents alternating signal current from flowing.

The parameter $h_{22} = i_2/v_2$ then from (3) it is necessary to cause $h_{21} i_1$ to be zero, hence i_1 must be zero and an open circuit input is once again the method by which, this is achieved.

The parameter $h_{21} = i_2/i_1$ then from (4) $h_{22} v_2$ needs to be zero; a short circuited output is once more the answer, v_2 becomes zero thus the term $h_{22} v_2$ becomes zero. This leads us quite naturally to the definitions of the h parameters which are as follows.

Definitions

$h_{11} = v_1/i_1$ slope of input characteristic for a constant output voltage.

$h_{21} i_1 = i_2/i_1$ slope of transfer characteristic for constant output voltage.

$h_{22} = i_2/v_2$ slope of output characteristic for constant input current.

$h_{12} v_2 = v_1/v_2$ slope of voltage feedback characteristic (v.f.r.) for constant input current.

It might be desirable to reinforce the following:

A very large capacitor across the output will allow correct d.c. conditions whilst preventing any signal variation, i.e., a constant voltage. A constant input current infers an open circuit input. This is obtained by inserting a large inductance in series with the input terminals. This allows correct d.c. biassing but prevents any change of current i.e., open circuit to a.c.

The definitions above, are more often expressed as follows.

h_{11} = input resistance with output short circuited to a.c.

$h_{21} i_1$ = current gain with output short circuited to a.c.

h_{22} = output conductance with input open circuit to a.c.

$h_{12} v_2$ = reverse Voltage Feedback Ratio with input open circuit to a.c.

h parameters can be measured and with reasonable care quite accurate results may be obtained. 1 KHz is a convenient frequency to choose when taking these measurements.

h_{11} = is the input resistance measured in ohms.

h_{21} = is a ratio and also has no dimensions.

h_{22} = is the output conductance measured in ohms.

h_{12} = is the v.f.r. and has no dimensions.

h PARAMETERS

Example

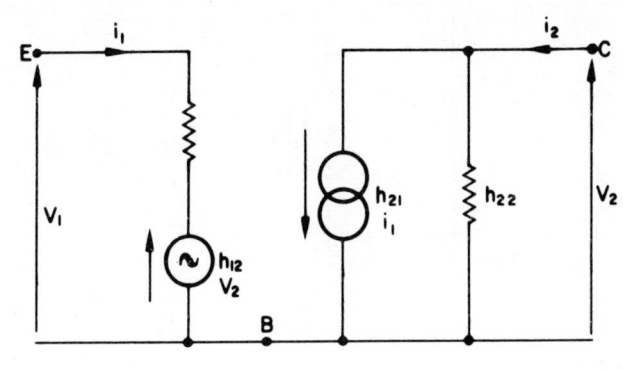

Fig. 22.1.4.

h_{11} input resistance

The equivalent circuit is shown in figure 22.1.4.

Method

Apply v_1 and calculate i_1. $h_{11} = v_1/i_1$. The generator $h_{12} v_2$ would act in opposition to v_1 and must be 'removed' to avoid this effect.

As $h_{12} v_2$ is not required we must short circuit the output to a.c. Hence v_2 is zero and $h_{12} v_2$ becomes zero also.

The circuit becomes as shown in figure 22.1.5.

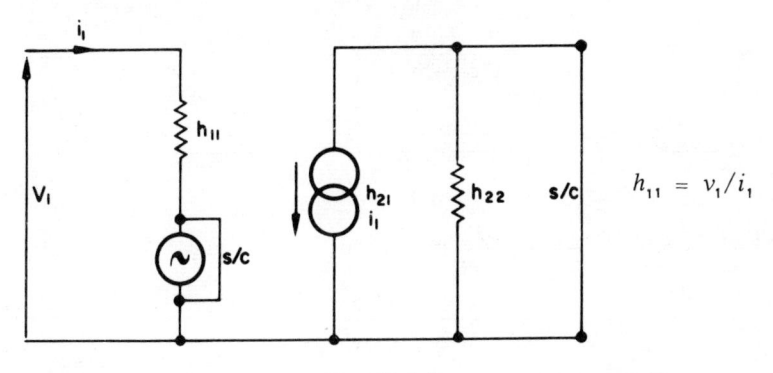

Fig. 22.1.5.

$h_{11} = v_1/i_1$

Alternatively, using an equivalent 'T' to solve for h_{11} (figure 22.1.6.).

The current through $r_b = i_1 (1 - \alpha)$ hence the p.d. across $r_b = (1 - \alpha) i_1 r_b$.

The effective resistance across points $XX = \dfrac{i_1 (1 - \alpha) r_b}{i_1} = (1 - \alpha) r_b$.

The input resistance is therefore $r_e + (1 - \alpha) r_b$.

(ignoring r_c as it is $\gg r_b$)

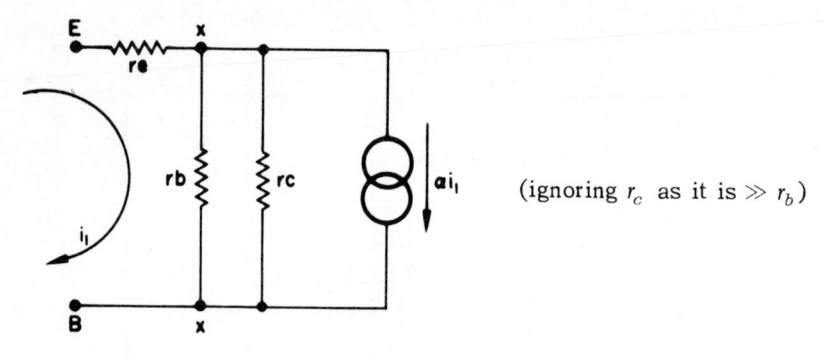

Fig. 22.1.6.

h_{21} current gain

We need to apply an input current i_1, short circuit v_2, so as to cause $h_{12} v_2$ to become zero, and determine i_{out}/i_{in}. The 'reduced' equivalent circuit is given in figure 22.1.7.

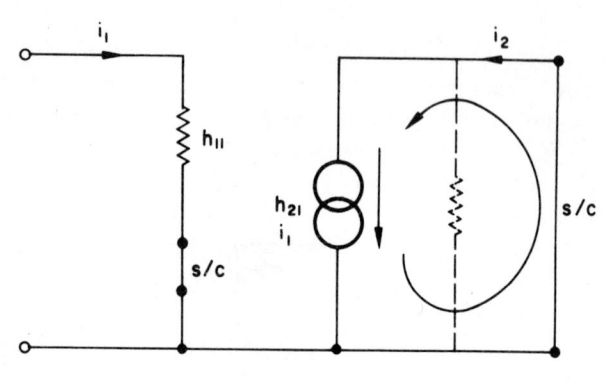

Fig. 22.1.7.

The current $h_{21 i_1}$ flows in the short circuit 'load', hence

$$h_{21} = \frac{i_2}{i_1} = \frac{h_{21 i_1}}{i_1} = h_{21}$$

h_{22} Output conductance $= i_2/v_2$.

The generator $h_{21} i_1$ is not required, as it will cause an opposing current in the output and the result will be inaccurate. The input must be open circuit in order to prevent i_1 from flowing. Thus $h_{21} i_1$ is open circuit as shown in figure 22.1.8.

Hence $h_{22} = i_2/v_2$.

h_{12} Voltage feedback ratio $= v_1/v_2$

Fig. 22.1.8.

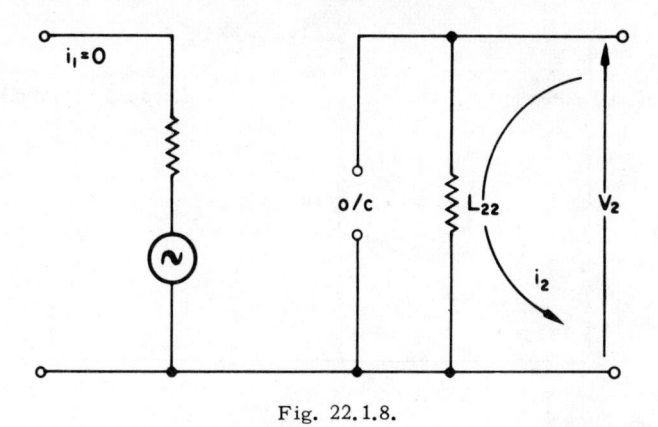

As i_1 would modify the voltage in the input due to $h_{12} v_2$, it must be zero. An open circuit input is required. The potential difference $i_1 h_{11}$ is therefore zero and does not affect the e.m.f. h_{12}. The equivalent circuit is given in figure 22.1.9.

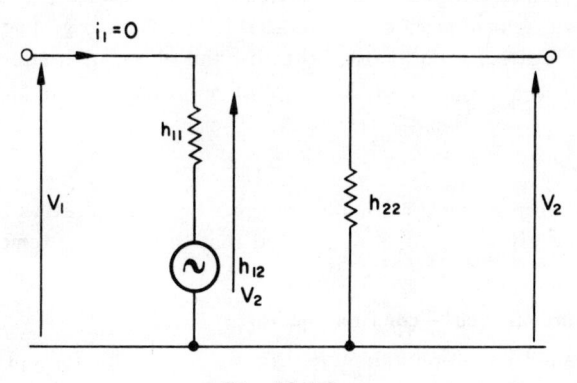

Fig. 22.1.9.

$h_{12} v_2 = v_1$ as $h_{12} v_2$ is an e.m.f., and appears across the open circuit input.

22.2. h parameters and equivalent 'T' circuits

Using an equivalent 'T' find for $h_{12} = \left(\dfrac{v_1}{v_2}\right)$

Figure 22.2.1. shows the simplified circuit using r parameters. The e.m.f. $h_{12} v_2 = v_1$ with an open circuit input.

By load over total, *as no current flows in r_e,*

$$v_1 = \frac{r_b}{r_b + r_c} \times v_2 \text{ or } \frac{v_1}{v_2} = h_{12} = \frac{r_b}{r_c} \text{ as } r_b \ll r_c.$$

2H

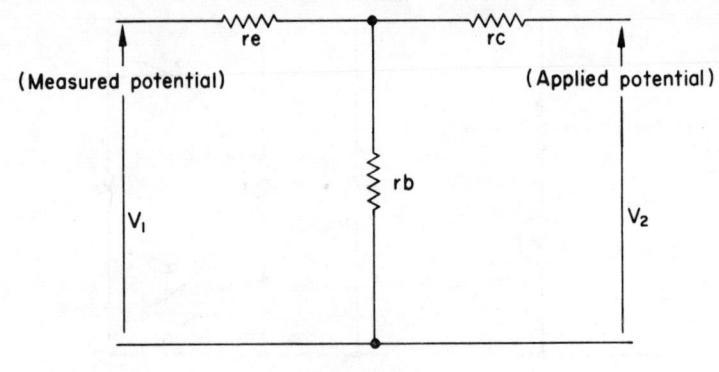

Fig. 22.2.1.

It may be seen that the voltage across r_b is dependant upon v_2.

Subsequently it may be seen that any change in v_2 (a.c. quantities) 'reflect' into the input circuit by the ratio r_b/r_c and is in the order of 8.10^{-4} for some low power transistors.

There are numerous possible variations to this 'theme'; using the grounded base as a reference, it is possible to examine grounded emitter and grounded collector equivalent circuits and obtain the appropriate h' and h'' parameters in terms of the original h parameters for the grounded base.

h' parameters

One or two examples on a method of converting h parameters to h' parameters will be given, but it is suggested that the reader attempts a few for himself.

h'_{22} output conductance — common emitter

From a grounded base equivalent h circuit, derive the appropriate h'_{22} parameter for a grounded emitter. The input (of the grounded emitter) must be open circuit for h'_{22}.

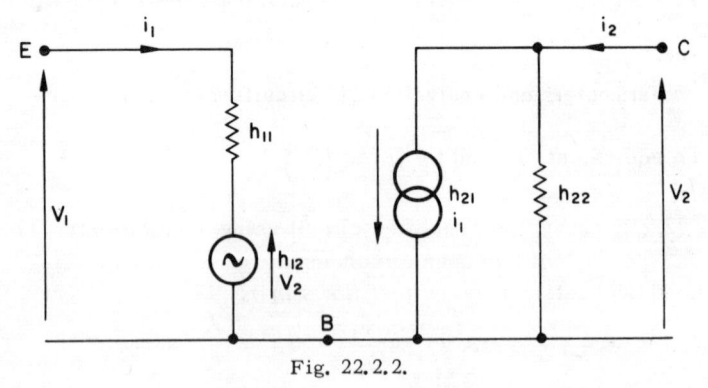

Fig. 22.2.2.

h PARAMETERS

The base therefore must have no external connection.
The circuit becomes

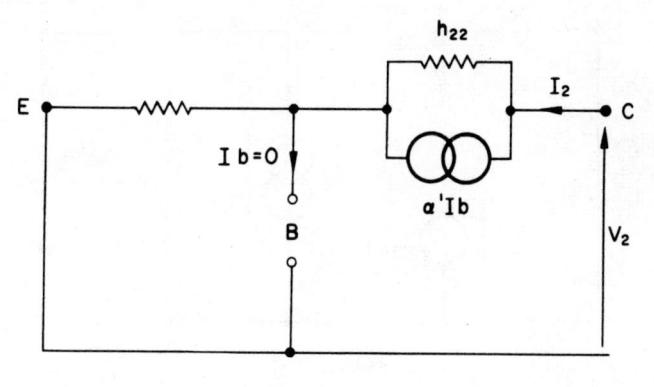

![Fig. 22.2.3 circuit diagram showing E, B, C terminals with h_{22}, current generator $a'I_b$, and $I_b = 0$]

Fig. 22.2.3.

The current generator becomes zero as the input current is zero.

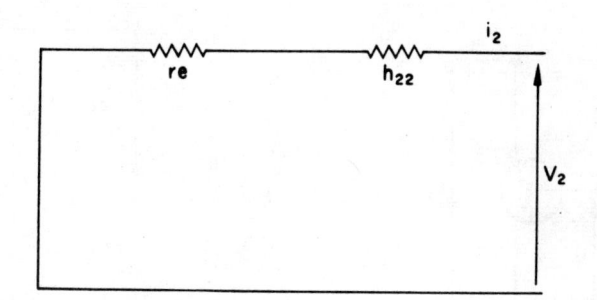

![Fig. 22.2.4 circuit diagram with r_e, h_{22}, i_2, V_2]

Fig. 22.2.4.

as $r_e \ll r_c$, this may be ignored.

$$h_{22} = \frac{i_2}{v_2} \text{ but } h_{22} = \frac{1}{r_c} \text{ and } r_c = \frac{R_c}{1 - a}$$

$\therefore \quad R_c = (1 - a) r_c$

and as $\quad h'_{22} = \frac{1}{R_C} \quad h'_{22} = \frac{1}{(1 - a) r_c}$

and as $\quad \dfrac{1}{(1 - a)} = 1 + a', \; h'_{22} = \dfrac{1 + a'}{r_c}$

and as $\quad \dfrac{1}{r_c} = h'_{22} \qquad h'_{22} = (1 + a') h_{22}$

h'_{12} v.f.r. for a common emitter

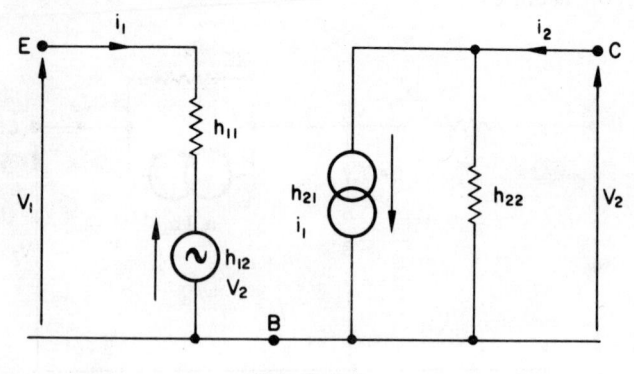

Fig. 22.2.5.

The input must be open circuit. Hence $h_{21} i_1 = 0$.
The circuit becomes

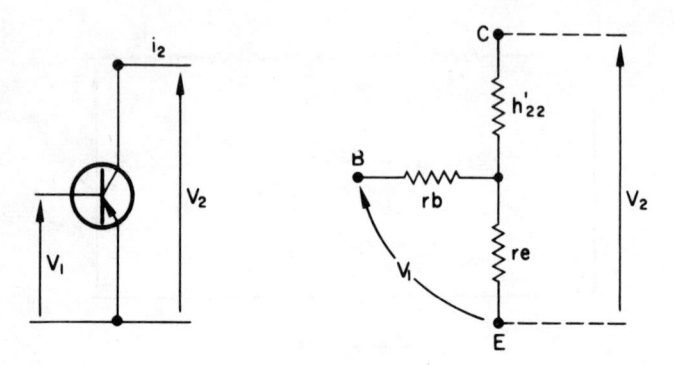

Fig. 22.2.6.

$$h'_{12} = \frac{v_1}{v_2} = \frac{v_2 \times r_e}{r_e + \dfrac{1}{h'_{22}}} = \frac{v_2 \, r_e \, h'_{22}}{r_e \, h'_{22} + 1}$$

$\therefore \quad \dfrac{v_1}{v_2} = \dfrac{r_e \, h'_{22}}{r_e \, h'_{22} + 1}$ but as $r_e \, h'_{22}$ is $\ll 1$

$$\frac{v_1}{v_2} = h'_{12} \doteq \underline{h'_{22} \, r_e}$$

Little more can be covered in a book of this size. It is left to the reader to practice other examples for himself. Answers can be checked against the following table (with acknowledgements to Mullards Limited).

22.3. Conversion from T-network parameters to h parameters

Common Base.

$h_{11} = r_e (1 - \alpha) r_b$

$-h_{21} = \alpha$

$h_{22} = 1/r_c$

$h_{12} = r_b/r_c$

Common Emitter.

$h'_{11} = (1 + \alpha') h_{11}$

$h'_{21} = \alpha'$

$h'_{22} = (1 + \alpha') h_{22}$

$h'_{12} = h'_{22} r_e$

Common Collector.

$h''_{11} = h'_{11}$

$-h''_{21} = 1 + h'_{21}$

$h''_{22} = h'_{22}$

$h''_{12} = \dfrac{1}{1 + h'_{12}}$

22.4. Practical measurements of h parameters

The following arrangement is typical of that required to derive individual h parameters for a given transistor.

Equipment required   OC71 Transistor (or similar type)
Valve Voltmeter.
Signal Generator.
Meters as shown in the Circuit Diagrams.

Circuit Diagrams. (Figures 22.4.1. and 22.4.2.)

(a) For measurement of h_{12} and h_{22}

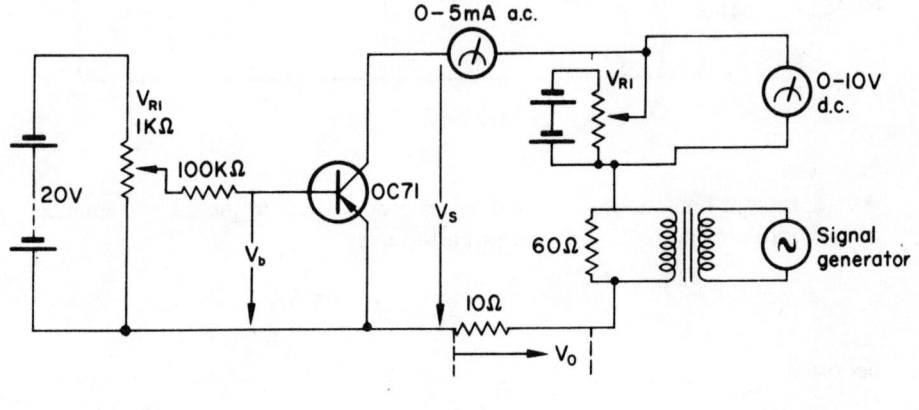

Fig. 22.4.1.

Method

The circuits should be connected as shown in (a). This configuration is correct for the measurements of h_{12} and h_{22}.

h_{12}. The signal generator should be set to an appropriate frequency, say 1K/cs. The signal generator output voltage should be set to an appropriate level such that v_s is 1V. v_s, of course, must be measured with the valve voltmeter. Set the voltage v_{CE} to 4V d.c. The collector current may be controlled by v_{R_1} in the base circuit. The collector current should be varied in increments of 0.1 mA, from 0 mA to 3 mA. v_o and v_b should be recorded for each value of I_c. As v_s is 1V, the voltage feedback ratio, may be expressed directly in terms of v_b. If v_b is 0.85 mV, then h_{12} is 0.85×10^{-3}. The decimal point may be rearranged so as to give a result more commonly met, the typical answer given could be written as 8.5×10^{-4}.

(b) For measurement of h_{11} and h_{21}

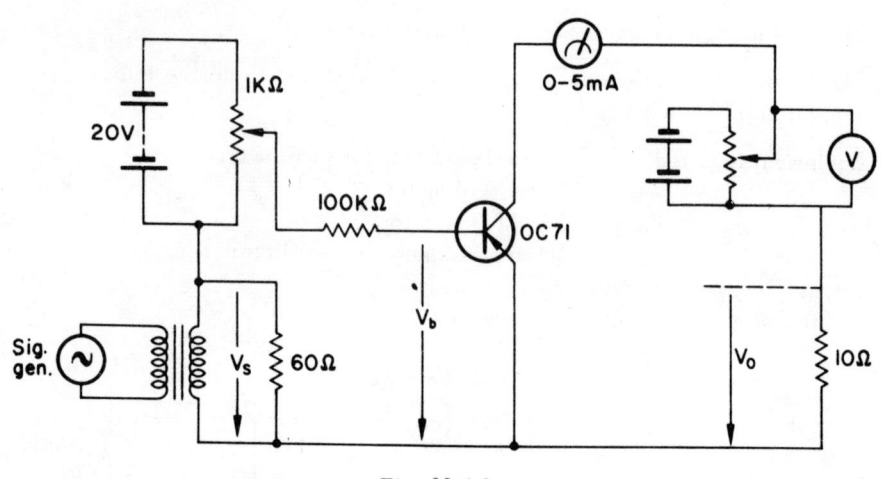

Fig. 22.4.2.

h_{22}.

As v_s is kept constant at 1V, and v_o is proportional to the a.c. component of the collector current, the task is quite easy.

$$v_o = 10 \times i_C \text{ therefore } i_C \text{ is } \frac{v_o}{10}. \text{ As } h_{22} \text{ is } \frac{i_C}{v_s} \; h_{22}$$

becomes $\dfrac{v_o}{10 \times v_s}$

As an example, if v_o is 0.82 mV, h_{22} becomes a tenth of this voltage. In this case, h_{22} would have a value of $0.082 \, m\Omega^{-1}$. Again this could be expressed as $82 \, m\Omega^{-1}$. This also is a typical value.

h_{21} **and** h_{11}

The circuit should be connected as in (b). The signal generator output should be set such that v_s has a constant amplitude of 100 mV. The set of readings for this experiment are similar to those of the previous example.

h PARAMETERS

For the same collector current values as before, the voltages v_o and v_b, should be recorded. The input level is 100 mV. The limiting resistor is 100 KΩ. The input current is assumed to remain constant at

$$\frac{100 \text{ mV}}{100 \text{ K}\Omega} = 10^{-6} \text{ A}.$$

The input resistance h_{11} is given as v_b divided by the input of $1 \mu A$.
If v_b was 1.0 mV, then the value of h_{11} would be

$$\frac{1.0 \text{ mV}}{1.0 \mu A} = 1.0 \text{ K}\Omega.$$

The current gain, alpha, is given as the ratio of output current to the input current.

v_o is the product of the a.c. component of the collector current and the 10 Ω resistor.

The parameter h_{21} may be shown to be $\dfrac{i_c}{i_b} = \dfrac{v_o \text{ (mV)}}{10 \times 10^{-6}}$

As an example, if v_o was 0.49 mV, then h_{21} would be

$$\frac{0.49 \cdot 10^{-3}}{10^{-6} \times 10} = 49.$$

This value is again a typical value.

The 60 Ω resistor was chosen so as to form a correct load for the signal generator, via the transformer. The 60 Ω met the requirements of the expression $R_L = n^2 R_g$. Where R_g is the output resistance of the signal generator. And n is the turns ratio of the transformer. It is advisable not to earth the equipment, apart from the signal generator.

The h parameters for the circuit shown in figure 22.4.3. is

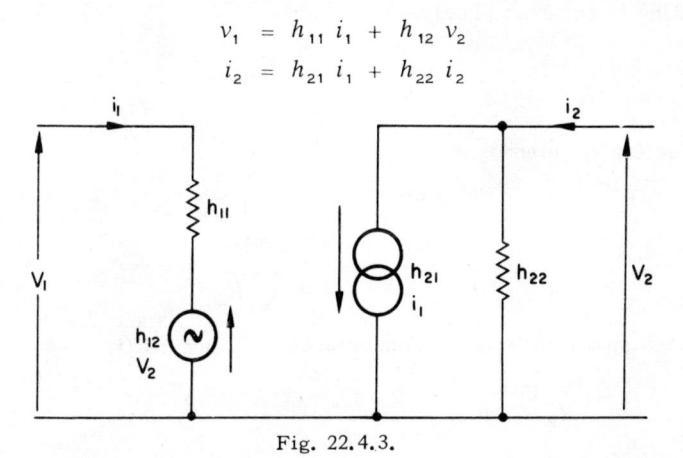

$$v_1 = h_{11} i_1 + h_{12} v_2$$
$$i_2 = h_{21} i_1 + h_{22} i_2$$

Fig. 22.4.3.

The following examples will allow for the source resistance of the input generator v_1. Let this source resistance be R_s. The same will apply when v_1 is applied where v_2 is replaced by a load R_L.

When looking into the input, we replace v_2 by a load resistor, R_L as shown in figure 22.4.4.

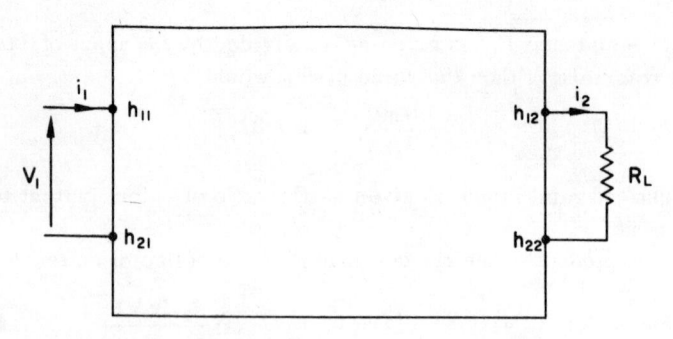

Fig. 22.4.4.

22.5. Input resistance with R_L connected

$$R_{in} = \frac{v_1}{i_1}.$$

When v_1 is applied, i_2 flows *into* R_L and will cause $-v_2$ to be developed.

$$v_1 = h_{11} i_1 - h_{12} i_2 R_L \qquad (1)$$
$$i_2 = h_{21} i_1 - h_{22} i_2 R_L \qquad (2)$$

where $-i_2 R_L = v_2$.

From equation (2) $\quad i_2 (1 + h_{22} R_L) = h_{21} i_1 \qquad (3)$
substituting in equation (1) gives

$$v_1 = h_{11} i_1 - \frac{h_{12} R_L h_{21} i_1}{1 + h_{22} R_L}$$

and collecting i_1 terms

$$v_1 = i_1 \left[h_{11} - \frac{h_{12} R_L h_{21}}{1 + h_{22} R_L} \right]$$

$$R_{in} \therefore = \frac{v_1}{i_1} = h_{11} - \frac{h_{12} h_{21} R_L}{1 + h_{22} R_L}$$

but expressing R_L in terms of conductance G_L. where $G_L = \frac{1}{R_L}$

$$R_{in} = h_{11} - \frac{h_{12} h_{21}}{G_L + h_{22}}$$

h PARAMETERS

22.6. Current gain

This is determined from equation (3) $\dfrac{i_2}{i_1} = \dfrac{h_{21}}{1 + h_{22} R_L}$

and in terms of G_L, current gain $= \dfrac{h_{21} G_L}{G_L + h_{22}}$

22.7. Voltage gain

From equation (3) $i_1 = \left\{ \dfrac{1 + h_{22} R_L}{h_{21}} \right\} i_2$.

Substituting in equation (1) gives

$$v_1 = \dfrac{h_{11} (1 + h_{22} R_L) i_2}{h_{21}} - h_{12} R_L i_2$$

Hence

$$v_1 = \dfrac{i_2 [h_{11} (1 + h_{22} R_L) - h_{21} h_{12} R_L]}{h_{21}}$$

$\therefore$

$$i_2 = \dfrac{h_{21} v_1}{[h_{11} (1 + h_{22} R_L) - h_{21} h_{12} R_L]}$$

and as $\quad v_2 = -i_2 R_L$ and voltage gain $= \dfrac{v_2}{v_1} = \dfrac{-i_2 R_L}{v_1}$

Voltage gain $= \dfrac{-h_{21} R_L}{[h_{11} (1 + h_{22} R_L) - h_{12} h_{21} R_L]}$

and in terms of G_L

Voltage gain $= \dfrac{-h_{21}}{h_{11} (G_L + h_{22}) - h_{12} h_{21}}$

22.8. Output admittance

The output admittance will be a function of the input source impedance R_S. In this example, v_2 will be applied and the input terminals will have R_s connected across them so as to allow for this impedance when calculating the output admittance.

$$v_1 = -i_1 R_S.\ \text{(as the 'input' current will now flow } out$$
of the network).

Rewriting the equations (1) and (2)

$$-i_1 R_S = h_{11} i_1 + h_{12} v_2 \qquad (4)$$

$$i_2 = h_{21} i_1 + h_{22} v_2 \qquad (5)$$

From (4) $\quad 0 = i_1 (h_{11} + R_S) + h_{12} v_2$

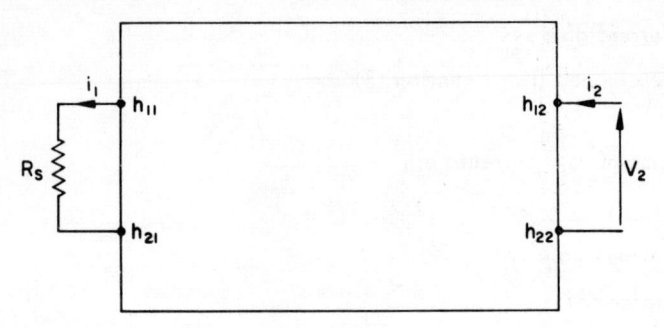

Fig. 22.8.1.

$$\therefore \quad i_1 = \frac{-h_{12} v_2}{h_{11} + R_S}.$$

Substituting in (5)

$$i_2 = \left[\frac{-h_{21} h_{12}}{h_{11} + R_S} + h_{22}\right] v_2$$

output admittance $Y_o = \dfrac{i_2}{v_2} = h_{22} - \dfrac{h_{21} h_{12}}{h_{11} + R_S}.$

22.9. Power gain

A useful method of determining the power gain is to first determine the maximum power the input generator can supply.

This will occur when the input resistance equals the source resistance. This is called the 'available power gain'.

The available power gain will occur when R_S is equal to the input resistance of the transistor. Hence $R_S = R_{in}$.

$$v_{in} = \frac{v_S R_{in}}{R_S + R_{in}} = \frac{v_S R_{in}}{2 R_{in}} = \frac{v_S}{2}$$

$$\text{Power in} = \frac{v_{in}^2}{R_{in}} = \frac{v_{in}^2}{R_S}$$

$$\text{hence power in} = \frac{v_S^2}{4 R_S}. \tag{6}$$

To derive the expression for power gain, the circuit shown in figure 22.9.1. is considered.

$$v_1 = v_S - i_1 R_S \text{ and } v_2 = -i_2 R_L.$$

$$v_S - i_1 R_S = h_{11} i_1 - h_{12} R_L i_2 \tag{7}$$

$$i_2 = h_{21} i_1 - h_{22} R_L i_2 \tag{8}$$

h PARAMETERS

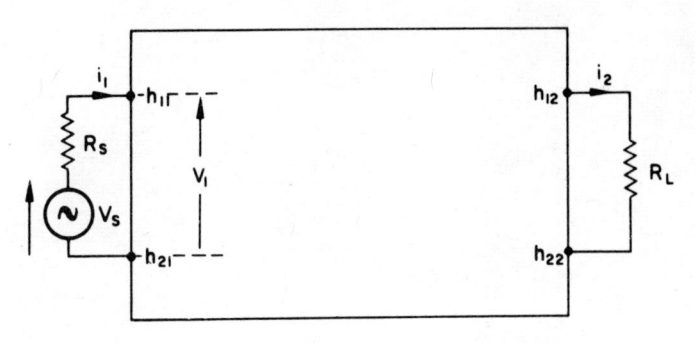

Fig. 22.9.1.

from (7) $\qquad v_S = (h_{11} + R_S) i_1 - h_{12} R_L i_2 \qquad (9)$

and from (8) $i_2 (1 + h_{22} R_L) = h_{21} i_1 \qquad (10)$

and from (10) $\qquad i_1 = \dfrac{(1 + h_{22} R_L)}{h_{21}} i_2$ and substituting this in (9)

$$v_S = \left[\dfrac{(h_{11} + R_S)(1 + h_{22} R_L)}{h_{21}} - h_{12} R_L \right] i_2$$

$\therefore \qquad v_S = \dfrac{(h_{11} + R_S)(1 + h_{22} R_L) - h_{12} h_{21} R_L}{h_{21}} i_2$

but power out $= i_2^2 R_L$

Power out $= i_2^2 R_L = \dfrac{h_{21}^2 v_S^2 R_L}{[(h_{11} + R_S)(1 + h_{22} R_L) - h_{12} h_{21} R_L]^2}$

and dividing by input power of $\dfrac{v_S^2}{4 R_S}$ gives power gain

available power gain $= \dfrac{4 R_S R_L h_{21}^2}{[(h_{11} + R_S)(1 + h_{22} R_L) - h_{12} h_{21} R_L]^2}$

and in terms of G_L

available power gain $= \dfrac{4 R_S G_L h_{21}^2}{[(h_{11} + R_S)(G_L + h_{22}) - h_{12} h_{21}]^2}$

CHAPTER 23

'H' parameters

23.1. H parameters (cascade circuits)

Technician engineers may often be called upon to deal with circuits that contain two or more transistors in cascade or cascode. These fall into a range of Compound Circuits.

Should we wish to analyse the function of compound circuits, we can use a 'compound' h parameter system. This we can call H parameters.

Although we will discuss compound circuits containing two transistors only in this chapter we should note that the approach can be extended for a number of interconnected stages.

We shall continue with the convention adopted earlier, that the collector current i_2, is positive when it enters h_{12}, but when the same current leaves h_{22}, we will consider the latter to be negative going.

The reader is advised to consider the following examples most carefully as on many occasions we will be discussing one current *leaving* the collector of say, V_{T_1} (this will be negative) and flowing *into* h_{11} of the next transistor V_{T_2} which according to our convention, is positive going.

The net effect will be that we shall consider the current flowing in a positive direction but the actual input signal current applied will be negative going input, i.e. i into h_{11} will be written as $(-)i$. This is shown in figure 23.1.1.

In this manner we can allow for phase shift in appropriate stages. However, the examples should help make this quite clear.

Note that we will follow the same rules as with h parameters in connection with short circuit and open circuit inputs and outputs.

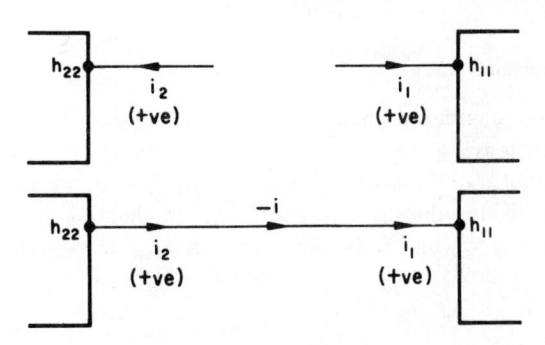

Fig. 23.1.1.

476   ELECTRONICS FOR TECHNICIAN ENGINEERS

Hence input to h_{11} when taken from h_{22} in previous stage is a negative going input and is written as $-i$.

We intend to consider the following circuit and derive the appropriate parameters H_{11}, H_{12}, H_{21}, and H_{22} for a two stage amplifier consisting of two grounded base amplifiers in cascade, as shown in figure 23.1.2.

Circuit diagram to a.c.

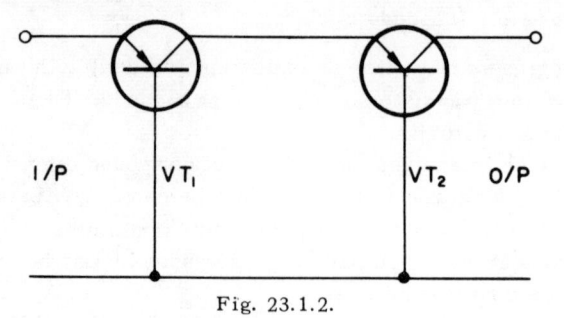

Fig. 23.1.2.

The equivalent h circuit is as shown in figure 23.1.3.

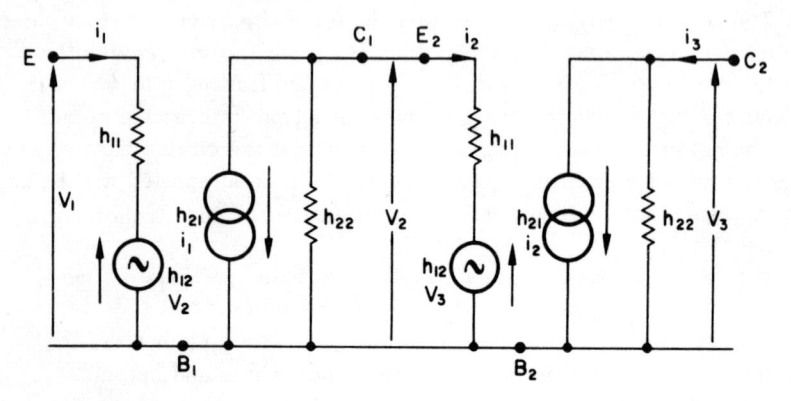

Fig. 23.1.3.

23.2. H_{11} (common base)

Note that i_2' is considered positive when entering h_{11}, but is considered negative, when leaving the collector of v_{T_1}. Consider H_{11}. We must short circuit the output. v_3 therefore $= 0$. Thus $h_{12} v_3 = 0$. v_2 is the potential due to $h_{21} i_1$, flowing through h_{11} and $1/h_{22}$ in shunt.

This current is i_2, but i_2 is negative. As $h_{21_{i_1}}$ is negative, v_2 is positive.

H PARAMETERS

$$v_2 = \frac{h_{21} i_1 \dfrac{h_{11}}{h_{22}}}{h_{11} + \dfrac{1}{h_{22}}} = \frac{h_{21} h_{11} i_1}{1 + h_{11} h_{22}}$$

but

$$H_{11} = \frac{v_1 - h_{12} v_2}{i_1} = \frac{h_{11} - h_{12} h_{21} h_{11}}{1 + h_{11} h_{12}}$$

(The feedback voltage generator opposes v_1 and must be accounted for when considering the effective input resistance.)

As the term in the denominator ($h_{11} h_{22}$) is $\ll 1$

$$H_{11} \doteq \underline{h_{11} - h_{12} h_{21} h_{11}}$$

Now consider the v.f.r. for the same circuit, as shown in figure 23.3.1.

23.3. H_{12} (common base)

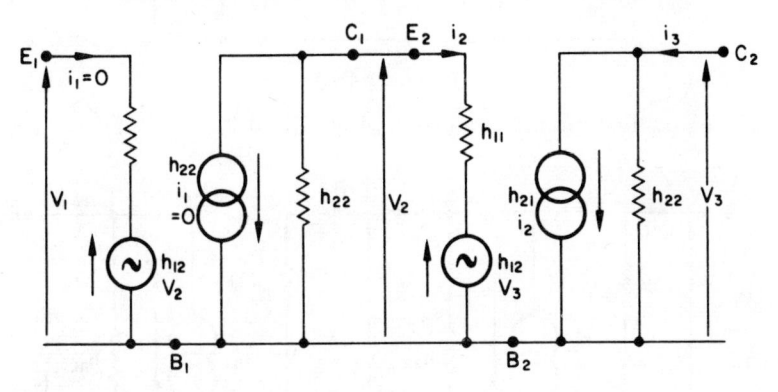

Fig. 23.3.1.

The input must be open circuited in order to cause $h_{21} i_1$ to become zero. $i_1 = 0$; therefore $h_{21} i_1 = 0$, v_3 is applied and $h_{12} v_3$ generates an e.m.f.

The voltage

$$v_2 = \frac{h_{12} v_3 \dfrac{1}{h_{22}}}{\dfrac{1}{h_{22}} + h_{11}} \quad \text{(from load over total)}$$

∴

$$v_2 = \frac{h_{12} v_3}{1 + h_{11} h_{22}} \quad \text{after multiplying top and bottom by } h_{22}.$$

The generator $h_{12} v_2$ generates an e.m.f. h_{12}, v_2.
The open circuit e.m.f. (as the input is open circuit)

$$= H_{12} = h_{12} v_2 = \frac{h_{12} h_{12} v_3}{1 + h_{11} h_{22}}$$

and ignoring h_{11}, h_{12} in the denominator as before.

$$H_{12} \doteq h_{12}^2 \, v_3$$

and as h_{12} is extremely small, h_{12}^2 is very much smaller.

A change in amplitude of v_3 will cause a change of v_1 (open circuit input voltage) of approximately $64 \times 10^{-8} \, v_3$ for the low power transistors discussed earlier.

23.4. H_{12} (common collector)

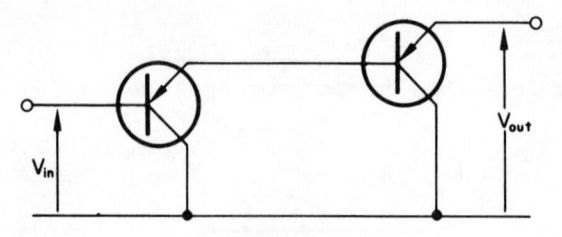

Fig. 23.4.1.

The equivalent circuit is shown in figure 23.4.2.

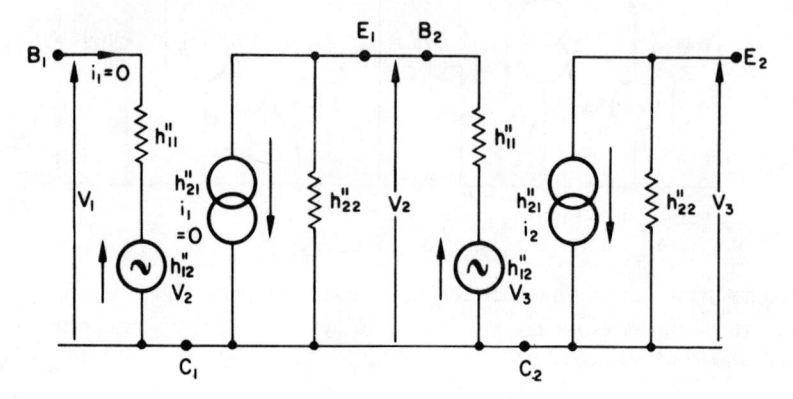

Fig. 23.4.2.

The input must be open circuit, i.e. $i_1 = 0$. Therefore $h_{21}'' \, i_1 = 0$.

The voltage $v_2 = \dfrac{h_{12}'' \, v_3 \, \dfrac{1}{h_{22}''}}{\dfrac{1}{h_{22}''} + h_{11}''} = \dfrac{h_{12}'' \, v_3}{1 + h_{22}'' h_{11}''}$

The voltage generator in the input circuit generates an e.m.f. h_{12}'', v_2.

This magnitude of e.m.f. appears across the input terminals as the input terminals have no external load connected.

$H_{12} = h''_{12} \, h''_{12} \, v_3$ once more ignoring the denominator.

∴ $H_{12} = (h''_{12})^2$ and is extremely small.

23.5. H_{21} (common base)

Consider now the current gain for the circuit (figure 23.5.1.).

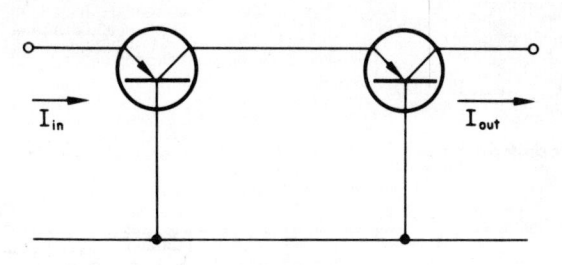

Fig. 23.5.1.

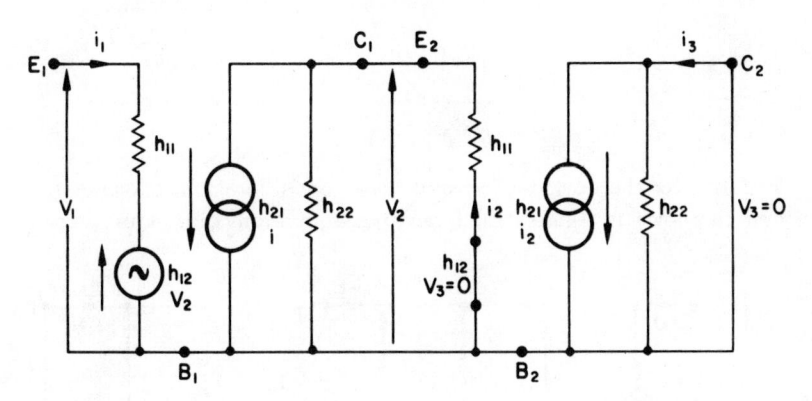

Fig. 23.5.2.

The equivalent circuit is shown in figure 23.5.2.

The compound current gain, using h parameter principles is obtained by short circuiting the output.

$$v_3 = 0 \quad \therefore \quad h_{12} \, v_3 = 0.$$

The current, i_2 entering h_{11} of the second transistor is negative due to $h_{21} \, i_1$ and i_2 being in opposite sense.

$$i_2 = \frac{-h_{21} \, i_1 \, \dfrac{1}{h_{22}}}{\dfrac{1}{h_{22}} + h_{11}} = \frac{-h_{21} \, i_1}{1 + h_{11} \, h_{22}} \quad \text{assuming an input } i_1 \text{ to } V_{T_1}.$$

The current generator $h_{21} \, i_2$ generates a current of magnitude

21

$$\frac{-h_{21} \, h_{21} \, i_1}{1 + h_{11} \, h_{22}}$$

and as the output is short circuit we can assume all this current flows in the short circuit load and is in fact, i_3.

The current gain $= \dfrac{o/p}{i/p} = \dfrac{i_3}{i_1} = \dfrac{(-h_{21})^2 \, i_1}{(1 + h_{11} \, h_{12}) \, i_1} \doteq \underline{(-h_{21})^2}.$

As $-h_{21} = \alpha$. Result is α^2.

23.6. H_{22} (common base)

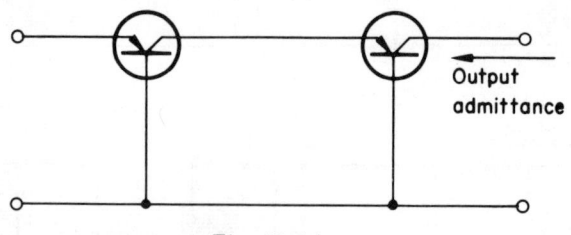

Fig. 23.6.1.

Finally, for the compound circuit under consideration, examine H_{22}. The circuit is shown in figure 23.6.1. and the equivalent circuit in figure 23.6.2.

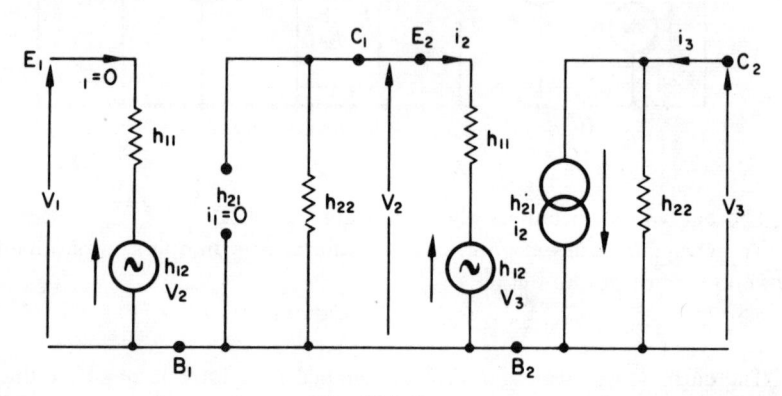

Fig. 23.6.2.

This is calculated by applying v_3 and calculating i_3. The output conductance $= i_3/v_3$.

We must open circuit the input so as to cause i_1 to become zero. If $i_1 = 0$, $h_{21} \, i_1 = 0$.

The current flowing in h_{11} (second transistor) $= \dfrac{-h_{12} \, v_3}{h_{11} + \dfrac{1}{h_{22}}}$

H PARAMETERS

(This is negative as the e.m.f. due to $h_{12} v_3$ is antiphase to the assumed direction of i_2).

Tidying up the equation for i_2, $i_2 = \dfrac{-h_{12} h_{22} v_3}{1 + h_{11} h_{22}}$

The generator $h_{21} i_2$ generates a current, $\dfrac{-h_{21} h_{12} h_{22} v_3}{1 + h_{11} h_{22}}$

The applied voltage v_3 also causes a current to flow, $v_3 h_{22}$. The output resistance is given as

$$\frac{v_3}{i_1 + i_2}$$

where i_2 represents i_3, and i_3 represents the current due to v_3,

$$\therefore H_{22} = \frac{i_3 + i_2}{v_3}$$

$$= \frac{v_3 h_{22} - h_{21} h_{22} h_{12} v_3}{(1 + h_{11} h_{12}) v_3} = \underline{h_{22} - h_{21} h_{22} h_{12}}$$

23.7. H_{21}. A simpler approach (any configuration)

This approach may be applied to a compound cascaded circuit irrespective of the individual configurations.

A general approach found to be most useful, is as follows:
For the first transistor, label each parameter as A_{11}, A_{12}, A_{21}, A_{22}. For the second, B_{11}, B_{12}, B_{21}, B_{22}, and so on for each transistor.

Consider the following circuit, in figure 23.7.1. using this method of attack.

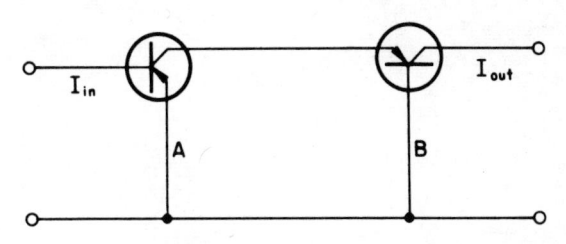

Fig. 23.7.1.

Transistor A is a grounded emitter, whilst B is a grounded base. The equivalent H circuit is as follows in figure 23.7.2.

Following the rules for the previous example an expression is obtained for H_{21}

$$H_{21} = -A_{21} B_{21}$$

but as
$$A_{21} \text{ is } h'_{21} \text{ and } B_{21} \text{ is } h_{21}$$
$$H_{21} = -h'_{21} h_{21} \text{ as the current gain.}$$

Therefore
$$H_{21} = \alpha' \cdot \alpha = \frac{\alpha^2}{1-\alpha} \text{ and if } \alpha = 0.98,$$

$$H_{21} = \frac{0.98 \times 0.98}{1 - 0.98} \simeq \underline{48.}$$

This method is less complicated as the suffixes are not included until the final result is obtained.

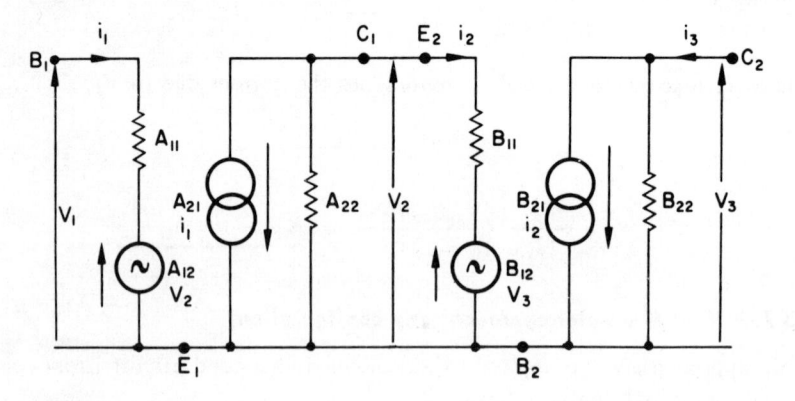

Fig. 23.7.2.

CHAPTER 24

M. O. S. T. devices

One common method of producing microelectronic devices is as follows.

A bar of silicon is cut into thin slices about 0.25 mm thick. These are put into a reaction chamber during which an epitaxial process occurs thus coating one side of the slice with silicon oxide.

These are then coated with a layer of 'photo-resist', a material which hardens after exposure to ultraviolet light. The slice is then subjected to ultraviolet light via a mask that contains patterns of parts of an electronic circuit.

The slice is then 'developed' and subsequently etched to leave exposed area which then undergo a diffusion process.

This is repeated many times, each time the circuit is growing more and more complete as different masks are used.

Once the circuit is complete, and there are about 400 on each slics, a thin layer of aluminium is evaporated on to the slice. Further etching removes the unwanted aluminium leaving a good electrical contact between components and provides a means of connecting external leads to the relevant parts of the circuit.

The M.O.S.T. is also manufactured as a discrete component and it is this discrete component we shall discuss in this chapter.

24.1. Metal oxide silicon transistors (M.O.S.T.)

Of the many problems that arise during the analysis of junction transistor circuits, the allowance for the base current predominates. If it were not for this, the analysis would be as simple as that for the thermionic valve. The transistor unfortunately presents a comparatively low resistive load to the input signal source. The transistor is a current amplifier whilst the valve is essentially a voltage amplifier. The need for a transistor that possesses some of the qualities of the valve and yet to retain some of the qualities of the transistor, has been obvious since transistors were first introduced. The following is a brief provisional introduction to one of the more recent n type devices and to some elementary circuit analyses.

Figure 24.1.1. shows a simple block schematic construction of a M.O.S.T. transistor (n type).

Both the source and the drain are heavily doped n-type regions and are diffused into the p-type silicon substrate. The aluminium control electrode, known as the gate, is insulated from the n-type channel by a layer of silicon dioxide (C).

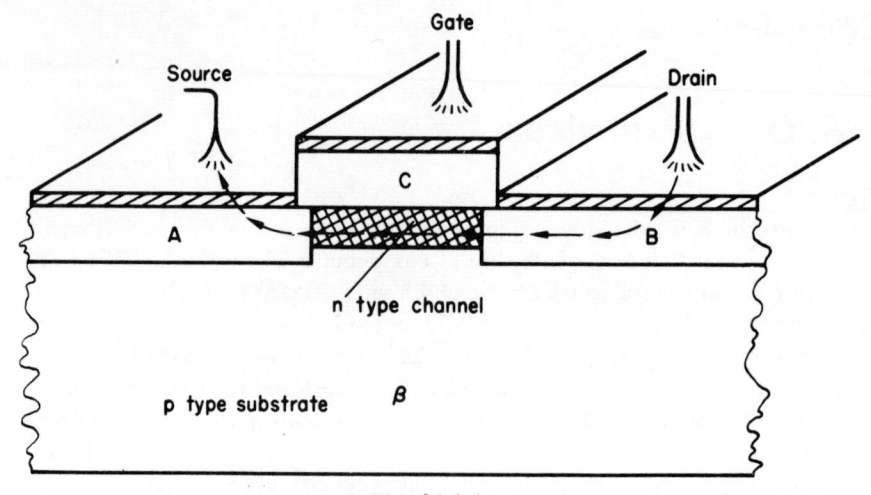

Fig. 24.1.1.

The gate electrode will form one plate of a capacitor whilst the other plate may be seen to be the *n*-type channel. The dielectric of the capacitor is the silicon dioxide insulator (*C*). The source electrode is connected to the *n*-type region (*A*), and the drain electrode is similarly connected to the *n*-type region (*B*). Figure 24.1.2. shows a typical amplifier circuit compared with a triode amplifier.

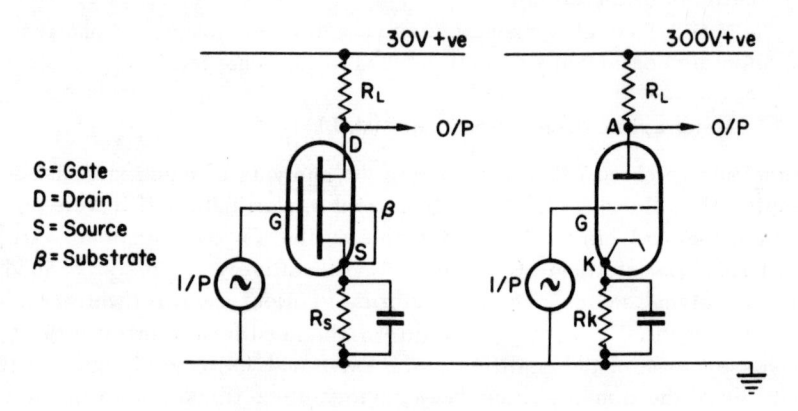

G = Gate
D = Drain
S = Source
β = Substrate

Fig. 24.1.2.

Electron current will flow by way of the channel to the drain from the source. Conventional current will therefore flow from the drain to the source. There is no *p*–*n* junction across which the current will have to flow, hence there will be no depletion layer to repel or attract the current that will flow in the path outlined. In the *n* type shown, the current carriers are electrons. With p types, the current carriers will be holes. A p–n junction will exist between the drain and the substrate and although this does not obstruct the signal flow path, may be very useful as we will see at a later stage. It will be sufficient for the time being, to consider that the substrate is to be connected to the outer case which may be assumed to be at the source potential. (The substrate is often connected internally to the case, and the source is connected externally to the substrate in most circuits.)

Figure 24.1.3. shows a positive potential applied to the gate. The effect of the input is illustrated.

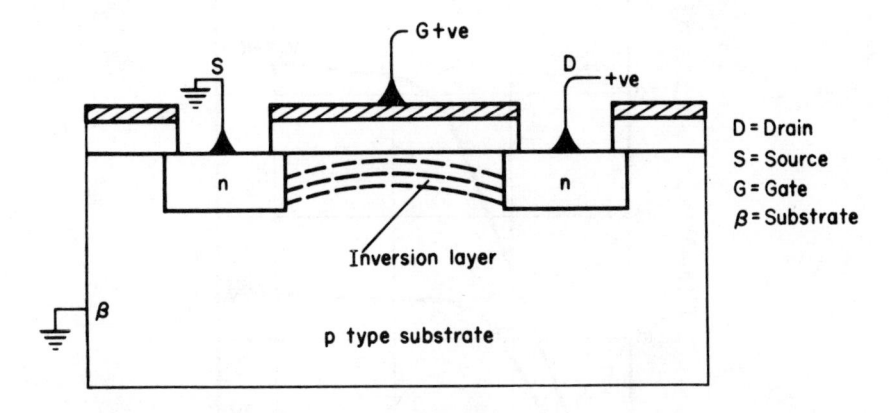

Fig. 24.1.3.

When a positive going input is applied to the gate, the 'upper plate' of the capacitor will be charged to the value of the input. The substrate will be charged to a negative value. The negative charge will result from the movement of the minority carriers in the substrate as holes are repelled from the capacitor. An n-type inversion layer which provides a conductive path will exist a little below the semiconductor surface. The inversion layer will be approximately proportional to the input potential. The higher the input, the thicker the inversion layer. The thicker the inversion layer, the lower the resistivity between the source and the drain, thus allowing a larger current to flow.

The n-type source is connected to an n-type drain by an n-type inversion layer. Even without the gate positively biassed, a drain current will flow depending upon the amount of doping in the substrate. The capacitance between the gate and the source will be in the order of 4 pF. The capacitance between the gate and the drain will be in the order of 0.5 pF. and the capacitance between the drain and source 3 pF. For the device under discussion the gate will require a potential of -5V with respect to the source before current flow ceases.

If considered as a low to medium frequency device, the capacitors may be ignored. The device is essentially a voltage amplifier unlike its predecessors which are current amplifiers. The input resistance of the 95 BFY, made by Mullards, is $10^{12}\,\Omega$. The control of the drain current is made possible by the field effect of the capacitor. This is also known as the 'insulated field effect'. The drain characteristics are shown in figure 24.1.4. These curves show the type of characteristics. It is not necessarily a typical curve. The 'cut off' is known as the 'pinch off' for these devices.

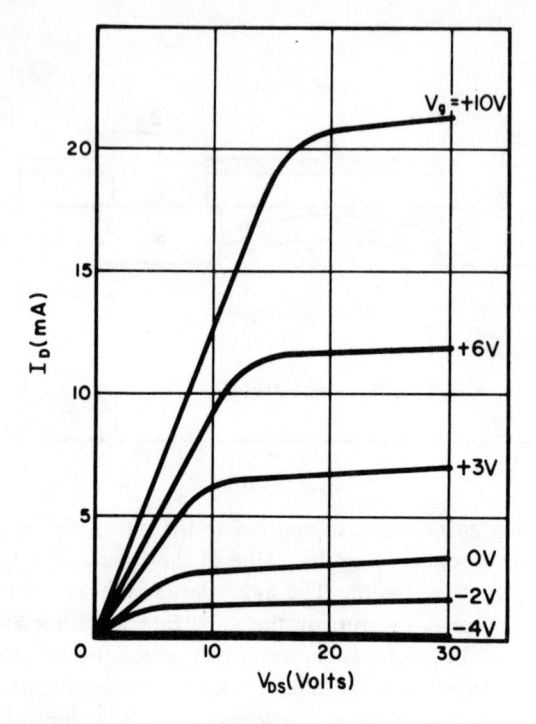

Fig. 24.1.4.

The I_D/V_{DS} characteristics may be seen to be very similar to those of a pentode. Note that for zero or small negative gate voltages, drain current will continue to flow. (This development model is one of three types. Each of the three types require different values of gate voltage to cause the drain current to become zero. Some types have a pinch off voltage that is positive.)

The source is normally considered connected to earth or chassis. The substrate is normally connected to the source. The substrate may be connected to a different potential resulting in a variation of the drain-substrate $p-n$ junction. This will also control the drain current in a manner not unlike that of the suppressor grid of a pentode. The substrate should not be allowed to become positive to the drain or source. If an input is applied to the substrate, and an input applied to the gate, the resultant drain current will be proportional to the product of the two voltage inputs.

Figure 24.1.5. shows the mutual characteristics for the device.

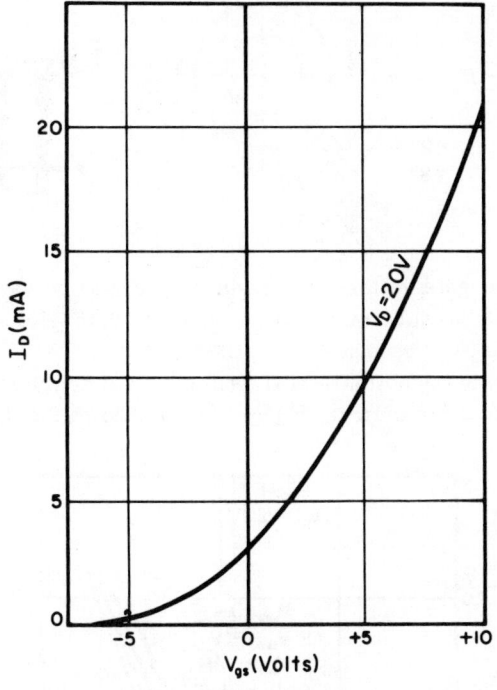

Fig. 24.1.5.

It should be noted that for any given constant value of gate potential, the drain current may be controlled by the substrate potential. These devices have a square law transfer characteristic and hence may be very useful for mixer circuits, etc. The very high input resistance will allow the devices to be used in many circuits that hitherto were only possible with other devices such as electrometer valves. As there is no base current to allow for, the calculations when using the devices in multivibrator and switching circuits, are much simplified.

There will be small drain leakage even when the device is 'cut off'. This will usually be considerably less than 50×10^{-9} amps. This quantity may be more readily expressed as 50 nA. For the device considered, the pinchoff is -5 V.

Because the pinch off voltage for the 95 BFY is negative, a small negative gate potential will allow conduction as may be seen from the I_D/V_{DS} characteristics in figure 24.1.4.

The physical dimensions of the 95 BFY are shown in figure 24.1.6.

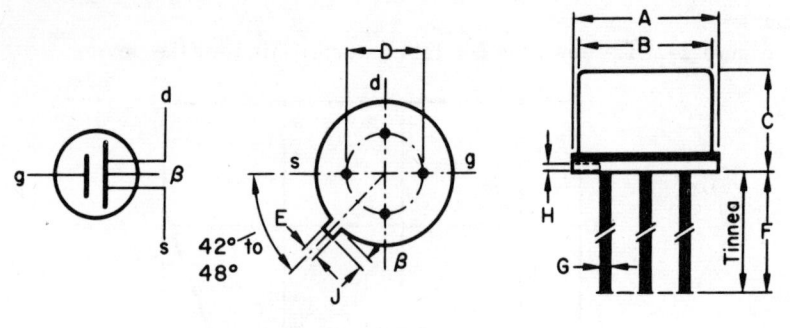

Fig. 24.1.6.

With a constant gate voltage, the drain current may be varied according to the potential of the substrate. With an input to both gate and the drain, the device may be used as a mixer-oscillator, etc.

The drain to gate capacitance will remain almost constant irrespective of the value of the drain current. Figure 24.1.7. shows the I_{DS}/V_{GS} characteristics.

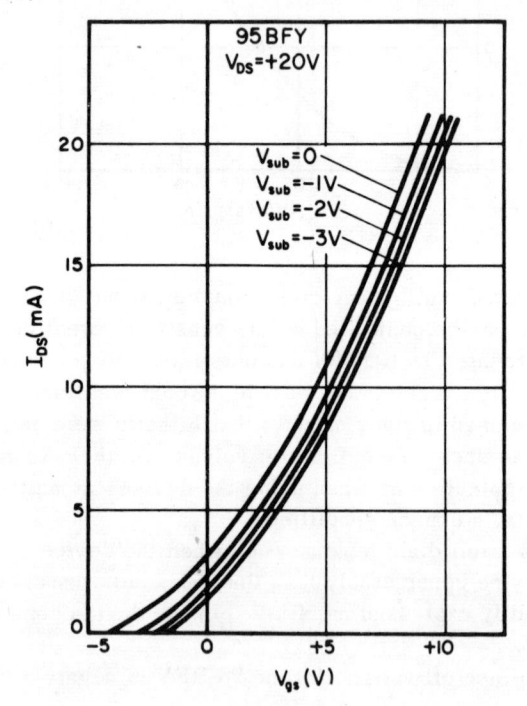

Fig. 24.1.7.

M.O.S.T. DEVICES

These characteristics show quite clearly that, for a constant gate voltage, the drain current will depend on the substrate potential.

24.2. A simple amplifier

Figure 24.2.1. shows a simple circuit diagram of an amplifier using a M.O.S.T. device.

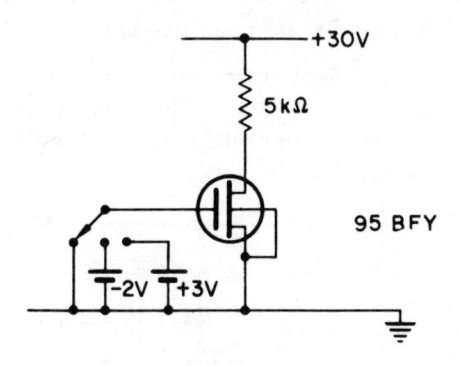

Fig. 24.2.1.

Figure 24.2.2. shows the drain characteristics with a load line drawn.

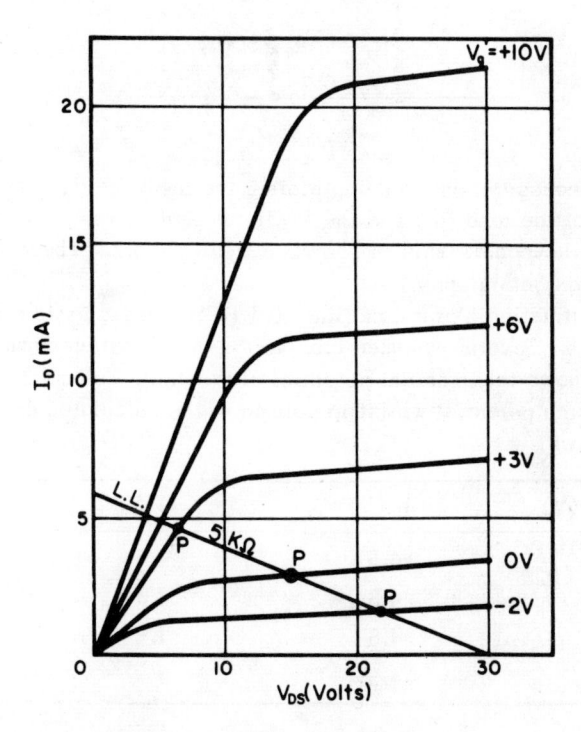

Fig. 24.2.2.

With zero bias (V_{GS}) the operating point, P, is seen to be (15, 3.). With S_1 set to position 2, the $-2\,V$ bias results in an operating point of (22, 1.6.). With the switch S_1 set to the third position, the 3 V bias results in an operating point, P, of (6.5, 4.7). It may be seen that the device conducts over a range of bias from a positive or a negative voltage. (Great care must be taken when plotting bias load lines as will be demonstrated later.)

24.3. Analysis of an amplifier with positive bias

Figure 24.3.1. shows a circuit of an amplifier employing a positive bias. The actual bias voltage (V_{gs}) will be smaller than the gate voltage as due to the normal source-follower action, the source will follow the gate and will be at a potential just below the gate.

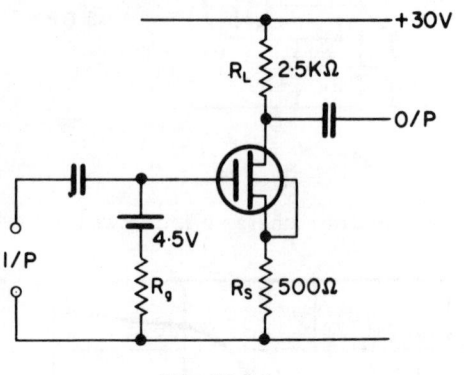

Fig. 24.3.1.

It will be necessary, once having plotted the load line, to refer the intersections of the load line and the family of gate curves, to a second graph of a similar scale. This is shown in figure 24.3.2. (These are shown as point $A-F$ on both graphs.)

Before constructing a bias load line, it will be necessary to draw up a table as follows. Several assumed bias voltages are written down in column 1. Column 2 shows the constant potential of the gate. Column 3 shows the calculated source potential whilst in column 4, the calculated drain current values are shown.

Assumed bias (V)	V_g (V)	V_s (V)	I_D (mA)
0	4.5	4.5	9.0
1.5	4.5	3.0	6.0
3.0	4.5	1.5	3.0
4.5	4.5	0	0

M.O.S.T. DEVICES

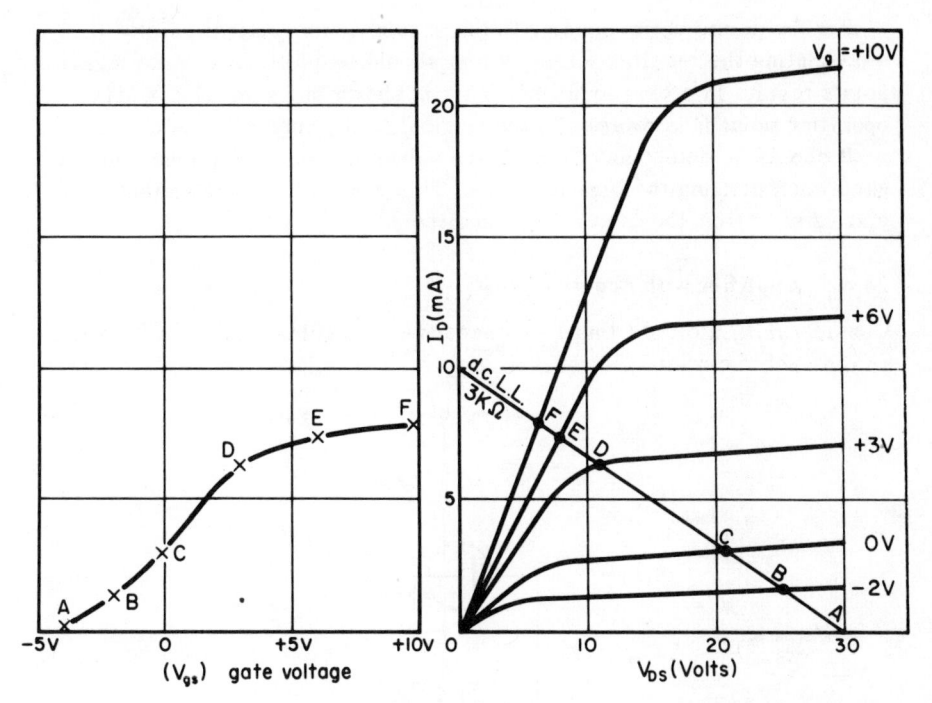

Fig. 24.3.2.

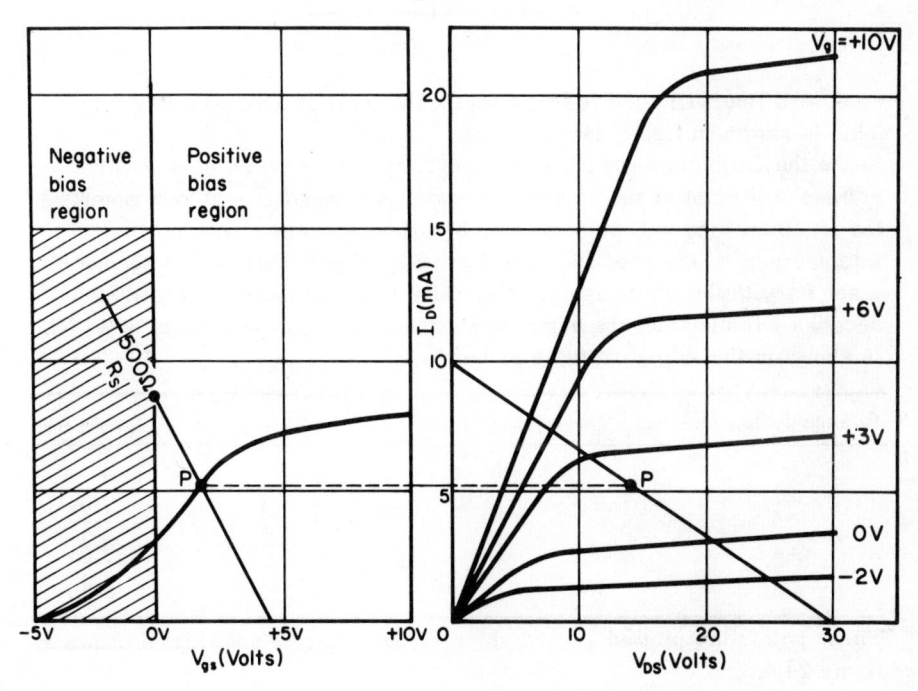

Fig. 24.3.3.

For the points corresponding to the assumed bias, further points representing the resultant drain current should be plotted. Joining these points results in a bias load line. This is shown in figure 24.3.3. The operating point P is referred to the original I_D/V_D characteristics.

It should be noted that the *negative* values of bias voltage were *not* used when plotting the bias load line. This was due to the fact that the bias is seen from the circuit to be positive.

24.4. Amplifier with negative bias

Consider now, a circuit that has a negative bias voltage. Figure 24.4.1. shows such a circuit.

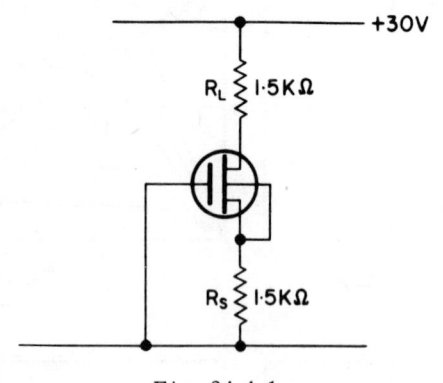

Fig. 24.4.1.

A load line will need to be drawn for the total of $3\,K\Omega$ and $30\,V$ h.t. This is shown in figure 24.4.2.

As the drain current will flow through the source resistor, a positive voltage will exist at the source electrode. As the gate is at zero potential, the resultant bias will be negative. It will be necessary to refer the intersections of the load line and the family of gate curves to a second graph as with the previous example. Once referred however, it will be necessary to consider the *negative* portion of the bias potentials only. This is shown in the following table as before.

Assumed bias (V)	V_g (V)	V_s (V)	I_D (mA)
0	0	0	0
−2	0	2	1.33
−4	0	4	2.66
−6	0	6	4.00

These points are plotted as with the previous example and this is shown in figure 24.4.3.

Fig. 24.4.2.

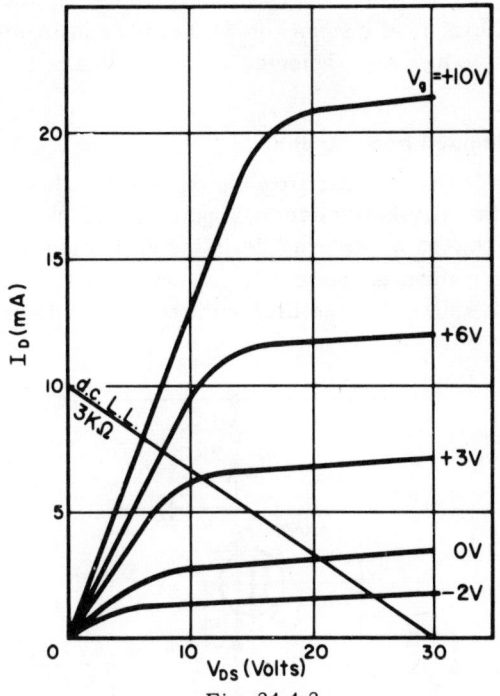

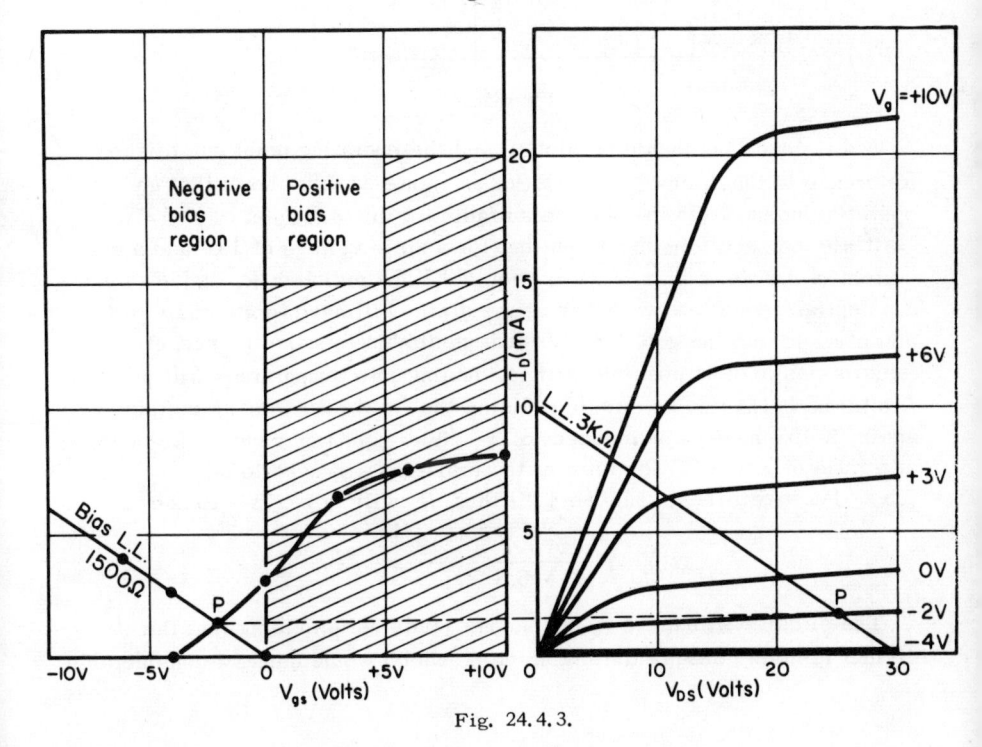

Fig. 24.4.3.

Note that the bias load line was drawn in the *negative* region and that the positive bias values were ignored. The various d.c. levels may be seen from the graph.

Positive bias obtained from the drain

Positive bias is easily obtained from the one supply. In a similar manner as was used in the transistor circuits, a potential divider may be connected from the drain to the gate. An example in the derivation of a simple amplifier resistor values is shown.

Figure 24.4.4. shows the amplifier with the gate voltage derived from the drain.

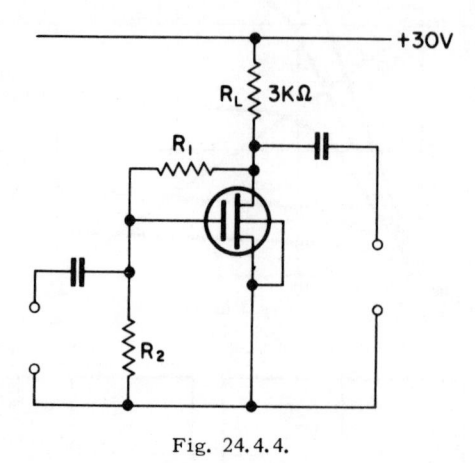

Fig. 24.4.4.

A d.c. load line should be plotted and the operating point established. Reference to the drain characteristics in figure 24.4.5. shows that an operating point of (15, 5.) will be suitable for this example.

It may be seen from the graph that for a drain voltage of 15V and a drain current of 5.0 mA, a bias of 2V is required. The resistors R_1 and R_2 form a potential divider and with 15V at the drain, will have to attenuate to this potential so that the gate has 2V. It is desirable to make the sum of the resistors as high as possible without running into temperature drift problems A total of 10 MΩ may be suitable. Using the 'load over total' principle once again, if 15V have to be reduced to 2V, the divider will need to attenuate at a ratio of 7.5 to 1. The ratio of the resistors will need to be 7.5 − 1 = 6.5 to 1. R_2 may be 1 MΩ thus R_1 will be 6.5 MΩ. Hence

$$V_g = \frac{15 \times 1}{1 + 6.5} = \frac{15}{7.5} = 2V.$$

The circuit will behave in the manner predicted providing that the divider resistors are not too low in value which would cause a considerable

M.O.S.T. DEVICES

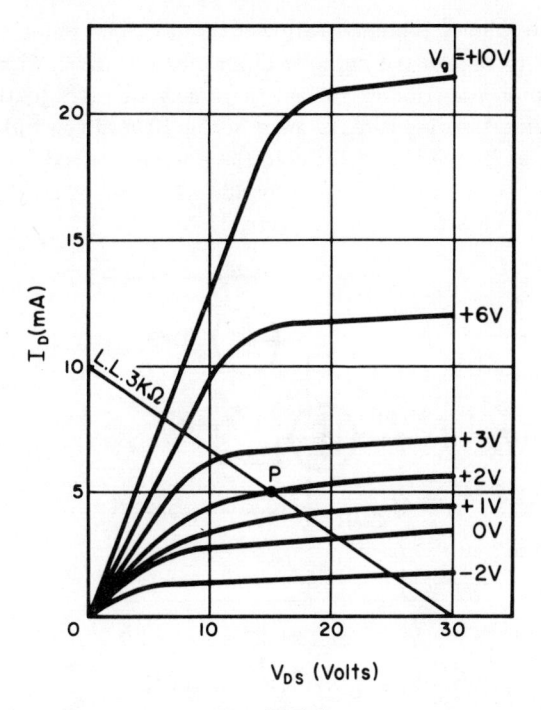

Fig. 24.4.5.

current to flow in the chain thus upsetting the drain potential. Making the resistors too high might lead to drift problems due to the temperature coefficient of the resistors. R_1 should be tapped and bypassed as with the comparable circuit using junction transistors, if a.c. degeneration is to be avoided.

Source followers

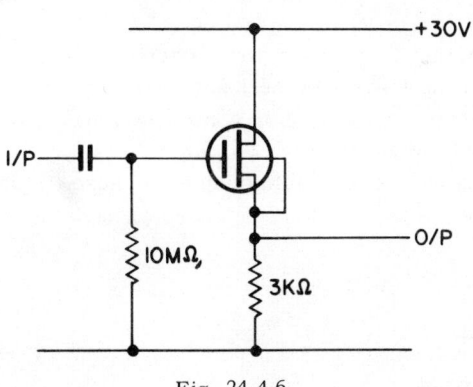

Fig. 24.4.6.

Figure 24.4.6. shows a source follower circuit. The function of the stage is similar to that of the cathode of emitter follower. With an M.O.S.T. input-resistance of a million megohms, there may be little justification in attempting to raise the 'input-resistance of the 10 MΩ even further by connecting the 'gate leak' (grid leak) to the source (cathode).

The gain may be increased simply by adding a further transistor to the collector circuit. This is shown in figure 24.4.7.

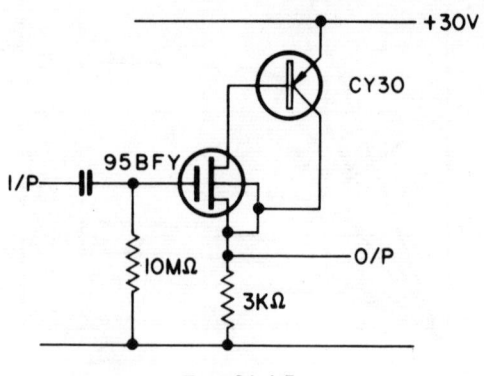

Fig. 24.4.7.

The output-resistance will be in the order of 200 Ω.

These devices may be used in a variety of circuits. Some of these are shown in a basic form in figure 24.4.8.

The M.O.S.T. has some of the qualities of both the valve and the transistor. It has a very high input resistance which can be compared very favourably with the grid circuit of a valve. The drain characteristics are similar to those of a pentode. With a high value of drain load resistor, a useful voltage gain may be obtained. It will probably be common practice to run some of these particular devices at about 1 mA.

Although at the time of writing these devices were in the early development stage, experiments carried out with one or two of the devices show quite clearly that they may become a serious rival for place of importance to both the valve and the junction transistor. The substrate provides a most useful means of gain control in a manner that will not seriously affect the input stage.

The 95BFY was selected for this chapter because its characteristics lent themselves most readily for the intended examples.

Xtal mike

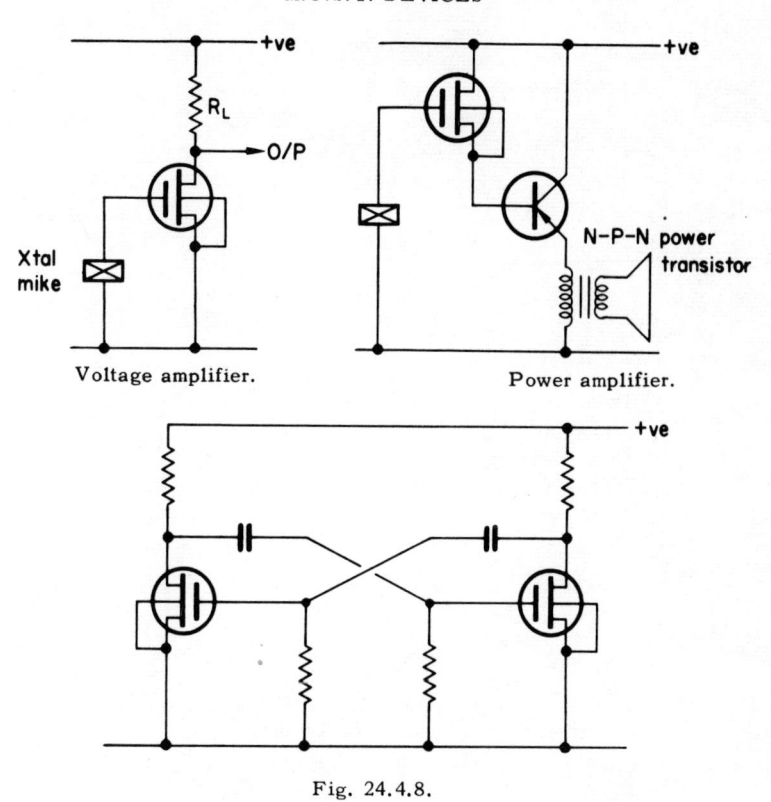

Voltage amplifier.

N-P-N power transistor

Power amplifier.

Fig. 24.4.8.

CHAPTER 25

Ladder networks and oscillators

In this chapter we will be looking at a particular type of ladder network. One or two very simple tests will be applied to the networks and formulae derived for the attenuation of the network.

Once we have dealt with simple resistive networks and having discussed a number of techniques, we will replace some resistors by reactive components.

These networks themselves may be useful but in this chapter their real value is perhaps in the manner in which they can be used to simulate other circuits. These other circuits are sometimes a little tricky to analyse quickly, but when using the ladder network approach, analyses are very greatly simplified.

A number of alternatives have been shown where practicable throughout this book which helps to allow more lattitude in practical every day life, it enables second and possibly third methods to be used to re-check answers obtained by other means.

These alternatives also help to broaden ones outlook and should convince the reader that the quickest way to approach a problem may include time initially set aside to think about the problem first of all and decide upon the best method for that particular task.

25.1. Ladder networks

A simple ladder type attenuator consisting of one stage of attenuation is shown in figure 25.1.1.

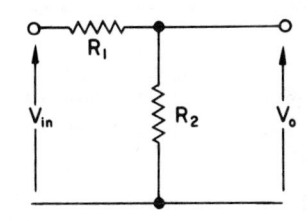

Fig. 25.1.1.

From load over total, $\dfrac{V_o}{V_1} = \dfrac{R_2}{R_1 + R_2}$. Then invert both sides of the equation.

And $\qquad \dfrac{V_1}{V_o} = \dfrac{R_1 + R_2}{R_2} = \dfrac{R_2}{R_2} + \dfrac{R_1}{R_2} = 1 + \dfrac{R_1}{R_2},$

but $1/R_2$ may be expressed as an admittance (Y). In order to establish a simple, general formula, we will write $1/R_2$ as Y. If, then, $1/R_2$ may be written as Y and R_1 written as Z (following the same arguments), then, as

$$\frac{V_1}{V_0} = 1 + \frac{R_1}{R_2},$$

this may be written as $V_1/V_0 = 1 + ZY$.
An example may be given, taking $R_1 = 10\Omega$ and $R_2 = 20\Omega$ and $V_1 = 3V$.

From $\dfrac{V_1}{V_0} = 1 + ZY$, then $\dfrac{V_1}{V_0} = 1 + \dfrac{R_1}{R_2} = 1 + \dfrac{10}{20} = 1.5$.

$V_0 = \dfrac{V_1}{1.5}$ or $\dfrac{2}{3} V_{in}$. $\therefore \dfrac{V_{in}}{V_0} = 1.5$. By load over total,

$V_0 = \dfrac{V_{in} \times 20}{30} = \dfrac{2}{3} V_{in}$ $\therefore \dfrac{V_{in}}{V_0} = 1.5$

in the same manner, and the answers are identical.

This rather simple example may be used to evaluate very complex problems, and should not be lightly dismissed.

A more complicated attenuator consisting of four stages is shown in figure 25.1.2. and the general formula for any number of stages is also given.

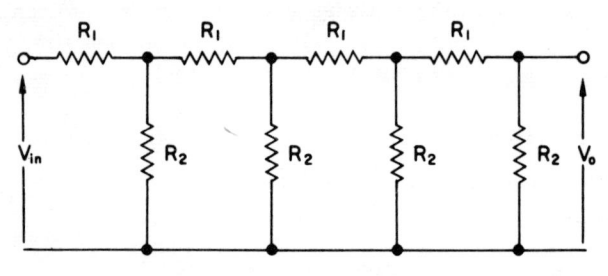

Fig. 25.1.2.

Note: This chapter deals solely with the special case where all resistors marked R_1 have the same value. The values of R_2 are also assumed to be constant.

$$\frac{V_1}{V_0} = 1 + AZY + BZ^2Y^2 + CZ^3Y^3 + DZ^4Y^4 + EZ^5Y^5 + \ldots 1.$$

If we simplify the product ZY and replace it by S, where S is a constant, the formula becomes

$$\frac{V_1}{V_0} = 1 + AS + BS^2 + CS^3 + DS^4 + ES^5 + \ldots 1.$$

LADDER NETWORKS AND OSCILLATORS 501

The coefficients of S are determined from a rather 'distorted' Pascal's Triangle. The triangle is shown in figure 25.1.3. and also the method of construction. The triangle itself need not be memorised, it is easy to construct, and after a little practise may be drawn in a minute or two.

How to Construct the Triangle.

Write down a number one.   1
This is the apex of the triangle.
Then write a figure one beneath the 1, above, and a light dot to the   1
left of the second number 1;   .1
The dot must now be replaced by a number (which will vary from line to line). You obtain the number by adding the figure above the dot to its right hand neighbour, in this case, $0 + 1 = 1$.
Then replace the dot by this figure 1.   1
    11

The dot in column two is now replaced by the sum of the number above the dot and its right hand neighbour; 2, in this instance.   1
    11
    .21

The remaining dot (far left hand side) is obtained in the same way and always becomes a figure 1.   1
    11
    121

Now place a 1 in the right hand column, and 3 dots. The dot nearest the right hand column is replaced by a number equal to the sum of the number above the dot and its right hand neighbour, $(2 + 1) = 3$ in this case.   1
    11
    121
    ... 1
    1
    11
    121
    ..31

The far left hand dot is *always* replaced by a figure 1, whilst the remaining dot becomes a $(2 + 1) = 3$ also.   1
    11
    121
    1331

This process may be continued indefinitely, but of course one would stop at the appropriate point once a sufficient number of coefficients has been obtained.

Each line consists of one more dot than the previous line. Each dot is replaced by a number corresponding to the sum of the figure immediately above the dot plus that figure's right-hand neighbour. The dots need not be drawn in practice; they are merely given here as a constructional guide. A complete triangle sufficient to obtain co-efficients for 4 stages is shown below in figure 25.1.3.

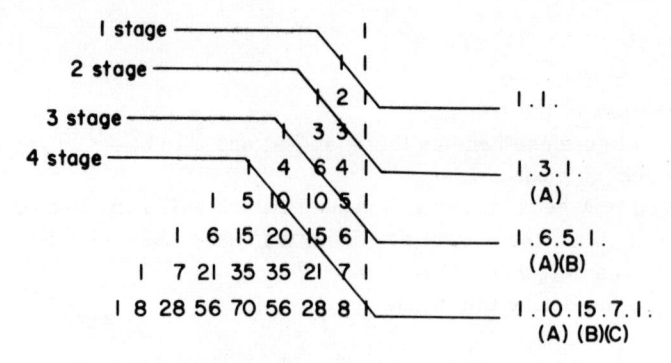

```
  I stage ─────────────┐    │
  2 stage ───────────┐ │ I  │
                     │ 2 │  │────── 1.1.
  3 stage ─────────┐ │ │ │  │
                   │ 3 3 │  │
  4 stage ───────┐ │ │ │ │  │
                 │ 4 6 4 │  │────── 1.3.1.
           I   5   10  10  5  I                    (A)
         I   6  15  20  15  6  I    ────── 1.6.5.1.
                                                   (A)(B)
       I   7  21  35  35  21  7  I
     I   8  28  56  70  56  28  8  I ────── 1.10.15.7.1.
                                              (A)(B)(C)
```

Fig. 25.1.3.

An example in the use of this technique is shown below.

25.2. The Wien network

The two relevant arms of a Wien bridge are shown below in figure 25.2.1.

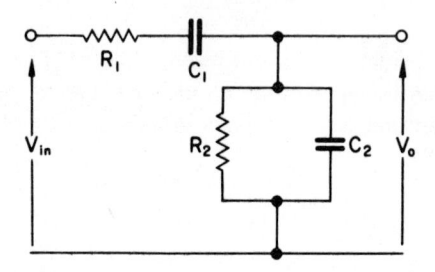

Fig. 25.2.1.

It is drawn as a single stage attenuator.

$$\frac{V_1}{V_o} = 1 + ZY$$

Where $Z = \left(R_1 + \dfrac{1}{j\omega C_1}\right)$ and $Y = \dfrac{1}{R_2} + j\omega C_2$

$$\frac{V_1}{V_o} = 1 + ZY = 1 + \left(R_1 + \frac{1}{j\omega C_1}\right)\left(\frac{1}{R_2} + j\omega C_2\right)$$

$$\therefore \quad \frac{V_1}{V_o} = 1 + \frac{R_1}{R_2} + j\omega C_2 + R_1 + \frac{1}{j\omega C_1 R_2} + \frac{C_2}{C_1}$$

LADDER NETWORKS AND OSCILLATORS

If we let $R_1 = R_2$ and $C_1 = C_2$

then
$$\frac{V_1}{V_0} = 1 + 1 + j\omega CR + \frac{1}{j\omega CR} + 1$$

Hence
$$\frac{V_1}{V_0} = 3 + j\omega CR - \frac{1}{\omega CR} \tag{1}$$

If we equate the j term to zero,

then $\omega CR = \dfrac{1}{\omega CR}$ ∴ $\omega^2 = \dfrac{1}{C^2 R^2}$

and $\omega = \dfrac{1}{CR}$ and $f = \dfrac{1}{2\pi CR}$

This is the frequency at which each arm has the same impedance. We need to know the attenuation and to see if there is any 180° phase shift. And considering the 'real' or active terms
$$\frac{V_1}{V_0} = 3$$
the attenuation is 3 and as there is no negative sign, V_0 is in phase with V_{in}.

A rather simpler example is given in figure 25.2.2. to illustrate a fourth application of the general formula. (By load over total).

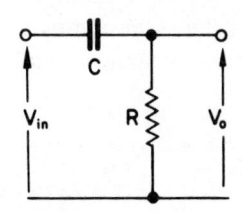

Fig. 25.2.2.

Now, $\dfrac{V_0}{V_1} = \dfrac{\text{Load}}{\text{Total}} = \dfrac{R}{R + \dfrac{1}{jwc}}$

and $\dfrac{V_1}{V_0} = \dfrac{R + 1/j\omega C}{R} = 1 + \dfrac{1}{j\omega CR}$

(by the formula)

$\dfrac{V_1}{V_0} = 1 + Zy = 1 + \dfrac{1}{j\omega C} \times \dfrac{1}{R}$ which of course is the same.

A final example in figure 25.2.3. using resistors only. This is a 2 stage problem. Figure 25.1.3. shows the co-efficient of $A = 3$.

$$Z = 10\Omega \quad Y = 1/5\Omega.$$

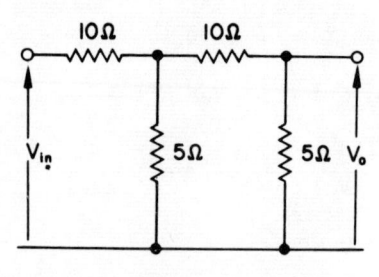

Fig. 25.2.3.

$\dfrac{V_1}{V_0} = 1 + 3ZY + Z^2Y^2$ (where '3' is obtained from the triangle)

$\dfrac{V_1}{V_0} = 1 + \dfrac{3 \times 10}{5} + \dfrac{(10)^2}{5^2}$

$\dfrac{V_1}{V_0} = 1 + 6 + 4 = 11$

∴ $V_0 = \dfrac{1}{11}$ of V_1 which is very useful for this type of network and very often much quicker than other methods.

The proof of the general formula is beyond the scope of this book, and consequently should be used only in practice and not when sitting for examinations requiring proof of any formulae used. As an exercise in handling various techniques, the following example is given in figure 25.2.4. It embraces many of the techniques outlined in previous pages.

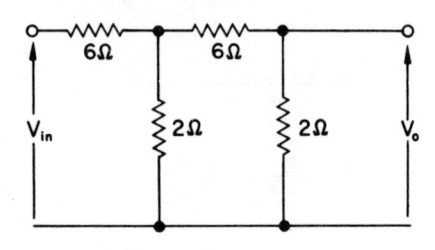

Fig. 25.2.4.

$\dfrac{V_1}{V_0} = 1 + 3 \dfrac{(6)}{2} + \dfrac{(6)^2}{(2)^2} = 1 + 9 + 9 = 19$

∴ $\dfrac{V_0}{V_1} = \dfrac{1}{19}$

Or, as an exercise, convert part of the circuit to a T network. (Figure 25.2.5.).

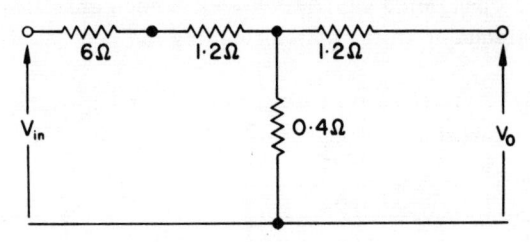

Fig. 25,2.5.

(as measured with a loss free voltmeter).

from load over total $\dfrac{V_o}{V_{in}} = \dfrac{0.4}{7.6} = \dfrac{1}{19}$ as before.

Note that the remaining 1.2Ω does nothing to modify our calculations as no current is flowing through it, as V_o is the open circuit voltage. Now by Kirchoff's laws: (figure 25.2.6.).

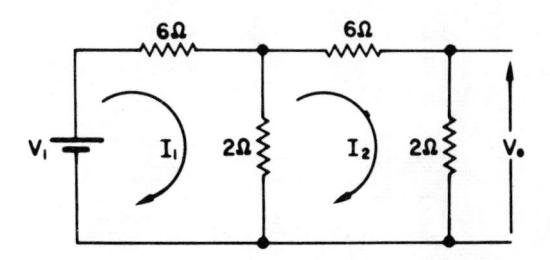

Fig. 25.2.6.

$$V_o = I_2 (2)$$

$$V_1 = I_1 (6 + 2) - I_2 (2) \therefore V_1 = 8I_1 - 1I_2 \quad (1)$$

$$0 = I_1 + I_2 (6 + 2 + 2) \text{ and } 0 = 2I_1 + 10I_2 \quad (2)$$

(1) $I_1 = \dfrac{V_1 + 2I_2}{8}$ and substituting in (2)

$$0 = \dfrac{-2(V_1 + 2I_2)}{8} + 10I_2$$

∴ $\quad 0 = \dfrac{-V_1}{4} - \tfrac{1}{2}I_2 + 10I_2 \quad \therefore \quad 0 = -V_1 - 2I_2 + 40I_2$

and $\quad 0 = -V_1 + 38I_2$ but as $V_o = 2I_2$

$\quad 0 = -V_1 + 19 V_o \quad \therefore \quad V_o/V_1 = 1/19.$

A final example, using nodal analysis is given.

Each node is considered separately, and in each case the current flowing into the node is equated to the current leaving the node. Let the nodes be (A) and (B).

The voltage V, is assumed (it disappears later but we need it in order to build up our equations). (Figure 25.2.7.).

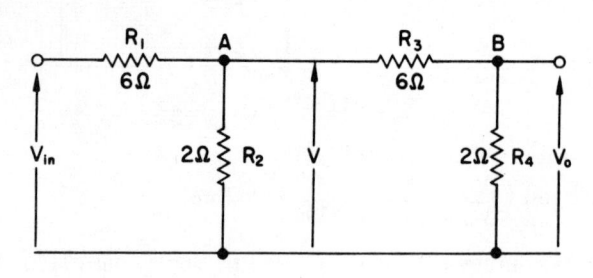

Fig. 25.2.7.

Node A   Current leaving = Current entering

$$\therefore \quad V\left[\frac{1}{R_1} + \frac{1}{R_2} + \frac{1}{R_3}\right] = \frac{V_1}{R_1} + \frac{V_o}{R_3} \quad (1)$$

Node B

$$V_o\left[\frac{1}{R_3} + \frac{1}{R_4}\right] = V\left[\frac{1}{R_3}\right] \quad (2)$$

Tidying up, (1)

$$V\left[\frac{1}{R_1} + \frac{1}{R_2} + \frac{1}{R_3}\right] = \frac{V_1}{R_1} + \frac{V_o}{R_3}$$

and (2)

$$V\left[\frac{1}{R_3}\right] = V_o\left[\frac{1}{R_3} + \frac{1}{R_4}\right]$$

divide (1) by (2) to eliminate V.

$$\frac{\frac{1}{R_1} + \frac{1}{R_2} + \frac{1}{R_3}}{\frac{1}{R_3}} = \frac{\frac{V_1}{R_1} + \frac{V_o}{R_3}}{V_o\left[\frac{1}{R_3} + \frac{1}{R_4}\right]}$$

then putting in known values:

$$\frac{\frac{1}{6} + \frac{1}{2} + \frac{1}{6}}{1/6} = \frac{\frac{V_1}{6} + \frac{V_o}{6}}{V_o\left[\frac{1}{6} + \frac{1}{2}\right]} \quad \text{(then obtain common denominator of 6)}$$

$$\frac{1 + 3 + 1/6}{1/6} = \frac{V\,1/6 + V_o\,1/6}{V_o\,(1 + 3)/6} \quad \text{(common denominator cancels)}$$

$$\therefore \quad 5 = \frac{V_1 + V_0}{4 V_0}$$

and $20 V_0 = V_1 + V_0$ $\therefore$ $20 V_0 = V_1 + V_0$ $\therefore$ $20 V_0 - V_0 = V_1$

and $19 V_0 = V_1$ $\therefore$ $\dfrac{V_0}{V_1} = \dfrac{1}{19}$ as before.

25.3. Phase shift oscillators

An oscillator consists of two major units, (a) a frequency determining network (F.D.N.) and (b) an amplifier.

In general, both units are connected as a single network where the F.D.N. attenuates the signal flowing through it, and the amplifier has a gain equal to the reciprocal of the loss in the F.D.N. The resultant signal fed back must be in phase with the input, and this is known as positive feed back (P.F.B.). This is shown in figure 25.3.1.

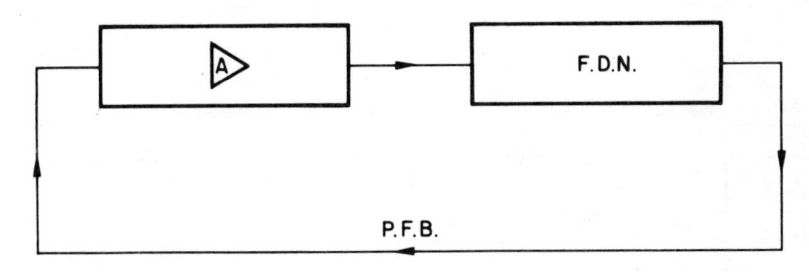

Fig. 25.3.1.

When analysing oscillators, one assumes say, a positive going input to the amplifier. If one stage of amplification only is used, then the amplifier output will normally be negative going. To cause oscillation, the output from the F.D.N. must be in phase with the assumed amplifier input, hence the F.D.N. must reverse the phase.

If the F.D.N. has an attenuation of 0 : 1, then the amplifier must have a gain of 10 times, thus causing the overall gain of the entire network to be unity.

If we use a single stage amplifier, having a 180° phase shift, then the F.D.N. must have a 180° phase shift also, in order to ensure the necessary overall zero phase shift (P.F.B.).

If we use a two stage amplifier, i.e. zero phase shift, then the F.D.N. should also have zero phase shift in order to keep the overall phase shift to zero. These are shown in figure 25.3.2.

There is a clear analogy between voltage and current networks; a similar analogy exists between valve and transistor networks. Simply, if network (1) is a voltage (or valve) network, then network (2) is the current (or transistor) analogy. Figure 25.3.3.

Single stage amplifier.

Two stage amplifier.

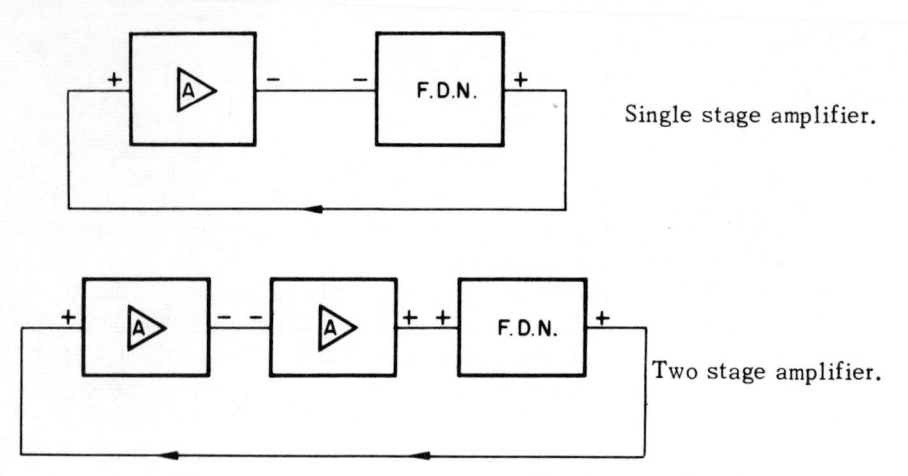

Fig. 25.3.2.

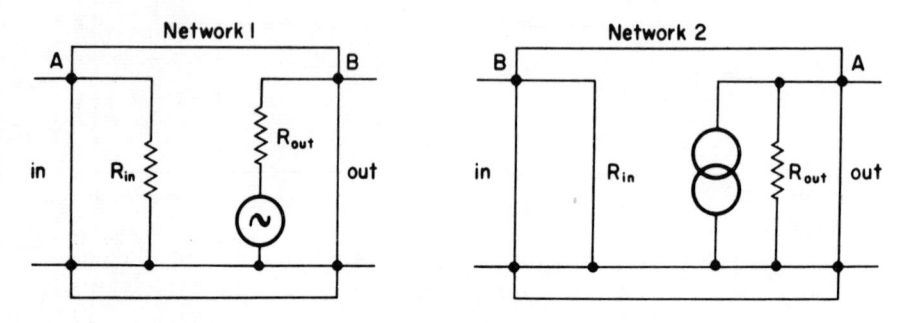

Fig. 25.3.3.

Requirements

Network 1.
R_{in} very high
R_{out} very low

Network 2.
R_{in} very very low
R_{out} very high

Note, that if we *reverse* the input and output terminals of network 2, and fulfil the above requirements, then by duality, the calculations we make for voltage networks apply equally to current networks.

Basic circuit diagram of R.C. phase shift oscillators

Note that the F.D.N. in the transistor circuit shown in figure 25.3.5 is reversed compared with the valve network shown in figure 25.3.4. The

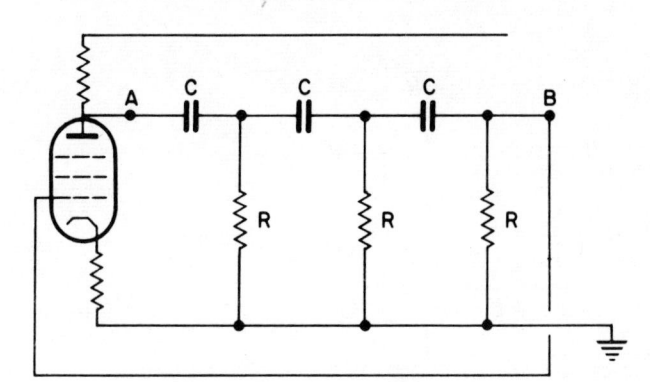

Valve.

Fig. 25.3.4.

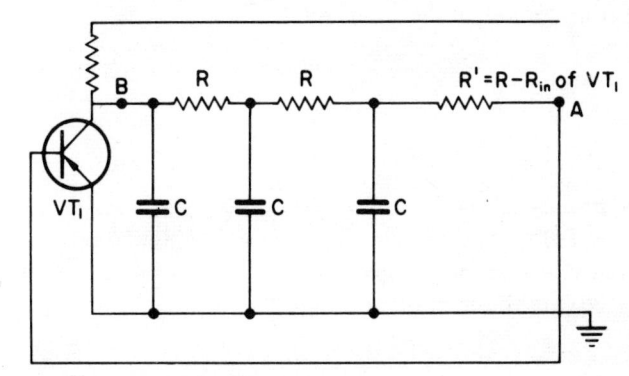

Transistor.

Fig. 25.3.5.

transistor base offers a virtual short circuit to the (A) terminal, while the grid offers an infinite resistance to the (B) terminal of the valve version.

Both F.D.N.'s are examined, using methods previously outlined. It is assumed that the d.c. conditions of valve and transistor have been decided upon.

Valve F.D.N. (Figure 25.3.6.).

Fig. 25.3.6.

Using the formula for ladder attenuators, it is required to find the attenuation and phase shift.

$$\frac{V_{in}}{V_0} = 1 + \frac{6}{j\omega CR} + \frac{5}{j^2\omega^2 C^2 R^2} + \frac{1}{j^3\omega^3 C^3 R^3}$$

$$\therefore \frac{V_{in}}{V_0} = 1 + \frac{6}{j\omega CR} - \frac{5}{\omega^2 C^2 R^2} - \frac{1}{j\omega^3 C^3 R^3} \quad (\text{where } j^2 = -1)$$

Collecting j terms and equating to zero for null.

$$\frac{6}{j\omega CR} - \frac{1}{j^3\omega^3 C^3 R^3} = 0 \quad \therefore \quad 6 = \frac{1}{\omega^2 C^2 R^2}$$

$$\therefore \quad \omega^2 = \frac{1}{6C^2 R^2} \text{ and } \omega = \frac{1}{\sqrt{6}CR} \text{ and } f = \frac{1}{2\pi\sqrt{6}CR}$$

substituting for ω^2 in the 'real' terms,

$$\frac{V_{in}}{V_0} = 1 - \frac{5}{\omega^2 C^2 R^2} \text{ gives } \frac{V_{in}}{V_0} = 1 - \frac{5}{\frac{1}{6C^2 R^2} C^2 R^2}$$

$$= 1 - 30 = -29.$$

The attenuation is 29:1 down (amplifier gain must be 29:1 up) and there is a 180° phase shift indicated by the minus sign.

As the valve has a 180° phase shift, the overall phase shift is zero, thus satisfying the requirements.

Transistor network. (Figure 25.3.7.).

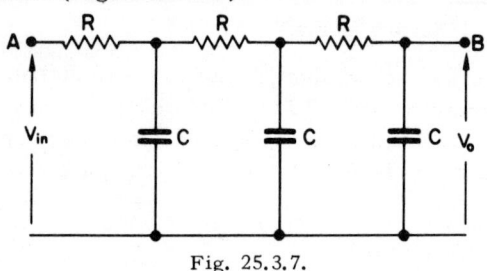

Fig. 25.3.7.

$$\frac{V_{in}}{V_0} = 1 + 6Rj\omega C + 5R^2 j^2 \omega^2 C^2 + R^3 j^3 \omega^3 C^3$$

and, as before, equating j terms to zero,

$$6R\omega C = \omega^3 C^3 R^3$$

$$\therefore \quad \omega^2 = \frac{6}{C^2 R^2} \text{ and } \omega = \frac{\sqrt{6}}{CR} \quad \therefore \quad f = \frac{\sqrt{6}}{2\pi CR}$$

and substituting ω^2 in real part of equation,

$$\frac{V_{in}}{V_o} = 1 - 5R^2C^2 \left\{\frac{6}{C^2R^2}\right\} = 1 - 30 = -29 \text{ as before.}$$

The gain of the transistor amplifier must be 29, and must have a phase shift of 180°, which it has in a common emitter configuration.

There are many variations to these basic circuits. These are offered as a preliminary step towards later, more complex, investigations into the theoretical and practical aspects of other C.R. sinusoidal oscillators the reader may expect to deal with at a later stage.

25.4. An analysis of a transistorised 3 stage phase shift network

The following analysis shows the loading effect of a transistor on a 3 stage network. The network components, when used with thermionic valves, may be fairly accurately determined. The frequency of oscillation, f_o, can also be quite accurately established.

In practice, these values work out quite well.

With junction transistors however, due to the shunt effect of the transistor output impedance, the network parameters will often have a value quite different to that of the predicted values, unless some allowance is made during the initial calculations.

The following circuit in figure 25.4.1. shows a transistor connected to a low pass network, where the generator, I_1 is the collector current generator, and the component R/n is the effective transistor amplifier output resistance. This might be the collector output resistance R_0, in shunt with the collector load resistor.

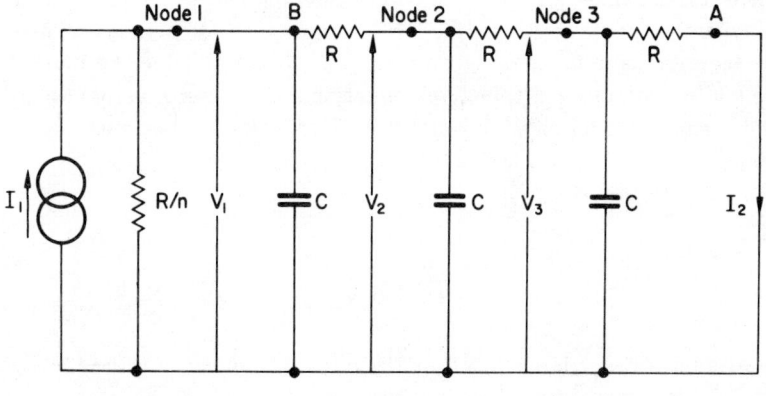

Fig. 25.4.1.

An extension of the method outlined earlier, when analysing electrical networks, will be employed. This approach is often called nodal analysis.

There are three nodes in the above circuit, whereas there was only one in the previous circuits. The method is simply to equate currents entering each node to the currents leaving each node.

Consider node 1, it can be seen that the current entering the node may be shown to be, $I_1 + V_2/R$, whilst the current leaving may be expressed as

$$V_1 \left[\frac{n}{R} + \frac{1}{R} + j\omega C \right]$$

Analysing each node separately.

$$V_1 \left[\frac{n}{R} + \frac{1}{R} + j\omega C \right] - \frac{V_2}{R} = I_1 \qquad (1)$$

Multiplying through by R, the equation becomes, where $K = j\omega CR$

$$V_1 [n + 1 + K] - V_2 = I_1 R \qquad (1)$$
$$V_2 [2 + K] - V_1 - V_3 = 0 \qquad (2), \text{ for node 2.}$$
$$V_3 [1 + K] - V_2 = -I_2 R \qquad (3), \text{ for node 3.}$$

From (3) $I_2 R = V_3$, $\therefore V_3 [2 + K] = V_2$ $\therefore V_2 = V_3 [2 + K]$
From (2) $\quad V_1 = I_2 R [(K + 2)(K + 2)] - I_2 R$.
$\therefore V_1 = I_2 R (K^2 + 4K + 3)$.
and in (1), $I_2 R (K^2 + 4K + 3)(n + 1 + k) - I_2 R (K + 2) = I_1 R$.
and $\dfrac{I_1}{I_2} = K^2 n + K^2 + K^3 + 4Kn + 4K + 4K^2 + 3n + 3 + 3K - K - 2$

As a single stage common emitter amplifier is employed and if the circuit is to oscillate, the network current feedback must be 180° out of phase with the input, or $I_2 = -I_1$. This was explained earlier, and no further mention need be made at this time. We now determine the null frequency by collecting j terms and equating them to zero, as usual, (where all odd terms are imaginary, or reactive). Thus $- K^3 = K(4n + 6)$

$$- K^2 = (4n + 6)$$

and as $K = j\omega CR$., $\omega_0 CR = \sqrt{4n + 6}$ and $f_0 = \dfrac{\sqrt{4n + 6}}{2\pi CR}$

The frequency, f_0, is modified as the term $(4n)$ appears in the root sign and should be allowed for, if greater accuracy is required.

The active, or real part, gives the attenuation,

Therefore, $\quad -5(4n + 6) + 3n + 1 - n(4n + 6) = \dfrac{I_1}{I_2}$

LADDER NETWORKS AND OSCILLATORS

$$\therefore \frac{I_1}{I_2} = -20n - 30 + 3n + 1 - (4n^2 + 6n)$$

and

$$\frac{I_1}{I_2} = \underline{-29 - 23n - 4n^2}$$

Although the first term is -29, as with previous examples, it illustrates the loading effect of R_0, as shown in the second and third terms.

If $n = 1$, the gain of the transistor amplifier should not be less than $(-29 - 23n - 4n^2) = 56$.

Therefore an a' greater than 56 is essential if the circuit is to oscillate if $n = 1$.

The frequency f_0, will be increased as the numerator is increased by the term $(4n)$ inside the root sign.

[If $n = 0$, R/n is infinity. The transistor therefore would not load the network and the expression for I_1/I_2 would revert to -29 as with previous examples.]

Increasing the number of stages results in a lesser attenuation, i.e., a value smaller than -29. The reader might derive the actual value for a four stage phase shift network.

CHAPTER 26

Zener diodes

26.1. Operating points

A particular series of diodes are carefully doped during manufacture so as to ensure that when a critical value of reverse bias is applied, the devices break down and allow current to flow. The current is limited only by both external resistance and the power dissipation of the device. These devices are known as Zener Diodes and are used in transistor circuitry in the same manner that neon stabilising tubes are in valve circuits. The voltage at which the large reverse current flows is called the zener voltage, V_z.

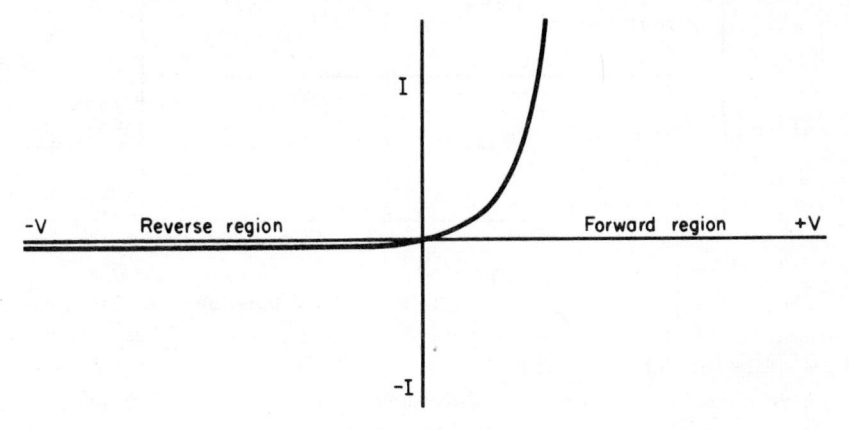

Fig. 26.1.1.

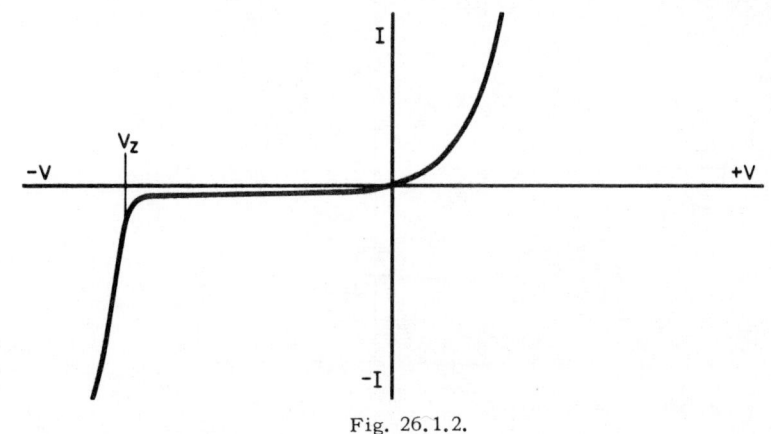

Fig. 26.1.2.

The forward characteristic of a typical zener diode is similar to the forward characteristics of a silicon diode.

The reverse bias region however, is quite different in that for a fairly low critical bias, a turn-over point exists.

These diodes must be operated well below the knee so as to maintain a stable voltage across itself. Operating on the knee will result in an indeterminate zener voltage, particularly with fluctuating supply voltage V.

The actual value of reverse bias at which the turnover occurs is determined during manufacture and is known as the Zener Voltage V_z.

These diodes have temperature coefficients which are often negative ($-ve$), positive ($+ve$) or zero according to their respective V_Z as shown in the figure 26.1.3.

V_Z (approximate)	<5V	5V	>5V
Temp coeff.	-ve	0	-ve

Fig. 26.1.3.

The diodes have a manufacturing tolerance and for diodes having a nominal V_Z of $-9V$, a voltage may be obtained in practice of 8.1 – 9.9V, i.e., a tolerance of $V_Z \pm 10\%$.

A simple reference voltage circuit consisting of a Zener diode and a series limiting resistor R_S is given in figure 26.1.4.

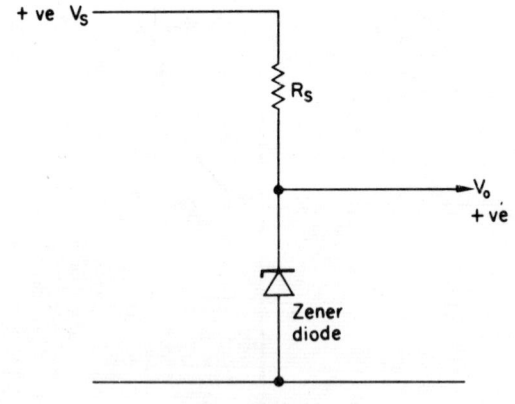

Fig. 26.1.4.

ZENER DIODES AND THEIR APPLICATIONS

A load line is plotted for R_s and an operating point established. Point P is seen to exist at the intersection of the device characteristic and the d.c. load line for R_s.

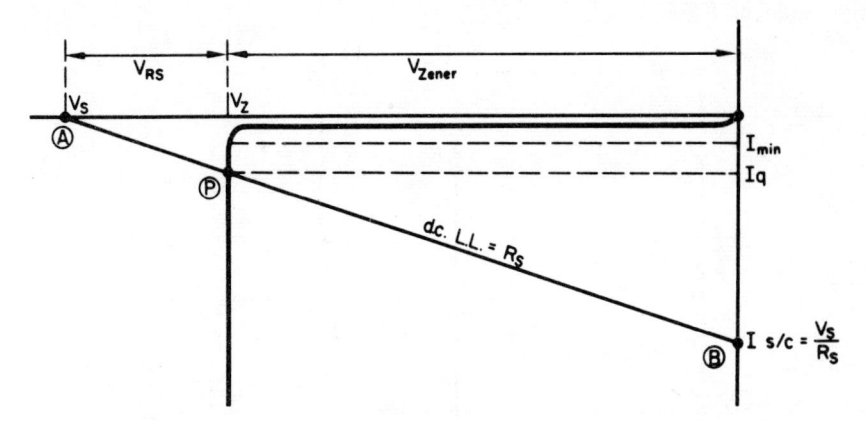

Fig. 26.1.5.

V_s is the d.c. supply voltage, V_Z is the 'steady' zener voltage and I_q, the 'steady' zener current. V_s/R_s is the value of zener current that would flow through R_s should the zener diode become short circuit; this would not normally occur of course, but needs to be calculated in order to position that end of the d.c. load line. Finally I_{min} is the minimum current that must flow through the zener diode to ensure that point P is well below the knee. The zener has an internal resistance r_Z of about $10 - 100\,\Omega$ once it is operating below the knee. We will deal with this later on and allow for voltage drops across r_Z in more advanced power supply units.

Suppose we decide to operate a 5V zener diode with an I_q of $200\,\mu A$, and with a supply voltage of say 20V,

$$R_s = \frac{(20-5)\,V}{200\,\mu A} = \frac{15\,K\Omega}{0.2} = 75\,K\Omega.$$

Point A in figure 26.1.5. would be $V_s = 20V$, and point B would be $20V/75\,K\Omega = 267\,\mu A$.

The operating point P would give an I_q of $200\,\mu A$.

An alternative method is to choose the point P. Draw a straight line from V_s — through point P — and stop at the point marked B on the characteristics.

The value of R_s required $= V_s/I$ or A/B in figure 26.1.5.

26.2. Voltage reference supply

This simple circuit would normally be used to supply a constant voltage across a load, as was the case with the neon stabiliser.

Figure 26.2.1. shows a basic stabilised circuit providing a load with its required voltage and current. The load current must now be considered in the equation for R_s.

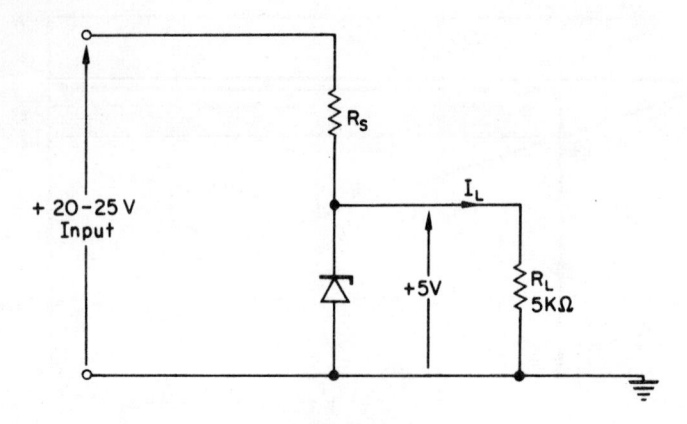

Fig. 26.2.1.

Suppose R_L was a 5 KΩ load, and required 1 mA load current (for a load supply of 5V).

The total current to flow through R_s = 1 mA + 200 μA = 1.2 mA.

For a 20V supply, $R_s = \dfrac{20 - 5}{1.2} = \dfrac{15}{1.2} = 12.5$ KΩ.

Should the supply voltage increase to an acceptable maximum of 25V, then the current through R_s, for a constant load current I_L, would be

$$I_s = \dfrac{(25 - 5)\,V}{12.5\,K\Omega} = \dfrac{20}{12.5}\ mA = 1.6\ mA$$

and as 1 mA is required for the load, 600 μA must flow through the Zener diode. It remains to ensure that the power rating of the zener diode is not exceeded.

$$P_z = I_z V_z = 0.6 \times 5 \times 10^{-3} = 3\ mW.$$

If the load were to be disconnected, or take no current for any other reason, all of the current would need to be taken up by the zener diode. The dissipation would then be $P_z = I_s R_z = I_z R_z = 8$ mW.

Zener diodes can be connected in series so as to give a number of different output voltages by summing their individual voltages, provided that the I_q flowing through the diodes is greater than the minimum required to cause all diodes to operate on the stable portion of their curves below the knee.

A particularly stable low voltage reference can be obtained with a circuit similar to that shown in figure 26.2.2.

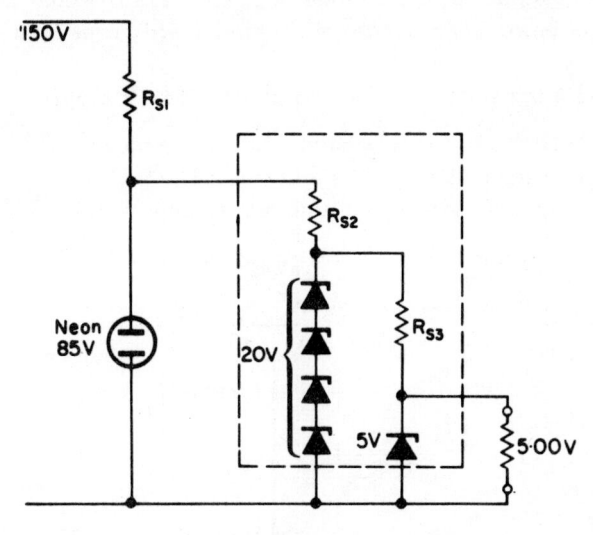

Fig. 26.2.2.

A particularly stable 5.00V supply for experimental and calibration purposes could be made with a circuit similar to that shown. In this instance, the diodes could be situated in a large solid copper container situated inside a temperature controlled oven. Zener diodes are prone to temperature changes, but with care, excellent stable supplies can be obtained.

The output would normally be referred to a standard cell, by 'potting down' the output and placing a sensitive galvonometer between both voltages. (Careful selection of both $+Ve$ and $-ve$ temperature coefficient diodes connected in series assist even further).

Example.

Suppose a 10V regulated d.c. supply of 10 mA was required from a d.c. source of 12 – 15V. An I_q of 200 µA and a V_z of 10V is assumed for a pair of 5V diodes in series.

The first step is to determine R_s for a supply of 12V.

$$R_s = \frac{(12 - 10)\,V}{(10 \cdot 2)\,mA} = \frac{2}{10.2} K\Omega = 196\,\Omega.$$

When the supply is increased to its known maximum of 15V, the current

$$I_s \text{ through } R_s = \frac{(15 - 10)\,V}{(196)\,\Omega} = 25.5\,mA.$$

When the load is not drawing current, all of the 25.5 mA must be taken up by the zener diode.

The power dissipated in the diode $= I_z V_z = I_s V_z = 25.5 \times 10 = 255$ mW. Many small power zener diodes will satisfactorily cope with 255 mW.

26.3. Using a transistor in a basic stabilised power supply unit

Figure 26.3.1. shows the simple basic circuit extended to include a transistor connected as an 'emitter follower'. The inclusion of the transistor reduces the load current through R_s by a factor of $(1 + a')$.

Fig. 26.3.1.

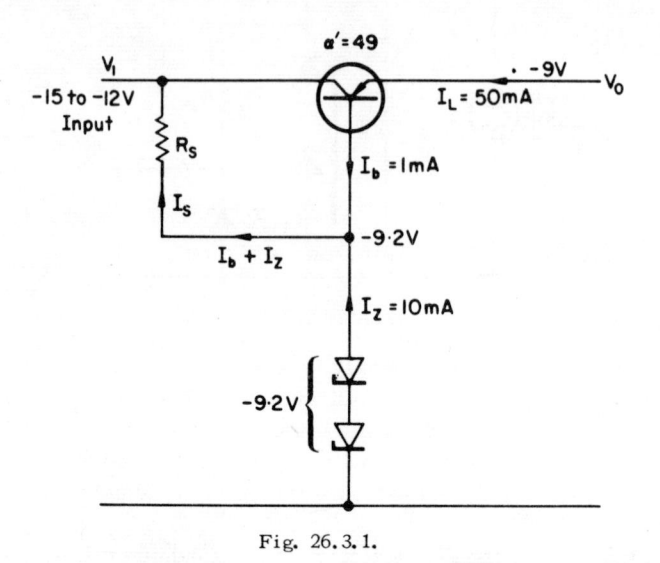

The first step is to select a suitable transistor.

The power dissipation under 'worst' conditions is determined as follows:

$$P_{TOT} = (V_1 \max - V_o)(I_L \max) = V_{ce}(\max) \times I_L(\max)$$
$$= (15 - 9.0)(50) = 6 \times 50 = 300 \text{ mW}.$$

Assuming we select a transistor capable of dissipating say 400 mW and an a' of 49, and that a V_{be} of 0.2V is required for the necessary base current.

The base current $= \dfrac{I_e}{1 + a'} = \dfrac{50 \text{ mA}}{50} = 1 \text{ mA}$.

The current in R_s will have two components, the base current and the current I_z. It is good practice to ensure that I_z is in the order of 10 times that of the base current.

As $I_b = 1\,\text{mA}$, I_z will be $10\,\text{mA}$. Hence for the minimum supply voltage,

$$R_s = \frac{(12 - 9.2)\,\text{V}}{11\,\text{mA}} = \frac{2.8\,\text{K}\Omega}{11} = 255\,\Omega.$$

The power dissipation of the zener must be checked.

$$P_z = V_z I_z \text{ and for } I_L = 0 \text{ (the worst condition)},$$

$$P_z = V_z I_s = \left[\frac{(12 - 9.2)}{R_s} - I_b\right] V_z$$

and assuming $I_b = 0$ at $I_L = 0$, $P_z = \left[\dfrac{2.8}{255}\right] 9.2 \doteq \underline{100.0\,\text{mW}}$

and is shared between the zener diodes in proportion to their respective zener voltages.

We shall be discussing an extension to this basic circuit in chapter 27.

CHAPTER 27

Composite devices

27.1. Silicon controlled-rectifiers (S.C.R.s)

If an n-p-n transistor V_{T_2}, is connected to a p-n-p- transistor V_{T_1}, as shown in figure 27.1.1., the resultant composite device will function as an electronic switch. When manufactured as a single component, it is known as a thyristor or silicon controlled-rectifier.

The switch can be 'closed' by a number of methods, two of which will be discussed here.

(1) The switch can be 'closed' by applying steep fronted trigger current pulse input, I_g, to the gate terminal, typified by V_s/R_s in figure 27.1.1. or (2) by raising the anode to cathode voltage, V_{AK}, to a suitable level.

This level is called the breakover voltage, V_{bo}.

Once 'closed', the switch cannot be 'opened' by removing the gate pulse. The function of the composite device can best be discussed by considering the two separate transistors shown in figure 27.1.1.

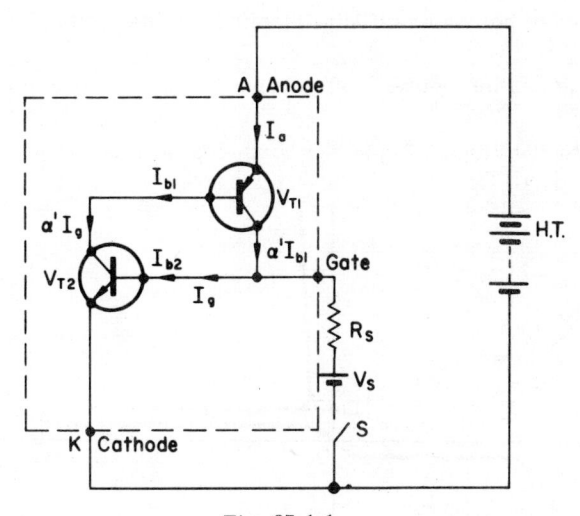

Fig. 27.1.1.

With a suitable h.t. supply connected, S_1 is closed. A gate current input of magnitude $I_g = V_s/R_s$ will flow into the n-p-n transistor, V_{T_2} causing it to conduct. The collector current of V_{T_2} will be of the magnitude of $a'I_{b2}$, where $I_{b2} = I_g$, and as the collector is directly connected to the base of the p–n–p transistor, V_{T_1}, current will be dragged from V_{T_1} base causing V_{T_1} to conduct also.

The resultant current flowing out of V_{T_1} collector will be a' times that of its base current, i.e. $a'I_{b1}$. But $I_{b1} = a'I_{b2}$, hence the current flowing from V_{T_1} collector will be $(a')^2 I_{b2}$. (This assumes each transistor has a similar value of a').

As $I_{b_2} = I_g$, the collector current of V_{T_1} will be $(a')^2 I_g$. The emitter current of V_{T_1} will be

$$\frac{I_{C1}}{a} = \frac{(a')^2 I_g}{a}$$

and will flow into the base of V_{T_2} causing this transistor to conduct heavily. The increase in V_{T_2} collector current will drag V_{T_1} further into conduction.

This process will continue and the cumulative effect causes both transistors to become bottomed.

Disconnecting V_s (the trigger input) and hence I_g, will have no effect upon the bottomed states of V_{T_1} and V_{T_2} as the emitter current from V_{T_1} entering the base of V_{T_2} will greatly exceed the relatively low level of external gate current. The currents will therefore be self maintained. The emitter of V_{T_1} is the anode of the composite device hence the current flowing in V_{T_1} emitter is the load current of the whole device.

The minimum current flowing into the anode of the composite device needed to ensure bottoming of both transistors is called the holding current, I_H.

The switch can be 'opened' by reducing the anode current, I_a, to a value less than I_H.

The S.C.R. functions externally in a similar manner to the Thyratron discussed earlier.

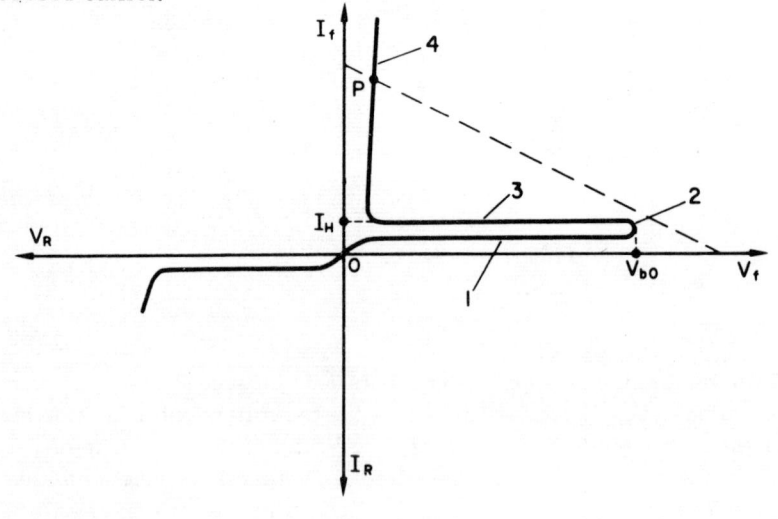

Fig. 27.1.2.

COMPOSITE SEMICONDUCTOR DEVICES 525

The name given to this component was derived from both THYRatron and transISTOR, i.e. Thyristor.

Figure 27.1.2. gives the I_a/V_{ak} characteristics of a thyristor.

The reader will no doubt remember that when we first discussed the I_a/V_{AK} characteristics of a Triode valve, we examined the curve for $V_{gk} = 0V$. In other words, we looked at a diode curve.

Such is the case here. With no input to the gate, we obtain a $p-n-p-n$ diode.

V_R is the reverse voltage, V_f the forward voltage. I_f is the forward current and I_R the reverse current. I_H is the holding current and V_{b_0} is the breakover voltage.

Point 1 on the curve is a relatively constant current similar to the saturation current of a normal diode. This constant saturation current is of the same order as the leakage current I_{co}'.

Increasing V_{ak} will have little effect upon this current until point 2 is reached. This point is V_{b_0} and once V_{ak} reaches this potential, internal multiplication processes occur and the current is increased.

The portion of the curve is a slight negative-resistance region. When the current reaches I_H, it rapidly traverses point 3 as this is a high negative-resistance region. Point 4 is reached when the collector junction becomes forward-biassed and the characteristics revert to those of a forward biassed diode.

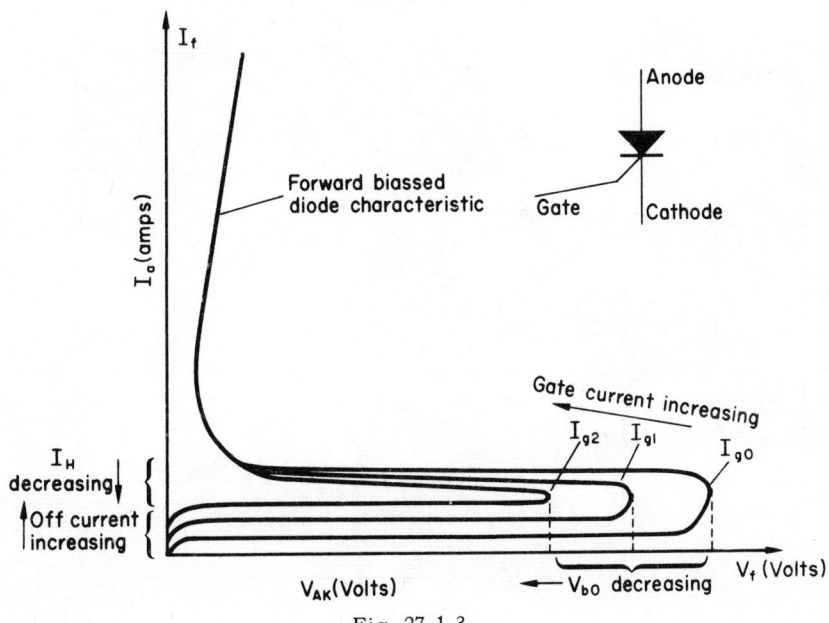

Fig. 27.1.3.

A load line is plotted for the load resistor R_L. The operating point cannot remain on the negative-resistance portion of the curve hence for a V_{AK} greater than V_{bo}, the operating point will rest as indicated at point P. The reverse region is similar to that of a reverse-biassed diode.

Figure 27.1.3. shows both the circuit symbol and the I_a/V_{ak} characteristics for a thyristor with gate current/inputs, i.e. as with the triode valve with values of V_{gk}, these characteristics are those of a $p-n-p-n$ triode.

Figure 27.1.3. shows clearly that the breakover voltage V_{bo}, is a function of the gate current input.

Increasing the gate current, I_g, lowers the breakover voltage and the holding current but increases the off current.

The thyristor is normally biassed well below V_{bo} and triggered on by a suitable gate current input pulse.

Whilst 'turning on', there is a delay of $1 - 5\,\mu S$ and delays of $8 - 30\,\mu S$ often occur whilst turning off. These delay times are measured as normal rise and decay times for the V_{ak} to fall from 90% to 10% of its initial off value.

The equivalent circuit of a thyristor is given in figure 27.1.4.

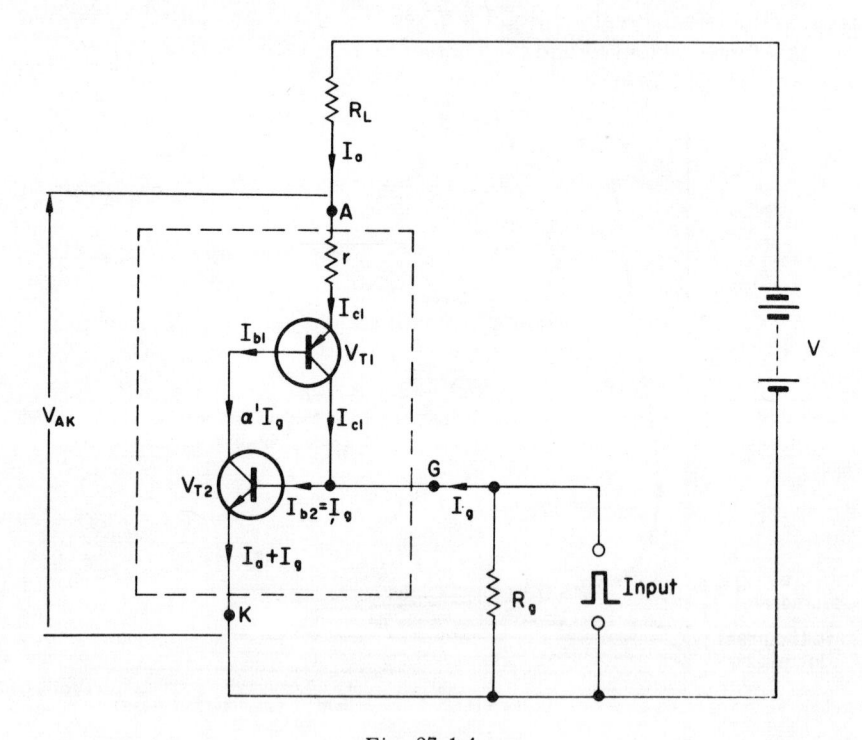

Fig. 27.1.4.

R_g is connected so as to provide a path to divert the leakage current I'_{co}. If R_g is not connected, the leakage current may enter the lower transistor base and cause it to conduct. A common value for R_g is about 1 KΩ. R_L is determined by

$$\frac{V - V_{ak}}{I_a}$$

where V_{ak} is the anode to cathode voltage and I_a the required anode, or load, current. R_L may often be in the order of fractions or units of ohms depending upon the circuit requirements and the supply voltage V.

With say, 100 mA gate current input determined by V_s/R_s applied to V_{T_2} base, a collector current of $\alpha' I_g$ will be taken from the base of V_{T_1}. The emitter current of V_{T_1} is approximately equal to $(\alpha')^2 I_g$ hence this is the anode current for the device as a whole.

Thyristors are high voltage high current devices and achieve large power gains.

Compared to power transistors which often have a current gain approaching unity and a limited capability to handle high voltage, the thyristor is a very useful device.

When fired, a small voltage in the order of a volt or so, is developed across the thyristor. This is accounted for in the equivalent circuit by considering the internal resistance, r.

A typical value for gate current inputs is 100 mA determined by about two volts and the series input resistor R_s. Anode currents of 70 A are now common.

The gate pulse should be steep fronted and have a sufficient duration to allow the load current to build up to just beyond I_H.

The gate should never be taken positive whilst the anode is negative. Heat sinks must be used as directed by the manufacturers.

Reference to figure 27.1.4. will show that $I_{b_2} = I_g + I_c$ also $I_{c_2} = I_{b_1}$ Hence $I_{e_2} = I_{e_1} + I_g$ but $I_g \ll I_e$, ∴ $I_{e_2} \simeq I_{e_1}$.

With the gate open and a positive anode voltage applied; the outer p–n junction will be forward-biassed whilst the centre junction will be reverse-biassed. Under these conditions a very small leakage current will flow, as with a normal reverse-biassed silicon junction diode but if R_g is connected, the device will not fire.

The device will fire if (a) the junction temperature is raised sufficiently or (b) the anode to cathode voltage is raised to a value sufficient to cause the centre junction to avalanche. When this occurs, the thyristor will conduct heavily whilst the amplitude of current flow is limited by external resistance.

Reference to figure 27.1.4. will show that

$$I_{b_2} = \alpha'_1 I_{b_1} \ldots\ldots (1) \text{ and } I_{b_1} = \alpha'_2 I_{B_2} \ldots\ldots (2)$$

2M

and rearranging (1)

$$I_{b_1} = \frac{I_{b_2}}{a'_1}$$

and substituting for I_{b_1} in equation (2),

$$\frac{I_{b_2}}{a'_1} = a'_2 I_{b_2}$$

hence

$$\frac{1}{a'_1} = a'_2$$

and therefore 'switch on' conditions occur when $a'_1 a'_2 = 1$. If $a'_1 = a'_2$, then 'switch on' condition occurs when $(a')^2 = 1$.

As the gate must never be made positive with respect to the cathode whilst the anode is negative with respect to the cathode, a zener diode connected as shown in figure 27.1.5. will ensure that this requirement is met.

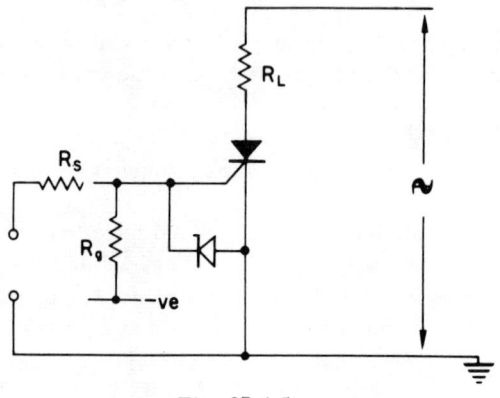

Fig. 27.1.5.

Figure 27.1.5. shows a thyristor with no gate input and a sinusoidal anode-to-cathode supply.

The 'turn on' is accomplished when the supply voltage reaches V_{b_0}. Turn off occurs when the supply voltage falls to a slightly negative value with respect to the cathode.

Figure 27.1.6. shows the waveforms for this circuit. The figure also shows the waveforms for two thyristors connected back to back.

COMPOSITE SEMICONDUCTOR DEVICES   529

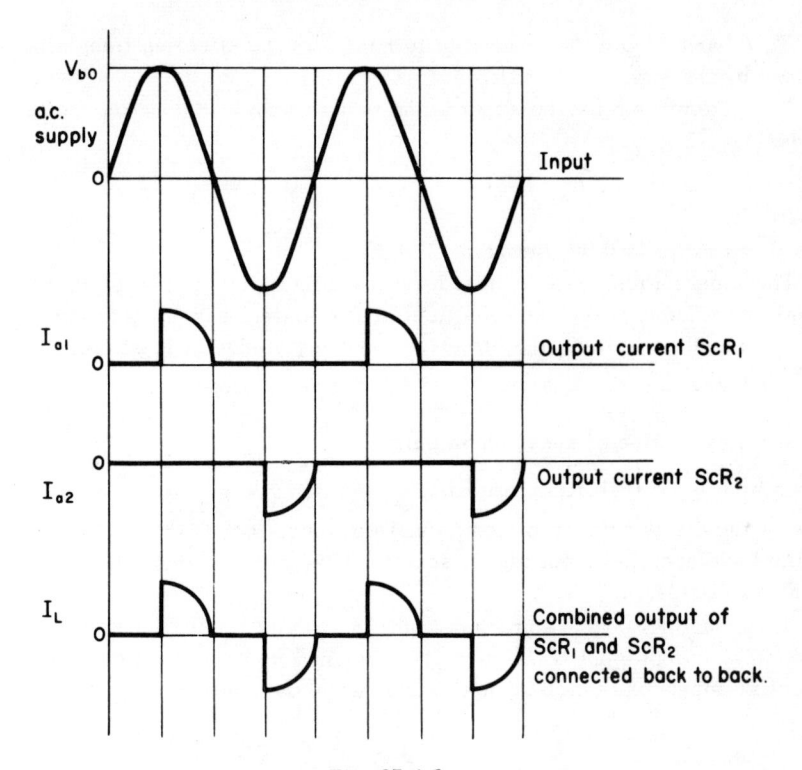

Fig. 27.1.6.

27.2. Super alpha pair. (Darlington pair)

Figure 27.2.1. shows a super alpha pair. This composite circuit has a very high input resistance, low input base current and a gain much greater than either of the transistor gain values when connected individually as common emitter amplifiers. Figure 27.2.1.

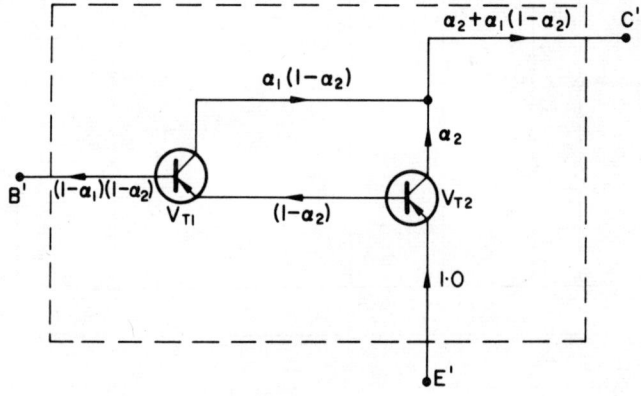

Fig. 27.2.1.

B', C' and E' are the composite terminals of the effective transistor formed by the super alpha pair.

The current gain for the super alpha pair as shown in common emitter mode

$$= \frac{\alpha_2 + \alpha_1(1 - \alpha_2)}{(1 - \alpha_1)(1 - \alpha_2)}$$

and if $\alpha_1 = \alpha_2 = 0.99$, the gain = 9.999.

The input current into B' is reduced by a factor $(1 - \alpha)$ times that of a single transistor, hence the effective input resistance is considerably increased. If $\alpha = 0.99$, the effective input resistance is increased by 100 times.

27.3. Application of super alpha pairs

(1) A high input-resistance amplifier

One of the disadvantages of some junction transistors is the loading effect of the base upon the input signal source. This might be the order of $10\,\text{K}\Omega - 20\,\text{K}\Omega$.

If we employ a super alpha pair and use techniques similar to the valve version of a high input-resistance amplifier in 18.13. we can raise the effective input-resistance to input signal sources appreciably.

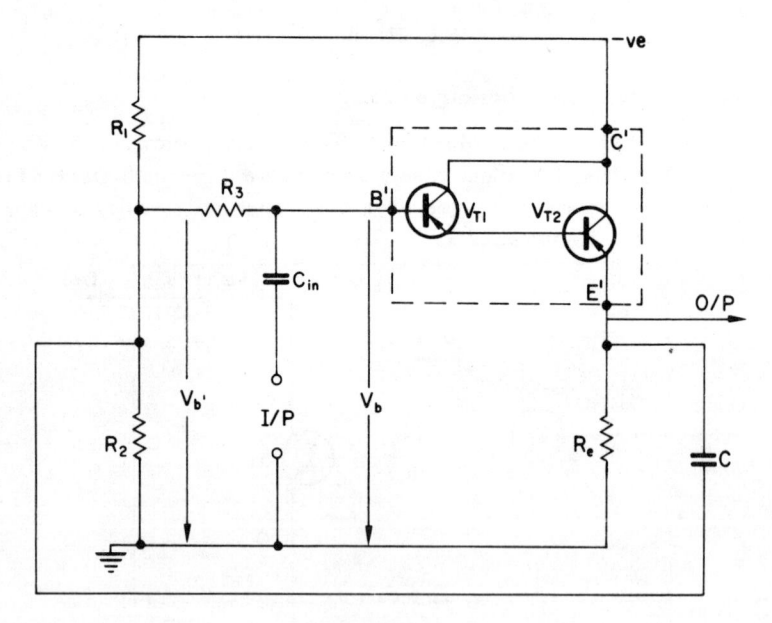

Fig. 27.3.1.

Figure 27.3.1. shows a simple amplifier using a super alpha pair. The latter is contained in the dotted 'black box'.

With 'C' open circuit and R_3 short circuit, the input resistances will be the shunt combination of R_1 and R_2. The input resistance to the super alpha pair would not appreciably effect the input resistance, as was shown in 27.2.

Suppose now we inserted R_3 as shown in figure 27.3.1. but left 'C' open circuit.

The input resistance would be increased slightly due to R_3 and would be $R_3 + R_1 /\!/ R_2$. Once more we ignore the small current flowing in V_{T_1} base.

If 'C' is now connected in circuit, and a signal v_b applied to the input shown, a signal v_b' will appear across R_e and will be fed back via 'C' to the junction of R_1, R_2 and R_3.

Should the signal fed back approach the amplitude of the input signal, and be in phase with the input, a very small potential will exist across R_3.

The effective input resistance will be approximately

$$\frac{v_b - v_b'}{I_b}$$

once more ignoring the input resistance to the super alpha pair and the relatively low ohmic value of R_1 in shunt with R_2.

A typical input-resistance value with a simple circuit such as this is in the order of $2\,\text{M}\Omega$.

The value of R_3 can be determined once the base current of V_{T_1} is chosen.

$$R_3 \doteq \frac{v_b - v_b'}{I_b}$$

Knowing V_{T_2} emitter potential, and initially assuming V_{T_2} base-emitter voltage to be zero, V_{T_2} base potential is known also.

As V_{T_1} emitter is directly connected to V_{T_2} base, the potential at V_{T_1} emitter is also known. Once more assuming initially a zero base emitter voltage for V_{T_1}, the base voltage V_b is also known.

Knowing V_b, and deciding upon the base current, R_3 can be determined.

Once R_3 is known, the required potential at the junction of R_1 and R_2 is easily determined by summing V_b and the small voltage drop across R_3. 'C' should have a reactance of about $1/10 - 1/20$ of the shunt value of R_1 and R_2 is parallel at the lowest frequency to be used.

27.4. Applications of super alpha pairs

(2) Transistorised regulated power supply units

The simplest form of regulation is to use a shunt connected zener diode in the same manner as a neon stabiliser is used with valve circuits. This circuit is usually used for low power requirements. The zener diode has an a.c. resistance in the order of $5 - 50\,\Omega$, R_Z. Figure 27.4.1. shows such a circuit.

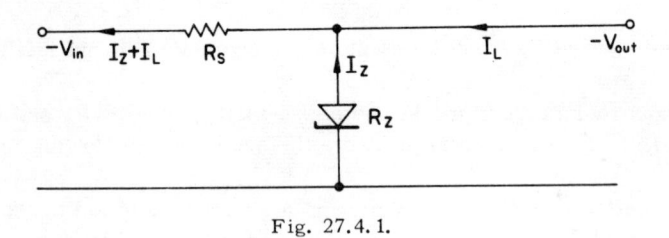

Fig. 27.4.1.

If the zener voltage, V_o, is significant compared to V_{in}, then the value of
$$R_s = \frac{V_{in} - V_o}{I_z + I_1}$$

where I_z is the steady zener current shown as I_q on the characteristics in figure 27.4.2.

A load line for R_s is plotted on the characteristics for the zener in order to establish the steady zener I_q. Any changes in load current must be taken up by the zener diode. A very small change in zener voltage will result as seen in figure 27.4.2. Should the load be disconnected, the zener diode will need to take up all of the load current. In order to reduce the change in zener current under steady or varying load conditions, a transistor may be added as in figure 27.4.2.

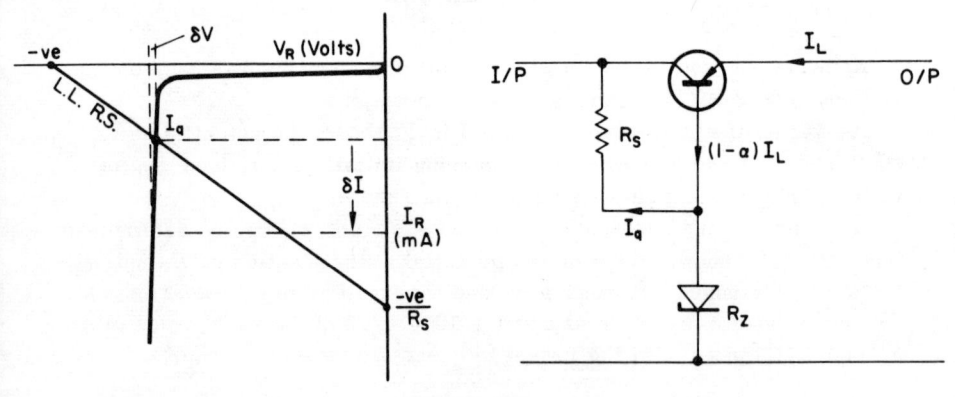

Fig. 27.4.2.

In this circuit the *change* in zener current due to a *change* in load current is reduced by a factor $(1 - a)$. If $a = 0.99$, the change in zener current would be $\delta I_L/100$, where δ represents the change in I_L.

The mutual conductance gm, of a transistor may be determined graphically. Figure 27.4.3. illustrates this.

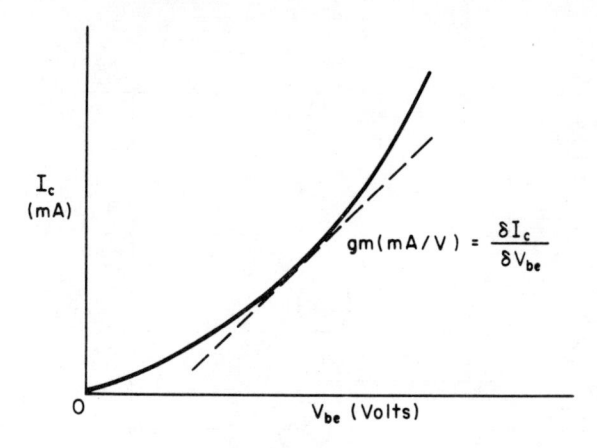

Fig. 27.4.3.

Example: A change in v_{be} of 125 mV caused I_c to change by 5 mA.

$$gm = \frac{\delta I_c}{\delta v_{be}} = \frac{5 \text{ mA}}{125 \text{ mV}} = 40 \text{ mA/V}.$$

Hence the gm is 40 mA/V.

Figure 27.4.4. shows a regulated power supply unit containing a super alpha pair. We intend to examine the change in output voltage for a change in input level, V_{in}, and to establish how effective the regulator is in reducing the effects upon the stabilised output for a given change in amplitude of the unregulated d.c. input voltage.

The circuit function is as follows:

An increase of amplitude in the unregulated d.c. supply will cause the stabilised output to rise.

A fraction of the rise in output is detected by the base of V_{T_3}. The collector current of V_{T_3} increases causing V_{T_3} collector to fall in a positive direction. Consequently the base of V_{T_1} is taken less negative and this positive going movement appears at the output of the super alpha pair. As this output is connected to the stabilised output, this positive going excursion tends to cancel the original negative going excursion of the stabilised output.

This brief description did not take into consideration several factors, one of which is the function of the zener diode. The following analysis is discussed in detail and results in the derivation of an expression for the stability factor S. A good regulated p.s.u. will have a very small value of S. This is defined on the following page.

An analysis of a transistorised regulated power supply unit

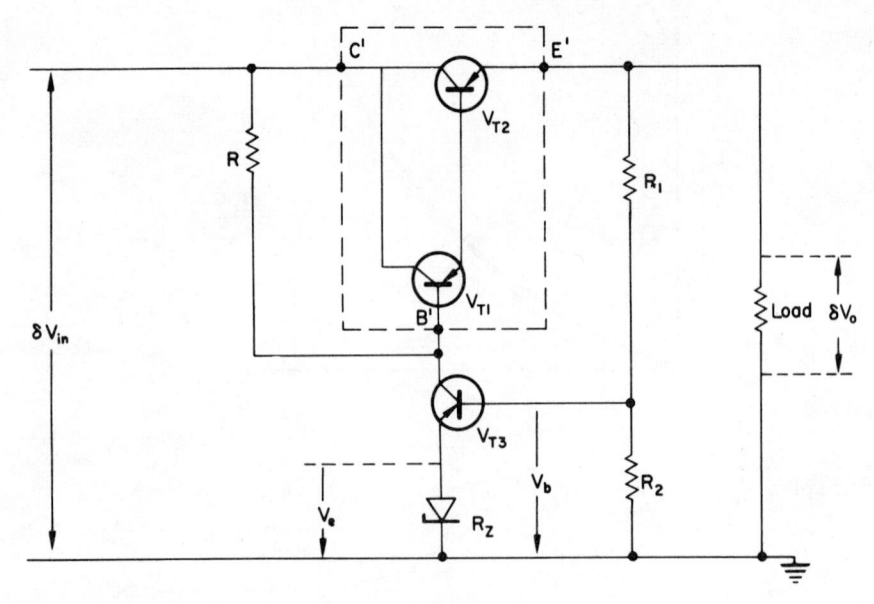

Fig. 27.4.4.

If the input tended to rise to a greater $-ve$ value, (δV_1), the output would tend to follow (δV_0). A fraction of the change of output level would be transmitted to V_{T_3} base (δV_b). This change,

$$\delta V_b = \delta V_0 \times \frac{R_2}{R_1 + R_2}.$$

A change in V_b will cause a change in collector current.

This change in V_{T_3} collector current (δI_c) in turn causes a change in potential across the zener diode. The change in potential across the zener diode $= \delta V_e$ $\therefore$ $\delta V_e = \delta I_c \cdot r_Z$

$$\delta V_{be} = \delta V_b - \delta V_e$$

$$\delta V_{be} = \frac{\delta V_0 \cdot R_2}{R_1 + R_2} - \delta I_c \, r_Z \text{ but } \delta I_c = gm \, V_{be}$$

$$\text{and } \delta V_{be} = \frac{\delta I_c}{gm}$$

hence $$\frac{\delta I_c}{gm} = \frac{\delta V_0 \cdot R_2}{R_1 + R_2} - \delta I_c \cdot r_Z$$

hence
$$\delta I_c = gm \cdot \frac{\delta V_0 \cdot R_2}{R_1 + R_2} - gm \, \delta I_c \cdot r_Z$$

Thus $\delta I_c \left[1 + gm \cdot r_Z \right] = \frac{gm \cdot R_2}{R_1 + R_2} \times V_0 \quad \therefore \delta I_c = \left[\frac{gm \, R_2}{R_1 + R_2} \times \frac{1}{1 + gm \, r_Z} \right]$

Since the high input resistance super alpha pair is employed, the change in collector current will pass through R. Hence

$$\delta V_1 \simeq \delta I_c \, R. \quad \therefore \delta I_c = \frac{\delta V_1}{R}$$

$$\therefore \quad \delta V_1 = \left(\frac{gm \, R_2 \cdot R}{R_1 + R_2} \right) \left(\frac{1}{1 + gm \cdot r_Z} \right) \delta V_0.$$

The stability factor is a measure of the change of output for a change in input and should be as small as possible.

The stability factor $\quad S = \dfrac{\delta V_0}{\delta V_1}$

$$S = \frac{(R_1 + R_2)(1 + gm \cdot r_Z)}{gm \cdot R_2 \cdot R}.$$

If $R_1 = R_2 = 1000\,\Omega$, $gm = 50\,\text{mA/V}$, $r_Z = 40\,\Omega$ and $R = 10\,\text{K}\Omega$.

Then
$$S = \frac{(2000)\left(1 + \dfrac{50}{1000} \times 40\right)}{\dfrac{50}{1000} \cdot 1000 \cdot 10\,000} = \frac{2000\,(3)}{500\,000} = \frac{6}{500} = \underline{0.012}$$

for a 1V change in input, the output will change by 0.012V, or 12 mV.

The effect on an input change in level is reduced at the load by approximately 83 times.

The output resistance of a stabilised power supply unit provides an indication of the likely fall in output voltage for a given increase in load current. One method of determining the output resistance is as follows :-

Connect a millivoltmeter suitably shunted with a shorting link, between the output of the p.s.u. in question and the output of a stable reference supply. The latter should produce a voltage of the same magnitude and polarity as the p.s.u. being investigated. A variable load should be connected across the output of the p.s.u. to be tested.

Both supplies should be switched on and allowed to warm up, if appropriate, and allowed to 'settle down'. Using a general purpose voltmeter, measure the voltage between both output terminals and adjust one p.s.u. until the voltmeter reading is zero on the low volts range. Remove the meter and then set the millivoltmeter to a high range and remove the shorting link.

Adjust the load until a given current flows. Note this current and the millivolt reading, if any, on the millivoltmeter. Increase the load current by a given value and note both this value and the millivoltmeter reading. The output resistance is given as the change in voltage output divided by the change in load current. Typical values of output resistance may be in the order of $0.001 - 1.0\,\Omega$ depending upon the circuit components, valves, and/or transistors.

This measurement is known as a voltage differential method and allows very small changes in quite high voltages to be recorded quite simply.

Example 1.

Determine the output resistance of a stabilised p.s.u. from the following measurements. With a load current of 20 mA, the millivoltmeter records a voltage differential of 5 mV. When the load current is increased to 50 mA, the millivoltmeter reads a differential of 8 mV. Ans. $0.1\,\Omega$.

Example 2.

(a) Calculate the reduction in input base current to a super alpha pair consisting of two identical transistors having an $\alpha = 0.95$, when compared to a single transistor connected alone.

(b) By what factor is the input resistance increased compared to a single transistor connected alone. Ans. (a) 0.05. (b) 20.

CHAPTER 28

Simple logic circuits

28.1. Transistorised multivibrator circuits

For each of the valve circuits previously considered, there is a transistor version. Figure 28.1.1. shows a free running multivibrator, the pulse duration, T, is given as approximately $T \simeq 0.7\,CR$. The base resistors must not be reduced to too small a value, or due to lack of proper bias, the transistor may be damaged. Consequently, the coarse variable would be the capacitor, C, whilst R may be varied, just a little, if a fine control is required.

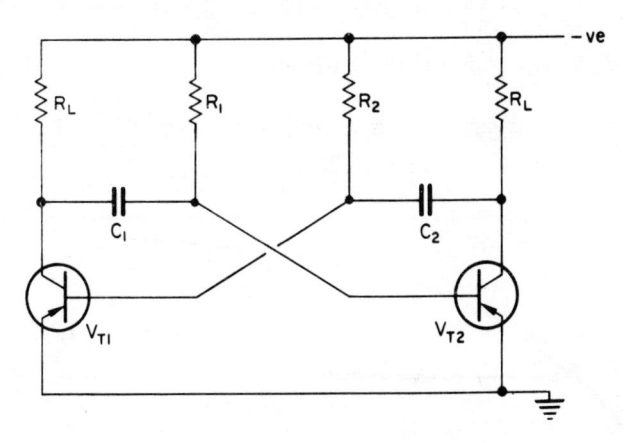

Fig. 28.1.1.

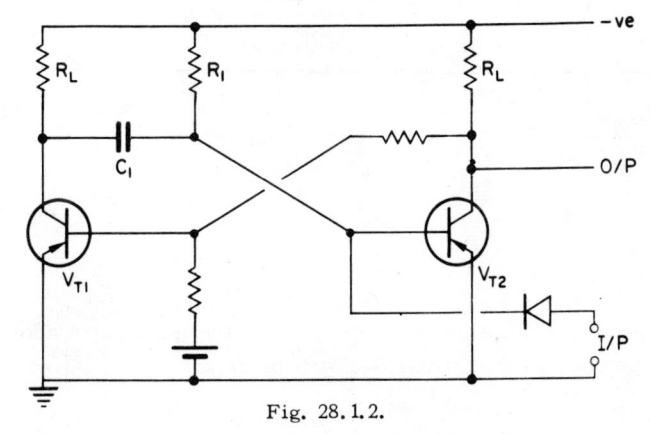

Fig. 28.1.2.

Figure 28.1.2. shows a monstable multivibrator. As one coupling component is a resistor and the other a capacitor, the circuit is one which has one stable state. It will 'flip' over when a pulse is applied to the input and will, after a time T, return to its original state. The pulse width of the output pulse is given as approximately $0.7\,C_1R_1$. When on, each transistor will have a $V_{ce} \simeq 0V$. The capacitor will be charged to the h.t. rail and when switched over, will take the base down by that amount. The base will want to climb $2 \times$ h.t. but will stop at the point when the base is zero. The actual climb will be equal to the h.t.

If we now insert known values in the expression $V_c = V(1 - e^{-t/CR})$ where $V_c = $ h.t. and $V = 2$ h.t., we have

$$1 = 2(1 - e^{-t/CR}) \quad \therefore \quad 1 = 2 - 2e^{-t/CR}$$

$\therefore \quad 1 = 2e^{-t/CR}$ and $\frac{1}{2} = e^{-t/CR}$

$\therefore \quad 2 = e^{t/CR}$ and taking $\log_e$s of both sides,

$$\log_e 2 = \frac{t}{CR} \text{ hence } t = CR \log_e 2.$$

Thus $\quad t = 0.693$ or $\simeq 0.7\,CR$ seconds.

The waveform is shown in figure 28.1.3.

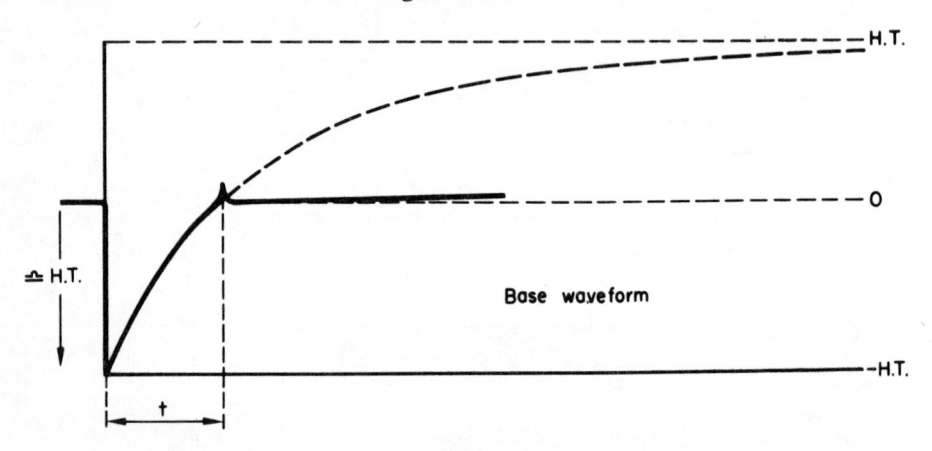

Fig. 28.1.3.

The output pulse duration is independant of the input pulse duration. In the absence of an input signal, V_{T_2} will be on and V_{T_1} off.

SIMPLE LOGIC CIRCUITS

A basic transistorised Schmitt trigger circuit.

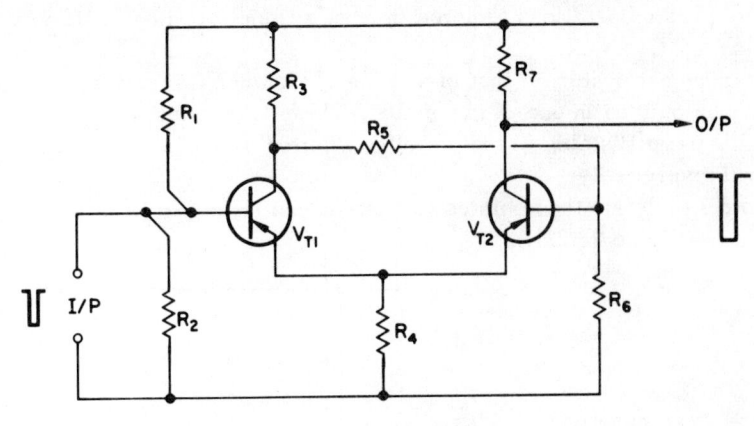

Fig. 28.1.4.

In the absence of a signal, V_{T2} is conducting and is 'bottomed'. The voltage across the transistor, V_{T2} will be in order of 0.2V. The base of V_{T1} is more positive than V_{T2} base, and as they share a common emitter resistor, V_{T1} will be considered to be off. When an input is applied, V_{T1} base is lifted, in a negative direction and is switched on. The flow of collector current in V_{T1} causes V_{T1} collector to fall in a positive direction. This fall is transmitted to V_{T2} base, via R_5, thus causing V_{T2} to become cut off. V_{T2} collector current rapidly falls to zero and a large negative going pulse appears at V_{T2} collector. This circuit is very sensitive to small voltage changes to V_{T1} base and produces a relatively large output voltage for a small input signal. A speed up capacitor may be connected across R_5 in the same manner as the valve circuit. When considering a circuit such as this, it may be noted that the bias voltage for V_{T2} (which is on) is determined by the divider chain consisting of R_3, R_5 and R_6, in the usual manner. A load line would have been plotted for R_4 and R_7. Note that the lower end of R_3 will be almost at the h.t. potential, as V_{T1} is off and apart from I_{co}, V_{T1} current will be zero. When V_{T1} is on, R_5 is at a potential determined by the divider R_3 and R_4 and the current through V_{T1}. There will be about 0.2 volts across V_{T1} at this time, but as a first approximation, may be ignored. The base current of V_{T2}, which will flow through R_5, must be sufficient to ensure that V_{T2} V_{be} is of the right order to cut V_{T2} off. If R_3 is the same ohmic value as R_7, the problem will be easier as there will be a reasonably constant emitter voltage, irrespective of which transistor is on, as when on, they would both be 'bottomed' in turn.

28.2. A brief introduction to simple digital systems

The three basic operations required in most systems are Logic, Memory and Counting.

The simplest system is that of a 'Binary', or where all elements in the system can only be in one of two states. These two states are stable — hence the name Bistable — and are often referred to as Binary 1 and Binary 0 respectively.

In order to define these states, a common emitter amplifier is to be examined. (Figure 28.2.1.).

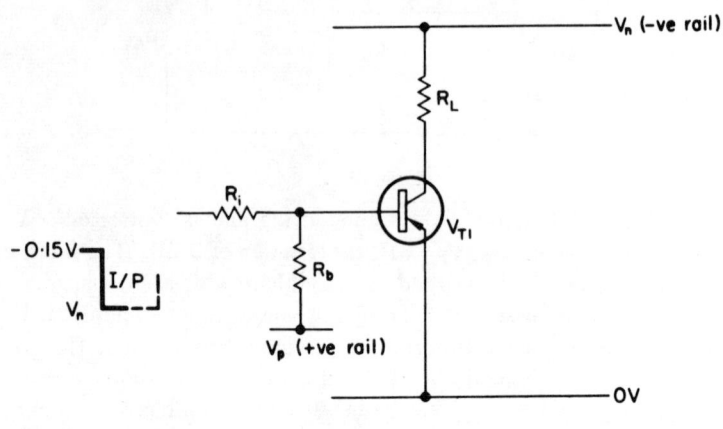

Fig. 28.2.1.

The bias resistor, R_b, is chosen such that with no input, the base is positive with respect to the emitter (say + 0.5V or so). The transistor, V_{T_1}, is then cut off. The output terminal will be at the $-ve$ rail potential, (V_n), as there is no appreciable current through R_L — apart from I_{co}, which is small.

When an input is applied (V_n), the base becomes negative with respect to the emitter. The transistor will conduct, and the value of I_c will be determined by the particular transistor and R_L. The limiting resistor R_i limits the drive to the base to a safe value, but allows the conducting transistor to 'bottom'.

The output potential will be at the 'bottoming potential' of the particular transistor, usually about $-0.2V$, V_{ce}. The output has fallen in a positive (less negative) direction for a negative going input so that this circuit may be termed an inverter.

There are two output levels: binary 0 corresponds to 0V to $-0.5V$, while binary 1 corresponds to V_n.

The current available in the binary 0 state depends upon the transistor characteristics, while for the binary 1 state the current is determined by

R_L. A typical range of output potential for binary 1 is from V_n to $0.7 V_n$, depending upon R_L and I_{co}. For V_n $-6V$ or more, a reasonable discrimination is obtained between the two binary state potentials.

Logic.

The three main logical operations are 'OR', 'AND' and 'NOT'. An 'OR' gate will produce a binary 1 output when any or all of its inputs are at binary 1. An 'AND' gate will produce a binary 1 output only when all of its inputs are at binary 1. A 'NOT' circuit will provide an inverted output, i.e. for a binary 1 input, a binary 0 output is produced. The 'NOT' function would be performed by the common emitter amplifier previously considered.

Memory.

If two common emitter inverter amplifiers are connected as shown in figure 28.2.2. in the form of a binary stage, as described in an earlier chapter, it has the ability to 'remember'.

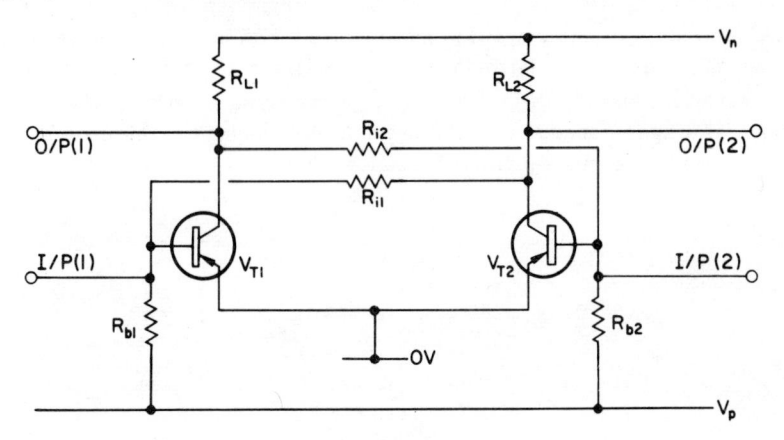

Fig. 28.2.2.

Operation of the binary stage.

Assuming an h.t. rail of $-6V$, the following sequences will apply. For a change of input as shown in figure 28.2.3. an output pulse of 6V is obtained.

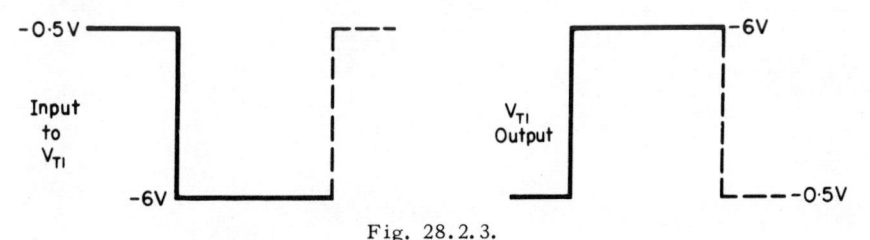

Fig. 28.2.3.

With a negative input to input 1, a $-6V$ output appears at output 2.
With a negative input to input 2, a $-6V$ output appears at output 2.
With a positive input to input 1, a 0V output appears at output 1.
With a positive input to input 2, a 0V output appears at output 2.
The action of the circuit is as follows:

For a negative input to V_{T_1} (input 1), V_{T_1} conducts. This conduction causes the V_{T_1} collector to fall to 'zero' volts, thus giving an output at output 1 of 0V. This fall in potential is transmitted to V_{T_2} base via R_{I_2}, which causes V_{T_2} to be cut off. The collector of V_{T_2} is then at $-6V$, as there is no current in R_{L_2}.

Using a similar argument, it may easily be seen that the roles of the transistors are reversed, when a negative input is applied to V_{T_2} base. A very important feature of this circuit is that once a signal has been applied, the circuit will remain in the state it acquires until a further signal is applied. The bistable then, has the ability to 'remember' the signal applied to it.

The outputs are both amplified and inverted with any one transistor and with an 'OR' circuit; the letter N is often prefixed so as to indicate the phase reversal, and the circuit may be referred to as a 'NOR' circuit.

Several stages connected in the appropriate manner may be used as a counter, as will be evident from an earlier chapter.

CHAPTER 29

Combined AND/OR gate

We will discuss further in this chapter, one or two very basic networks often used in 'Logic' circuits.

The field of logic design is primarily that of system design, i.e. once a number of 'building bricks' or modules are designed, these can be used in a variety of ways so as to produce a large number of different systems.

We will therefore content ourselves in this chapter with a very brief look at one or two very simple networks.

29.1. Simple logic circuits

Diodes may be 'seen' as electronic switches, that is they can be either 'on' or 'off' — the switch is either closed or open.

When transistors are used in these circuits and switched 'on', the voltage across the transistor is often very small. When off, the voltage across the transistor is often the full supply.

It follows then that we can often obtain an output from such a circuit, equivalent to almost the full supply voltage.

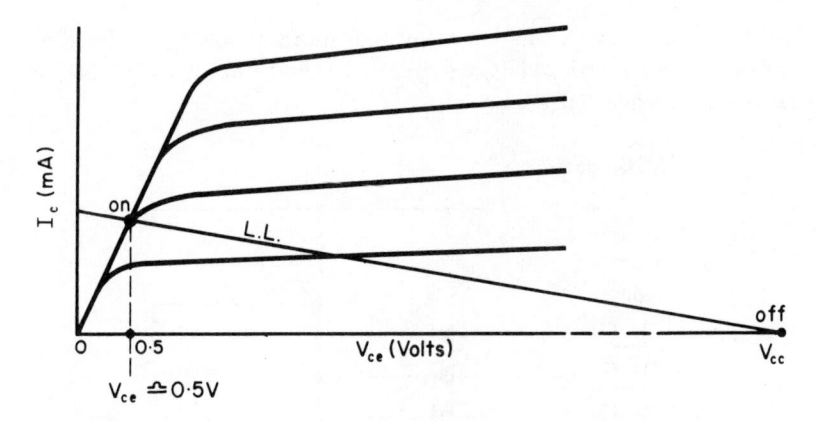

Fig. 29.1.1.

It has been shown that when forward biassed, a diode is 'ON'. In effect, the diode is conducting as much as external resistance will allow, and a very small forward voltage is developed between anode and cathode. For all practical purposes, the slight forward drop may be ignored and the diode regarded as virtually short circuit.

When reverse biassed, very little current flows, and the diode may then be considered as an 'open circuit'. Figure 29.1.2. shows the I_a/V_a characteristics for a diode.

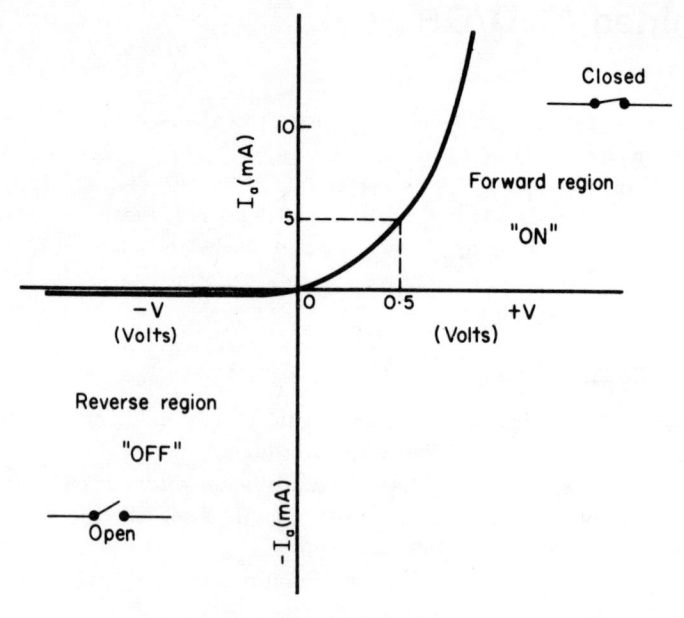

Fig. 29.1.2.

Sufficient basic theory has been covered in this respect, and for the present exercise we will consider a diode, as short circuit when 'ON' and as open circuit, when 'OFF'.

29.2. Simple 'AND' gate.

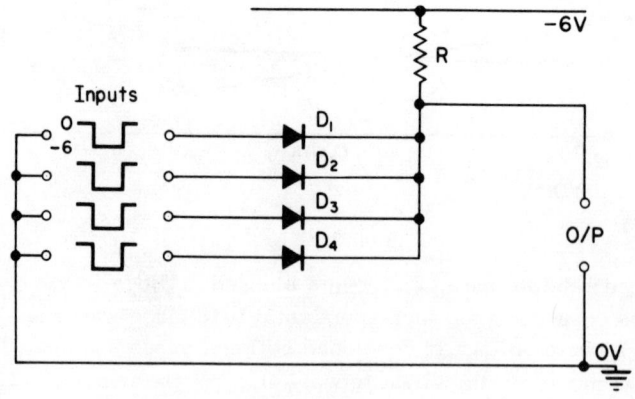

Fig. 29.2.1.

ANALYSIS OF A SIMPLE COMBINED AND/OR GATE

The circuit in figure 29.2.1. consists of four diodes, each separately connected to an input but with a common output terminal. If the input levels were all zero, the output level will be zero also due to the sum of the diode currents through R. In this case, all the diodes will be 'ON'. If the input to D_1, say, were to rise to $-6V$, D_1 will be 'OFF'. The current through R due to the remaining 3 diodes will be sufficient to cause the output to remain at 0V. If D_1 and D_2 were both 'OFF' due to a $-6V$ input, the output will remain at 0V due to the current through R flowing through D_3 and D_4.

To cause the output to rise to $-6V$, it may be seen that all diodes must be 'OFF', thereby causing the current in R to become 'zero'. When this state occurs, the output potential rises to $-6V$ as there is now no drop across R.

Therefore, for an output of $-6V$ to appear at the output, it is necessary to apply an input signal of $-6V$ to D_1 and D_2 and D_3 and D_4. This, therefore, is an 'AND' circuit. The order of diode currents may be 20–50 μA reverse, when OFF and 0.5–1.0 mA when ON.

29.3. Simple 'OR' gate circuit

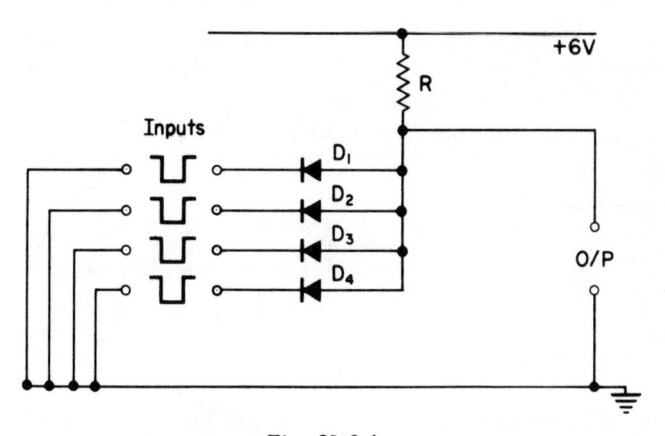

Fig. 29.3.1.

With no input (0V) applied to all diodes, the output is at zero potential due to the diode currents in R. If, say a $-6V$ input is applied to D_1, the diode will conduct heavily and the output will fall to $-6V$, as D_1 is now 'short circuit'. The output ($-6V$) will now appear on each diode anode; thus D_2, D_3, D_4 are OFF.

If the input had been applied to D_1 OR D_2 OR D_3 OR D_4 the result would have been the same. This, then, is an 'OR' circuit.

For a signal of $-6V$ to appear at the output, a $-6V$ input should be applied to D_1 or D_2 or D_3 or D_4.

In practice, the output will be −6V − (forward drop) and will be in order of $(-6 + 0.5)V = -5.5V$ or so. The diodes are usually silicon types, which produce a 0.5V positive shift or forward drop.

29.4. Coincidence gates

We have seen that when applying a rectangular pulse to a $C.R.$ coupling network, we may differentiate the 'waveform', provided that the time constants are correctly chosen.

We also remove any d.c. level that may have been present, and may, if required, restore the d.c. level with a diode.

Consider the following circuit in figure 29.4.1.

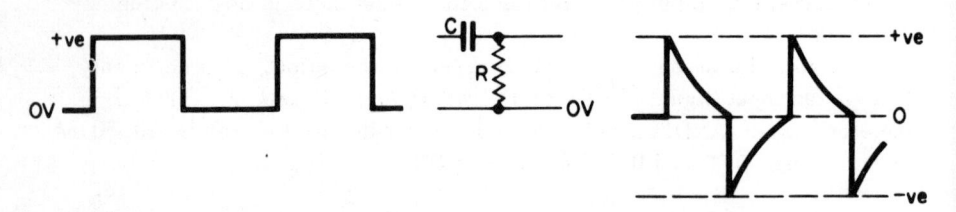

Fig. 29.4.1.

The pulse is differentiated, and is symmetrical about 0V, so that consequently there is no d.c. level. To restore the d.c. level, we simply connect a diode across R. The d.c. level is restored by removing the negative going component, as shown in figure 29.4.2.

Fig. 29.4.2.

The negative portion of the waveform causes the diode to conduct, and effectively short circuits the signal to earth. If we reverse the diode, the reverse is the case, and a train of negative pulses is obtained across both R and the reverse biassed diode in shunt.

It will now be convenient to examine a circuit embracing many of the features associated with diodes as shown in figure 29.4.3.

The waveforms at points A, B and C as well as diode inputs, are shown.

It is clear that the only time that all inputs are positive is between T_8 and T_9. During this time interval, all diodes are OFF. The potential at 'A' rises from 0V to +5V as at this time; there is no diode current in R_1 and the p.d. across R_1 is reduced to zero.

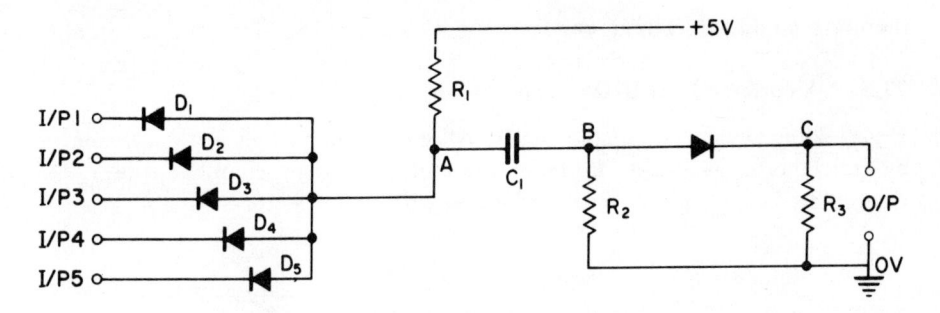

ANALYSIS OF A SIMPLE COMBINED AND/OR GATE

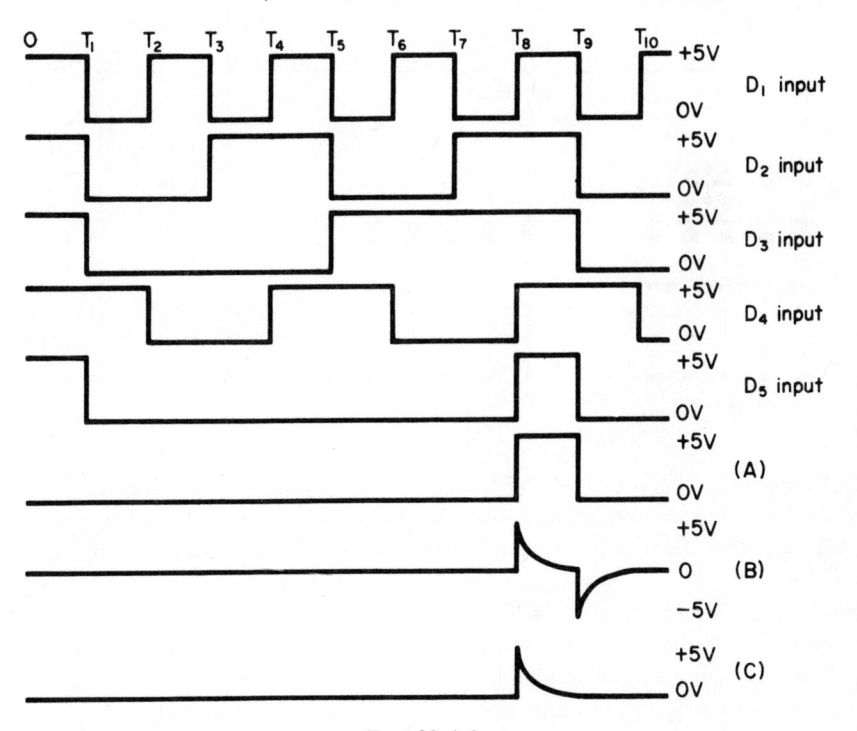

Fig. 29.4.3.

At the instant all diodes are off, point A rises to 5V. The positive going pulse is differentiated by C_1, R_2, and appears at point B. The d.c. level is now lost as during differentiation of the pulses, there is an equal positive and negative potential across R_2. The leading edge at T_8 is positive going, and causes D_6 to conduct; consequently the signal current in R_3 produces a positive spike at the output.

When the lagging edge of the pulse appears at D_6 anode, the diode is OFF and practically open circuit, so that no current flows in R_4 and therefore no output voltage exists.

29.5. A combined 'AND/OR' gate circuit

Figure 29.5.1. shows a simple 'AND/OR' gate diode circuit and a common emitter invertor amplifier. In the absence of a signal to any diode, the transistor, V_{T_1}, should be non conducting with its collector sitting at V_{cc} (– 6V).

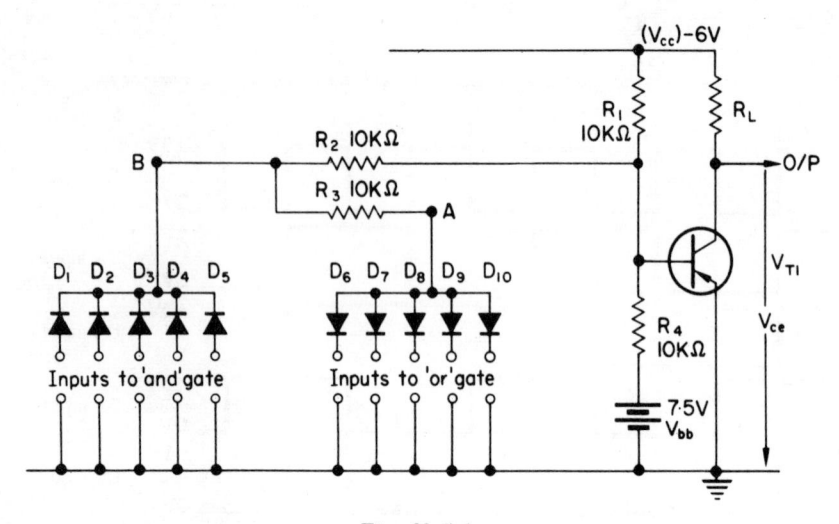

Fig. 29.5.1.

Despite the simplicity of the basic circuit, it does contain a great deal of useful information relating to simple static switching.

The diodes D_1 to D_5 form the 'AND' gate whilst the diodes D_6 to D_{10} form the 'OR' gate. These gates may be considered as switches. When any one diode is conducting, it is 'ON' and is the electronic equivalent of a closed switch. When the diode is 'OFF' it is electronically, an open switch.

The following simple analysis will enable the d.c. levels in the circuit to be established and in particular, the base-emitter voltage of the transistor, V_{be}.

The base current due to the leakage current, $I_{co}{'}$, will be disregarded as any such base current flowing in the base circuit will be 'swamped' by the external current flowing. The 10 KΩ resistors are not necessarily ideal or even optimum values but they do enable a simple analysis to be made.

The input signal will be in one of two states, either zero potential or at V_{cc} (– 6V). *It is important to note that when no signal is applied, the input terminal is at earth potential.*

The input signal is shown in figure 29.5.2.

ANALYSIS OF A SIMPLE COMBINED AND/OR GATE

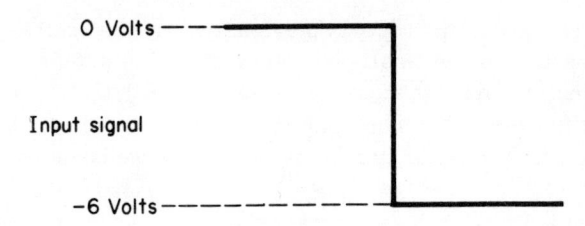

Fig. 29.5.2.

No signal condition.

With no signals applied, the battery V_{bb}, will cause a current to flow that will place a positive potential at the anodes of the 'OR' gate diodes and a positive potential at the cathodes of the 'AND' gate diodes. Reference to figure 29.5.1. will show that the 'OR' gate diodes are forward biassed and are 'ON' whilst the 'AND' gate diodes are reverse biassed and are 'OFF'.

Figure 29.5.3. shows the circuit under no signal conditions. All of the diodes are replaced by switches.

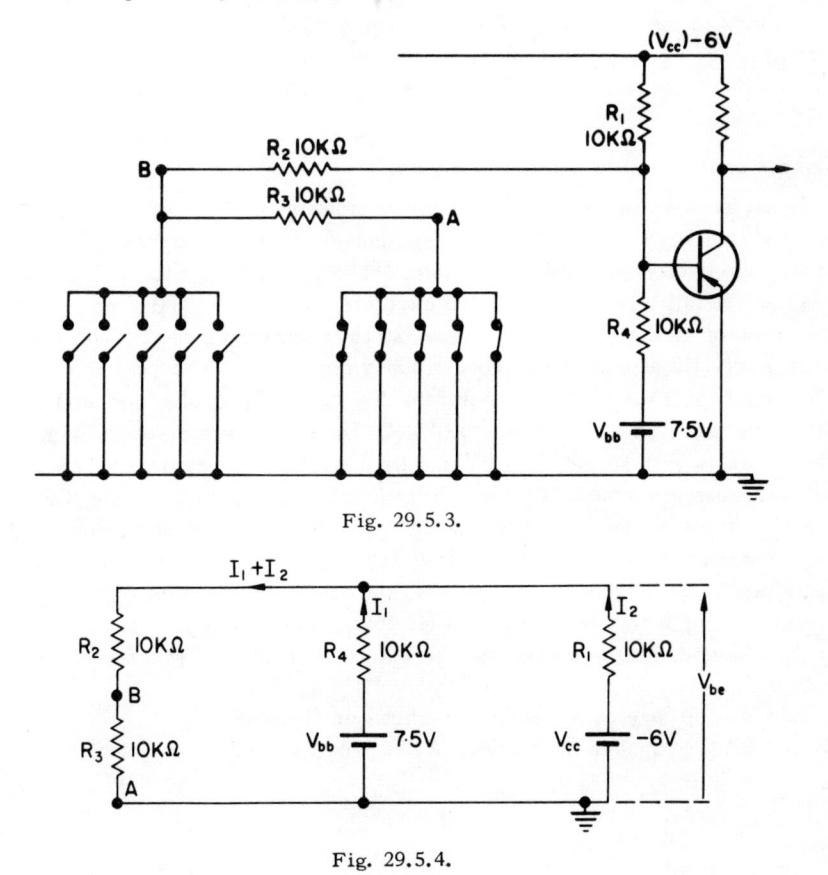

Fig. 29.5.3.

Fig. 29.5.4.

By using the analytical methods previously described earlier, the base-emitter potential may easily be determined. The circuit is redrawn in a more familiar form for this exercise in figure 29.5.4.

Using the formula in the same manner as for simple electrical networks, derived in an earlier chapter, the base-emitter voltage becomes,

$$V_{be} = \frac{\dfrac{V_{bb}}{R_4} + \dfrac{V_{cc}}{R_1}}{\dfrac{1}{R_1} + \dfrac{1}{R_4} + \dfrac{1}{R_2 + R_3}}$$

$$\therefore V_{be} = \frac{\dfrac{7.5}{10} - \dfrac{6}{10}}{\dfrac{1}{10} + \dfrac{1}{10} + \dfrac{1}{20}} = \underline{0.6V.}$$

As the base is sitting at 0.6V, point B will be at 0.3V. Point A will be very slightly positive by an amount determined by the forward drop of the diodes in the 'OR' gate. The 'AND' gate diodes are effectively open circuit. V_{T_1} is non conducting with its base sitting at 0.6V. (The forward drop of the 'OR' gate diodes may be ignored). The collector will be at $-6V$.

Signals applied to the 'OR' gate diodes.

When an input is applied to any one or more of the 'OR' gate diodes, point A is dragged down to the input potential of $-6V$. The 'OR' gate diodes are even more 'ON' than before. When point A falls to $-6V$, point B attempts to fall also. The negative excursion of point B is limited by the forward drop of the 'AND' gate diodes as they commence to conduct and point B is held at a potential very slightly negative with respect to earth. This small potential may be ignored in this case. All of the 'OR' diodes that have received an input will be 'ON'. Those that had no such input will be cut off as with point A at $-6V$ and their cathodes at zero volts, they will be reverse biassed. The change in level of point A due to the 'OR' input(s) is prevented from reaching the base of the transistor by the 'AND' gate diodes which are now conducting and holding point B at approximately earth potential. Figure 29.5.5. shows this condition. In order to clarify the circuit, an input will be shown applied to D_6 and D_8 only.

The base-emitter potential may be established in the same manner as before.

The circuit may be simplified as shown in figure 29.5.6.
(Note that the input in series with R_3 is 'short circuited' to earth).

ANALYSIS OF A SIMPLE COMBINED AND/OR GATE

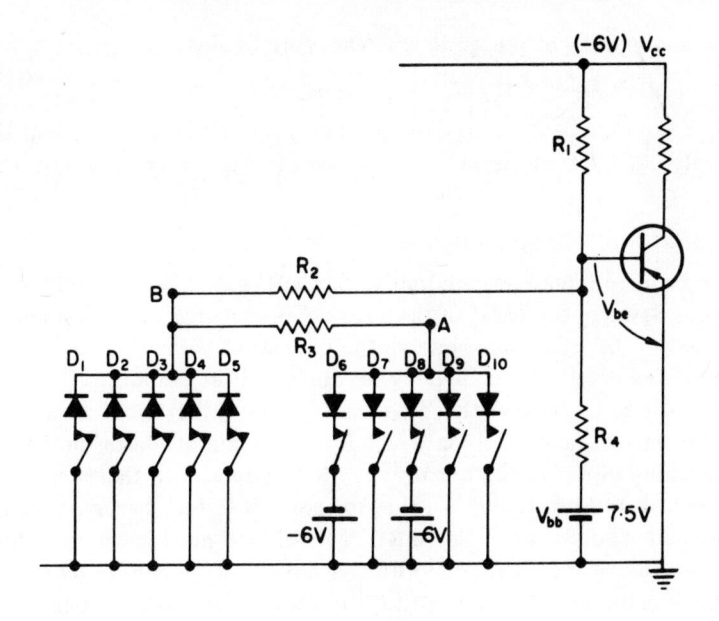

Fig. 29.5.5.

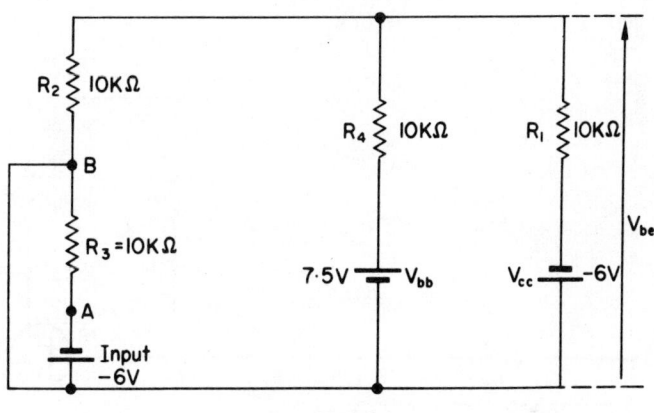

Fig. 29.5.6.

$$V_{be} = \frac{\dfrac{7.5V}{10} - \dfrac{6}{10}}{\dfrac{1}{10} + \dfrac{1}{10} + \dfrac{1}{10}} = \underline{0.5V.}$$

The base is at a positive potential and therefore the transistor is still non conducting. The collector remains at $V_{cc} = (-6V)$.

Point A is at $-6V$ also, but cannot affect the transistor due to the short circuit action of the 'AND' gate. The position would remain exactly the same for any number of inputs to the 'OR' gate diodes.

Input applied to the 'AND' gate diodes only.

Should any number of inputs be applied to the 'AND' gate diodes, they will merely cut off even further those diodes that are reverse biassed due to V_{bb}.

Inputs applied to both gate circuits.

If one or more inputs are applied to the 'OR' gate diodes, point A will be dragged down to the level of the input. This change of level cannot be transmitted to the transistor base as the change at point B is restricted by the conduction of all of the 'AND' gate diodes. If at the same time, an input was applied to one of the 'AND' gate diodes, that diode would be cut off and become an open switch. Point B will remain at 'earth' potential due to the latching effect of the remaining 'AND' gate diodes that are still conducting. It follows therefore, that for point B to fall due to an input to the 'OR' gate circuit, *all* of the 'AND' gate diodes need an input. When this occurs, all of the 'AND' gate diodes are cut off – behave as open switches – and point B is able to fall to a level sufficiently negative to drag the transistor base down into conduction. Figure 29.5.7. shows two inputs to the 'OR' gate and an input to all of the 'AND' gate diodes. The diodes are again replaced by switches.

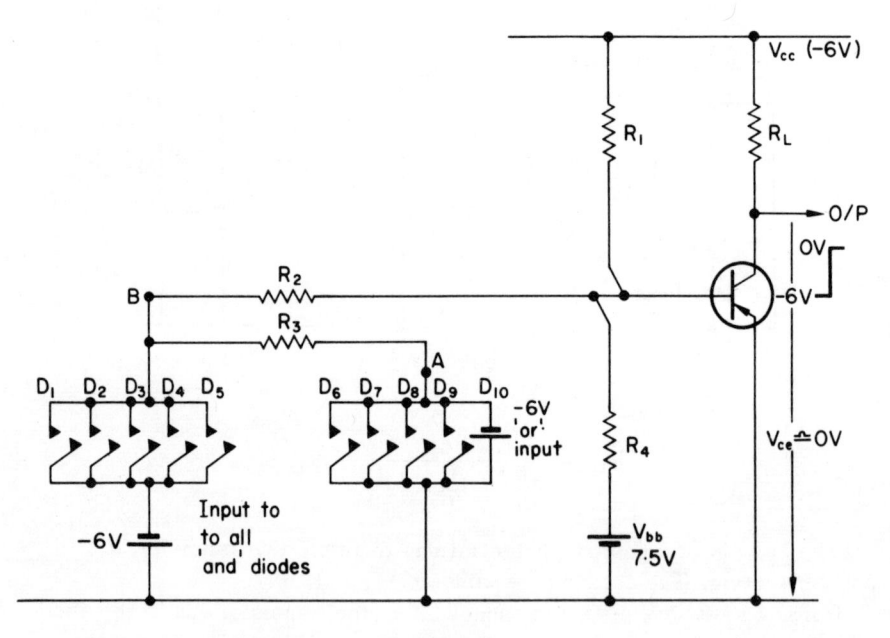

Fig. 29.5.7.

ANALYSIS OF A SIMPLE COMBINED AND/OR GATE

The equivalent circuit shown in figure 29.5.8. enables the base-emitter voltage to be evaluated.

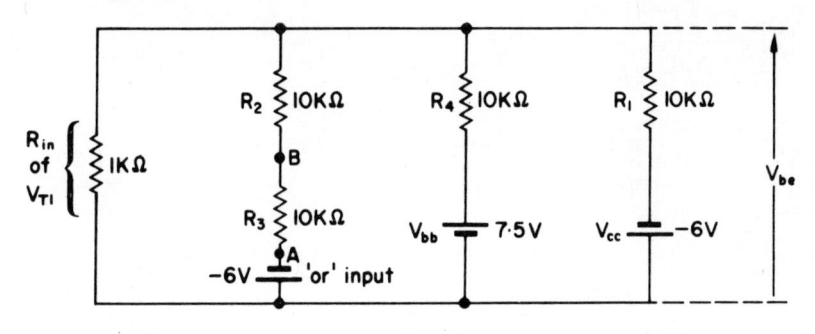

Fig. 29.5.8.

$$V_{be} = \frac{\frac{\text{'OR' input}}{R_3 + R_2} + \frac{V_{bb}}{R_4} + \frac{V_{cc}}{R_1}}{\frac{1}{R_1} + \frac{1}{R_2 + R_3} + \frac{1}{R_4} + \frac{1}{R_{in}}} = \frac{\frac{-6}{20} + \frac{7.5V}{10} - \frac{6}{10}}{\frac{1}{20} + \frac{1}{10} + \frac{1}{10} + \frac{1}{1}} = -0.12V.$$

With the base-emitter potential at $-0.12V$ the transistor will conduct. When conduction occurs, the input resistance of V_{T_1}, usually about 1000Ω, will be presented across the gating circuits. This is allowed for in the expression above. Its value will *not* affect the *polarity* of V_{be} but will affect the amplitude of course.

Summary.

An output signal will be obtained only when an input is applied to one or more 'OR' gate diodes and an input signal is applied to all of the 'AND' gate diodes *at the same time*. The output from the transistor in this circuit will have a value equal to that of the input and may therefore be used to trigger a further stage.

Suitable combinations of this circuit and binary stages can result in many useful practical examples.

CHAPTER 30

Analogue considerations

30.1. Laplace terminology

It was shown earlier in this book that the impedance of simple series circuits could be expressed in terms of a $\pm jb$. For instance a series $L-R$ circuit could be written as $R + j\omega L$.

Non-sinusoidal signals, e.g. squarewaves, pulses, sawtooth waveforms, etc, may be expressed in a form of a series. The series could contain a number of sinusoidal quantities of differing frequencies.

It is often necessary to use non-sinusoidal voltages in electronic equipment and for non-sinusoidal quantities, the expression $j\omega$ becomes invalid. As the term ω, when expanded, becomes $2\pi f$, it can be valid for one frequency only, i.e. f is a constant and could not apply for say, pulse waveforms.

One method of overcoming the difficulties outlined is to use Laplace terminology. For the purpose of dealing with the following simple analogue computers, it will be initially sufficient for the reader to know how to write down expressions for circuit impedance that are valid for non-sinusoidal quantities without needing to know how to derive the expressions from first principles, although the subject should be learnt at a later stage.

Normal impedance may be expressed as a $\pm jb$ where b may be ωL or $1/\omega C$. To write the corresponding Laplace expression one simply writes S for $j\omega$ wherever the latter occurs.

Example. A series $L-R$ and $C-R$ circuit.

Normal expression. $R + j\omega L$ or $R + \dfrac{1}{j\omega C}$

Laplace expression. $R + SL$ or $R + \dfrac{1}{SC}$.

The latter expressions using Laplace terminology are known as 'operational impedances', and are valid for pulse quantities.

Consider the following table of examples in figure 30.1.1.

The expression in column 3 in row 6 is given as

$$\dfrac{1}{\dfrac{1}{R} + SC}$$

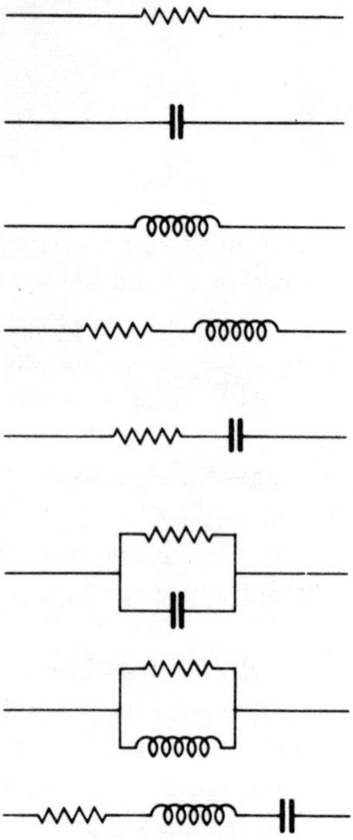

Circuit network	a + jb expression	Operational impedance	
—\/\/\/—	R	R	(1)
—‖—	$\dfrac{1}{jWC}$	$\dfrac{1}{SC}$	(2)
—⏝⏝⏝—	jWL	SL	(3)
—\/\/\/—⏝⏝⏝—	$R + jWL$	$R + SL$	(4)
—\/\/\/—‖—	$R + \dfrac{1}{jWC}$	$R + \dfrac{1}{SC}$	(5)
R ∥ C	$\dfrac{1}{\frac{1}{R} + jWC}$	$\dfrac{1}{\frac{1}{R} + SC}$	(6)
R ∥ L	$\dfrac{1}{\frac{1}{R} + \frac{1}{jWL}}$	$\dfrac{1}{\frac{1}{R} + \frac{1}{SL}}$	(7)
—\/\/\/—⏝⏝⏝—‖—	$R + jWL + \dfrac{1}{jWC}$	$R + SL + \dfrac{1}{SC}$	(8)

Fig. 30.1.1.

This may be simplified by re-arranging as follows:

$$\frac{1}{\frac{1}{R} + SC} = \frac{R}{1 + SCR}$$

after multiplying top and bottom by R.

This is a common form of expressing the operational impedance for a shunt circuit having R and C in parallel.

30.2. Operational amplifiers

Operational amplifiers have a very high gain and often work down to d.c. When d.c. operated, special precautions need to be taken to ensure minimum d.c. drift and one must include provision for zero setting. Push pull circuits, often long tailed pairs, are frequently employed throughout the whole of the amplifier when it is d.c. connected.

A simple operational amplifier complete with 'input' and 'feedback'

SIMPLE ANALOGUE CONSIDERATIONS 557

resistors R_1 and R_0 is shown in figure 30.2.1.

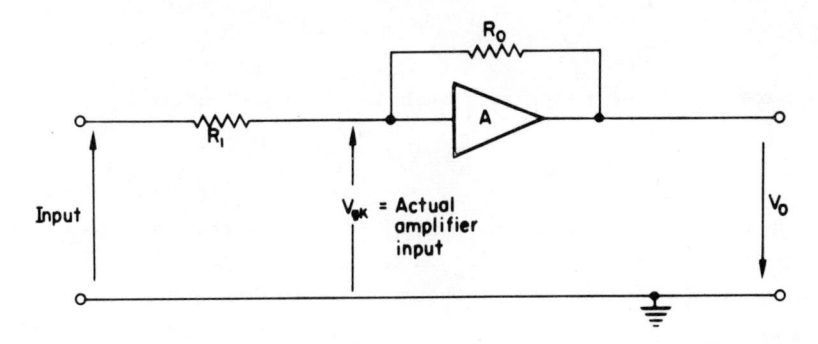

Fig. 30.2.1.

A resistor R_1 is connected between the input terminal and the 'actual amplifier' input, i.e. the grid of the first valve.

A feedback resistor R_0 is connected between the output and the 'actual amplifier' input. Provided that the open loop gain is very very high, when feedback is connected as shown, the actual effective input voltage v_{gk}, is very very small. This actual input v_{gk} is virtually zero in this type of system and this point is called a 'virtual earth'. The amplifier has a high input-resistance.

As v_{gk} (for a valve amplifier) is virtually zero, it will not affect the current flowing through R_1 resulting from the input signal v_1.

Figure 30.2.2. typifies operational amplifiers although R_1 and R_0 are often replaced by more complex components or networks.

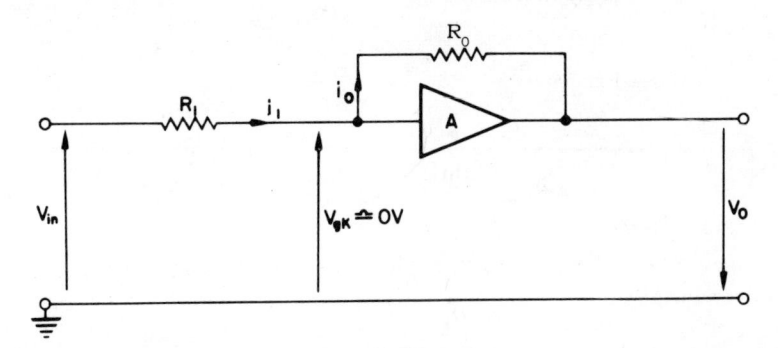

Fig. 30.2.2.

As the input current i_1 will flow through R_0, $i_1 = i_0$.

$$i_1 = \frac{v_1 - v_{gk}}{R_1} \text{ and } i_0 = \frac{v_{gk} - v_0}{R_0}$$

but as $v_{gk} \simeq 0V$,

$$i_1 = \frac{v_1}{R_1} \text{ and } i_0 = \frac{-v_0}{R_0}$$

hence the gain of the system as a whole may be expressed as

$$\frac{v_0}{v_1} = \frac{-R_0}{R_1}.$$

This is known as the transfer function. It is seen therefore that the gain is determined by R_0 and R_1. The $-ve$ sign indicates the 180° phase shift of the amplifier. This is also accounted for by the direction of the arrow indicating v_0.

If R_0 and R_1 have the same value of resistance, i.e., $R_0 = R_1$, the gain of the system will be unity. When this is the case, the circuit is simply a phase invertor. If R_0 is increased and is greater than R_1, the feedback is reduced and the gain will be greater. If R_0 is less than R_1, the gain will be less than unity.

We can extend the circuit a little by providing a number of inputs, each with its associated input series resistor. This is shown in figure 30.2.3.

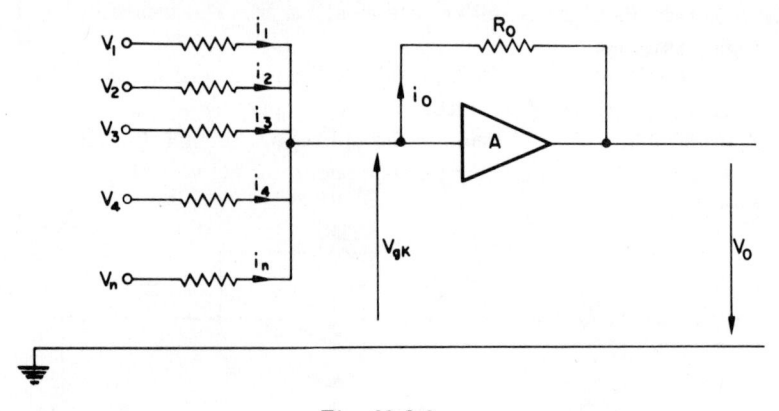

Fig. 30.2.3.

All input currents will flow through R_0 as was the case with i_1 in the previous example.

Hence $i_1 + i_2 + i_3 + i_4 + i_n = i_0$.

Thus $\dfrac{v_1 - v_{gk}}{R_1} + \dfrac{v_2 - v_{gk}}{R_2} + \dfrac{v_3 - v_{gk}}{R_3} + \dfrac{v_4 - v_{gk}}{R_4} + \dfrac{v_n - v_{gk}}{R_n} = \dfrac{v_{gk} - v_0}{R_0}$

but as before, we can ignore v_{gk}, therefore,

SIMPLE ANALOGUE CONSIDERATIONS

$$\frac{v_1}{R_1} + \frac{v_2}{R_2} + \frac{v_3}{R_3} + \frac{v_4}{R_4} + \frac{v_n}{R_n} = \frac{-v_0}{R_0}$$

then $v_1 \cdot \left(\dfrac{R_0}{R_1}\right) + v_2 \cdot \left(\dfrac{R_0}{R_2}\right) + v_3 \cdot \left(\dfrac{R_0}{R_3}\right) + v_4 \cdot \left(\dfrac{R_0}{R_4}\right) + v_n \cdot \left(\dfrac{R_0}{R_n}\right) = -v_0$

showing that v_0 is the sum of all input voltages, the individual amplitudes of which when summed will be determined by the ratio of R_0 to the individual input series resistors.

If all resistors were of the same value,

$$-v_0 = v_1 + v_2 + v_3 + v_4 + v_n$$

and the output voltage would be the direct sum of all inputs.

Example

$R_1 = 2\,\text{M}\Omega.$ $R_2 = 1\,\text{M}\Omega.$ $R_3 = 0.5\,\text{M}\Omega.$ $R_0 = 1\,\text{M}\Omega.$

$v_1 = 1\text{V}.$ $v_2 = 2\text{V}.$ $v_3 = 3\text{V}.$

Thus $\qquad -v_0 = v_1 \cdot \left(\dfrac{R_0}{R_1}\right) + v_2 \cdot \left(\dfrac{R_0}{R_2}\right) + v_3 \cdot \left(\dfrac{R_0}{R_3}\right)$

∴ $\qquad -v_0 = 1 \cdot \left(\dfrac{1}{2}\right) + 2 \cdot \left(\dfrac{1}{1}\right) + 3 \cdot \left(\dfrac{1}{0.5}\right)$

∴ $\qquad -v_0 = \dfrac{1}{2} + 2 + 6 = 8.5\text{V}.$

Hence $\qquad v_0 = -8.5\text{V}.$

If all resistors were equal in value,

$$v_0 = -6\text{V}.$$

This circuit is a simple summing invertor.

30.3. Difference amplifier

Should we require v_0 to be proportional to the difference between two inputs, we need only to charge the polarity of one input and add as in the previous example.

Example

We need the difference between 6V and 4V. Using the summing invertor, we would apply 6V and −4V. The addition of the inputs would be 6 − 4 = 2V. Figure 30.3.1. shows this principle.

The 4V input is phase inverted and applied to the summing invertor and effectively added to the 6V input.

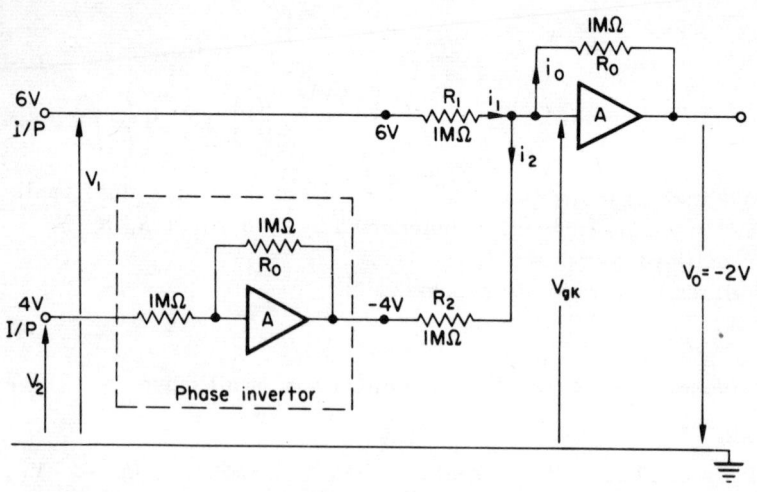

Fig. 30.3.1.

Both i_1 and i_2 will flow through R_0 in the summing invertor.

Hence
$$i_1 + (-i_2) = i_0.$$

$$\frac{6V - v_{gk}}{1\,M\Omega} + \frac{-4V - v_{gk}}{1\,M\Omega} = \frac{v_{gk} - v_0}{1\,M\Omega}$$

thus
$$\frac{6V}{1\,M\Omega} + \frac{-4V}{1\,M\Omega} = \frac{-v_0}{1\,M\Omega}$$

therefore
$$-v_0 = (6 - 4)\,V$$

∴
$$v_0 = -2V \text{ as required.}$$

This principle applies for any number of inputs.

30.4. Servomechanisms

A 'differential' is an integral part of a remote position control servo-mechanism. It notes the difference between the command input and the actual output. The output is sampled and fed back to the differential and any difference (error) is inverted and passed through the system to correct the error.

The difference amplifier performs the 'differential' function.

The feedback resistor R_0 may be replaced by other components; a capacitor, or a combination of components depending upon the required function.

In order to allow for any type of component between output and input, it is usual to write R_0 as Z_0 and for the input series components, $Z_1 \cdot Z_2$, etc.

SIMPLE ANALOGUE CONSIDERATIONS   561

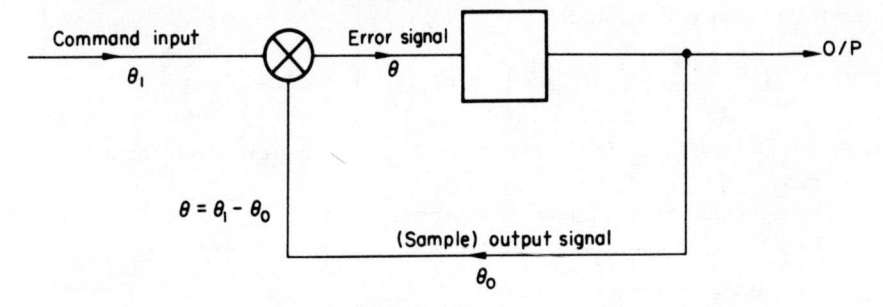

```
Command input ─────⊗─── Error signal ──[  ]──────•────▶ O/P
         θ₁          │        θ                   │
                     │                            │
      θ = θ₁ - θ₀    │                            │
                     └──────(Sample) output signal┘
                                   θ₀
```

Fig. 30.4.1.

We can then say that
$$\frac{v_0}{v_1} = \frac{-Z_0}{Z_1}$$

where Z_0 and Z_1 can be complex or simply resistors.

30.5. Summing integrator

Let us now consider an operational amplifier where Z_0 is a capacitor as shown in figure 30.5.1.

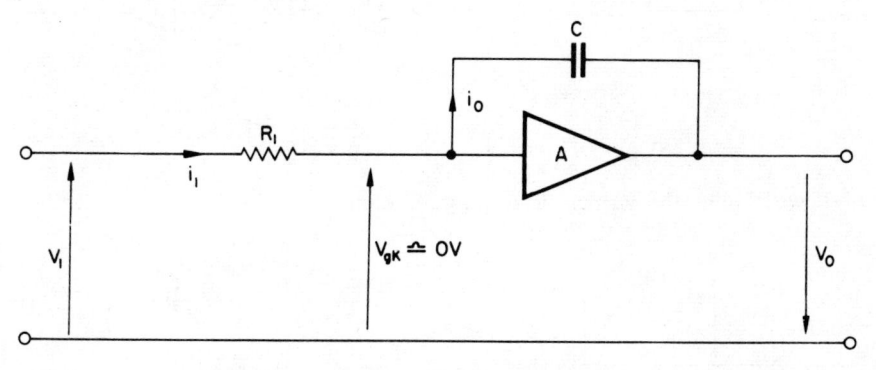

Fig. 30.5.1.

$$i_1 = i_0 \quad \therefore \quad \frac{v_1 - v_{gk}}{R_1} = C.d/dt\,(v_{gk} - v_0)$$

and ignoring v_{gk},

$$\frac{v_1}{R_1} = -C\,d/dt\,(v_0) \quad \therefore \quad \frac{v_1}{R_1 C} = -d/dt\,(v_0)$$

hence
$$v_0 = \frac{-1}{CR_1} \int v_1\,dt.$$

and for a number of inputs,

$$v_0 = \frac{-1}{CR_1} \int v_1 \, dt \; \frac{-1}{CR_2} \int v_2 \, dt \; \frac{-1}{CR_3} \int v_3 \, dt \ldots \frac{-1}{CR_n} \int v_n \, dt.$$

The output voltage v_0 is the scaled sum of the integrals of all input voltages.

The terms shown are integrated with respect to t.

The product of $C.R.$ is also in terms of t.

$1\mu F \times 1 M\Omega = 1 \sec$ and is unity.

When $R = 1 M\Omega$, $C = 1\mu F$ and for 3V input, $v_0 = -3V$.
When $R = 0.1 M\Omega$, $C = 1\mu F$ and for 3V input, $v_0 = -30V$.
When $R = 10 M\Omega$, $C = 1\mu F$ and for 3V input, $v_0 = -0.3V$.

Let us compare two operational amplifiers and with a common input, examine their respective outputs.

Fig. 30.5.2.

$$v_0 = \frac{-1}{0.1} \times 7 = -70V \quad v_0 = \frac{-1}{0.1} \times 7 = -70V.$$

If we change the values of R_1 in each case,

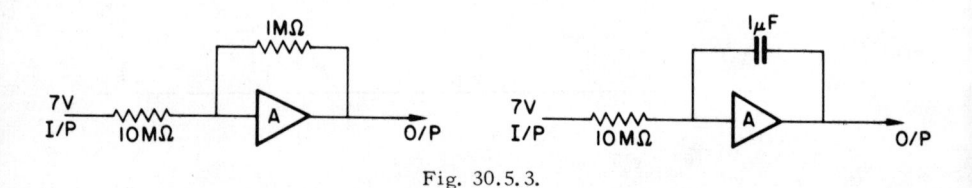

Fig. 30.5.3.

$$v_0 = \frac{-1}{10} = -0.7V \quad v_0 = \frac{-1}{10} \times 7V = -0.7V.$$

and finally,

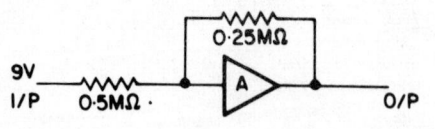

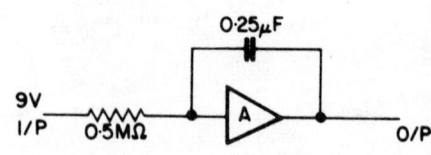

Fig. 30.5.4.

$$v_0 = \frac{0.25}{0.5} \times 9 = \underline{-4.5V} \qquad v_0 = \frac{-1}{.125} \times 9 = \underline{-72V.}$$

Note that $v_0 \, \alpha \, \frac{R_0}{R_1}$ in the summing invertor.

Note that $v_0 \, \alpha \, \frac{1}{CR}$ in the summing integrator as $Z_0 \, \alpha \, \frac{1}{C}$.

Let us now deal with the case where both Z_0 and Z_1 consist of shunt networks and to employ operational impedance technique.

The circuit is as shown in figure 30.5.5.

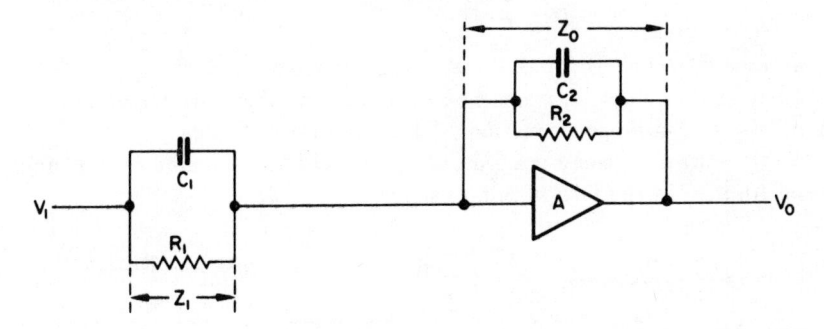

Fig. 30.5.5.

$$Z_1 = \frac{1}{\frac{1}{R_1} + SC_1} = \frac{R_1}{1 + SC_1 R_1}$$

and similarly

$$Z_0 = \frac{1}{\frac{1}{R_2} + SC_2} = \frac{R_2}{1 + SC_2 R_2}$$

hence

$$\frac{v_0}{v_n} = \frac{-Z_0}{Z_1} = \frac{1 + \frac{R_2}{SC_2 R_2}}{1 + \frac{R_1}{SC_1 R_1}} = \left[\frac{1 + SC_1 R_1}{1 + SC_2 R_2} \right] \cdot \frac{R_2}{R_1} .$$

The function of a system may often be expressed in the form of a differential equation. A combination of the operational amplifiers shown can be used to solve for the variable and to display it on say, a pen recorder or an oscilloscope. Electronic analogies may be constructed quite easily for other systems and enables one to see the behaviour of non-electronic systems and allows easy adjustment to component values in order to achieve the desired result. An equation of the form $(D + 3D + 2)x = f(y)$ might be encountered. $f(y)$ is the input to a system and the equation $(D^2 + 3D + 2)x$ describes the system response to the input. This can be solved mathematically but an example will be given in the use of these operational amplifiers to solve the problem.

30.6. Simple analogue computer

Consider the need for a simple analogue computer to display for 'x' in a system described by the differential equation $(D^2 + 3D + 2)x = f(y)$.

The first step is to re-arrange the equation and write the highest order differential co-efficient on the l.h.s. taking all other terms to the r.h.s.

$$D x = f(y) - 3Dx - 2x.$$

We now commence to 'build' our simple computer circuit.

We must then draw as many input terminals as there are terms on the r.h.s. of the re-arranged equation. (Three in this example).

We then draw as many integrators in cascade as indicated by the order of the l.h.s. terms. (Two in this example).

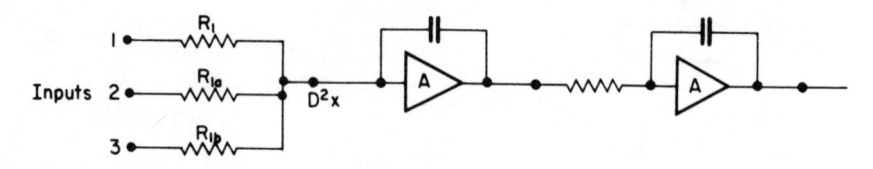

Fig. 30.6.1.

We have to apply to inputs 1, 2 and 3, the terms shown on the r.h.s. of the equation, i.e. $f(y)$ and $-3Dx$ and $-2x$ respectively. The algebraic sum of these terms must equal D^2x (from our equation) therefore we can write D^2x at the summing junction of R_1, R_{1a} and R_{1b}.

The next step is to write the l.h.s. term on the circuit at the input to the first amplifier (point A on circuit in figure 30.6.2.).

We can see that the output from the first integrator will be the integral of the input to that stage and we can therefore write $-Dx$ at point B. The same principle applies for the second integrator, if $R_2 = 1\,M\Omega$ and $C_2 = 1\,\mu F$, then we will have x at point C. (Figure 30.6.2.).

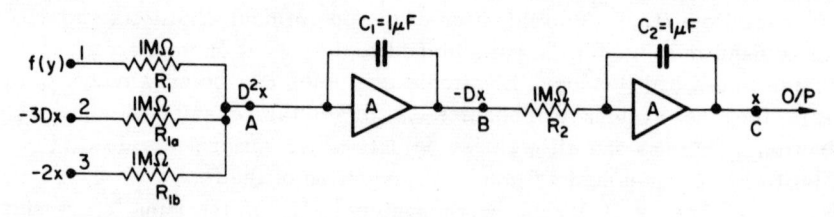

Fig. 30.6.2.

SIMPLE ANALOGUE CONSIDERATIONS   565

We have therefore obtained the required output, x. Input terminal 2 requires $-3Dx$. We have $-3Dx$ available at point B but before we can transmit this to the input terminal 2, we must amplify it by 3 to give us $-3Dx$.

If we use an operational amplifier for this task, we will also reverse the phase. It will be necessary then to follow this with an invertor to restore the signal to $-3Dx$.

A simple operational amplifier connected at point C, to give a gain of -2 will suffice for the feedback to satisfy the input terminal 3.

Figure 30.6.3. shows the complete system including all feedback loops and component values.

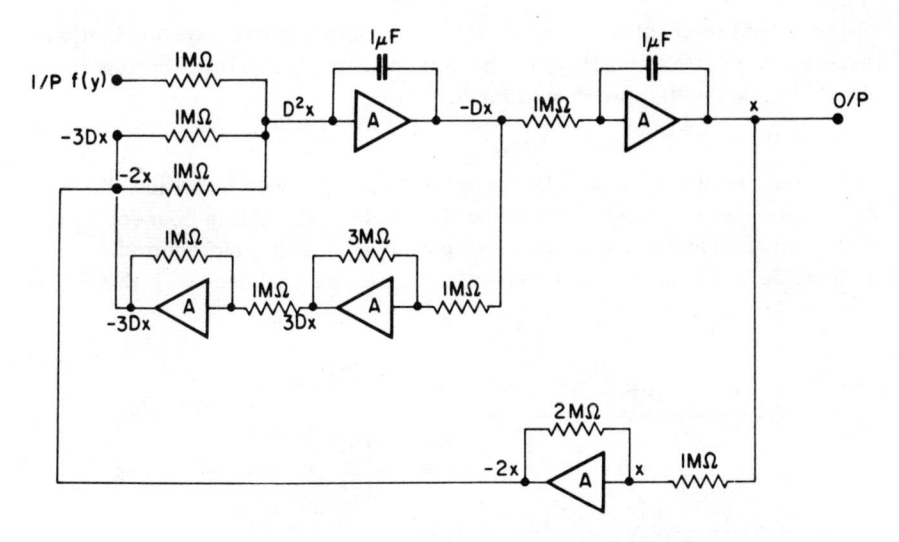

Fig. 30.6.3.

A little time taken on these systems will often reveal a much more economical use of components and operational amplifiers. No attempt was made in this example to reduce the number of components. The reader might try for himself.

30.7. Application to a simple servomechanism system

Let us now consider a simple servomechanism system. (Figure 30.7.1.).

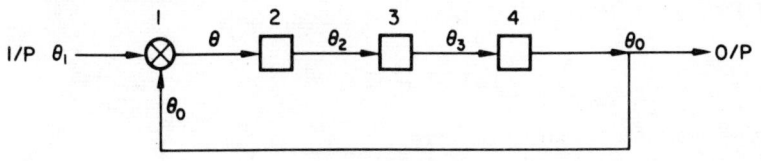

Fig. 30.7.1.

θ_1 is the input or command signal. θ_0 is the actual output operation (which should agree with the input command). θ_0 is fed back (this is often just a fraction of the output only) to the differential.

θ is the error signal and is the difference between input and output, i.e. $\theta = \theta_1 - \theta_0$.

An analysis of the system might result in the following equations.

$$\theta = \theta_1 - \theta_0.$$
$$S\theta_2 = 4\theta$$
$$S\theta_3 = 2(\theta_2 - \theta_3) + 6S\theta_2.$$
$$\theta_0 = 8\theta_3.$$

We intend to construct, step by step, a simple analogue system that will become an electronic analogy of the servomechanism system, the function of which, can be displayed on a c.r.o.

Step 1. Block 1. The differential.

We require $\theta = \theta_1 - \theta_0$. Hence we need a difference amplifier. We can use a summing amplifier and reverse the phase of θ_0 before connecting it to the input. This will give us an output of $\theta_1 + (-\theta_0)$ as required, (Figure 30.7.2.) but will be reversed in phase. We will deal with this further, later on.

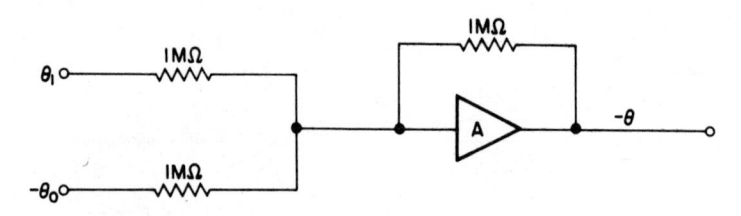

Fig. 30.7.2.

Step 2. Block 2.

Given that $S\theta_2 = 4\theta$, the transfer function is $\dfrac{O/P}{I/P} - \dfrac{\theta_2}{\theta} = \dfrac{4}{S}$

The transfer function of an integrator, Z_0/Z_1, is $-1/SCR$ and may be compared with $4/S$.

Hence $\dfrac{4}{S} : \dfrac{-1}{SCR}$ shows that $4 = -1/CR$.

If $C = 1\mu F$, $R = 250\,K\Omega$. And as a check,

$$\frac{1}{1 \times 0.25} = 4$$

and the requirements are satisfied.

Our circuit will then become as shown in figure 30.7.3.

Fig. 30.7.3.

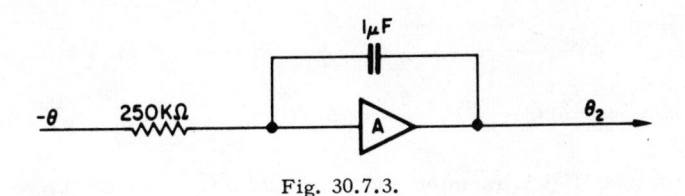

Step 3. Block 3.

$$S\theta_3 = 2(\theta_2 - \theta_3) + 6 S\theta_2.$$

Re-arranging both sides of the equation to collect like terms,

$$\theta_3 (S + 2) = \theta_2 (2 + 6S)$$

hence the transfer function,

$$\frac{O/P}{I/P} = \frac{\theta_3}{\theta_2} = \frac{2 + 6S}{S + 2}.$$

This transfer function is in the form given earlier for shunt networks.

Comparing this expression with the general expression derived earlier

$$\frac{2 + 6S}{S + 2} \quad \text{and} \quad \frac{-R_2}{R_1}\left[\frac{1 + S\,C_1 R_1}{1 + S\,C_2 R_2}\right]$$

Letting $R_1 = R_2 = 1\,\text{M}\Omega$ for convenience, we have

$$\frac{2 + 6S}{S + 2} \quad \text{and} \quad \frac{-1}{1}\left[\frac{1 + S\,C_1 R_1}{1 + S\,C_2 R_2}\right]$$

and

$$\frac{2[1 + 3S]}{2[1 + S/2]} \quad \text{and} \quad -1\,\frac{[1 + S\,C_1 R_1]}{[1 + S\,C_2 R_2]}$$

we can see that $3S$ corresponds to $S\,C_1 R_1$ and $S/2$ corresponds to $S\,C_2 R_2$.

Hence $\quad 3 = C_1 R_1$ and as $R_1 = 1\,\text{M}\Omega$, $C_1 = 3\,\mu\text{F}$

and $\quad 1/2 = C_2 R_2$ and as $R_2 = 1\,\text{M}\Omega$, $C_2 = 0.5\,\mu\text{F}$.

The circuit may now be drawn, complete with component values as in figure 30.7.4.

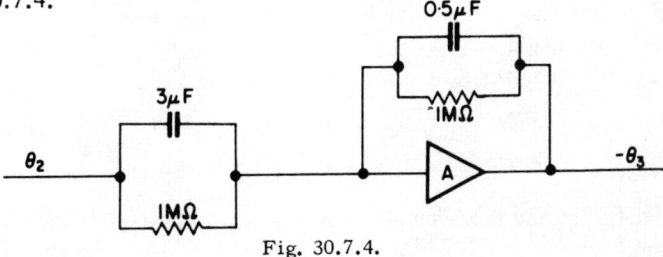

Fig. 30.7.4.

Step 4. Block 4.

Given $\qquad\qquad\theta_0 = 8\theta_3$.

The transfer function, $\qquad\dfrac{\theta_0}{\theta_3} = 8$.

An invertor amplifier is required with a gain of 8 (figure 30.7.5.).

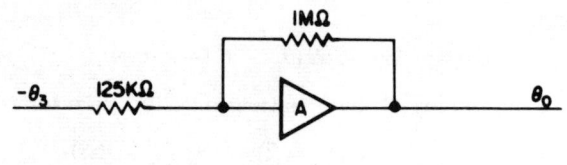

Fig. 30.7.5.

We need $-\theta_0$ for applying to the differential therefore an invertor must be connected between this output and the differential input.

The complete circuit of the whole system is shown in figure 30.7.6.

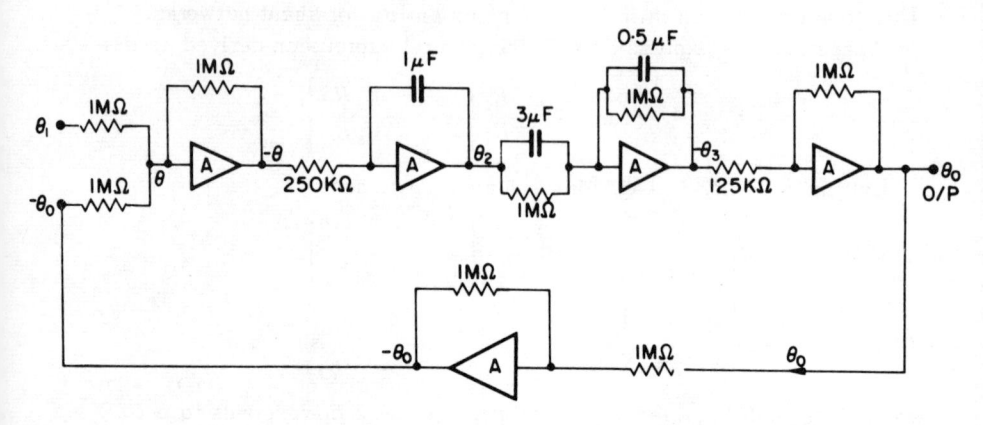

Fig. 30.7.6.

θ_1 can now be applied and the output, θ_0, can be visually examined.

30.8. Solving simultaneous differential equations

A similar system may be employed to solve for say, both x and y in a simultaneous equation.

Consider the simultaneous equation,

$$(D^2 - 4)y + (D + 1)x = f(t) \qquad (1)$$
$$(D + 1)x + (D + 4)y = 0. \qquad (2)$$

We must follow the rules outlined previously. The equations should be re-arranged.

SIMPLE ANALOGUE CONSIDERATIONS 569

$$D^2y = f(t) + 4y - Dx - x. \qquad (1)$$
$$D^2x = -x - Dy - 4y. \qquad (2)$$

We need 2 integrators for both equations, four inputs for equation (1) and three inputs for equation (2).

We will construct one system to solve for x and another to solve for y. We will apply one input of $f(t)$, all other 'inputs' will be taken from the circuit by way of feedback. This is shown in figure 30.8.1.

Once $f(t)$ is applied, the output signals x and y can be displayed for visual examination.

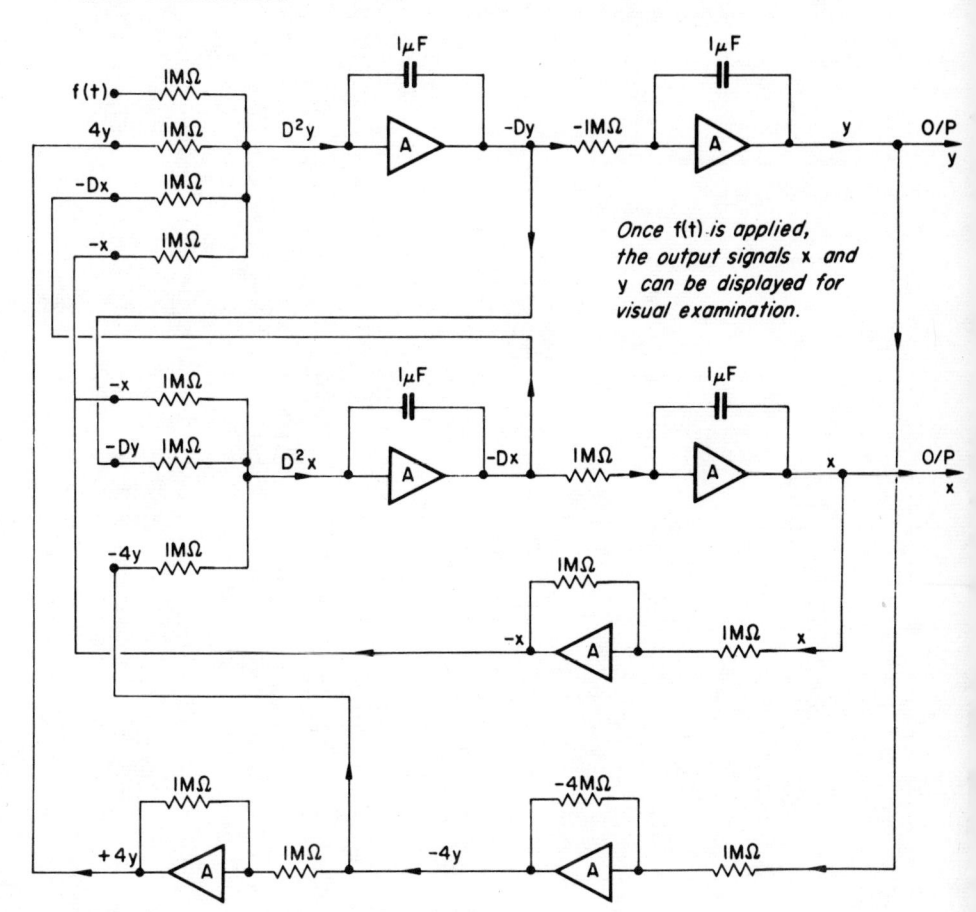

Fig. 30.8.1.

CHAPTER 31

Sawtooth generation

Sawtooth generators provide a waveform whose instantaneous value is continually changing at a constant rate.

The waveform may be obtained by charging, or discharging, a capacitor with a current whose rate of flow is constant.

The examples of the Puckle timebases demonstrate this principle nicely and practical use is made of the Pentode and its constant current characteristics.

The modified Miller circuit uses the initial portion of the charging curve where the rate of 'climb' is linear. Very high amplification of this tiny but linear portion of the curve provides an output signal that is linear within normal limits. These limits are also discussed in terms of % deviation from perfect linearity.

The modified Miller may be also represented by an operational amplifier with a capacitor connected between input and output. This was shown as an Integrator in the last chapter.

The modified Miller circuit waveform suffers from a disadvantage in that its otherwise linear waveform has a voltage step at the commencement of the linear voltage 'rundown'. This step can be eliminated and the means of so doing will also be discussed.

There are many practical applications for the sawtooth waveform although perhaps the most common is its application to the c.r.o. where it provides the means of obtaining a linear timebase, or sweep which is seen as a horizontal trace on the screen.

31.1. A modified Miller sawtooth generator

The Miller effect in a triode is exploited in this circuit. The function of the circuit is to provide a sawtooth waveform of controlled amplitude and duration t.

Figure 31.1.1. shows a basic modified Miller sawtooth generator and figure 31.1.2. shows a simplified typical output from such a circuit.

The portion of the waveform, $a-b$, is known as the sweep voltage, and should be a straight line. It is often applied to the 'X' plates of a c.r.t. so as to provide a means of deflecting (or sweeping) the 'spot' in a linear manner across the face of the tube.

The deflection will depend upon the relative potential between the X_1 and X_2 plates, and the linearity of the movement of the spot will depend upon the 'straightness' of the sweep voltage waveform, $a-b$.

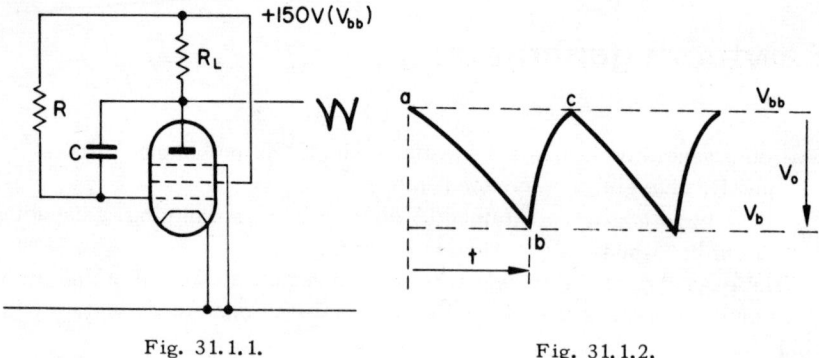

Fig. 31.1.1.   Fig. 31.1.2.

The portion $b-c$ is known as the flyback and should be vertical, returning to its maximum potential V_{bb} instantaneously.

This is difficult to achieve and a compromise is the usual result.

The sawtooth waveform may, for a given setting of an oscilloscope, be a 'single shot' or repetitive depending upon the requirement at the time. In more complicated circuits, a gating arrangement is built in to allow the sawtooth generator to provide either a single waveform or a series of repetitive waveforms as required, depending upon the gating circuit.

The suppressor grid of a pentode valve may be used to cut off all flow of anode current. Most valves if they are to do this, need a suppressor potential very much more negative than the control grid although some are specially made with suppressor characteristics very similar to that of the control grid. For example, normal pentodes with a bias of say $-5V$ needed to cut off all anode current, might require $-80V$ to do the same (with the control grid at normal bias potential).

If we remove this negative potential and take the suppressor up to cathode potential, anode current will flow (providing the control grid is not biassed back beyond cut off).

Alternatively we can bias the suppressor from a d.c. source, thus cutting off anode current, and apply a positive going pulse to the suppressor overcoming the suppressor bias which drags it up to cathode potential for a given duration, thus allowing anode current to flow for the same duration. This pulse would be a gating pulse.

The sawtooth generator in figure 31.1.1. basically operates as follows: when the valve is cut off, no anode current will flow and the instantaneous anode potential V_b, will be equal to the supply, V_{bb}. This is shown as point A in figure 31.1.2.

The grid is returned via R to a positive potential (V_{bb} in this example) and as it is positive with respect to the cathode, current will flow from

V_{bb} — through R — into the grid. This is known as grid current. The grid current flowing causes a large voltage drop across R and consequently the grid is just a few volts positive above the cathode potential. The grid and cathode behave as a diode, this results in a low input resistance to the grid and is in the order of $1\,K\Omega$. The grid behaves as the anode of the diode. When the valve is in grid current, large anode current flows. The valve under these conditions will not behave as an amplifier as, when positive to the cathode, the grid has no control over anode current value. The voltage drop across R is partly self-compensating, if the supply V_{bb} were higher, grid current would increase through R and a larger voltage drop would result. The drop across R is $\simeq V_{bb}$.

During grid current, the anode potential will fall rapidly. This negative going excursion is transmitted via C to the grid. Once the grid has fallen by a volt or so, it is within its normal bias region — controls anode current — and the rapid fall in anode potential is arrested. This rapid fall is usually unwanted in most circuits.

The grid however soon experiences a rising potential as C begins to discharge via R. As the grid potential rises, an inverted and amplified charge occurs at the anode and it falls at a controlled rate. This negative going anode voltage is once more transmitted via C to the grid and eventually the grid is taken sufficiently negative to cut the anode current off and the cycle is complete.

In the following examples, we will discuss in more detail, more practical circuits and derive component values. We will discuss other relevant factors that govern the amplitude and duration of the output waveforms from sawtooth generators.

31.2. A modified Miller with suppressor gating

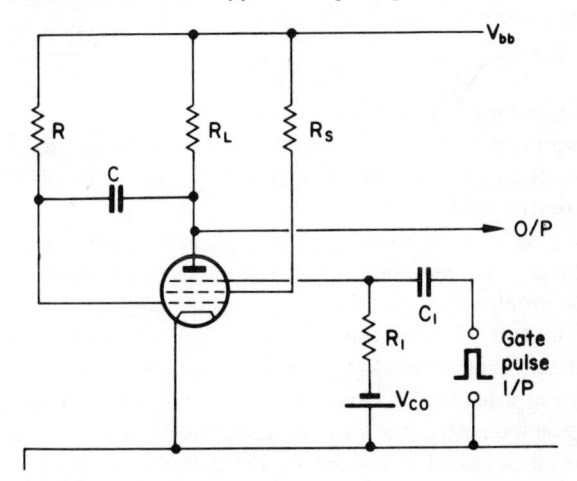

Fig. 31.2.1.

The Miller timebase generator utilises the 'Miller effect' and by connecting an external capacitor between anode and grid, a linear sweep voltage is produced. Its action is as follows:

In the absence of a positive going gating pulse to the suppressor, the suppressor is biased sufficiently negative so as to ensure complete cut off of anode current. R_1 provides a load for the gating pulse input and C_1 blocks the d.c. from V_{co} reaching the input pulse generator.

The screen grid is connected to V_{bb} via the screen resistor, R_s. The value of the maximum screen current is given from the expression,

$$I_{g2} = \frac{V_{bb} - V_{g2}}{R_s}.$$

Therefore R_s is chosen to limit I_{g2} to a safe value. This information is easily obtained from the valve data book. As no anode current is flowing, the screen current will be at a maximum.

The control grid is connected to V_{bb} via the resistor, R. The control grid will be at a potential slightly positive with respect to the cathode. Grid current will be flowing at a value given

$$I_{g1} \doteqdot \frac{V_{bb}}{R}.$$

The path for the grid current will be via R and the grid – cathode diode. The resistance of the diode, R_d, will be in the order of $1000\,\Omega$. The control grid potential may be determined by the expression

$$V_{g1} = \frac{V_{bb} \cdot R_d}{R + R_d}.$$

As R will be very much greater than R_d, R_d may be ignored in the denominator. Hence

$$V_{g1} \doteqdot \frac{V_{bb}}{R}$$

which may often be in the order of a volt or so.

The timing capacitor C is connected between the anode and the grid. The grid is held in grid current at almost zero volts whilst the anode will be at the potential of V_{bb}.

The capacitor C will, therefore, be fully charged to V_{bb}.

At the instant that the positive gating pulse is applied to the suppressor grid, anode current will flow. The anode potential will rapidly fall by a few volts. The fall in anode potential will drag the control grid down by the same amount due to the capacitor, C. The control grid, once it is dragged down to a few volts negative, will arrest the increase in anode current and consequently, the fall in anode potential.

The valve, now out of grid current, will behave as an amplifier. The

SAWTOOTH GENERATION

effective capacity between grid to cathode will be $C_{gk} = C_{gk} + C_{ag}(1 + A)$ and as C_{gk} will be very much smaller than the term $C_{ag}(1 + A)$, it may be ignored. The time constant in the grid circuit will now be $T.C. = C(1 + A).R$. The negative potential at the control grid can only be maintained whilst the capacitor remains fully charged. This is not possible and as the capacitor commences to discharge, the control grid potential will begin to rise towards V_{bb}. The rate at which the grid will rise is determined by the time constant, $C(1 + A).R$. The capacitor, C, discharge path is through R_L, the anode load, the valve and the resistor, R. The screen current will have fallen, as much of the screen current prior to the gating pulse, will now be flowing in the anode circuit.

Gating pulse. Suppressor grid.

Anode waveform. — Rapid fall, Slow rundown, Flyback, V_{bb}, Time constant $C.R_L$, V_b

Control grid. — Rapid fall, Time contant $(1+A)CR$

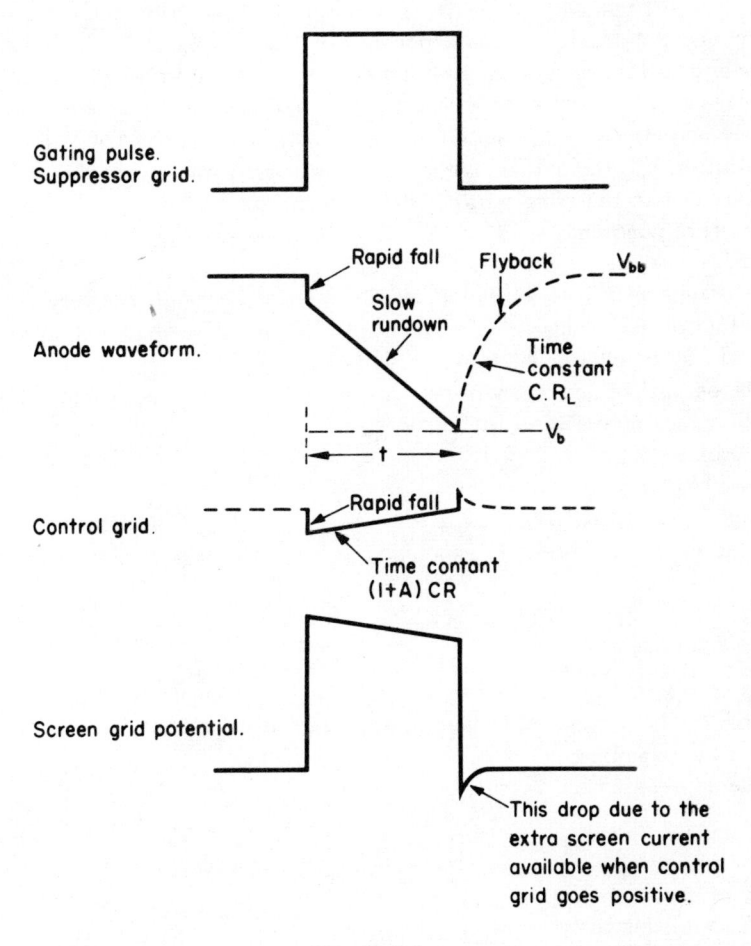

Screen grid potential. — This drop due to the extra screen current available when control grid goes positive.

Fig. 31.2.2.

The control grid will rise towards V_{bb} in an exponential manner. This rise will be amplified by the amplifier thus causing the anode potential to fall at the rate of grid potential rise, multiplied by the gain of the stage. The grid can only rise a few volts before it reaches the cathode potential and grid current will flow once more.

It would be desirable to ensure that the duration of the gating pulse is less than the complete run down for the anode. This will prevent the anode from bottoming as the grid reached the cathode potential.

Assuming that the gating pulse is removed before the anode waveform bottoms, the anode current will fall to zero. The anode potential will rapidly rise to V_{bb} and the control grid will be dragged up by C. The grid cannot rise very far above the cathode as it will be held a fraction of a volt positive to the cathode due to the grid current that will flow through R.

The grid potential will rise exponentially towards V_{bb}. The actual climb can only be a few volts as any further rise will run the valve into grid current and the grid will be latched by grid current to a level just above the cathode potential. As the actual rise is but a few volts and the 'aiming' potential is V_{bb}, the actual rise is but a very small fraction of the maximum. The rate of rise therefore will be very linear.

The time constant is $C(1 + A) \cdot R$. But as A is $\gg 1$, the 1 may be ignored.

The stage will have a gain of $-A$. The anode potential will vary $-A$ times that of the change in grid potential. The instantaneous potential of the grid may be expressed as $v_g = V_{bb}(1 - e^{-t/ACR})$.

The output voltage v_o, will be $-A \cdot v_g \quad v_o = V_{bb} - v_b$, where v_b is the instantaneous potential of the anode.

Therefore $V_{bb} - v_b = -A v_g \therefore v_b = V_{bb} + A \cdot v_g$. But $v_g = V_{bb}(1 - e^{-t/ACR})$.

Thus $v_b = V_{bb} + A V_{bb}(1 - e^{-t/ACR}) \therefore v_b = [V_{bb}(1 + A(1 - e^{-t/ACR})]$.

If the exponential term is expressed as a series, it becomes

$$v_b = V_{bb}\left[1 + A\left(1 - 1 - \frac{t}{ACR} + \frac{t^2}{2(ACR)^2} \quad \frac{-t^3}{3(ACR)^3} + \ldots\right)\right]$$

As only a fraction of the exponential rise will be embraced, the rise will linear. It is reasonable therefore, to consider the first, and linear, term of the expansion.

The expression for v_b will become

$$v_b = V_{bb}\left[1 + A\left(1 - 1 - \frac{t}{ACR}\right)\right]$$

which simplifies to $v_b = V_{bb}\left(1 - \dfrac{t}{CR}\right)$.

At $t = 0$, $v_b = V_{bb}$. At $t = CR$, $v_b = 0$.
The output will not fall to zero as the anode will at worst, fall only to the bottoming potential of the valve.

The deviation from perfect linearity may be expressed as a % error between the first and second terms of the series. The % error may be expressed as

$$\frac{\dfrac{t^2}{2(ACR)^2}}{\dfrac{t}{(ACR)}} \times 100\% = \frac{50\,t}{ACR}.$$

It has been shown that $v_b - V_{bb}(1 - t/CR)$. It can be reasoned in a more elementary analysis, that provided the sweep or anode fall, is linear and the gain is very high, the expression for the output voltage is $v_o = I.T./C$. As $v_o = I.T./C$ then $V_{bb} - v_b = I.T./C$. But $I = V_{bb}/R$. Therefore

$$V_{bb} - v_b = \frac{V_{bb} \cdot t}{CR} \quad v_b = V_{bb}\left(1 - \frac{t}{CR}\right)$$

which agrees with the result obtained in this example.

Should the anode have fallen to its minimum value, the removal of the gating pulse would cause the Miller to revert to its original state. Anode current would cease. The screen current would rise to its maximum value whilst the anode would be at V_{bb}. The grid would be once more in grid current and held a fraction of a volt positive with respect to the cathode. The time taken for the anode to rise from bottoming to V_{bb} will be determined by the time constant $(R_L + R_d)C$. The valve will not be an amplifier and therefore the time constant of the anode circuit will be $(R_L + R_d).C$. The grid cathode diode, R_d, will be very much less than R_L, therefore the time constant may be regarded as $C.R_L$.

It is important to ensure that the suppressor is never driven positive to the cathode and a diode should be suitably connected from the suppressor to earth. Some simple circuits use the screen grid waveform to switch the suppressor grid, hence the whole circuit is free running.

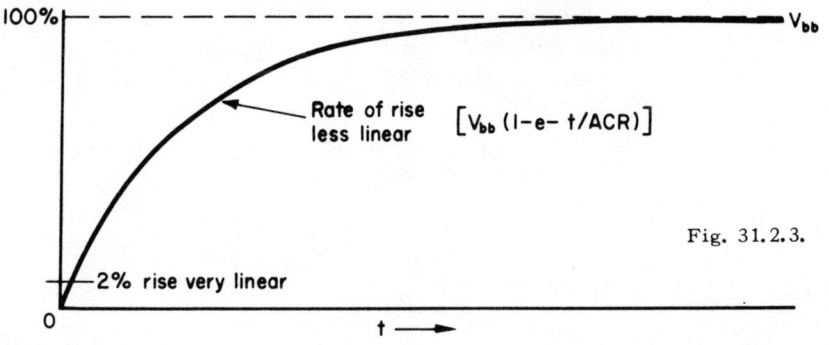

Fig. 31.2.3.

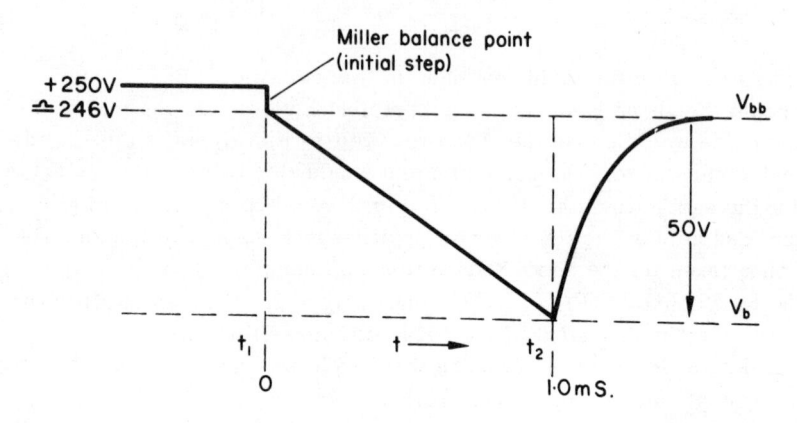

Fig. 31.2.4.

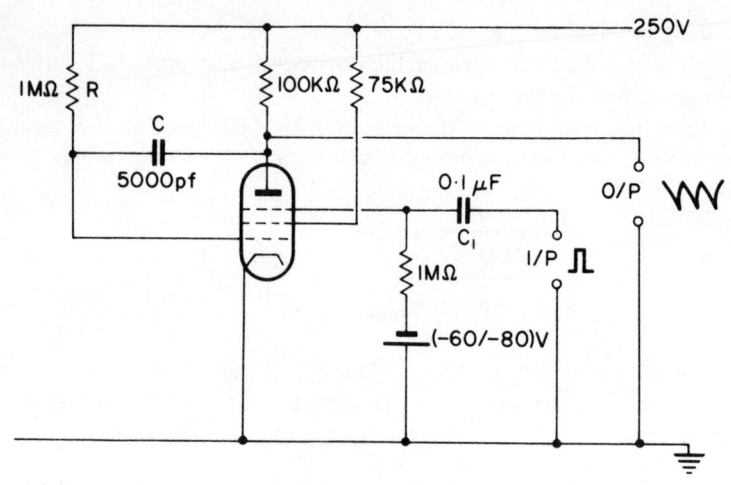

Output waveform

Example.

A pentode valve having a 250V supply, is to be used as a Miller sawtooth generator. The screen grid requires 100V at a current of 2.0 mA. The valve has a *gm* of 2.0 mA/V. A positive input pulse is to be applied to the suppressor grid for a period of 1.0 mS. A 50V output from the anode is required at a linearity represented by a % error of 0.05%. The grid current must be limited to 250 μA. Determine the values of all resistors. The value of R is obtained from the expression

$$R = \frac{250\,\text{V}}{250\,\mu\text{A}} = 1.0\,\text{M}\Omega.$$

SAWTOOTH GENERATION

The screen resistor will need to be

$$R_s = \frac{(250 - 100)\text{V}}{2\,\text{mA}} = \underline{75\,\text{K}\Omega}.$$

The value of C may be derived from the expression, $v_b = v_{bb}(1 - t/CR)$. Hence

$$C = \frac{V_{bb}\,t}{R(V_{bb} - v_b)} = \frac{10^{-3} \cdot 250}{10^6 \cdot 50} = 5.10^{-9} = \underline{5000\,\text{pF}}.$$

From the expression $\% = \dfrac{50\,t}{A.C.R.}$ $A = \dfrac{50\,t}{\%\,C.R.}$ $A = \dfrac{50.10^{-3}.10^9.10^{-6}}{0.05 \quad 5.1} = \underline{200}.$

From the expression for the stage gain,

$$R_L = \frac{A}{gm} = \underline{100\,\text{K}\Omega}.$$

The capacitor C_1 should have a reactance at the repetition frequency which is very much lower than the resistor to which it is connected.

The completed circuit is shown in figure 31.2.4.

Placing a resistor in series with the charging capacitor, C, will modify the initial step in the waveform. A large resistor will accentuate the step (for subsequent use in current waveforms for magnetic deflection).

The modified Miller circuit can also be shown as an operational amplifier as in figure 31.2.5.

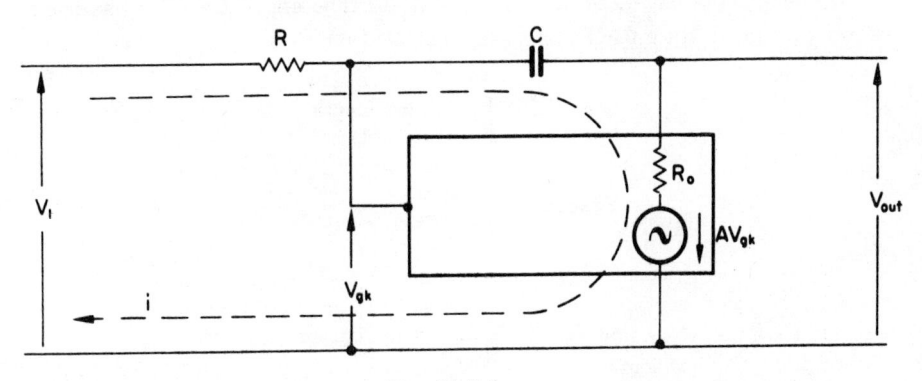

Fig. 31.2.5.

The input current $i \doteq v_1/R$. This current flows through R, to the left hand side 'plate' of the capacitor C and through the amplifier output resistance R_0.

$$v_c = \frac{1}{C} \cdot q.$$

Therefore the rate of change of v_c

$$= \frac{1}{C}\frac{dq}{dt} \div \frac{1}{C} \cdot \frac{v_1}{R}$$

hence the rate of change of p.d. across the output $\div \frac{-1}{C} \cdot \frac{v_1}{R}$

This may be written as $\frac{dv_0}{dt} = \frac{-1}{CR} \cdot v_1$

and after integrating,

$$[v_0]_{t_1}^{t_2} = -\frac{1}{CR}\int_{t_1}^{t_2} v_1 \, dt.$$

This means that the change of output over an interval t_1 to t_2 in figure 31.3.4. is proportional to the integrated input signal. Inversion occurs and is indicated by the minus sign.

We have already discussed the rapid fall in anode potential just prior to the sweep or rundown. This fall, or step, may be reduced or even eliminated by the inclusion of a further component.

31.3. Eliminating the 'Miller Balance Point' step

Let us refer to figure 31.2.4. and considering the 'step voltage output' only, derive an expression for its amplitude. v_0 therefore is, for this example, not the sawtooth, but *the step only*. The capacitor C is assumed short circuit as its p.d. cannot change instantaneously.

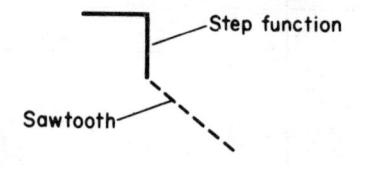

Fig. 31.3.1.

$v_0 = iR_0 - Av_{gk}$ and $i = \dfrac{v_1 - v_0}{R}$ whilst $v_{gk} = v_1 - iR$

$\therefore v_0 = \dfrac{R_0 v_1 - R_0 v_0}{R} - Av_1 + \dfrac{Av_1 R - Av_0 R}{R}$

$\therefore v_0(R) = R_0 v_1 - R_0 v_0 - Av_0 R$

$\therefore v_0 R = v_1 R_0 - v_0[R_0 + AR] \quad \therefore \quad v_0[R + AR + R_0] = v_1 R_0$

hence $v_0[R_0 + (1+A)R] = v_1 R_0$

$\therefore v_0 = \dfrac{v_1 R_0}{R_0 + (1+A)R}$

SAWTOOTH GENERATION

The amplifier components include R_o ($R_L // r_a$) but R_L would be difficult to vary in order to reduce the step without causing other difficulties. R is the one component in the expression that could be increased so as to reduce v_o (the step voltage) but is the resistor that forms part of the c.r. timing portion of the circuit and again might prove too inconvenient to vary.

A practical solution is to insert a resistor in series with the timing capacitor C as shown in figure 31.3.2.

If we consider figure 31.3.2., with r connected as shown, we can derive an expression for the amplitude of v_{out} in terms of the gain A, output resistance R_o, the timing circuit resistance R and r. The reader is reminded that v_o is the step one, and not the sawtooth waveform.

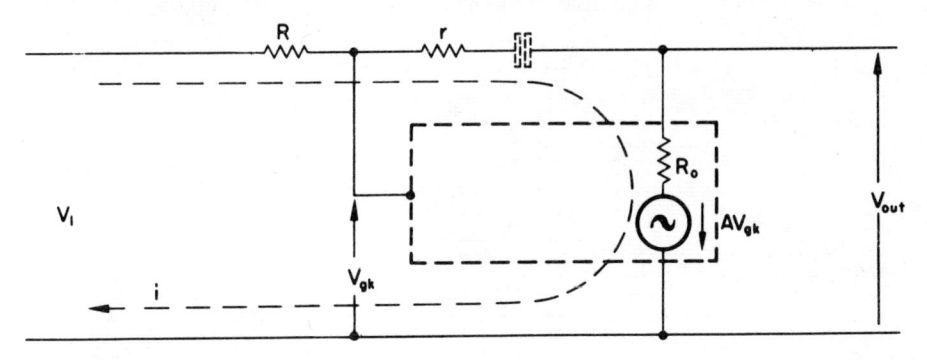

Fig. 31.3.2.

$$v_o = iR_o - Av_{gk} \qquad i = \frac{v_1 - v_o}{R + r} \qquad v_{gk} = v_1 - iR$$

$$\therefore \quad v_o = \frac{(v_1 - v_o)R_o}{R + r} - A(v_1 - iR)$$

$$\therefore \quad v_o = \frac{v_1 R_o - v_o R_o}{R + r} - Av_1 + AiR$$

$$\therefore \quad v_o = \frac{v_1 R_o - v_o R_o}{R + r} - Av_1 + \frac{A(v_1 - v_o)R}{R + r}$$

$$\therefore \quad v_o = \frac{v_1 R_o - v_o R_o}{R + r} - Av_1 + \frac{Av_1 R - Av_o R}{R + r}$$

$$\therefore \quad v_o = \frac{v_1 R_o - v_o R_o}{R + r} - \frac{(R + r)Av_1}{R + r} + \frac{Av_1 R - Av_o R}{R + r}$$

$$\therefore \quad v_o [R + r] = v_1 R_o - v_o R - rAv_1 - Av_o R$$

$$\therefore \quad v_o [R + r] = v_1 R_o - v_o R + v_1 AR - v_o AR$$

581

$$\therefore \quad v_0 [R + r + R_0 + AR] = v_1 [R_0 - Ar]$$
$$\therefore \quad v_0 [r + R_0 + (1 + A)R] = v_1 [R_0 - Ar]$$
$$\therefore \quad v_0 = \frac{v_1 [R_0 - Ar]}{r + R_0 + (1 + A)R}$$

and as v_0 must be zero to satisfy our requirements,

$$0 = v_1 [R_0 - Ar]$$
$$\therefore \quad R_0 = Ar \text{ hence } r = \frac{R_0}{A}$$

Now the output resistance R_0 consists of r_a and R_L in shunt,

$$R_0 = \frac{r_a R_L}{r_a + R_L}$$

and the gain

$$A = \frac{\mu R_L}{r_a + R_L}$$

then

$$r = \frac{R_0}{A} = \frac{\frac{r_a R_L}{r_a + R_L}}{\frac{\mu R_L}{r_a + R_L}} = \frac{r_a}{\mu} = \frac{1}{gm}.$$

Therefore, if a resistor, r, having a value equal to $1/gm$ of the amplifier valve, is inserted as shown, the initial few volts step will be eliminated.

It might be prudent to remind the reader that increasing the value of r above $1/gm$, i.e., $r > 1/gm$, the waveform will be of little value as a linear sawtooth, but will provide a waveform of the shape required for 'current' deflection in a system that uses electromagnetic coils.

31.4. Puckle timebase (1)

Figure 31.4.1. shows a simple Puckle timebase. This circuit uses a pentode as a constant current device. V_1 produces a constant current, determined by the screen grid potential, V_{g2}.

With 'C' initially uncharged, V_1 anode will be at h.t. potential once the supply is switched on. V_1 will conduct and draw its anode current from the lower 'plate' of the capacitor C.

As current (positive charges) are drawn from the 'lower plate' of C, the potential of the 'lower plate' will fall.

This fall in potential is of course, the output sawtooth waveform developing. V_2 is a gas filled triode and in this simple circuit, has its grid

SAWTOOTH GENERATION

connected to a variable positive potential.

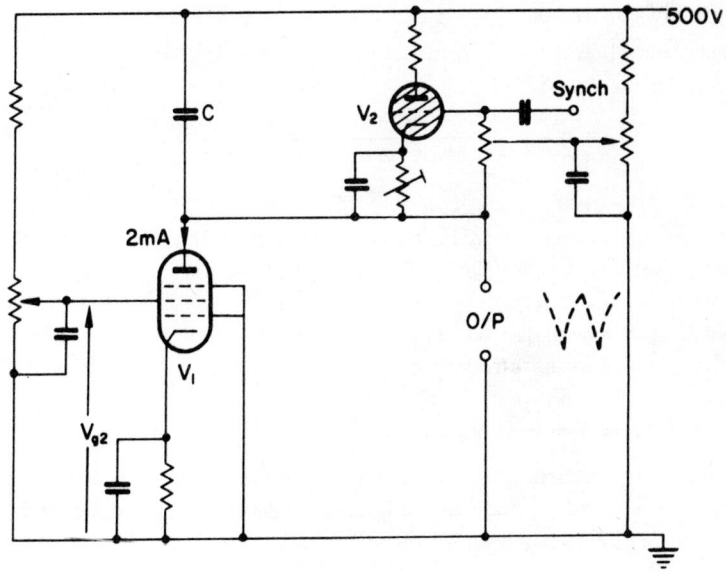

Fig. 31.4.1.

As V_1 anode falls, it drags V_2 cathode down with it. This process continues until V_2 cathode is at a potential such that its V_{AK} divided by the control ratio of V_2, is equal to the predetermined bias needed to fire the device.

When the device fires, it presents a 'short circuit' across C and discharges it, thereby raising V_1 anode to the h.t.

The cycle is then complete, i.e. one 'cycle' of the sawtooth waveform including flyback, has been completed.

Synchronisation may be affected by injecting an appropriate signal to V_2 grid so as to drag the grid sufficiently positive at a predetermined point in time during the rundown, thus initiating the flyback.

Example.

Let $C = 0.02\mu\text{F}$. Assume that V_2 has a control ratio of 33.3 (i.e. conduction, or firing will occur when $V_{AK}/V_{gk} = 33.3$.). Suppose we require to display exactly 2 cycles of a 600 c/s waveform on an oscilloscope for which this circuit is the timebase. V_1 is set to provide 2 mA anode current. Determine the grid potential of V_2 which will satisfy these requirements.

The period of timebase waveform corresponding to 2 cycles of a 600 Hz waveform = 2/600 secs = 3.33 mS.

The potential V_c. after 3.33 mS $= \dfrac{I.T.}{C} = \dfrac{3.33 \cdot 10^{-3} \cdot 2 \cdot 10^{-3}}{0.2 \cdot 10^{-6}} = 333V$.

The cathode voltage of $V_2 = V_k = 500 - 333 = 167V$.

The thyratron should fire at this instant and with $V_{AK} = V_C = 333V$, and for a control ratio of 33.3,

$$V_{gk} = \dfrac{V_{AK}}{\text{control ratio}} = \dfrac{333}{33.3} = 10V.$$

$V_K = 167V$. Hence $V_g = V_K - V_{gk} = 167 - 10 = 157V$.

Therefore, if V_g is set to 157V, the timebase will produce a sawtooth waveform that will enable 2 c/s of a 600 Hz waveform to be displayed on a c.r.o.

This circuit has a limited upper frequency of about 40 Hz – 100 Hz, depending upon the thyratron.

31.5. A Puckle timebase using a hard valve (2)

A variation on the circuit is that a 'hard' valve may be used instead of the thyratron. A second pentode and a triode is used in this second example, the circuit diagram for which is shown in figure 31.5.1.

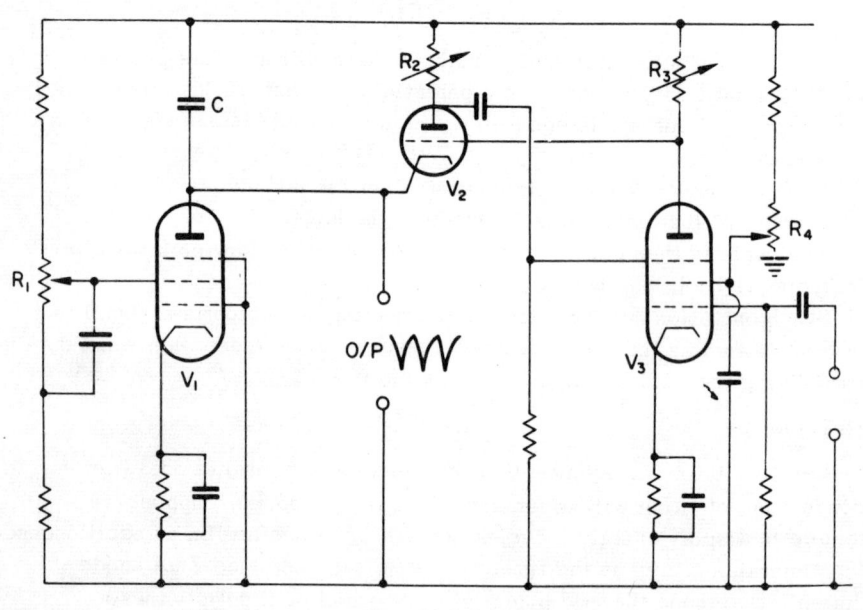

Fig. 31.5.1.

V_1 determines the amplitude of constant current 'through' C.

V_2 terminates the sweep and initiates the flyback, as it discharges C as V_3 goes into grid current.

V_3 is a synch amplifier.

Initial conditions. $\bar{V}_2$ cut off. V_3 conducting. V_1 conducting.

Once the supply is switched on, V_1 anode falls at a linear rate, taking with it the cathode of V_2. Once the cathode is at a sufficiently low potential, V_2 will be operating within its grid base (normal bias conditions).

V_2 will commence to pass current, causing a negative going impulse to be developed across R_2. This impulse is transmitted via C_2 to the suppressor grid of V_3.

As the suppressor is dragged down, V_3 anode current falls. This reduction in anode current causes a positive going impulse to develop across R_3, which in turn drags V_2 grid up with V_3 anode.

When this occurs, V_2 passes much more current and the large negative going impulse across R_2 is fed to V_3 suppressor. These effects are cumulative and V_2 is rapidly taken into grid current.

Once V_2 is in grid current, it presents a 'short circuit' across 'C' thus discharging C through $\dot{V}_2$ and R_2 and at the same instant, initiating the flyback.

The first cycle is complete.

R_1 is the fine frequency control, coarse ranges being determined by the value of C. R_2 is the trigger threshold control and should have the smallest value possible so as to keep the $(C.R_2)$ time constant small, thus flyback will occur rapidly.

R_3 is the course amplitude control and will determine the 'length' of the timebase. This is achieved by predetermining the instant of flyback.

Fine amplitude control is by means of R_4.

We have seen that certain techniques covered earlier enabled us to derive formulae for triode valve circuits.

One basic equation we met was

$$I_a = \frac{v_{ak}}{r_a} + gm \cdot v_{gk}$$

(the reader will see that both terms contribute to the anode current).

We need an expression for zero anode current, i.e. for complete cut off, therefore we can substitute zero for I_a in the expression, hence

$$0 = \frac{v_{ak}}{r_a} + gm \cdot v_{gk}.$$

If we divide both sides of the equation by gm, we get

$$0 = \frac{v_{ak}}{\mu} + v_{gk}.$$

Therefore, for cutt off,

$$v_{gk} = -\frac{v_{ak}}{\mu} \quad \text{or} \quad \frac{v_{ak}}{\mu} = -v_{gk}.$$

We will use this expression in the following example, the skeleton circuit diagram of which is shown in figure 31.5.2.

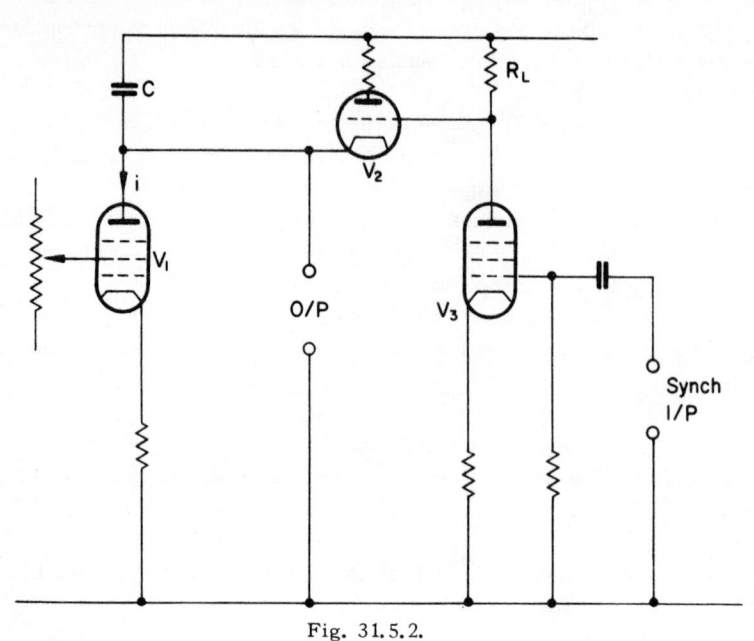

Fig. 31.5.2.

Example (2)

V_1 is set to pass 2 mA. V_2 has a μ of 20. V_3 has a gm of 5 mA/V. R_L of V_3 = 20 KΩ.

We require a 110V output signal and wish to display exactly 5 c/s of a 100 Hz waveform during the sweep.

In the free running state, the sweep duration will, for 110V output, be greater than we need to exactly display our waveform. We want to apply a synchronising input which, after amplification, will suddenly drag V_2 grid up — causing V_2 to conduct — thus terminating the sweep at the exact instant we are displaying 5 complete cycles of our waveform.

Problem: Determine the amplitude of the synch input pulse.

Free running condition.

The duration of sweep for an output of 110V is given by

SAWTOOTH GENERATION

$$t = \frac{Cv}{i} = \frac{10^{-6} \; 110}{2.10^{-3}} = 55 \, \text{mS} \tag{1}$$

Just prior to $t = 55 \, \text{mS}$, V_2 will be sitting with a bias of

$$\frac{-v_{ak}}{\mu} = \frac{-55V}{20} = -2.75V \, . \, v_{gk}.$$

V_2 grid potential = V_3 anode potential = h.t. $- 110 - 2.75 = 187.25V$.

Thus V_3 anode will be 112.75V below the h.t. (2)

Synchronised condition.

The required sweep time $= 5 \times 10 \, \text{mS} = 50 \, \text{mS}$. V_2, v_{ak} increases at a linear rate of 2V/mS, therefore at $t = 50 \, \text{mS}$, the output voltage = 100V (negative going). This 100V fall in potential will also exist across V_2. Hence just prior to conduction, V_2 will require a bias of

$$v_{gk} = \frac{-v_{ak}}{\mu} = \frac{-100V}{20} = -5V.$$

Therefore V_2 grid potential = V_3 anode potential = $300 - 100 - 5 = 195V$.

Thus V_3 anode will be sitting at 105V below h.t. (3)

We can see then, that to 'stop' the sweep 5 mS before it would normally stop, and to give us 110V output at $t = 50 \, \text{mS}$, the anode potential of V_3 will have to rise from (2) 112.75V below h.t. to (3) 105V below.

The change in anode potential required is therefore 7.75V.

V_3 has a stage gain of $gm . R_L = 100$.

Allowing for stage inversion, we need to apply an input synch signal of $7.75/100 = 77.5 \, \text{mV}$, negative going input pulse.

ANSWERS TO PROBLEMS

1. (a) Name the units of electric current, pressure and resistance.
   (b) State Ohms law in terms of V, I and R.
   (c) An electric device passes 2A when connected to a supply of 240V. Calculate the resistance of the device.

   Ans. (c) 120Ω

2. A 2Ω resistor is connected in series with two other resistors which are connected in parallel. The latter having values of 4Ω and 6Ω respectively. What voltage would need to be applied to the complete network in order to maintain a current through the 2Ω resistor.

   Ans. 8.8V

3. Three resistors are connected in parallel. They have values of 30Ω, 15Ω, and 10Ω respectively. When connected to supply of unknown voltage, a total current of 1A flows. Calculate the supply potential.

   Ans. 5V

4. (a) Explain what is meant by the internal resistance of a cell.
   (b) With the aid of a sketch, show how you would determine the value of the internal resistance of a cell.
   (c) The e.m.f. of a cell is measured and found to be 5V. When a 8.99Ω load resistor is connected, the p.d. across the load is found to be 4.495V. Calculate the internal resistance of the cell.

   Ans. 1.01Ω

5. A lamp takes a current of 0.5A at its normal operating voltage of 12V. When the lamp is used with a 24V supply a further component is required. Sketch the complete circuit and calculate the value of the additional component.

   Ans. 24Ω

6. A 2.4Ω resistor is required in an electrical circuit. If a 6Ω resistor only is available, what value of resistor needs to be connected in parallel to give an effective resistance of 2.4Ω.

   Ans. 4Ω

7. (a) Sketch a graph representing a 12Ω resistor connected across a supply variable from 0 − 36V.

   (b) Modify the graph to show the effects of connecting a second resistor of 6Ω in series with the first.

   (c) From the modified graph, state the current that would flow at a supply potential of 18V, 9V, 4.5V, is connected.

   Ans. 1A, 0.5A, 0.25A

8. Two identical 200V, 50W lamps are connected in series with a 250V supply. How many similar lamps need to be connected across one of the lamps in order to cause the other to operate at its rated condition.

   Ans. 3 lamps

9. A resistor 'R' is connected in series with a pair of resistors connected in parallel. The latter having values of 7Ω and 3Ω respectively. What value of 'R' would be necessary to maintain a current of 0.15A through the 7Ω resistor when a supply voltage of 5V is connected across the complete network.

   Ans. 7.9Ω

10. (a) What is meant by 'load over total'.

    (b) Derive the expression 'load over total' from basic Ohms law.

    (c) Derive the expression for 'load over total' for current.

    (d) How does the term 'load' vary in (b) and (c).

    (e) A series circuit consisting of a 12Ω and 18Ω resistor is connected across a 60V battery having zero internal resistance. Using 'load over total' calculate the p.d. across the 18Ω resistor.

    (f) A shunt circuit consists of a 12Ω and 18Ω in parallel. The total current flowing through their resistors is 3A. Using 'load over total' for currents, calculate the current flowing in the 18Ω resistor.

    Ans. (e) 36V. (f) 1.2A

11. Show from first principles that the effective resistance of a circuit consisting of R_1 and R_2 in shunt is given by

$$\frac{R_1 R_2}{R_1 + R_2}$$

Hence determine the effective resistance of the circuit for the following values of R_1 and R_2.

PROBLEMS AND ANSWERS 591

R_1 9 KΩ	R_2 1 KΩ	Ans.	900 Ω
R_1 8 KΩ	R_2 2 KΩ	Ans.	1600 Ω
R_1 7 KΩ	R_2 3 KΩ	Ans.	2100 Ω
R_1 6 KΩ	R_2 4 KΩ	Ans.	2400 Ω

12. A battery having an e.m.f. of 100V and negligible internal resistance is connected across a network comprising a 10 Ω fixed resistor in series with R_L. R_L is a variable resistor from 0 − 20 Ω. Calculate the power across R_L as it is varied in 2 Ω steps. Draw a graph of P/R_L where P is power in W. The maximum power value occurs in R_L when it has a particular value, what is this value.

Ans. 10 Ω

13. Calculate the value of R_7 to maintain a current of 2A through R_4.

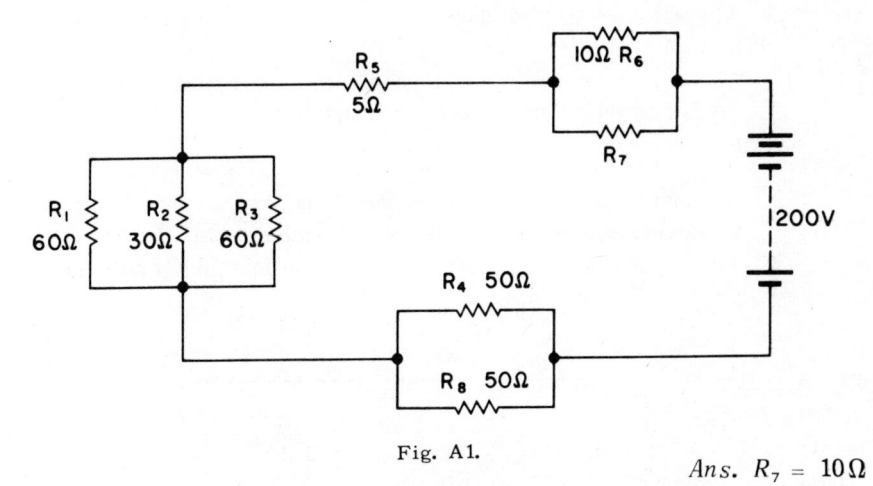

Fig. A1.

Ans. $R_7 = 10 \, \Omega$

14. A transformer has a primary consisting of 10 000 turns and two secondary windings each having 100 turns. If 100V is applied to the primary winding and a 1 Ω load connected across both secondary windings, what current will flow in the primary?

Ans. 20 mA

15. (a) Calculate the reactance of a 10 Henry inductor at 50 Hz, 100 Hz and 200 Hz.

(a) $\begin{cases} 3140 \, \Omega \\ 6280 \, \Omega \\ 12560 \, \Omega \end{cases}$

contd.

592   ELECTRONICS FOR TECHNICIAN ENGINEERS

15 contd
- (b) Calculate the reactance of a $0.159\,\mu\text{F}$ capacitor at 50 Hz, 100 c/s and 200 Hz. (b) $\begin{cases} 20\,\text{K}\Omega \\ 10\,\text{K}\Omega \\ 5\,\text{K}\Omega \end{cases}$
- (c) Sketch a graph showing X/f and determine the resonant frequency when the components are connected in series.

  (c)   126 Hz
- (d) If the inductor contained $5\,\Omega$ resistance, what is the total circuit impedance in (c).

  (d) $5\,\Omega$

16. A 500W lamp requires 250V. Determine the resistance of the filament under normal operating conditions.

    Ans. $125\,\Omega$

17. A transformer having 2000 turns on the primary is connected to 100V.
    The secondary winding is loaded with a $1\,\Omega$ resistor into which 1A is flowing. Determine the secondary turns.

    Ans. 20

18. A transformer has one primary and 3 secondary windings. Each secondary winding has 50 turns and each delivers 2A into a load. If the primary current is 300 mA on load, determine the number of primary turns.

    Ans. 1000

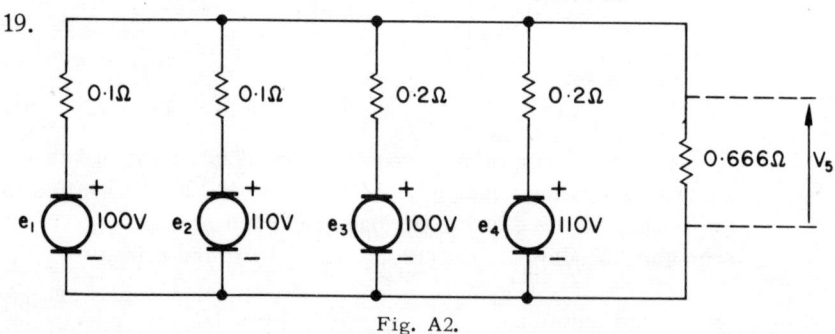

19.

Fig. A2.

Four generators are connected in shunt as shown in the diagram. Calculate, v_5, the terminal p.d. across the load resistor.
Any formulae used should be derived from first principles.

Ans. 100V

20.

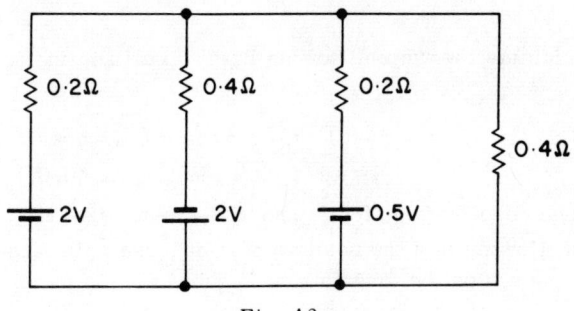

Fig. A3.

Calculate (a) the terminal p.d. across the load resistor, and (b) the current flowing in the 0.5V cell.

Ans. (a) 0.5V
(b) zero

21. Calculate the p.d. across the load resistor.

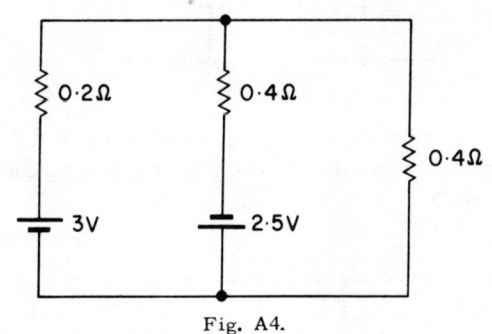

Fig. A4.

Ans. 0.875V

22.

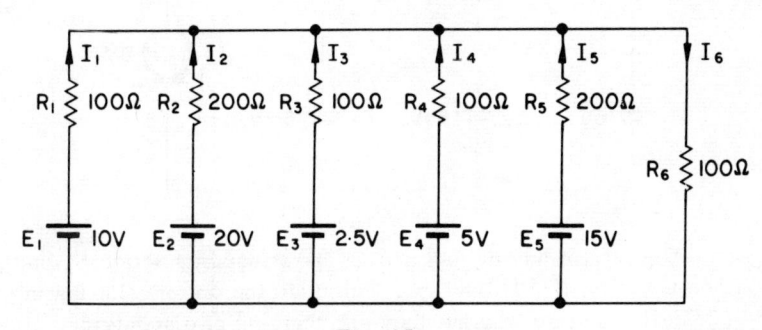

Fig. A5.

(a) Derive a formula to be used in the solution of the problem.
(b) Calculate the terminal p.d. across R_6.

contd.

594   ELECTRONICS FOR TECHNICIAN ENGINEERS

22. contd
   (c) Calculate the current flowing in each cell and in R_6.

   Ans. (b) 7V

   (c) I_1 = 30mA I_2 = 65mA I_3 = −45mA

   I_4 = −20mA I_5 = 40mA I_6 = 70mA

   (The negative sign for I_3 and I_4 indicates that these currents are flowing into the positive pole of these cells and not leaving as shown on the diagram).

23.

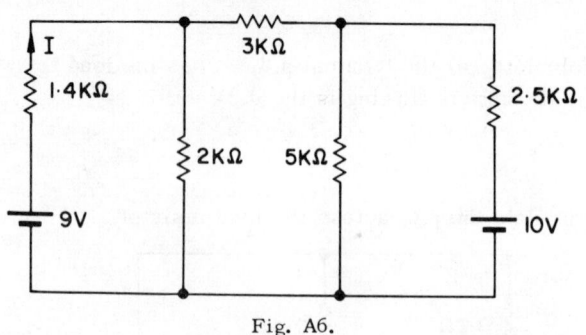

Fig. A6.

Calculate the current flowing in (or out) of the 9V cell. (The current will be said to flow out if it flows as indicated on the diagram).

Ans. 2.5A (flowing out)

24.

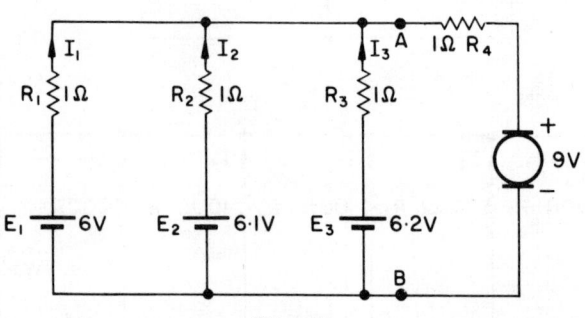

Fig. A7.

A generator having an e.m.f. of 9V is used as shown to charge the cells, E_1, E_2 and E_3. Calculate the current flowing in each cell and state whether they are charging or discharging.

Ans. E_1 = −0.825A (charging)
     E_2 = −0.725A (charging)
     E_3 = −0.625A (charging)

contd.

24. contd

(Hint)

As there is no load resistor, the formula should read,

$$v_{AB} = \frac{\frac{E_1}{R_1} + \frac{E_2}{R_2} + \frac{E_3}{R_3} + \frac{E_4}{R_4}}{\frac{1}{R_1} + \frac{1}{R_2} + \frac{1}{R_3} + \frac{1}{R_4} + \frac{1}{\infty}} \quad \text{where } \frac{1}{\infty} = 0$$

where the load resistor is shown having an infinite value.

25. Two generators are connected in shunt. A load is connected in parallel with the shunt combination. Generator 1 has an e.m.f. of 105V and an armature resistance of $0.02\,\Omega$. Generator 2 has an e.m.f. of 102V and an armature resistance of $0.04\,\Omega$. The load has a resistance of $0.04\,\Omega$. Each generator contributes to the load current. Calculate these currents and the p.d. across the load.

Ans. Generator 1 = 1350A
Generator 2 = 600A
Load current = 1950A
Terminal p.d. = 78V

26.

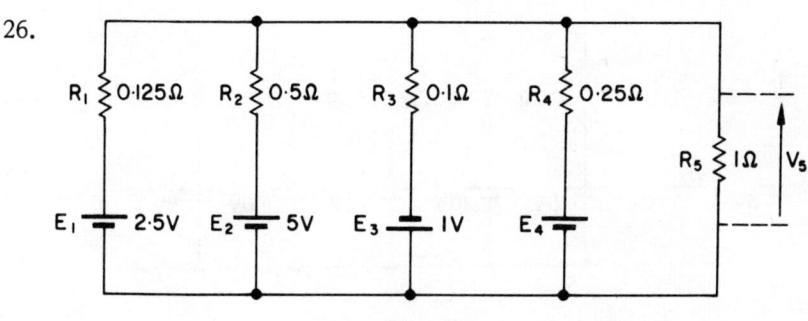

Fig. A8.

(a) Calculate the value of the cell E_4 such that 2A will flow through R_5.

(b) How much current flows in the cell E_4.

(c) Explain the magnitude of current in the cell E_4 and show why it is very much greater than the load current through R_5.

Ans. (a) $E_4 = 7.5V$
(b) 22A

27.

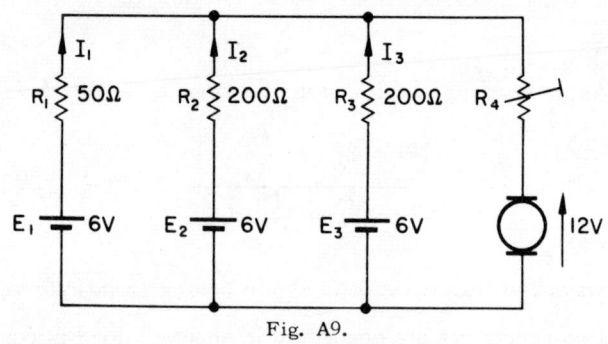

Fig. A9.

It is required to charge the cells from the generator. Calculate

(a) the value of R_4 so that the cell E_1 is charged at 48 mA.

(b) Calculate the charging current in the other two cells.

(c) What relationship is the total cell currents with respect to the generator current.

$Ans.$ (a) $50\,\Omega$

(Hint: Note that if the cell E_1 is being charged, I_1 will be negative).

28.

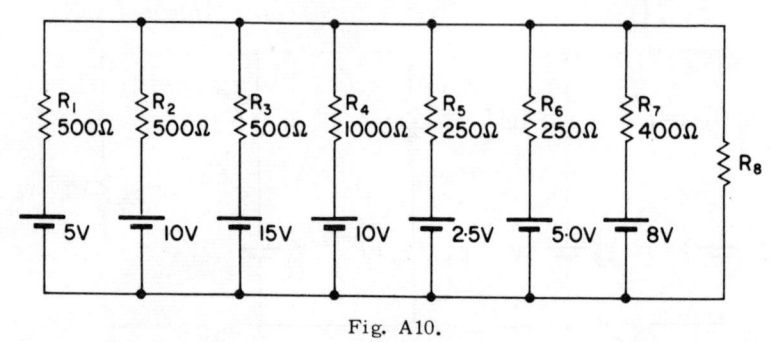

Fig. A10.

(a) A number of cells are shunt connected as shown. Calculate the value of the load resistor R_8 such that a p.d. of 6V will exist across it.

(b) Calculate the current in each cell.

(Hint: Express every resistor in KΩ (i.e. $500\,\Omega = 0.5\,$K). The currents will therefore be in mA. The arithmetic will be much easier.

$Ans.$ $400\,\Omega$ or $0.4\,$KΩ

29.

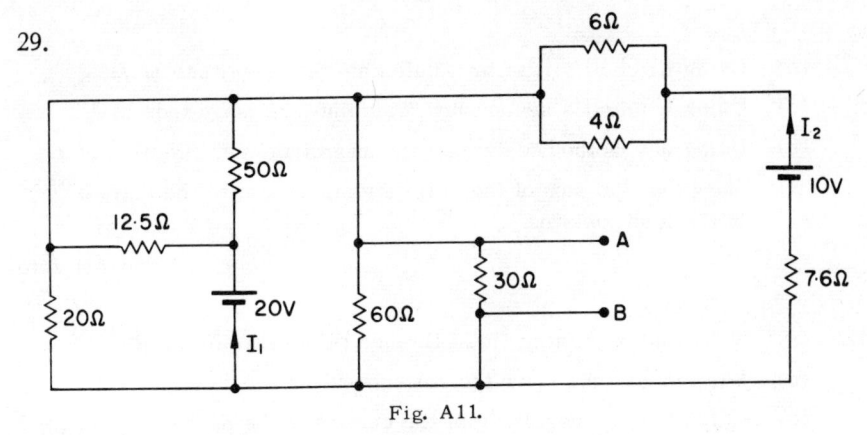

Fig. A11.

(a) Show that I_2 is zero.
(b) Calculate the p.d. across points $A - B$.
(c) Calculate the magnitude of I_1 and the current in the 30Ω resistor.

(Hint: If a shunt circuit consists of a 7Ω and 21Ω in parallel, the resultant effective resistance may often be derived easily by comparing one resistor value to the other. If the larger is exactly divisible by the smaller, the smaller may be represented by several of the larger in parallel. Example: 7Ω may be represented by three 21Ω resistors in shunt. The final effective resistance of 21Ω in shunt with 7Ω is equal to $1 + 3 = $ four 21Ω resistors in parallel. The resultant will be $21\Omega/4$).

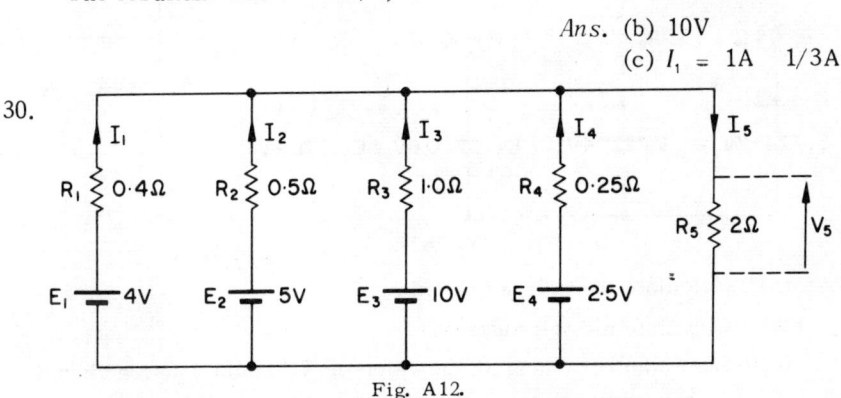

Ans. (b) 10V
(c) $I_1 = $ 1A   1/3A

30.

Fig. A12.

(a) Derive a formula to express V_5 in terms of all other quantities.
(b) Derive an expression for I_1.

contd.

598   ELECTRONICS FOR TECHNICIAN ENGINEERS

30. contd
   - (c) Using Kirchoff's first law, calculate the magnitude of I_1.
   - (d) Using Kirchoff's second law, calculate the magnitude of I_1.
   - (e) Using any method, calculate the magnitude and direction of I_4.
   - (f) Show that the sum of the cell currents equate to the current (I_5) in the load resistor.

   Ans. (c) and (d) zero
   (e) −6A

31. (a) Show that no current flows through the load resistor R_3.
   - (b) Explain (a) and show the current path in the circuit.
   - (c) Calculate the magnitude of the current in the path shown in (b).

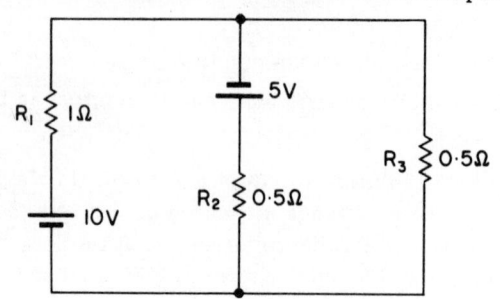

Fig. A13.

Ans. (c) 10A

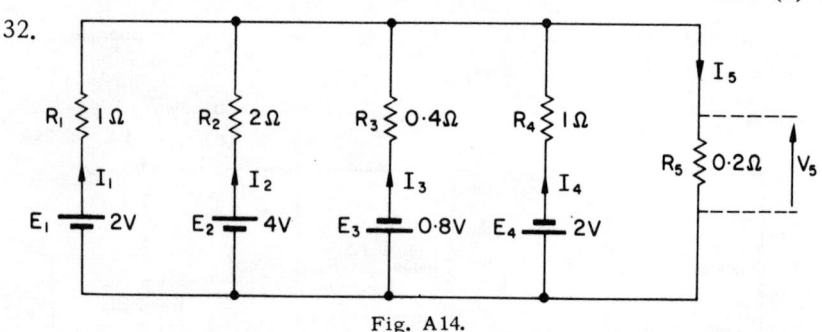

32.

Fig. A14.

   - (a) Calculate the p.d. across R_5.
   - (b) Calculate all cell currents.
   - (c) State whether changing the value of R_5 would cause a change in V_5.
   - (d) Justify your statement in (c) mathematically.
   - (e) Show that $I_1 + I_2 + I_3 + I_4 = I_5$.

   Ans. (a) zero. (b) 2A, 2A, −2A, −2A

33.

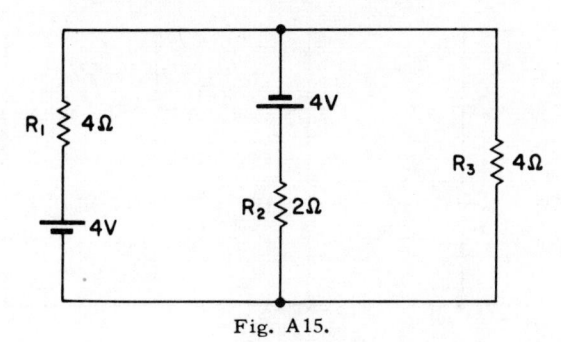

Fig. A15.

Derive a formula to solve the following problem.

(a) Calculate the p.d. across the load resistor R_3.
(b) Reverse both cells and calculate the p.d. across R_3.
(c) Repeat (a) and (b) for R_3 having a value of 0.8Ω.
(d) Calculate the value of R_2 that would cause the p.d. to become zero.

Ans. (a) −1V
(b) 1V
(c) −0.5V, 0.5V
(d) 4Ω

34.

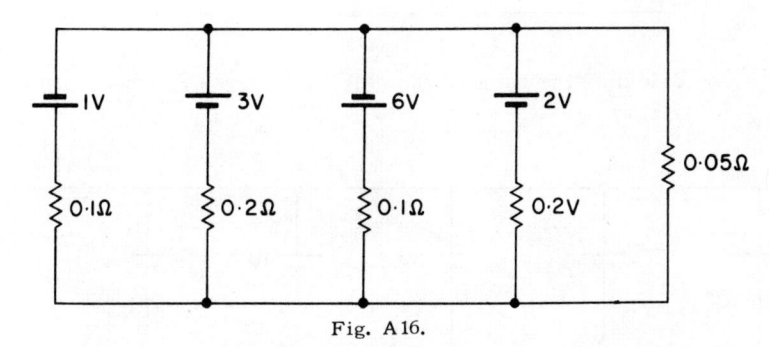

Fig. A16.

(a) Calculate the potential across the load resistor 0.05Ω.
(b) Calculate the current flowing in the 6V cell.

Ans. (a) −0.9V

(b) −69A

35.

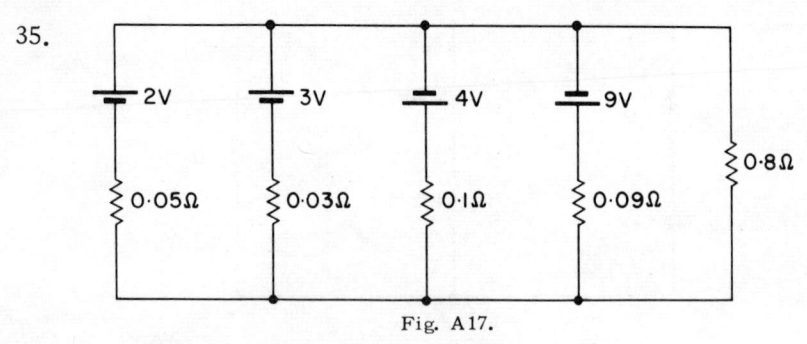

Fig. A17.

Calculate the current flowing in the 0.8Ω load resistor.

Ans. zero

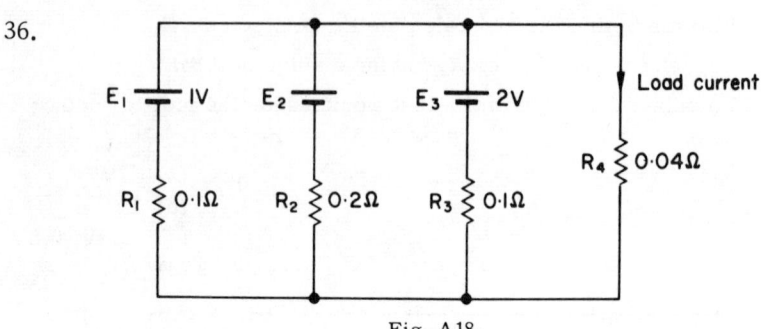

36.

Fig. A18.

Calculate the value of the cell E_2 to cause 25A load current to flow.

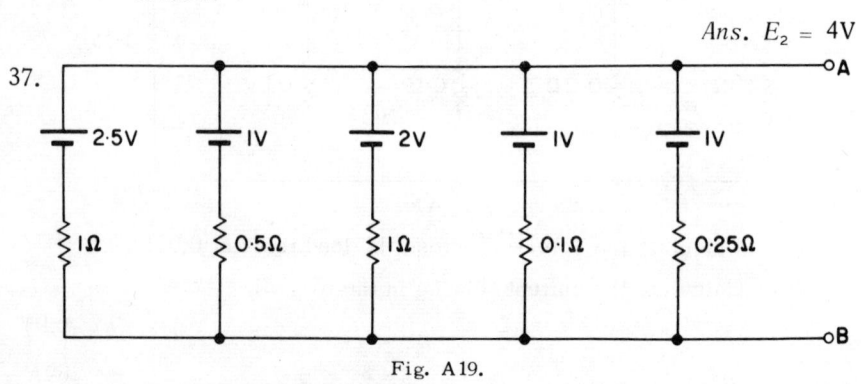

Ans. $E_2 = 4V$

37.

Fig. A19.

Calculate the p.d. across the points $A - B$ with
(a) the circuit as shown and
(b) with a 0.2Ω load resistor connected across $A - B$.

Ans. (a) 0.25V
(b) 0.2V

38.

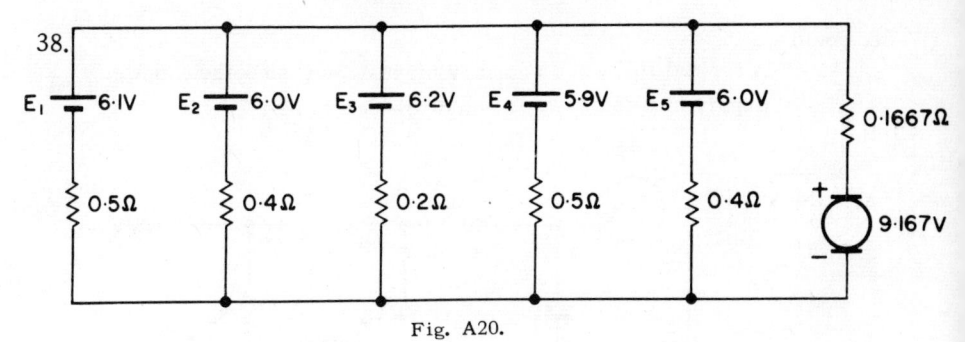

Fig. A20.

(a) Calculate by Kirchhoff's first law, the current flowing into each cell from the generator and the generator current.

(b) Repeat (a) using Kirchhoff's second law.

$$\text{Ans. } (E_1) \ 1.8\text{A}$$
$$(E_2) \ 2.5\text{A}$$
$$(E_3) \ 4\text{A}$$
$$(E_4) \ 2.2\text{A}$$
$$(E_5) \ 2.5\text{A}$$

Current from generator 13A

A typical half wave power supply unit employing resistance – capacitance smoothing is shown in the circuit diagram (Fig. A21).

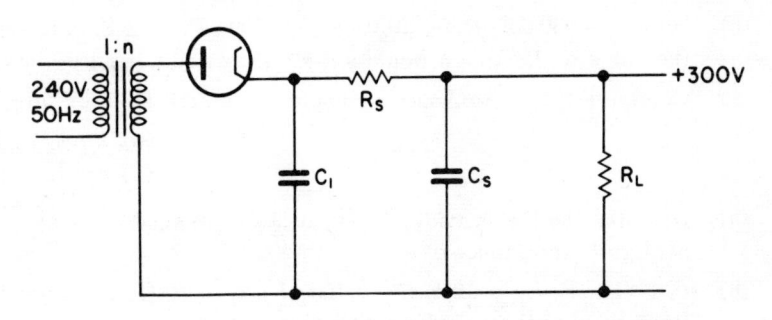

Fig. A21.

A typical full wave power supply unit is also shown in figure A.22.

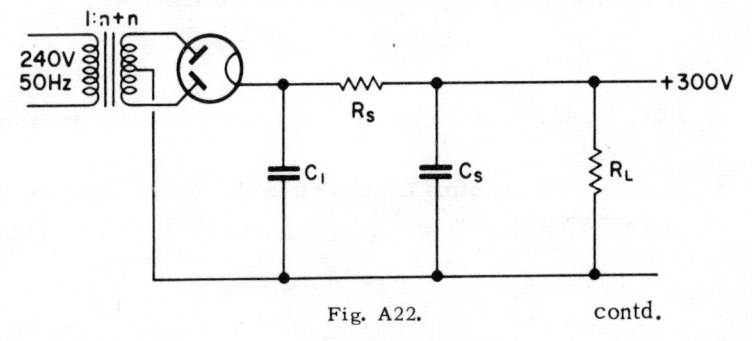

Fig. A22. contd.

38. contd

A typical full wave capacity input, $L - C$ smoothed, power supply unit is shown in figure A.23.

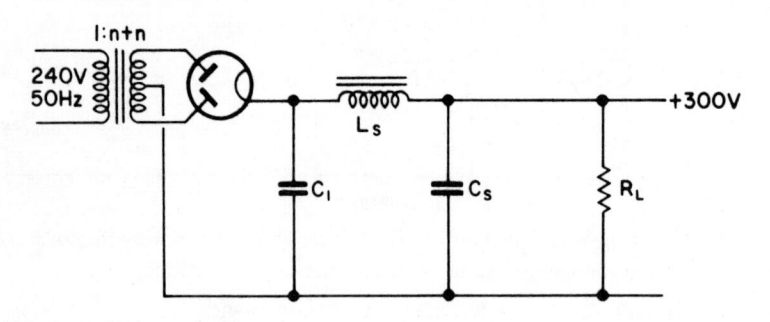

Fig. A23.

39. (a) Describe, with the aid of sketches, the p.s.u. shown in figure A21.
    (b) For $n = 1$, $C_1 = 10\,\mu\text{F}$, $C_s = 100\,\mu\text{F}$, $R_L = 30\,\text{K}\Omega$. Calculate the value R_S to satisfy the circuit d.c. requirements.

    Ans. $R_S = 2.6\,\text{K}\Omega$

40. (a) Describe with the aid of sketches, the function of the diode and C_1 in the p.s.u. shown in figure A22.
    (b) With $C_1 = 30\,\mu\text{F}$, $R_L = 10\,\text{K}\Omega$, $C_S = 100\,\mu\text{F}$, $n = 2$, calculate the value of R_S which would satisfy the circuit requirements.
    (c) Calculate the approximate pk ripple across the load resistor, R_L.

    Ans. (b) $R_S = 12.2\,\text{K}\Omega$
    (c) approx. 6 mV pk

41. (a) Describe the p.s.u. shown in figure A23. (Assume L has negligible resistance).
    (b) $C_1 = 10\,\mu\text{F}$, $L_s = 10\,\text{H}$, $C_S = 100\,\mu\text{F}$, $R_L = 2\,\text{K}\Omega$, evaluate the turns ratio of the transformer necessary to satisfy the circuit requirements.
    (c) Calculate the pk ripple across the load resistor R_L.

    Ans. (b) $1.12 + 1.12$
    (c) app. 192 mV pk

42. (a) Sketch a full wave bridge rectifier circuit. Describe the diode functions.
    (b) Sketch a full wave p.s.u. complete with a bridge rectifier and smoothing components.

    contd.

42. contd
   - (c) Explain one advantage of a bridge rectifier system.
   - (d) With an a.c. input of 14.14V r.m.s., a d.c. load current of ¼A and a reservoir capacitor of 500 μF, determine the d.c. output voltage and pk ripple voltage.

   Ans. 17.5V d.c.
   2.5V pk

43. Design a simple half wave p.s.u. (as shown in figure A21) to give an output voltage of 275V average d.c. at a load current of 55 mA. The ripple should be < 100 mV pk. Describe each step clearly and show all working.

44. Design a simple full wave p.s.u. to give an output of 12V average d.c. at a load current of 4A and a ripple of ⩽ 40 mV pk. Show all workings and explain each step.

45. Design a simple full wave p.s.u. employing capacity input, $L - C$ smoothing to give an output of 275V at 100 mA. The ripple should not exceed 50 mV pk at full load current.

46. (a) Show mathematically, the relationship between the output ripple and the load current.
    (b) What other factors affect the ripple voltage?
    (c) Explain p.i.v. and the precautions necessary in designing a p.s.u.
    (d) Design a full wave p.s.u. operating from a standard British mains supply, to give 200V at 100 mA and a maximum pk ripple of 100 mV. (Use actual diode valve characteristics allowing for the p.i.v. etc).

47. A simple thermionic diode is connected as a half wave rectifier. The transformer has a step up ratio of 1:5 and is connected to a standard 240V, 50 Hz mains supply.
    (a) Calculate (a) the average d.c. output voltage and
    (b) The approximate peak inverse voltage (p.i.v.).
    (c) If 20V is dropped across the diode at peak load current, what ratio should the transformer be to ensure the same d.c. output as in (a).

   Ans. (a) 540V
   (b) 1696V
   (c) 1:5.06

48. A half wave p.s.u. employing a $16\,\mu F$ as a reservoir, delivers an average load current of 32 mA at an average voltage of 300V. Calculate (a) the transformer ratio required when connected to a normal 240V, 50 Hz supply.

Ans. 1 : 1.06

49. A full wave rectifier system operating from a 50 Hz supply is filtered with two $L - C$ filters in cascade. The value of each $C = 8\,\mu F$ and $L = 10\,H$.

(a) To what approximate percentage is the fundamental ripple reduced?

(b) If the peak ripple at the rectifier output is 20V, calculate the final ripple voltage.

Ans. (a) 0.1%
(b) 20 mV pk

50. An $L - C$ filter is connected to the output of a full wave rectifier system. The inductance has a value of 16 H. Calculate the value of C so as to reduce the fundamental ripple to 2%.

Ans. $8\,\mu F$

51. A half wave rectifier employing an $8\,\mu F$ capacitor as a reservoir, is connected to a transformer having a ratio of 1 : 1.13 which in turn is connected to a standard 240V, 50 Hz mains supply. Calculate the average d.c. output for an average load current of

(a) 24 mA,
(b) 48 mA,
(c) 72 mA.

(Comment upon your answers and refer to the effective source impedance.)

Ans. (a) 270V
(b) 240V
(c) 210V

52. A high voltage p.s.u. requires to deliver 1200V at 10 mA. If the system is full wave — and operates from a 50 Hz mains supply — calculate the values of R and C in the filter for an average unfiltered voltage from the rectifier of 1500V. The fundamental ripple is to be reduced to 10%.

Ans. $R = 30\,K\Omega$
$C = 0.53\,\mu F$

53. A 1:1 + 1 transformer has its primary connected to a normal 240V, 50 Hz mains supply. The secondary winding is connected to a full wave rectifier stage which has a 20 μF reservoir. A resistor (R) is connected in series with the rectifier cathodes and the load. The load consists of two series connected 150V neon stabilisers across which a 10 KΩ load resistor is connected. The neon stabilisers normally require a tube current of 20 mA. Calculate the value of R to satisfy these conditions.

Ans. 530 Ω

54. A low voltage full wave power supply unit is required to deliver 12V at 2A. A 2000 μF reservoir capacitor and a series connected cut out having a d.c. resistance of 1 Ω is employed.
If the unit is to be connected to 240V, 50 Hz mains supply, determine the turns ratio of the transformer. *Ans.* 1 : 17.85 − 17.85 (step down).

55. Sketch a power supply unit employing a bridge rectifier. Describe, with the aid of sketches, the current flow on alternate half cycles and explain how a undirectional load current results. Show that the ripple voltage is proportional to the load current and inversely proportional to the value of the reservoir capacitor.

56. A simple rectifier is connected to a normal 240V, 50 Hz supply. A 5 KΩ load resistor is connected as a load. Calculate

(a) the peak load current.

(b) the average load current when a 10 K is connected in shunt with the first load.

Ans. (a) 67.8 mA pk
(b) 21.6 mA average
(c) 32.4 mA average

57. Consider a power supply circuit employing a reservoir capacitor.

(a) Show, by $Q = C.V. = I.T.$, that the ripple voltage is proportional to the load current and inversely proportional to the reservoir capacitor value.

(b) For the p.s.u. shown in figure A22, let $C_1 = 10 \mu F$. $R_L = 15 K\Omega$. Calculate the 'rise and fall' across C_1.

(c) Let $R_L = 7.5 K\Omega$. $C_1 = 10 \mu F$ as before, calculate the 'rise and fall' as in (b).
Do the results in (b) and (c) verify your statements in answer to (a)?.

58. Consider figure A21. Let $C_1 = 20\,\mu\text{F}$. $R_S = 150\,\Omega$ $n = 1$. $R_L = 5\,\text{K}\Omega$.
   (a) Show that these values conform to the circuit requirements.
   (b) Show whether it is possible to reduce R_L indefinitely and explain your reasons.

59. (a) Explain why it is necessary for a meter to have a low resistance when used as an ammeter and a high resistance when used as a voltmeter.
   (b) Draw a diagram of a single range voltmeter using a milliameter which has an R_m of $1000\,\Omega$ and a f.s.d. of $500\,\mu\text{A}$.
   (c) Calculate the value of the series resistor (R_s) in (b) for a f.s.d. voltage range of 1V, 10V, 100V, 1000V.
   (d) Using the same milliameter, draw a basic single range ammeter.
   (e) Calculate the value of the shunt resistor (r.s.h.) for a f.s.d. value of current equal to 1 mA, 10 mA, 1A.

   Ans. (c) $1\,\text{K}\Omega$, $19\,\text{K}\Omega$, $199\,\text{K}\Omega$, $1999\,\text{K}\Omega$
   (e) $1\,\text{K}\Omega$, $1/19\,\text{K}\Omega$, $5/9.995\,\Omega$

60. Draw a multirange meter having five ranges as follows: 1V, 10V, 1 mA, Ω range.
   The basic movement has an R_m of $100\,\Omega$ and a f.s.d. of $100\,\mu\text{A}$.

61. Draw a 5 range ammeter employing a universal shunt. The basic movement has an R_m of $500\,\Omega$ and a f.s.d. of 1 mA. The ranges should be: 5 mA, 50 mA, 1A, 10A.
   (Any formulae used for a shunt should be derived from first principles).

62. A milliameter requires 2 mA to cause full scale deflection. It has a resistance of $100\,\Omega$.
   What total current would need to flow through the network if a $200\,\Omega$ resistor is connected across the meter terminals in order to cause full scale deflection.

   Ans. 3 mA

63. Calculate, using 'load over total', the voltage at points A, B, C and D.
   (a) before the meter is connected and when
   (b) the meter is connected as shown in the diagram.
   Explain why the meter readings differ from these without the meter connected.

   contd.

63. contd

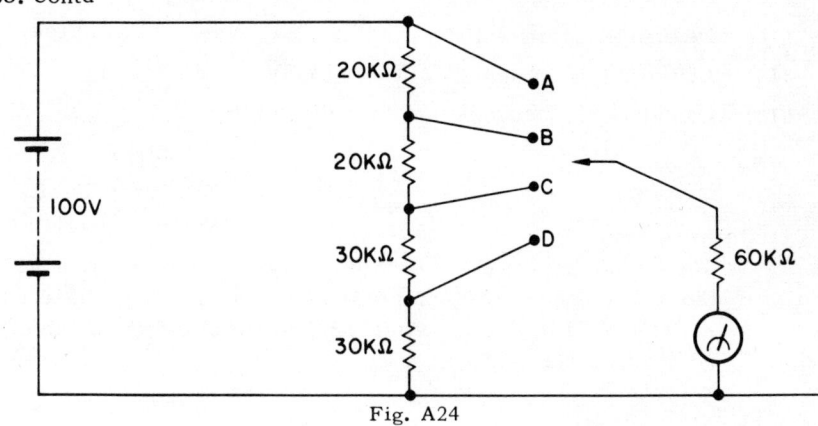

Fig. A24

Ans. (a) 100V, 80V, 60V, 30V
(b) 100V, 63.2V, 42.8V, 22.2V

64. Draw a diagram of a test rig which will enable the resistance (R_m) of a moving coil movement to be determined. Assume values for the components shown in your diagram and explain each step.

(a) Explain the difference between e.m.f. and p.d.

(b) Explain how you would use your explanation in (a) to calculate the internal resistance of a cell.

A typical electronic device has the following static characteristics.

V_a V	0	50	100	150	200	250	300	350	400	450	500
I_a A	0	4	20	40	60	80	93	96	97	98	99

65. (a) Determine the d.c. resistance of the device, whose static characteristics are shown, at V = 50, 100, 150, 200, 300, 500V.

(b) Draw the graph of I_a/V_a according to the table given.

(c) Compare the d.c. resistance at V = 50 and 500V and comment on the differences.

Ans. (a) 12.5 KΩ
5 KΩ
3.75 KΩ
3.34 KΩ
3.22 KΩ
4.12 KΩ
5.04 KΩ

66. (a) Draw a graph, according to the table given, of I_a/V_a.
    (b) Determine the current flow at $V = 125V$.
    (c) Determine the current flow at $V = 325V$.
    (d) Calculate the d.c. resistance for (b) and (c).

    Ans. (b) 30 mA
    (c) 95 mA
    (d) 4.17 KΩ, 3.42 KΩ

67. The typical device has a load resistor connected in series with the device and the supply. For a supply of 400V d.c. and a load resistor of 5 KΩ, determine the supply current and the potential across the load resistor.

    Ans. 47 mA, 233V

68. Draw the I_a/V_a graph for the typical device according to the table given.
    (a) When used with a d.c. supply of 400V and a series load resistor of 5 KΩ, determine the potential across R_L and the device potential.
    (b) Determine the device resistance at the device potential in (a) and show that when this is added to R_L, the total circuit resistance agrees with the voltage distribution as in (a) by load over total.

    Ans. (a) $V_{RL} = 233V$. $V_D = 167V$

69. (a) Construct a graph of I_a/V_a according to the table given.
    (b) The device is to be used in series with a 10 KΩ load resistor. The supply is to be varied from 0 − 400V, modify the static characteristics and produce a dynamic curve.
    (c) For supply potentials of 100V, 150V, 200V, 250V, 300V, 350V, 400V, determine from the modified graph, the current flowing at each potential.

    Ans. (c) 4.8 mA
    8.5 mA
    12 mA
    16 mA
    20 mA
    24 mA
    28 mA

PROBLEMS AND ANSWERS

70. Draw a graph of I_a/V_a according to the table of characteristics for the typical device. When used with a variable load resistor and a constant 300V supply, determine from the graph
    (a) the load current,
    (b) V_{RL} and
    (c) V_D for values of R_L 3 KΩ, 4 KΩ, 6 KΩ, 10 KΩ.

    Ans. (a) 45.5 mA, 38.2 mA, 29 mA, 20 mA
    (b) 136.5V, 152.8V, 176V, 200V
    (c) 163.5V, 147.2V, 124V, 100V

71. Draw a graph of I_a/V_a for the typical device from the table of static characteristics. The supply voltage is to be varied according to the following table.

V (volts)	0	200	284	346.8	400	346.8	284	200	0
T (secs)	0	0.3	0.45	0.6	0.9	1.2	1.35	1.5	1.8

   (a) Sketch, to scale, the supply voltage from 0 − 1.8 seconds.
   (b) Determine the supply current at each time interval shown and produce a sketch of current using the same time scale.
   (c) A 4 KΩ load resistor is connected in series with the device − produce a dynamic graph of I_a/V_a and repeat (b) drawing this to the same scale as in (b).
   (d) Comment upon the results in (b) and (c) making particular reference to the similarity of either current graph to the voltage graph.

72. Draw a graph of I_a/V_a according to the static characteristics of the device in the table given.
    Produce in dynamic curve for a series load of 2 KΩ, 5 KΩ, 10 KΩ, 20 KΩ and show that the series circuit is more dependent upon R_L the greater its value and that the current flowing is less dependant upon the variations is the device resistance.

73. Draw a graph of I_a/V_a according to the static characteristics of the device whose characteristics are given in the table. The device is to be used in conjunction with a 10 KΩ load resistor.
    (a) Determine the total circuit d.c. resistors, by adding the 10K to the device resistance at V = 100, 200, 300, 400V.
    (b) Produce a dynamic curve for the device plus load resistor and determine graphically, the total circuit resistance at V = 100, 200, 300, 400V.

contd.

73. contd Ans. (a) 15 KΩ, 13.34 KΩ, 13.22 KΩ, 14.12 KΩ
(b) 20 KΩ, 16.5 KΩ, 15 KΩ, 14.3 KΩ

74. An electronic device has the following static characteristics.

V_a (volts)	0	100	150	200	250	300	350	400
I_a (mA)	0	8	18	52	88	97	97	97

With a 5 KΩ load connected in series with the device and a d.c. supply of 500V, determine (a) I_a, (b) V_{RL} (c) V_D.

Ans. (a) 58.5 mA
(b) 292.5V
(c) 207.5V

75.

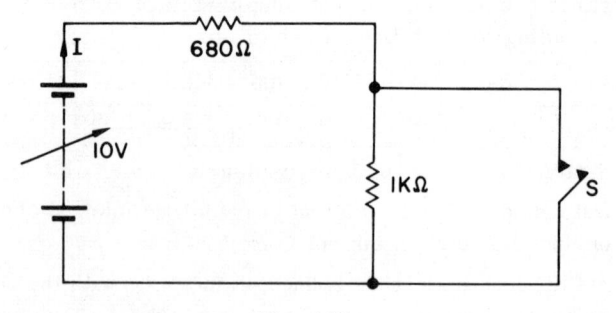

Fig. A25.

Draw a graph of I/V (when the (a) switch S is closed, and (b) when S is open).

76.

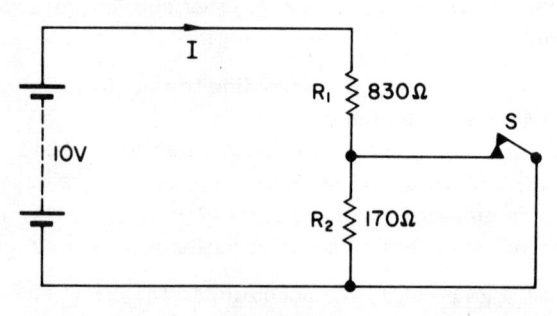

Fig. A26.

(a) Draw a graph whose slope will be determined by the 830Ω.
(b) With the switch 's' opened, draw a load line for the 170Ω and determine graphically, I, V_{R1} and V_{R2}.

Ans. (b) I = 10 mA
V_{R1} = 8.3V
V_{R2} = 1.7V

77.

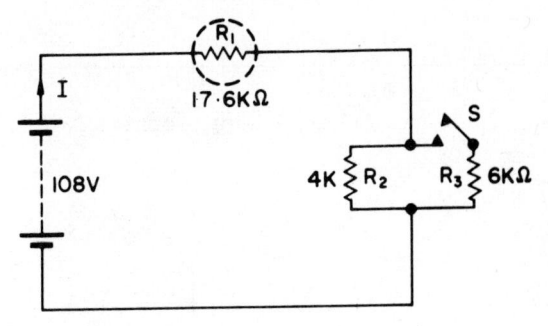

Fig. A27.

(a) R_1 is a device having a resistance of $17.6\,K\Omega$. It has a series load resistor of $4\,K\Omega$. Draw an I/V graph for R_1 and by load line technique, derive the current I, V_{R_1} and V_{R_2}.

(b) With the switch 'S' closed, draw a new load line and determine I, V_{R_1} and V_{R_2}.

Ans. (a) 5 mA, 88V, 20V
(b) 5.4 mA, 95.04V, 12.96V

78.

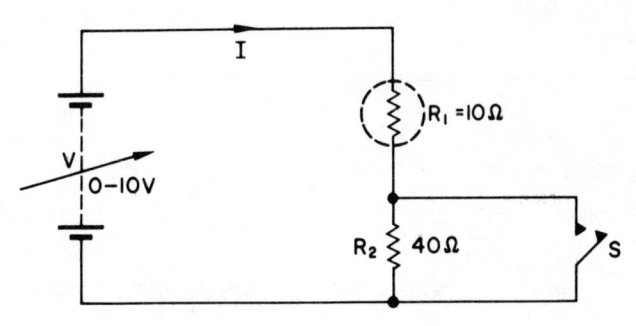

Fig. A28.

(a) Draw a graph of I/V for the 10V cell and R_1.
(b) Determine I when $V = $ 0V, 5V, 10V, when S_1 is closed.
(c) Determine I when $V = $ 0V, 5V, 10V, (when S_1 is open) using load line technique.
(d) Modify the graph in (a) to allow for R_2 in series with R_1 and determine I when $V = $ 0, 5, 10V with S_1 open as in (c).

Ans. (b) 0, 0.5A, 1A
(c) 0, 0.1A, 0.2A
(d) 0, 0.1A, 0.2A

79.

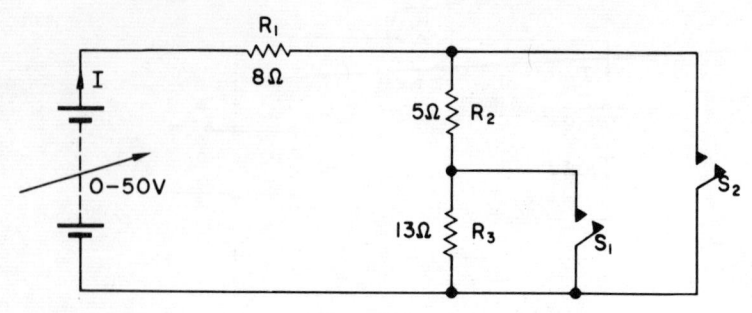

Fig. A29.

(a) With S_1 and S_2 closed, draw a graph of I/V for the diagram shown.
(b) With S_1 closed and S_2 open, modify the graph in (a).
(c) With both switches open, modify the graph drawn in (a).
From each graph, determine the current when $V = 26V$ in (a), (b) and (c).

Ans. (a) 3.25A
(b) 2.0A
(c) 1.0A

80.

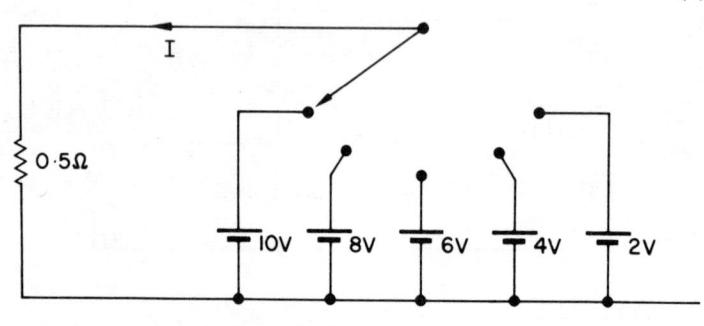

Fig. A30.

(a) Plot a graph of I/V for each of the switch positions shown.
(b) Assume that a 1.5Ω resistor is inserted in series with the 0.5Ω. Modify the graph and determine I for each position.

Ans. (b) 0, 1A, 2A, 3A, 4A, 5A

81. (a) Explain how you would produce a dynamic graph from a static graph of a linear device. Explain why the current depends more upon the load resistor the greater its value.
A sketch should be made to assist your explanation.
(b) Repeat (a) for a non-linear device.

PROBLEMS AND ANSWERS

82. Sketch a graph of V/I and explain how the slope of the curve determines R.

    Sketch a graph of I/V and explain how the slope of the curve determines $1/R$. Show that the slope of a load line for any given series load resistor determines $-1/R_L$. Supplement your explanation with sketches.

83. A 100V battery is variable from 0 – 100V. A device having a linear resistance of 20 Ω is connected across the supply. Draw a graph of I/V.

    A 20 Ω load resistor is connected in series with the device, modify the graph and show that for any value of voltage, the current is half that of the device alone.

84. Draw a graph of I_a/V_a from the table given in (74). Modify the static graph and produce a dynamic graph when a 5 KΩ load resistor is connected in series with the device.

    (a) Determine from the dynamic graph, the total resistance at $V = 150, 200, 250V$.

    (b) Determine the resistance of the device from the static graph at $V = 150, 200, 250V$, and add to the 5 KΩ load resistor.

    (c) Compare and comment upon, both results in (a) and (b).

85. A voltage amplifying valve has I_a/V_{ak} characteristics as shown in figure 8.6.2.

    Determine the operating point for each of the following combinations of R_L and R_K when the amplifier is connected to a 300V line.

    $R_L = 9.5$ KΩ $R_K = 500$ Ω   Answer $I_a = 5.4$ mA. $V_{ak} = 244$V

    $R_L = 9.75$ KΩ $R_K = 250$ Ω   Answer $I_a = 8.1$ mA. $V_{ak} = 220$V

86. A gas filled diode requires 150V to cause it to strike. Once it is in conduction, a burning voltage of 100V exists when a recommended 5 mA current is flowing through the tube. When used on a 300V supply, and with a 10 KΩ load resistor connected across the device.

    (a) Determine the value of R_s and

    (b) Calculate the highest value for R_s that will allow the device to strike in this circuit.

$$\text{Ans. (a) } 13.34 \text{ KΩ}$$
$$\text{(b) } 20 \text{ KΩ}$$

87. (a) Describe the step taken to determine gm and r_a for a triode.
    (b) Determine gm, r_a and μ from the I_a/V_{ak} characteristics in figure 8.6.2.
    (c) Calculate μ from $gm \times r_a$ and compare with the value obtained graphically in (b).
    (d) Explain the reasons for any difference in answers obtained in (b) and (c).

INDEX

Admittance, 366.

Alternating current, 53.
   magnification factor, 59.
   mean value, 56.
   reactance, 56.
   resonance, 58.
   r.m.s. value, 55.

Amplifiers, 129.
   a.c. load lines, 148.
   amplification, 131.
   bias load lines, 133, 134, 147.
   capacitive load, 166.
   efficiency, 450.
   inductive load, 146.
   load lines, 130, 133, 147.
   operating point, 130, 133, 147.
   transformer coupled, 155.

Argand diagram, 367.

Binary counter, 245, 267.

Black boxes, 16, 17, 189.
   four terminal devices, 18, 19.

Bridge.
   rectifiers, 90.
   Wein, 378.

Capacitors, 45.
   parallel connected, 49.
   parallel plate, 51.
   phase shift, 47.
   reactance, 354.
   series connected, 50.

Cathode follower, 184.
   input impedance, 167, 277.

Clipper circuit, 210.

Delay line, 291, 297.
   equations, 295.
   pulse generator, 302.

Digital circuits, 540.
   'AND' gate, 544.
   binary stage, 540.
   'OR' gate, 545.

Diode pump, 287.

Diodes, 63.
   full wave, 72.
   grid current in triodes, 573.
   p.s.u., 68, 72, 77.
   rectifiers, 65.
   voltage drop, 86.

Filters, 75, 80, 81.

Flow diagrams, 199.

Frequency response, 141, 374.

Frequency selective amplifier, 378.

Generators, voltage and current, 20.

Graphs.
   composite, 5, 67.
   current/voltage, 4.
   non-linear, 10, 67.
   static characteristics, 5.
   voltage/current, 4.

Inductors, 42.
   impedance, 354.

Kirchhoff's laws, 24, 32.

Load lines, 8.
   plotting, 130.
   positioning, 9, 12.
   restricted graphs, 13, 147.

Load over total, 2, 3.

Long-tailed-pairs, 187.

Metal rectifiers, 87.

Meters, 93.
   a.c. ranges, 101.
   current, 95.
   Ohms, 104.
   protection circuit, 107.
   universal shunt, 96.
   voltage, 93.

Miller effect, 165.

Multivibrators.
  cathode-coupled, 285.
  direct-coupled, 273.
  free-running, 232.
  monostable, 271.

Ohm's law, 1.

Oscillator.
  frequency determining network, 507.
  phase shift, 507.
  relaxation, 230.

Pentode, 172.

Power supply units, 68, 72, 77.
  regulated, 82.
  stabilised, 329, 336.

Pulse waveforms, 221, 222.
  lagging edge, 222.
  leading edge, 222.
  mean level, 226.
  rectangular, 221.
  rise time, 222.
  sawtooth, 221, 570.

Rectifiers, 65.
  bridge, 90.
  metal, 87.
  voltage doubler, 91.

Resistance.
  input, 16, 420, 529, 530.
  internal, 15.
  output, 19.

Resistors, 41.
  paralled, 3, 7.
  series, 2, 6.

Schmitt trigger circuit, 217.

Servomechanisms, 560.
  differential, 560, 566.
  summing integrator, 561.

Silicon controlled-rectifiers, 523.

Stabiliser circuits, 119.

Step function, 222.

Substrate, 483.

Theorems.
  Norton's, 29.
  reciprocity, 29.
  superposition, 28, 30.
  Thevinin's, 29, 31.

Thyratron, 122.
  control ratio, 125.
  mutual characteristics, 124.

Timebase synchronisation, 583, 587.

Time constant, 44.

Transformer.
  coupled output, 153.
  design, 389.

Transistors.
  gain, 419.
  h parameters, 457.
  H parameters, 475.
  junction, 411.
  mutual conductance, 533.
  metal-oxide-silicon-type, 483.

Triode valve, 109.
  anode dissipation, 137.
  grid current, 137.
  curves (I_a/V_a), 111, 115, 116, 117, 118, 133, 139, 147.
  parameters, 112, 145.

Twin Tee network, 380.

Virtual earth, 557.

Voltage reference tube, 118.

Zener diodes, 515.
  dissipation, 511.